Whether countries can transition to a greatly decarbonized society without significantly sacrificing economic performance or energy security remains one of the most pressing problems facing humanity. This book offers a fresh yet pragmatic examination of the economic and political constraints involved in this Herculean task. Important reading.

Benjamin K. Sovacool, Professor of Energy Policy, University of Sussex

Professor Looney has compiled a volume that is sure to become a classic for policymakers on the options for transitioning to an era of energy security, affordable prices and environmental preservation. While the country chapters afford best policies and roadmaps for every category of country, Professor Looney's synthesis is their beacon.

Hossein Askari, Iran Professor, George Washington University

Climate change is a game changer in the global energy landscape. Energy policies have moved beyond the binary focus on supply security in consumer nations and demand security in producer states. Transitioning away from carbon fuels will be necessary, yet challenging, while they are relatively abundant in the wake of the shale boom. There will be great variations between countries depending on their resource endowments, energy mixes, fiscal abilities and policy preferences. This Handbook offers invaluable theoretical insights and case studies to better understand the dialectic between climate and energy, a dialectic that will play a pivotal role in 21st century societies.

Eckart Woertz, Senior Research Fellow at CIDOB (Barcelona Centre for International Affairs) and Scientific Advisor to the Kuwait Chair at SciencesPo, Paris

Every nation and every community is facing critical decisions regarding the relative emphasis it will place on energy security, affordability, and sustainability. There is no better guide to these complex challenges than the *Handbook of Transitions to Energy and Climate Security*. Not only does it assess the policy frameworks of a broad spectrum of carbon-producing and carbon-consuming countries, but it also examines the global pressures that are so pivotal to the energy debate.

Michael T. Klare, Professor of Peace and World Security Studies, Hampshire College

This superlative volume reshapes the energy-security debate into an energy trilemma discussion, where physical access at reasonable prices is balanced with environmental concerns and financial burdens. Robert Looney, whose pioneering contributions to economic development studies are peerless, has magisterially harnessed the skills of experts to shed light on a global transition period, as developed and developing societies confront challenging tradeoff options. Readers will be grateful for the no-nonsense comprehensive options raised by this leading theoretician, who envelops objective scholarship with empathy, especially towards those who must struggle with conflicting objectives.

Joseph A. Kéchichian Senior Fellow, King Faisal Center for Research and Islamic Studies, Riyadh

Handbook of Transitions to Energy and Climate Security

The chapters in this volume flow from a central theme: the transformation of historic patterns of energy use resulting from two dramatic developments, the increased potential availability of energy in many parts of the world on the supply side, and on the demand side, increasing concerns over the harmful effects on the environment of the use of fossil fuels. The international framework for policy evaluation is outlined, followed by country studies demonstrating the manner in which states have adapted to the dramatic changes in the global energy environment, with sections examining carbon-producing countries (including Canada, Mexico, Russia and Saudi Arabia), intermediate carbon-producing/consuming countries (including Brazil, China, Indonesia and the United Kingdom), carbon-consuming countries (including India, Japan, Pakistan, Thailand and Turkey) and carbon reduction countries (including Denmark, France and Germany).

Robert E. Looney is a Distinguished Professor in the National Security Affairs Department at the Naval Postgraduate School, Monterey, California. He specializes in issues relating to economic development in the Middle East, East Asia, South Asia and Latin America. He has published 22 books, including *Economic Policymaking in Mexico: Factors Underlying the 1982 Crisis*, Duke University Press and *Iraq's Informal Economy: Reflections of War, Sanctions and Policy Failure*, The Emirates Center for Strategic Studies and Research. Professor Looney has also edited three Routledge handbooks, *Handbook of US-Middle East Relations* (2009), *Handbook of Oil Politics* (2012) and *Handbook of Emerging Economies* (2014). He is the editor of the Routledge series *Europa Emerging Economies*.

Handbook of Transitions to Energy and Climate Security

Edited by Robert E. Looney

LONDON AND NEW YORK

First published 2017
by Routledge

2 Park Square, Milton Park, Abingdon, Oxfordshire OX14 4RN
711 Third Avenue, New York, NY 10017

Routledge is an imprint of the Taylor & Francis Group, an informa business

First issued in paperback 2018

Europa Commissioning Editor: Cathy Hartley
Editorial Assistant: Eleanor Simmons

British Library Cataloguing in Publication Data
A catalogue record for this book is available from the British Library

Library of Congress Cataloging in Publication Data
Names: Looney, Robert E., editor.
Title: Handbook of transitions to energy and climate security / editor, Robert E. Looney.
Description: Abingdon, Oxon ; New York, NY : Routledge, 2017. | Includes index.
Identifiers: LCCN 2016022090 | ISBN 9781857437454 (hardcover)
Subjects: LCSH: Energy security. | Climatic changes–Government policy. | Energy policy. | Environmental policy.
Classification: LCC HD9502.A2 H2574 2017 | DDC 333.79–dc23
LC record available at https://lccn.loc.gov/2016022090

ISBN: 978-1-85743-745-4 (hbk)
ISBN: 978-1-857-43974-8 (pbk)

Typeset in Bembo
by Taylor & Francis Books

For Lauren and Caitie and the wonderful life that lies ahead

Contents

List of illustrations

Figures

Tables

Preface

For decades energy has been central to economic growth and improved standards of living for countries around the world. For most consuming countries, an overriding issue has been energy security – access to adequate supplies of energy at affordable prices. On the other hand, energy exporting countries have been primarily concerned with security of demand – the sustainment of strong, assured sales at stable, albeit elevated, prices.

Until quite recently, the dynamic between the energy producers and consumers remained relatively stable. Many of the world's energy trade and production patterns had not changed significantly from those that evolved out of the tumultuous 1970s which saw the rise of OPEC together with widespread energy nationalizations. However, a number of these patterns have begun breaking down in recent years, beginning with the shale oil and gas boom starting in the USA. For many of the consuming countries, energy supplies are either considerably more abundant or potentially so with the development of fracking and other techniques capable of expanding unconventional fuels. For the producing countries, the re-entry of Iraq into world oil markets together with the possible lifting of sanctions on Iran, the US removal of its crude oil export ban and the collapse in global oil prices starting in 2014 is likely to cause frictions within OPEC not seen for years.

At the same time, there has been rapidly increasing awareness in the scientific community, and within governments and the public at large of the impact of fossil fuels on the environment, especially increased levels of carbon dioxide released from either the burning or processing of certain hydrocarbons. Increased public concern over the causes and consequences of climate change is beginning to be translated into policies and measures designed to reduce carbon emissions and hence the use of certain fossil fuels.

Increased energy availabilities and political pressures to stem the factors contributing to global warming have converged to fundamentally alter the energy-security debate. For the present and for at least the next decade, energy will not be a scarce resource in the physical sense. Similarly, due to the development of efficient energy markets over the past decade, consuming countries are not, barring a major conflict, in much jeopardy of being cut off from foreign supplies. What will be scarce is non-fossil fuel – alternative energies that do not significantly contribute to climate change.

New energy vulnerabilities will come to the fore during the transition phase to lower carbon emitting fuels over the next several decades. Already visible are the alterations of global energy trade patterns and production mixes as many of the consuming countries accelerate their movement away from oil and coal to less environmentally damaging fuels. However, given the relatively slow development of green technologies and their delivery into major energy grids, periodic shortfalls of cleaner energies will present countries with difficult choices between energy security, climate security and energy affordability.

Preface

The central question posed in the current volume concerns the manner in which countries have managed their energy priorities during the early transition years to a reduced carbon environment, thus assuring improved climate security. Have energy security and climate security been treated by groups of countries as competing or complementary goals? Despite past worldwide climate treaties, are there still a number of settings where energy security and energy affordability are preferred over climate security? Are there specific country settings where one set of trilemma priorities is much more likely to occur, or least likely to be found?

To answer these questions, the volume is divided into two main parts. The first part describes the current environment and constraints under which countries are setting their energy priorities during these transition years. The second part consists of a series of country studies that provide insights as to how nations have been resolving their energy goals and objectives.

Clearly, a book of this scope and sheer length could not have come to completion without the contributions of many individuals. In addition to the volume's many contributors, special thanks go to my colleagues at the Naval Postgraduate School – Bob Springborg, James Russell, Nazeen Barma and Robert McNab, whose help and encouragement proved invaluable. Thanks also to Malik Nassem Abbas, Aminuddin Albek, Brandon Brown, Ellen Canup, Andromeda Windra Cipadi, W. W. Peterson Dumalo Fernando, Michael Lebrun, Daniel Nesmith, Amy Roznowki and Chaudhry Seed Uliah, students in my energy security seminar who helped me design the volume's framework. Greta E. Marlatt of the Naval Postgraduate School Knox Library went far beyond the call of duty to keep me informed of the latest oil developments throughout the course of the manuscript – a task only she could perform. Thanks also to Felicity Watts for a stellar job of copy-editing. Most of all, thanks go to Cathy Hartley, Europa Commissioning Editor, who conceived of the original study, provided ongoing guidance and most importantly provided good cheer and positive encouragement throughout.

Contributors

Robert E. Looney is a Distinguished Professor in the National Security Affairs Department at the Naval Postgraduate School, Monterey California. He received his PhD in Economics from the University of California, Davis. He specializes in issues relating to economic development in the Middle East, East Asia, South Asia, and Latin America. He has published 22 books including: *Economic Policymaking in Mexico: Factors Underlying the 1982 Crisis*, Duke University Press, and *Iraq's Informal Economy: Reflections of War, Sanctions and Policy Failure*, The Emirates Center for Strategic Studies and Research. He has also edited three previous Routledge handbooks: *Handbook of US-Middle East Relations* (2009), *Handbook of Oil Politics* (2012) and *Handbook of Emerging Economies* (2014). He is the editor of the Routledge series *Europa Emerging Economies*. Dr Looney is on the board of editors of *International Journal of World Peace* and *Journal of Third World Studies*. In addition, he has over 300 articles appearing in numerous professional journals, and is a regular contributor to the Milken Institute Review and Foreign Policy's Democracy Lab. As an international consultant, Dr Looney has provided advice and assistance to the governments of Iran, Saudi Arabia, Japan, Mexico, Panama and Jamaica as well as the World Bank, International Labour Office, Inter-American Development Bank, Stanford Research Institute, Rand Organization, and the International Monetary Fund.

Chris Bataille has been involved in energy and climate policy analysis for 20 years, as a modeller, analyst, writer, project manager and executive. He is currently an associate researcher at the Institute for Sustainable Development and International Relations (IDDRI.org) in Paris. At IDDRI he is the lead editor of a special issue of *Climate Policy* on the Deep Decarbonization Pathways Project (www.deepdecarbonization.org), as well as helping manage the DDPP, an international study to assess the feasibility and cost of restraining global temperature increases to +2°C, involving a 90% reduction in Canada's GHG emissions by 2050 to hit a global per capita target of 1.7 t/cap. Chris was also a co-author of the Canadian chapter of the DDPP, in co-operation with Carbon Management Canada. Dr Bataille is an Adjunct Professor at Simon Fraser University in Vancouver, and board member and co-chair of Ecotrust Canada. Dr Bataille was formerly founding managing partner of Navius Research Inc., a Vancouver-based energy policy consulting firm and executive director of MK Jaccard & Associates Inc., another energy policy consulting firm. He has managed many projects, including several large national climate change and energy policy studies for Natural Resources Canada, the National Round-table of the Environment and the Economy, Environment Canada, the Pembina Institute and the David Suzuki Foundation, the Ontario Ministry of the Environment and the Ontario Power Authority. He has published peer reviewed articles in *The Energy Journal* (EJ) and *Climate Policy*, has edited a special edition of the EJ on hybrid energy economy modelling, co-wrote a chapter of the most recent *International Handbook of Energy Economics*, and has written a number of public policy publications, including 'Pricing Greenhouse Gas Emissions: The impact on Canada's competitiveness' for the C.D. Howe Institute.

Mehmet Efe Biresselioglu is an Associate Professor of Energy Security and Policy in the Department of Political Science and International Relations, and Head of Sustainable Energy Division at Izmir University of Economics, Turkey. He is also acting as a Steering Committee Member of European Energy Research Alliance's (EERA) Joint Programme on Economic, Environmental and Social Impacts of Energy Policies and Technologies, and the Head of Izmir Circle of Mediterranean Citizens' Assembly (ACIMEDIT). He also taught as a Visiting Lecturer on Energy Security and Politics at Den Haag University, University of Aalborg, Budapest College of Management and University of Minho. Previously, he acted as a Lecturer at the EU's INOGATE Programme; Adviser to the Chairman of Ankara Chamber of Industry; Senior Visiting Research Fellow in the Norwegian Institute of International Affairs' Energy Programme; non-resident Junior Fellow at the Finnish Business and Policy Forum and Management Trainee at Deniz Investment's headquarters. He completed his PhD at IMT Institute for Advanced Studies, Lucca, Italy and also received his Doctor Europaeus degree with the approval of the Confederation of EU Rectors. He received his MA in European Studies from Jean Monnet Centre of Excellence at the University of Turku, Finland and his BA in Political Science and Public Administration from Bilkent University, Turkey. His research interests lie in the areas of energy security, energy politics, energy economics, geopolitics and sustainable energy. He has published several articles related to energy issues in leading journals. He is also the author of *European Energy Security: Turkey's Future Role and Impact* (Palgrave Macmillan, 2011) and the editor of *Enerji Güvenliği Perspektifinden Türkiye'ye Bakış* (EHAEY, 2015). He writes a monthly column on energy security and politics for *Energy World*, a Turkish periodical.

John Calabrese teaches US foreign policy at American University. He also serves as a Scholar at the Middle East Institute where he directs the Middle East-Asia Project (MAP). He is the Book Review Editor of *The Middle East Journal* and was Editor of *MEI Viewpoints*. He is the author of *China's Changing Relations with the Middle East* and *Revolutionary Horizons: Iran's Regional Foreign Policy*. He has edited several books and has written numerous articles on the international relations of the Middle East.

Luigi Carafa is research fellow at the Barcelona Centre for International Affairs (CIDOB), Spain, where he is in charge of energy and climate change. He is also co-convener of Barcelona Climate Futures, a new forum that engages experts and leaders of society to identify key challenges and actionable solutions under the Paris climate agreement. Previously he was post-doctoral researcher at the University of Cambridge, researcher at the University of Toulouse I, and visiting lecturer at the Zentrum für internationale Studien, Technical University of Dresden. His current work focuses mainly on low-carbon energy transitions in emerging economies, clean electricity in the Middle East and North Africa, as well as international climate negotiations. His research interests are in problems of energy policy and finance applied to developing countries. He has provided technical advice, contributed or participated in policy processes such as UN Rio+20, the Mediterranean Solar Plan and the UNFCCC COP21. He performed fieldwork in Morocco, Tunisia, Jordan, Turkey and China. He holds a PhD in Political Science from the University of Toulouse, and an MA in European Politics from the College of Europe, Bruges.

John S. Duffield is Professor of political science and director of academic assessment at Georgia State University, where he teaches and conducts research on international politics and the politics of energy and climate change. Educated at Williams College (BA), the University of Cambridge (MA) and Princeton University (MPA and PhD), he taught at the University of Virginia and the University of Georgia before joining the faculty at Georgia State in 2002. He is the author of *Power Rules: The Evolution of NATO's Conventional Force Posture* (Stanford, 1995), *World Power Forsaken: Political Culture, International Institutions, German Security Policy after Unification* (Stanford, 1998) and *Over a Barrel: The Costs of US Foreign Oil Dependence* (Stanford, 2008). His most recent book is *Fuels Paradise: The Search for Energy Security in Europe, Japan, and the United States* (Johns Hopkins, 2015).

Matthias Duwe is Head of Climate at the Ecologic Institute, where his research and work focuses primarily on EU energy and climate policy, including its connection to national policy and the international regime. Actively involved in the UN climate negotiations for well over a decade, he is an expert on the role of civil society in climate policy and its decision-making processes. He served as the Director of Climate Action Network Europe (CAN Europe) from 2005 to 2011 and played an active role in key political processes at the EU level including the European Climate Change Programme. He has also acted as a moderator for the International Carbon Action Partnership (ICAP) Summer School series on Emissions Trading. Matthias Duwe studied at Carl von Ossietzky University in Oldenberg, Germany with an emphasis on environmental politics and planning. He holds an MSc in development studies from the University of London.

Gonzalo Escribano is the Director of the Energy Programme at the Elcano Royal Institute for International Studies, Madrid, and professor of applied economics at the Spanish Open University (UNED). He holds a degree and a PhD in Economics from the Universidad Complutense de Madrid. His research agenda focuses on the international political economy of energy, including the external dimension of Spanish and European energy policies, the geopolitics and geo-economics of energy in the Mediterranean and Latin America, the transnational deployment of renewable energies, and the linkages between energy resources and economic development. Among his most recent international energy-related publications are: *Energy Security for the EU in the 21st Century: Markets, Geopolitics and Corridors* (Routledge, 2012), and several book chapters and articles in journals including *Energy Policy*, *Renewable and Sustainable Energy Reviews*, *European Journal of Political Economy*, *Global Policy*, *Turkish Studies*, *Global Governance* and *Mediterranean Politics*.

Nick Evans is a Research Assistant at the Ecologic Institute, where he supports work on EU and international climate policy. He completed his undergraduate studies in psychology at Davidson College, North Carolina and is currently pursuing a Master's in environmental policy and planning through the Environmental Policy Research Centre at the Freie Universität Berlin.

Fabio Farinosi is an environmental economist by training and a researcher in the fields of natural resource management, energy, disaster risk reduction, and climate change adaptation and mitigation strategies. He recently earned a PhD from the Science and Management of Climate Change programme at the Ca' Foscari University of Venice in Italy. Fabio Farinosi is currently post-doctoral research fellow in the Sustainability Science Program at Harvard University. In the last few years, he has been involved in research projects about natural hazards impact assessment, integrated water resource management, risk analysis, policy assessment and implementation. He worked as researcher at the Fondazione ENI Enrico Mattei and the Euro-Mediterranean Centre for Climate Change. There, he was involved in several research projects including natural hazard risk and impact analysis, policy assessment and implementation, water management, and analysis of climate change adaptation strategies. Among the other experiences abroad, he spent a semester at the United Nations Environment Programme – Regional Resource Centre for Asia and the Pacific in Bangkok, where he carried out a study on community driven rural development and community based adaptation to climate change in Thailand.

Francesco Femia is Co-Founder and Director of the Center for Climate and Security, where he leads the Center's policy development, analysis and research programmes, and facilitates the primary forum for climate and security dialogue in the US national security community. He has written, published and spoken extensively on the security implications of climate change, water stress and natural resource mismanagement in Syria and North Africa, including in the seminal report 'The Arab Spring and Climate Change', and in the SAIS Review of International Affairs, among others. He is also a regular commentator on how militaries and intelligence communities address climate change risks. He previously served as Program Director at the Connect US Fund, where he directed programmes ranging from international climate policy, to mass atrocity prevention and response. At the Fund, he founded and facilitated the US Climate Leadership Group, a multi-stakeholder effort involving policy institutes and donors in the national security and development sectors. He has over a decade of experience conducting research and policy development on the intersection of climate change, national and international security. Francesco has written for the *SAIS Review of International Affairs*, *Angle Journal*, *Defense News*, the Reuters Foundation, the *National Journal*, the *Bulletin of Atomic Scientists*, *Climate Progress* and *e-International Relations*, and is frequently cited on climate and security issues, including in the G7-commissioned 'A New Climate for Peace' report, and the UK Foreign and Commonwealth Office's 'Climate Change: A Risk Assessment'. He also serves on the advisory board of the Nuclear Security Working Group.

Benjamin Görlach is an environmental economist and Senior Fellow at the Ecologic Institute in Berlin, Germany. His main areas of work are the evaluation of environmental policy instruments and assessment of their performance as well as the economic valuation of environmental goods and services. He has authored and co-authored numerous reports and articles on the European climate policy instrument mix, in particular regarding the functioning of economic instruments for climate mitigation. Since 2009 he has facilitated and taught courses on Emission Trading to participants from emerging economies all across the globe in co-operation with the International Carbon Action Partnership (ICAP) Summer School series. Benjamin Görlach studied economics in Freiburg (Germany), Maastricht (the Netherlands), and Dublin (Ireland) and holds an MSc in international economic studies from the University of Maastricht. In 2006, he was awarded a prestigious Marshall Memorial Fellowship from the German Marshall Fund of the United States.

Peter R. Hartley is the George and Cynthia Mitchell Professor of Economics at Rice University. He is also a Rice Scholar of Energy Economics at the Baker Institute. He has worked on energy economics issues for 35 years, focusing originally on electricity, but also including work on natural gas, oil, coal, nuclear and renewable energy. He has also published in other areas including theoretical and applied issues in money and banking, business cycles, and international finance. He received his PhD from the University of Chicago and was an Assistant Professor of Economics at Princeton University. Originally from Australia, he has also held visiting appointments at Monash University, Melbourne University, the Australian National University and the University of Western Australia.

Peter Karnøe is professor in Innovation and Sustainable Development at Aalborg University, Department of Planning and Development, Denmark, and is associated with the Centre for Design, Innovation and Sustainable Transition (DIST). He has a background in Science, Technology, and Innovation Studies and has extended that perspective with Market and Valuation Studies and studied the co-creation of markets for renewable energy and clean-tech. Theoretically he takes a relational view on the complexes of sociotechnical entanglements that make up technological systems through the making and re-making of agencies and their arrangements. He has fostered agency-based conceptions like 'Distributed entrepreneurship through bricolage or breakthrough', 'Path creation (versus path dependency thinking)'. His current research is on the 'Innovative re-making of markets for a wind power dominated renewable energy system', which takes a performativity and interventionist approach to the re-construction of sociotechnical complexes that constitute energy technologies, and their markets and actors.

Marcus D. King is John O. Rankin Associate Professor in the Elliott School of International Affairs. Dr King also directs the Elliott School's Master of Arts in International Affairs Program. He has taught a number of courses on energy and environmental security including research methods and a capstone workshop. Dr King joined the Elliott School from the research staff of CNA Corporation's (Center for Naval Analyses). Previously, he was globalization planning fellow at Georgetown University. Dr King held Presidential appointments in the cabinet offices of the US Secretary of Energy and Defense, where he represented the USA in negotiation of treaties including the Kyoto Protocol. He has published widely, including book chapters and journal articles on the security implications of resource scarcity and energy policy. His current book project examines ties between water scarcity and violent extremists groups. He holds a PhD in International Relations and an MA in Law and Diplomacy from the Fletcher School at Tufts University.

R. Andreas Kraemer is Founder and Director Emeritus of the Ecologic Institute in Berlin, Germany and Founding Chairman of the Ecologic Institute US in Washington, DC. He is currently Senior Fellow at the Institute for Advanced Sustainability Studies (IASS) in Potsdam, Germany, the Centre for International Governance Innovation (CiGi), Waterloo, ON, Canada, Visiting Scholar (2015) at the Massachusetts Institute of Technology (MIT) Center for Energy and Environmental Policy Research (CEEPR), and Visiting Assistant Professor of Political Science and Adjunct Professor of German Studies at Duke University. His research focuses on the role and functions of science-based policy institutes or 'think tanks' in theory and practice in different political systems, the interactions among policy domains and international relations, and global governance on environment, resources, climate and energy. R. Andreas Kraemer is also Manager of the Konrad von Moltke Fund and co-Chairman of the Advisory Boards of OekoWorld, which sets global investment criteria for ethical and 'green' investment funds or mutual trusts, and Oekom Research, a rating agency specializing in corporate and governmental or 'sovereign' debtors' credit risk, ethics and sustainability.

Michael Kugelman is the senior associate for South and Southeast Asia at the Woodrow Wilson Center, where he is responsible for research, programming and publications on the region. His main area of specialization is Pakistan, India and Afghanistan and US relations with each of them. Michael Kugelman writes monthly columns for Foreign Policy's South Asia Channel and monthly commentaries for War on the Rocks. He also contributes regular pieces to the *Wall Street Journal*'s Think Tank blog. He has published op-eds and commentaries in the *New York Times, Los Angeles Times*, Politico, CNN.com, Bloomberg View, The Diplomat, Al Jazeera, and The National Interest, among others. He has been interviewed by numerous international major media outlets. He has also produced a number of longer publications on South Asia, including the edited volumes *Pakistan's Interminable Energy Crisis: Is There Any Way Out?* (Wilson Center, 2015), *Pakistan's Runaway Urbanization: What Can Be Done?* (Wilson Center, 2014), and *India's Contemporary Security Challenges* (Wilson Center, 2013). He has published policy briefs, journal articles and book chapters on issues ranging from Pakistani youth and social media to India's energy security strategy and transboundary water management in South Asia.

Caroline Kuzemko is Assistant Profesor in International Political Economy at the University of Warwick. She was previously a Senior Research Fellow and member of the EPSRC funded project 'Innovation and Governance' (IGov). IGov investigates complex inter-relationships between energy governance and innovative, sustainable practice change in energy systems, with an emphasis on demand management. In particular, she was responsible for comparing governing for innovations between Germany and the UK. From October 2011 to October 2012 she worked as a post-doctoral research fellow in the 'Energy and Environment' workpackage of the EU funded GR:EEN project at the University of Warwick focusing on EU energy and climate policy, and EU-Russia energy relations. She is also a Visiting Fellow at the Centre for the Study of Globalisation and Regionalisation (CSGR) at the University of Warwick. She has expertise in international political economy with specialized interests in energy and climate policymaking, governance norms and institutional change. She is the founder and convener of an academic network 'Political Economy of Energy in Europe and Russia' (PEEER) and the co-convener of the PSA special group 'Antipolitics and Depoliticisation' (APDSG).

Peter Maslanka is an analyst for the US Navy's Littoral Combat Ship programme. An earlier publication of his appears in the *Journal of Energy Security*. He previously worked as a graduate intern at the US Embassy in Jakarta, and also as an analyst at two Asia-focused political consultancies in Jakarta and Washington, DC. Peter Maslanka was a US-Indonesia Society language fellow at Universitas Gadjah Mada in Indonesia. He holds a master's degree in International Relations from Seton Hall University.

Julia Nesheiwat assumed the position of Presidential Deputy Envoy for Hostage Affairs in August 2015. Previously, Dr Nesheiwat was a Deputy Assistant Secretary of State in the Bureau of Energy Resources, where she focused on the nexus of national security, economic development and climate change. Her work in national security, bilateral and multilateral diplomacy, climate change and environment was closely linked to global energy issues for the 21st century: advancing energy security by promoting access to secure, reliable, and ever-cleaner sources of energy. Dr Nesheiwat also served as a visiting Professor at the US Naval Postgraduate School's National Security Affairs Department. She has held numerous government positions including Chief of Staff and Senior Adviser to the Under Secretary for Economic Affairs and US Special Envoy for Eurasian Energy Security, where she focused on energy security issues for Europe and Central Asia. Dr Nesheiwat also served as a Council on Foreign Relations Fellow in Japan, and on the Governing Advisory Council for the World Economic Forum. She earned a PhD from Tokyo Institute of Technology in Japan; an MA from Georgetown University, Washington, DC; and a BA from Stetson University in Florida.

Lydia Powell is head, ORF (Observer Research Foundation) Centre for Resources Management and Senior Fellow, ORF, New Delhi, India, and has been with the Foundation for over 12 years working on policy issues in energy, water and climate change in the Indian context. Her current interests include energy security, energy access, carbon constraints and their impact on India's energy security and efficiency of coal based power generation in India. Ms Powell contributes commentary and analysis to a weekly newsletter on the Indian energy sector. Her most recent paper was on India's approach to climate change negotiations. Ms Powell has also worked for Norsk Hydro and for Orkla, two of Norway's largest conglomerates whose interests include energy. Ms Powell has three postgraduate degrees: two on Energy Management from Norway and one in Solid State Physics from India.

James A. Russell serves as Associate Professor in the Department of National Security Affairs at the Naval Postgraduate School, where he teaches courses on Middle East security affairs, terrorism and national security strategy. From 1988–2001 Dr Russell held a variety of positions in the Office of the Assistant Secretary of Defense for International Security Affairs, Near East South Asia, Department of Defense. During this period he travelled extensively in the Persian Gulf and Middle East working on US security policy. His articles and commentaries have appeared in a wide variety of media and scholarly outlets around the world and he blogs at www.lobelog.com/author/james-russell. His articles include 'Nuclear Reductions and Middle East Stability: Assessing the Impact of a Smaller Nuclear Arsenal', *Nonproliferation Review* 20: 2 (Summer 2013), 263–268 and 'Counterinsurgency American Style: Considering David Petraeus and 21st Century Irregular War', in *Small Wars and Insurgencies* 25:1, 69–90. His latest books are *Military Adaptation in Afghanistan*, which he edited with Theo Farrell and Frans Osinga (Stanford University Press, 2013) and *Innovation, Transformation and War: US Counterinsurgency Operations in Anbar and Ninewa Provinces, Iraq, 2005–2007* (Stanford University Press, 2011). Professor Russell holds a Master's in Public and International Affairs from the University of Pittsburgh and a PhD in War Studies from the University of London.

Jack D. Sharples is a Lecturer in Energy Politics and Energy Law at the European University at St Petersburg, Russian Federation, where he also teaches EU Politics, and is an active researcher in the International Energy Centre. His academic research focuses on the political economy and geopolitics of energy relations between Russia, the EU and Ukraine, particularly in the sphere of natural gas. His previous research on the interplay between Russian environmental policy and energy policy was published in the article 'Russian approaches to energy security and climate change', in the journal *Environmental Politics*. Dr Sharples is the author of the Gazprom Monitor reports for the European Geopolitical Forum (Brussels) – a report published in monthly and annual editions analysing the foreign activities of the Russian state-owned gas-exporting company, Gazprom. Dr Sharples received his PhD from the Centre for Russian, Central and East European Studies (CRCEES) at the University of Glasgow, UK, having written his PhD thesis on state-business relations in the Russian gas sector.

Adam Simpson is Director of the Centre for Peace and Security within the Hawke Research Institute and Senior Lecturer in the International Relations programme within the School of

Communication, International Studies and Languages, University of South Australia. He is Adjunct Research Fellow at the Centre for Governance and Public Policy, Griffith University, and is currently a Visiting Research Scholar at the Centre for Southeast Asian Studies, Kyoto University. He previously taught at the University of Adelaide, where he remains an Associate in the Indo-Pacific Governance Research Centre. His research adopts a critical perspective and is focused on the politics of the environment and development in South-East Asia. He has published in journals such as *Pacific Review, Third World Quarterly* and *Environmental Politics*. He is the author of *Energy, Governance and Security in Thailand and Myanmar (Burma): A Critical Approach to Environmental Politics in the South* (Routledge, 2014) and is lead editor of the forthcoming *Routledge Handbook of Contemporary Myanmar*.

Morena Skalamera recently completed her PhD in Political Science and International Relations at the University of Trieste, Italy, which dealt with European-Russian energy co-operation and was funded by the International University Institute for European Studies (IUIES). Her dissertation was primarily focused on understanding the lack of binding institutionalization of the EU-Russia energy relationship despite the high degree of interdependence between the two sides. Morena Skalamera holds a BA and MA in Political Science and International Relations from the University of Trieste. During her tenure at the Belfer Center, she has been conducting research on a new project on the Sino-Russian gas relationship and China's disruptive rise in energy and geopolitics. Her areas of expertise and interest include energy co-operation between the EU and Russia, global energy governance, geopolitical and strategic issues arising from the unequal distribution of global energy resources (in particular, natural gas), the role of technological breakthroughs, Sino-Russian energy co-operation and the making of the USA's foreign and security policies. In summer 2013 she taught a seminar on the Geopolitics of Energy at the Peking University School of Government in Beijing (China). In the academic year 2015–16 she is also Visiting Fellow at Moscow's Higher School of Economics.

Mattijs Smits is Assistant Professor at the Environmental Policy Group of Wageningen University, the Netherlands. He holds degrees from four different universities on three continents: a BSc and MSc from the University of Utrecht and Wageningen University, and PhD degrees from the University of Sydney and Chiang Mai University (as part of a cotutelle arrangement). His research and teaching focus on (renewable) energy policy and politics, environment, sustainability, (rural) development and carbon markets. During his academic and professional career, he spent extended periods living and working as researcher and consultant in South-East Asia, notably in Laos, Thailand and Vietnam. He has published in journals such as *Geoforum, Energy Policy, Water Alternatives* and *Forum for Development Studies*. He is the author of *Southeast Asian Energy Transitions: Between Modernity and Sustainability* (Ashgate, 2015).

Robert Springborg is Visiting Professor in the Department of War Studies, King's College London and non-resident Research Fellow of the Italian Institute of International Affairs. Until October 2013 he was Professor of National Security Affairs at the Naval Postgraduate School and Programme Manager for the Middle East for the Centre for Civil-Military Relations. From 2002 until 2008 he held the MBI Al Jaber Chair in Middle East Studies at the School of Oriental and African Studies in London, where he also served as Director of the London Middle East Institute. Prior to this he was Director of the American Research Center in Egypt. From 1973 until 1999 he taught in Australia, where he was University Professor of Middle East Politics at Macquarie University. He has also taught at the University of California, Berkeley, the University of Pennsylvania and elsewhere. Professor Springborg's publications include

Mubarak's Egypt: Fragmentation of the Political Order; Family Power and Politics in Egypt; Legislative Politics in the Arab World (co-authored with Abdo Baaklini and Guilain Denoeux); *Globalization and the Politics of Development in the Middle East* (co-authored with Clement M. Henry); *Oil and Democracy in Iraq; Development Models in Muslim Contexts: Chinese, 'Islamic' and Neo-Liberal Alternatives* and several editions of *Politics in the Middle East* (co-authored with James A. Bill). He co-edited a volume on popular culture and political identity in the Gulf that appeared in 2008. He has published in the leading Middle East journals and was the founder and regular editorialist for *The Middle East* in London, a monthly journal that commenced publication in 2003. He has worked as a consultant on Middle East governance and politics for the United States Agency for International Development, the US State Department, the UNDP and various UK government departments, including the Foreign and Commonwealth Office, the Ministry of Defence and the Department for International Development, and has advised various intelligence organizations in the USA.

Jens Stissing Jensen is assistant professor at the Department of Development and Planning, Aalborg University, Denmark, and is associated with the Centre for Design, Innovation and Sustainable Transition (DIST). Coming from a background in philosophy and environmental planning, his current research interests span sectoral transformation processes, transitions of socio-technical systems, urban transformation and the relation between economic institutions and socio-technical transitions. His main focus is on the epistemic policies of such transitions, i.e. the politics involved in describing and delineating societal system and sector as objects of governance. His research addresses empirical domains such as construction, the built environment, energy, water, mobility and cities. Most of his latest research focuses on cities as contexts for transitions processes, due to the ongoing strategic work that goes into managing the ambiguities and tensions among the various systems and practices that characterize such contexts.

Caitlin Werrell is Co-Founder and President of the Center for Climate and Security, where she leads the Center's policy development, analysis and research programmes, and facilitates the primary forum for climate and security dialogue in the US national security community. She has written and published extensively on the security implications of climate change, water stress and natural resource mismanagement in Syria and North Africa, including in the seminal report 'The Arab Spring and Climate Change', and in the *SAIS Review of International Affairs*, as well as on the potential for new technologies such as additive manufacturing to address climate risks. Her primary research interests include climate change, water policy and international security. She has spent over a decade investigating the intersection of security, natural resources, conflict and co-operation. Caitlin Werrell has experience in international and domestic climate and water policy, including as co-founder of the MAP Institute for Water & Climate, a Senior Associate at AD Partners, and as Director of International Programs at EDN. Caitlin has written for *Angle Journal, Defense News*, the Reuters Foundation, the *National Journal*, the *Bulletin of Atomic Scientists, Climate Progress* and *e-International Relations*. She holds a master's degree from the University of Oxford, where she focused on transboundary water issues, concluding with a field study on water conflict and co-operation in Cyprus. Caitlin Werrell holds a BA in Environmental Politics from Mount Holyoke College. Caitlin also serves on the advisory board of the Nuclear Security Working Group.

Duncan Wood is the director of the Mexico Institute at the Wilson Center. Prior to this, he was a professor and the director of the International Relations Programme at the Instituto Tecnológico Autónomo de México (ITAM) in Mexico City. He has been a member of the

Mexican National Research System, an editorial advisor to both *Reforma* and *El Universal* newspapers, and is a member of the editorial board of *Foreign Affairs Latinoamerica*. In 2007 he was a non-resident Fulbright Fellow and, between 2007 and 2009, he was technical secretary of the Red Mexicana de Energía, a group of experts in the area of energy policy in Mexico. He has been a Senior Associate with the Simon Chair and the Americas Program at the Center for Strategic and International Studies (CSIS) in Washington, DC. His research focuses on Mexican energy policy, including renewable energy, and North American relations. He studied in the UK and Canada, receiving his PhD in political studies from Queen's University, Canada, and is a recipient of the Canadian Governor General's Visit Award for contributions to the Mexico–Canada relationship.

Xu Yi-Chong is a professor in the Centre for Governance and Public Policy and the School of Government and International Relations, Griffith University. Her research expertise is in the areas of energy security, international organizations, and nuclear policy in China and elsewhere. She holds a PhD from the University of Alberta, Canada. Her areas of interest include energy security, international organizations, and China and regional studies. Xu Yi-Chong has published numerous professional papers and has written eight books, including *The Politics of Nuclear Energy in China* (Palgrave Macmillan), *Electricity Reform in China, India and Russia: The World Bank Template and the Politics of Power* (Edward Elgar) and *Powering China: Reforming the Electric Power Industry in China* (Ashgate).

Abbreviations

CEO	Chief Executive Officer
CO_2	Carbon dioxide
COP	Congress of Parties (to the United Nations Framework Convention on Climate Change)
EU	European Union
EU-28	The 28 member states of the European Union
G7	Group of Seven
GDP	Gross domestic product
GHG	Greenhouse gas
IMF	International Monetary Fund
LNG	Liquefied natural gas
m.	million
MENA	Middle East and North Africa
MW	megawatt(s)
NGO	Non-governmental organization
OECD	Organisation for Economic Co-operation and Development
OPEC	Organization of the Petroleum Exporting Countries
UAE	United Arab Emirates
UK	United Kingdom
UN	United Nations
UNDP	United Nations Development Programme
UNFCCC	United Nations Framework Convention on Climate Change
US(A)	United States (of America)

Part I
The policy setting

1

Introduction

Robert E. Looney

Overview

With increased concern over global warming, countries are finding energy security at affordable supplies is no longer simply a matter of diversifying energy sources and expanding use of the cheapest type of energy. Energy security and affordable energy now come with a cost – greenhouse gases that contribute to climate change. With environmental sustainability as an additional energy goal, countries are often faced with the hard choice of improved energy security coming at the expense of either increased energy costs, or reduced climate security.

Focusing on these energy choices the theoretical underpinning of the volume is the idea of an energy trilemma,[1] and the associated policy constraints facing countries as they attempt to achieve their main energy priorities. The trilemma implies that given a spectrum of different energy types, their associated costs, and their varying impact on climate security, countries will be forced to prioritize the goals of energy security, energy equity and climate security – high levels of all three will be very difficult to achieve at any one time. By opting for the top two, countries will likely find a deterioration in the third. For example, if clean, renewable energy costs more to generate than conventional power, improved energy security (more domestic sources of energy) and an improved environment (fewer greenhouse gas emissions) would result in higher energy costs

In terms of anticipating the likely success of global efforts at combatting global warming through voluntary cut-backs in greenhouse gasses, it's useful to know if there is a predictable pattern of energy goal tradeoffs. Specifically, are countries in certain energy settings more likely to prioritize the same two energy goals at the expense of the third?

One can easily think of different national settings/environments where the trilemma tradeoffs may be fairly similar. For example, the major energy exporters automatically have less difficulty achieving high levels of energy security and energy equity. Other countries with limited domestic fossil fuel supplies can easily improve their environmental sustainability, and with it improved energy security due to a reduction in energy imports.

While a number of country groupings are conceptually possible, four provide logical starting points. On this basis, countries examined in this volume were placed in one of four groups: (1) carbon producing countries, (2) carbon producing and consuming countries, (3) carbon consuming countries. Several of the carbon consuming countries have shown a clear stated preference for the environment. Hence these were placed in a fourth group, the carbon reduction countries.

Energy priorities are also likely to be greatly affected by the global policy environment countries find themselves. The volume's first section identifies many of these elements and discusses their relevance in affecting national energy choices.

Policy environment

A number of recent dramatic developments in energy markets and technologies are producing forces and consequences that are not yet completely understood. However, as Marcus King (Chapter 2) notes, some global trends are discernable and must be taken into account if countries are to design strategies to balance their preferred mix of energy security, climate security and economic competitiveness (energy availability). As he correctly points out, the development of clean energy technologies will be undertaken by many countries to maintain competitiveness while advancing the other primary goals of energy and climate security. Complicating the energy policy-making in most countries is the fact that rapid changes in clean energy technologies in such areas as nuclear energy, cleaner fuels and the possibility of geoengineering of climate will no doubt alter the impacts of trends currently underway. One thing is clear, however, countries successful in managing their energy trilemmas will be the ones able to rapidly adapt to this changing environment.

How well are most countries and international bodies prepared for the consequences of climate change? Francesco Femia and Caitlin Werrell see a large gap between the climate risks nations and peoples face, and the capacity and political will to respond to these risks. Given past failures they suggest the necessity of improving, augmenting, and possibly even creating new international, regional, national and sub-national structures for addressing climate change. But this is just a start. They feel serious responses will require that nations, and international institutions, place climate change at the top of the international security agenda, and find ways of collaborating on reducing those risks. This isn't wishful thinking. Their well-researched essay shows why an international climate and security imperative is both necessary, and achievable.

The energy trilemma is not a universal constant that will be with us for the indefinite future. As Peter Hartley shows, the trilemma is purely a transition phenomenon since reducing fossil fuel combustion should increase energy security while also reducing potentially harmful climate change. Although we have two policy goals, they should be treated as one, since one policy instrument can simultaneously further both goals. The trilemma problem arises in the transition because the costs of reducing CO_2 emissions are currently high, especially for developing countries. Eventually these costs, through research and development, will come down and at that time policies to force reduced fossil energy consumption would be unnecessary. He notes that in the transition policies aimed at encouraging basic research to lower the cost of new energy technologies, limiting the harmful consequences from climate change, or contending better with damaging weather events would yield far greater expected benefits for a comparable level of expected costs.

The costs of renewable energy will eventually come down, but not before a number of mistakes are made and money lost. While there are a number of good case studies to illustrate this point, one of the best involves a solar power project, the Desertec vision, in Northern Africa, Luigi Carafa and Gonzalo Escribano (Chapter 5) delve deeply into the factors that led to the failure of the project. They conclude that in the future the project's troubles will require a rethinking of regional and industrial cooperation around a more inclusive narrative of sustainable energy development

Regional energy cooperation is increasingly seen as a means of achieving increased energy security, lower energy costs and a coordinated approach to achieving greater environmental sustainability. As one might imagine however, policy coordination across member countries with different energy needs, resource endowments, and priorities can be challenging. The two final essays in this section illustrate the difficulties and benefits of regional cooperative efforts.

In the first, focused on the EU, Benjamin Görlach, Matthias Duwe, and Nick Evans (Chapter 6) show there are a number of dynamics at work in EU climate and energy policy, which are not necessarily aligned. This has resulted in tension between unified EU targets on the one hand, and starkly different views of EU Member States about their future energy supply on the other hand.

In the second, Julia Nesheiwat (Chapter 7) ends the section on a more positive note. She sees regional coordination in energy systems not as a zero-sum game, but one that can benefit the national interest of all parties. Of course, there are always challenges and risks involved, but she shows that given the mutual problems of climate change and energy security, there is good reason for countries to cooperate. The integration of regional energy markets increases diversification of supply and delivery, cost-savings, and energy efficiency. Working together, countries can reduce climate change effects while promoting sustainable energy solutions and global energy security for the future.

With this background on energy trends, security concerns, climate developments, technological change and efforts at regional coordination the country case studies that follow show how many of these factors have played out in a broad spectrum of national settings. A number of interesting patterns emerge, leading to a set of predictions concerning where efforts at combatting climate change are likely to be the most concerted, and those where only limited progress can be expected.

Country studies

The first group of countries (Table 1.1) consists of carbon producing countries. In these countries energy production easily outstrips domestic energy usage enabling significant amounts of energy to be exported to international markets. Each country has consistently scored low in energy sustainability, in part because their abundant energy has simply enabled improved energy security and to a lesser extent, energy affordability to progress easily. However, just because a country has abundant fossil fuel supplies, energy security, as in the case of Saudi Arabia, is not automatically assured if the country becomes overly dependent on fuel exports. In addition to Saudi Arabia, the countries in this section include three other oil exporters, Canada, Russia, and Mexico. South Africa is the lone coal producer.

James Russell (Chapter 8) correctly assumes that it is unrealistic to assume a truly effective global accord to limit carbon emissions will be possible without the agreement of Saudi Arabia (and its Gulf State neighbours). Unfortunately, to date there has been little enthusiasm in that part of the world for a concerted effort to limit greenhouse gas emissions. The Kingdom's energy priorities largely lie in providing the domestic population with low cost energy (energy equity in Table 1.1).

Professor Russell notes that while Saudi Arabia has done a good job in slowing down the world's progress towards a climate accord, the Kingdom will face increasing international pressure to do its part in limiting carbon emissions. In this sense, the Kingdom is fighting a losing battle over climate change and its transition towards a more environmentally sustainable environment will be largely forced upon it, either indirectly through international condemnation, or directly through lower international oil prices as countries concerned over global warming transition to cleaner types of energy. Ultimately the shift away from oil internationally will force the Saudis to gradually diversify into more environmentally friendly forms of energy – solar and even nuclear. Professor Russell concludes that Saudi Arabia can survive for a time through oil revenues, but its days as a profligate welfare-spending state are slowly but surely coming to an end.

Canada and its oil stands are another country that has faced growing international pressure to cut back or even abandon production. Chris Bataille (Chapter 9) warns that, as in the case of Saudi Arabia, global climate security cannot be achieved solely by convincing fossil fuel exporting regions to stop producing. If pressure on one supplier is successful, as it seems to have been for the oil sands, there is a long list of potential suppliers waiting to take up the slack. Dr Bataille concludes that the best way to real climate security is to create real and perceived alternatives to fossil fuel consumption, and apply the necessary carrots (subsidies, development support) and sticks (technology performance regulations, carbon pricing) to encourage the great majority of firms and consumers to transition to low carbon options. As in the case of Saudi Arabia, this action will create a falling demand for oil, forcing Canada to limit oil sand development, and inducing the country to further develop green energies such as the country's hydro energy potential.

Russia is another country where energy security and affordability take precedence over environmental sustainability (Table 1.1). As Jack Sharples (Chapter 10) shows, the result has been little in the way of reduced greenhouse gas emissions. Unfortunately, the international community has, in contrast to Saudi Arabia and Canada, little leverage in inducing an energy path in Russia less associated with greenhouse gas emissions. Specifically, natural gas already provides the source fuel for half of Russia's electricity generation, with non-CO_2 emitting hydro and nuclear power providing a further third.

As Dr Sharples argues, the country's latent renewable potential will not be developed any time soon because at the present time it is simply not commercially profitable. In addition, shifting to renewables would require feed-in tariffs and electricity prices that are not politically and economically viable at present. In any case, with a low priority given to environmental sustainability together with abundant fossil fuel supplies an energy 'transition' in Russia is not expected in the medium-term future. The only significant development that may impact Russia's CO_2 emissions is the increasing use of nuclear power in place of coal for electricity generation.

Historically, Mexican energy policy has favoured energy security and equity to the detriment of energy sustainability. As a result, the country is the world's 10th largest emitter of greenhouse gas (GHG) emissions. However, Duncan Wood (Chapter 11) finds today that the country is assuming a leadership role in the international battle against climate change. He also finds that in recent years the Mexican government has initiated meaningful changes in national legislation aimed at reducing the country's GHG emissions. At the same time, the country has been successful in building a viable renewable energy industry.

Dr Wood feels the country's recent energy initiatives have defied those sceptics who argued that an oil producing state could not be taken seriously as a climate change leader. However, as he cautions, achieving the country's ambitious GHG reduction targets will require an enormous commitment of time, resources and political capital over a long period of time and across several administrations, a feat that has traditionally proved difficult in Mexico. Mexico's shift in energy priorities is too recent to have had a significant impact on the elements of the energy trilemma, but as Dr Wood notes, there is the hope that in the short term thanks to the modernization of the energy sector, in particular electricity generation results will soon be readily apparent.

The final country in the energy producer group is South Africa, a major coal producer. For years the country successfully implemented a 'minerals-energy-complex' development model, whereby cheap energy was able to give the country's mines a competitive advantage. However, as Robert Looney observes (Chapter 12) in recent years this energy paradigm has come under increasing criticism as being no longer economically or environmentally sustainable. Part of the problem stems from years of underinvestment by the state power company, Eskom. By 2008, demand for electricity began to noticeably outstrip supply, but because new investment in generating capacity was not forthcoming for several years the country experienced sharply rising power tariffs and frequent periods of load-shedding, brownouts and even blackouts.

To alleviate the situation, the South African government finally made the pragmatic decision to turn to the private sector and its capacity to quickly deliver renewable energy to the nation's grid. So far the experiment has been a huge success and the government is planning to gradually cut back coal usage, both on the grounds that this is the surest way in the near term to expand the country's electricity generating capacity, but also as a way of protecting the country's fragile environment from the effects of climate change.

Summing up this group of countries, one finds energy policies set to serve national interests and not necessarily taking into account global concerns over climate change. No doubt, when energy transitions come to Saudi Arabia, they will be largely induced by outside forces. Canada has active pro-environmental groups, so there is a much greater chance of a shift towards environmentally friendly energy sources. On the other hand, Russia seems immune to such pressures suggesting little hope for a significant reduction in greenhouse gas emissions over the foreseeable future. Finally, Mexico and South Africa represent cases where previously successful carbon based development strategies were becoming unsustainable, forcing a transition to a lower carbon energy mix. In Mexico's case it was choking pollution and a bankrupt state oil company, while in South Africa's case severe energy shortages, brought on by an under-funded, state electricity company, were an increasing strain on the economy.

Countries in the second group are both major carbon producing and consuming countries. This group has the highest level of energy security and relatively low energy equity scores (Table 1.1). On the other hand, it is more of a transition grouping with the UK and Brazil elevating a sustainable environment along with energy security above energy equity.

Traditionally, China has placed a high priority on energy security, and to a lesser extent to energy equity. The result has been an extremely low level of energy sustainability – an average ranking of 127 over the 2013–2015 period. As Xu Yi-chong observes (Chapter 13) however, China's energy policies are rapidly changing with China expanding its low-carbon energy sources significantly since 2003. The government's intent in this regard is to ensure adequate energy supplies while minimizing environmental and climate change threats. While the West's perception of policy making in China is one of a monolithic, smoothly functioning bureaucracy capable of rapid transformation, the reality on the ground is very different.

In particular Dr Xu Yi-chong argues China's energy transition will be difficult to implement because of: (a) fragmented government agencies competing for agendas, (b) the government lacking the capacity to adopt coherent and consistent policies and to implement them accordingly, and (c) slow creation of an operational legal and regulatory system. In addition, China's leaders face the challenge of balancing long- and short-term development along with balancing the diverse interests of urban and rural population, coastal and interior regions, and the elite and the masses. Needless to say the story of China's energy transition is far from over.

The United States energy picture is unique in a number of ways. As Robert Looney argues (Chapter 14) despite the fact that the US is the world's second largest consumer of energy (after China), and also a major energy producer, the country has never formulated what might be

Table 1.1 Sample country characteristics

Country group	*World energy council trilemma country rankings*			*WEC Evaluation*			*Trilemma priorities*
	Energy security	*Energy equity*	*Environmental sustainability*	*Security*	*Equity*	*Sustainability*	
Carbon producing							
Saudi Arabia	54	8.7	123	B	A	D	equity, security
Canada	1	2	62.3	A	A	C	security, equity
Russia	6.3	47.3	103.7	A	B	D	security, equity
Mexico	32	50.3	76.3	B	B	C	security, equity
South Africa	38.3	83.3	129	B	C	D	security, equity
Average	*26*	*38*	*99*				*security, equity*
Carbon producing and consuming							
China	19.3	87.3	127.3	A	C	D	security, equity
US	7.7	1	88	A	A	C	equity, security
UK	8	20	19.3	A	B	A	security, sustain
Brazil	33	83.3	17.7	B	C	A	sustain, security
Indonesia	17	78.7	95	A	C	C	security, equity
Egypt	53.7	64.7	75	B	C	B	security, equity
Average	*23.1*	*55.8*	*70.4*				*security, equity*
Carbon consuming							
Japan	64.3	18.7	41	C	A	B	equity, sustain
Thailand	94.3	76	105	C	C	D	equity, security
Pakistan	61.7	104	90	B	D	C	security, sustain
India	68.3	106.3	122	B	D	D	security, equity
Jordan	114.7	61.3	103.7	D	B	C	equity, sustain
Turkey	66	77	72.7	C	C	C	security, sustain
Average	*78*	*74*	*89*				*equity, security*

Country group	*World energy council trilemma country rankings*			*WEC Evaluation*			*Trilemma priorities*
	Energy security	*Energy equity*	*Environmental sustainability*	*Security*	*Equity*	*Sustainability*	
Carbon Reduction							
France	42	9.7	10.7	B	A	A	equity, sustain
Denmark	3.7	43	10.7	A	B	A	security, sustain
Germany	27.7	33	33.7	B	B	B	security, equity
Italy	67.3	44	22.7	C	B	A	sustain, equity
Dominican Republic	114.7	93.7	55	D	C	B	sustain, equity
Costa Rica	57	70	2	B	B	A	sustain, security
Nicaragua	103.7	97.7	71	D	D	B	sustain, equity
Average	*59.5*	*55.8*	*29.4*				*sustain, equity*

Notes: Trilemma data from World Energy Council Trilemma Data Base. Values are the average country ranking for 2013, 2014 and 2015 considered a comprehensive energy policy. Other policy critics contend that the US actually has too many individually focused plans and programmes. These are more often than not limited in scope, with little consideration given to their impacts outside their intended beneficiaries. The result has been a morass of competing and conflicting outcomes.

Interestingly, despite the country's failure at energy planning and the penchant for selecting conflicting policies that often negate each other the country appears, thanks to a series of fortuitous developments, to be on a successful path of transition towards a sustainable increase in climate and energy security.

On the surface, the United Kingdom appears to have managed its energy transition extremely well through the adoption of a number of best practice policies. The transition began with the innovative Climate Change Act of 2008 which committed the country to legally binding, long-term carbon emissions reduction targets and a series of carbon budgets. However, as Caroline Kuzemko (Chapter 15) observes, a closer look suggests that below the surface, the country has not made sufficient progress in policy implementation to effect significant system change. Specifically, she finds the Act's climate targets may be unrealistic because they do not provide enough impetus and direction for political and market actors to respond with supportable innovations that can facilitate profound system change.

In sharp contrast to most emerging/developing countries, Brazil scores the highest on environmental sustainability (Table 1.1), with energy security coming in at a significantly higher ranking than energy equity. In looking at the country's energy sector Fabio Farinosi (Chapter 16) questions whether the country's successful energy transition and pattern of robust economic growth can continue without significant improvements in energy equity. In this regard he finds a number of obstacles will have to be overcome. In particular, he feels the country will have to find more ways of introducing competition into the country's energy sector. To date the Brazilian authorities have shown reluctance to fully opening the national market to international competition. The benefits of increased competition would bring down energy prices thus providing a stimulus to the economy. More importantly, increased economic growth and freer movement of international capital into the energy sector would enable it to better meet the needs of the country.

Economically, Indonesia is a dynamic country with a rapidly growing population. Historically the country has opted for energy security and energy equity at the expense of the environment. As Peter Maslanka notes (Chapter 17), because the country's energy is growing 7% per year, it is becoming increasingly difficult to achieve and ensure energy access for the population while providing adequate energy for the country's economic sectors. An energy transition is not taking place, nor will the country begin addressing climate security for the foreseeable future.

As is the case with Indonesia, policy makers in Egypt have given preference to energy security and affordability at the expense of energy sustainability. Still, as Robert Springborg shows (Chapter 18), while Egypt's per capita carbon dioxide emissions are well below the average of OECD countries, they are high by the standards of lower middle income countries and rising at one of the world's fastest rates, doubling between 1990 and 2010. Egypt's energy efficiency of production has not improved over the past two decades, making it one of the world's least energy efficient producers.

Professor Springborg sees little in the way of an Egyptian energy transition to a reduced carbon environment. Instead, the Sisi government's resolution of the energy trilemma has been along the lines of what might be expected from a country lacking democratic checks and balances. Specifically, there is no energy transition. Instead government has opted for an unsustainable growth model that ignores the environmental consequences of its efforts to accelerate economic growth and expand the country's energy supply.

The next country grouping includes a group of nations that have limited domestic sources of energy, and yet are large consumers of energy. Of the four country groupings this set of countries has by far the lowest level of energy security and energy equity (Table 1.1). These countries also have, next to the carbon producing countries, the lowest level of energy sustainability. There is quite a bit of variation across these countries, however, with Japan, Pakistan, Jordan and Turkey showing a preference for energy sustainability.

Japan has been a global leader in energy and climate reform for decades. As Julia Nesheiwat notes (Chapter 19), it is no accident that Japan hosted the Kyoto Protocol, which is currently the most significant framework for mitigating climate change. Dr Nesheiwat feels that by consistently leading efforts towards a greener globe, Japan had been established as an exemplary energy model for transformation, especially through its use of nuclear technology.

While many nuclear power stations will eventually come back on-line, following the disaster of March 2011 and the shutting down of the country's nuclear plants, the share of power generated from this source will be considerably less than that anticipated before 2011. As Dr Nesheiwat shows, although the nuclear accident represented a serious blow to the country's energy transition to a reduced carbon environment, the country has not been deterred from this goal. Currently, plans are for a significant expansion in renewable energy to gradually replace nuclear and coal sources of power generation.

As is the case with many emerging economies, Thailand, especially in the last several years, has emphasized energy security and energy affordability at the expense of energy sustainability and climate security. This has been particularly the case with the post 2014 military regime. However, as Adam Simpson and Mattijs Smits argue (Chapter 20) the country's democratic history has spawned a dynamic civil society with large groups of environmental activists pressing for a more sustainable environment. After examining several case studies, the authors feel that even under a military government there is a chance for these groups to be successful in their quest for cleaner energy production.

For years, Pakistan has been on the verge of a major energy crisis with demand for energy continually outrunning supply. In addition, the country is highly vulnerable to the effects of climate change. Unfortunately, as Michael Kugelman (Chapter 21) observes, in its efforts to ease its energy woes, the country risks worsening its climate vulnerability. While it might be reasonable to conclude a successful energy transition is out of the question Mr. Kugelman finds reason for optimism. First, the government is well aware of the situation and has taken encouraging steps to address the country's energy and climate problems. In addition, he finds with assistance from international donors, the country has a number of policy interventions that should help mitigate the current difficulties and even put the country on a sustainable path in the not too distant future.

Historically India has been primarily concerned with the country's energy security, with energy affordability or environmental sustainability receiving considerably less attention. Things are changing and as Lydia Powell (Chapter 22) observes, several somewhat parallel, but contradictory, transitions are underway. First there is the low carbon transition taking place through an increasing share of renewable energy in the country's energy mix. Second an energy access transition is attempting to provide growing amounts of affordable petroleum fuels and cheap electricity generated using coal.

Ms Powell's assessment suggests both transitions will likely remain incomplete. As she observes, ironically the continued co-existence of both high and low carbon growth and the co-existence of energy poverty and energy affluence offer yet another example that for everything that is true in India, the opposite is also true.

Over the years Jordan's precarious energy situation has led many observers to predict a major calamity for that country's economy. John Calabrese (Chapter 23) documents the government's attempts to put the country on a more secure energy path. As he shows, the past decade has been marked by the unfurling of a comprehensive energy strategy and the enactment of legislation and policy reforms designed to facilitate its implementation. In addition, the government has launched a number of new energy projects. However, as an energy poor country in a violent, unstable part of the world, energy security needs have been so paramount that concerns of environmental sustainability, at least for the present, have not entered into the country's energy plans in any significant way.

In the last several years Turkey has given top priority to energy security. As in the case of Indonesia, Turkish energy demand is increasing due to rapid economic growth and population increase. It is expected that energy needs will increase by 80% by 2023. Mehmet Efe Biresselioglu (Chapter 24) notes that the overall Turkish energy dependency level is 74%, with much higher rates for oil and natural gas. As a result, the government has made increased energy security through diversification as its prime energy policy goal.

However, as in the case of India, Turkish energy policy has a fundamental contradiction. The country's diversification efforts have not only focused on suppliers, but also on resources. Hence, there are ambitious targets to decrease the share of natural gas in electricity generation. There are two main policy options for this decrease. The first is to increase the share of renewables and the second is to increase the share of coal in electricity generation. The intention is to implement these two policy options simultaneously. Professor Biresselioglu suggests there needs to be a balance between these resources in order to improve Turkey's position in terms of the energy trilemma. Specifically, he advocates these policy options need to be supported by the development of domestic resources, including nuclear energy.

The countries of the final group are also characterized as carbon consuming, but with the significant difference from the previous group in their determination to decarbonize. Each country[2] has placed a priority on sustainable energy, with Italy, Dominican Republic, Costa Rica and Nicaragua scoring the best on the energy sustainability dimension. As a result, this group of countries as a whole scores by far the highest of the four country groupings on the energy sustainability dimension.

France's energy transition began in the 1970s with massive investments in nuclear power. As John Duffield (Chapter 25) observes, nuclear power helped greatly to reduce both France's imports and consumption of fossil fuels, which in turn considerably lowered the country's CO_2 emissions. In addition, the substantial economies of scale made possible by such a large programme and a high degree of standardization meant that the cost of generating electricity from nuclear plants was kept relatively low in comparison with other fuels and other countries. However, these advantages did not come without a cost. Inexpensive nuclear power discouraged the development and introduction of new renewable sources of power. Similarly, an excess of generating capacity and the resulting low power rates discouraged efforts to improve energy efficiency. Currently with increasing concerns over nuclear safety and the escalating costs of new nuclear facilities, the continued dominance of nuclear power in France will be much less certain.

Professor Duffield concludes that France will eventually have to develop a new formula for managing the energy trilemma. The outlines of such a formula have been suggested by the energy law adopted in mid-2015. As in many other developed countries, greater emphasis will have to be placed on renewable sources of energy as well as advances in energy conservation and efficiency. This may result in higher energy costs, at least until technological developments result in lower generation and distribution costs.

Denmark has pioneered many of the new green technologies as part of the country's efforts at decarbonization. The country's new green industries have thrived enabling the country to pursue improvements in energy security without raising the costs of energy to levels experienced in other Western European countries.

Despite past successes in the renewable area, Peter Karnøe and Jens Stissing Jensen (Chapter 26) see some dark clouds on the horizon. In essence the country's energy discussion has shifted from the technical to the political. Specifically, a new Danish right-wing minority government elected in June 2015 has introduced the notion of 'Green realism' in its climate policy. 'Green realism' is based upon the argument that given the significant gains already made by the country in reducing greenhouse gas emissions, the country can take a break from further progress in the area, especially given the material and economic consequences of integrating the increasing amount of wind power generated electricity in the electricity/energy system. What lies ahead for the country's energy transition? Given this surprising development, the authors develop a model capturing the country's political/economic dynamics to get a sense of the changes likely to occur over the next few years.

German energy policy is rather unique in that no one area stands out. Over the period from 2013–2015, Germany had nearly identical scores across the energy trilemma ranking an average 27.7 in energy security, 33 in energy equity and 33.7 in environmental sustainability (Table 1.1). This pattern has not occurred by chance, but as Andreas Kraemer (Chapter 27) shows is a consequence of the country's energy policies or Energiewende. As he notes, energy transitions are much cheaper and easier to traverse than is commonly believed. Kraemer argues the Energiewende provides valuable lessons for other parts of the world. That is energy transformations can produce short-term benefits that outweigh their costs. Furthermore, those costs are not higher than maintaining the old, non-sustainable energy system, and the costs are coming down as experience accumulates.

Italy is another country that ranks the highest on the sustainable energy dimension of the energy trilemma. Morena Skalamera and Fabio Farinosi (Chapter 28) argue that this pattern is likely to continue to transition toward an energy mix largely consisting of a combination of renewables and natural gas. Renewables have been influenced by the EU's 20–20–20 targets which Italy is not only expected to meet but highly exceed.[3]

Barring major and, for the moment, unexpected changes in attitudes to nuclear, the contributions of wind and solar to Italian power generation will continue growing. However, as they show there is considerable uncertainty over the pace and composition of the country's future energy mix. The country's depressed economic conditions stemming from the 2008–2009 international financial crisis and the ongoing eurozone crisis have greatly eroded the government's resolve to continue subsidizing these technologies as generously as they have in the recent past.

The final country chapter is actually a broad survey of the manner in which the countries of the Caribbean and Central America have dealt with their energy problems and the manner in which several of the larger countries, Dominican Republic and Haiti in the Caribbean and Costa Rica and Nicaragua in Central America, are transitioning to a more sustainable environment with improved energy security.

In his survey of the region Robert Looney (Chapter 29) notes the difficulty of making sweeping generalizations concerning energy transitions. Even in a regional setting where countries share a number of similarities such as small domestic markets with limited fossil energy reserves, a great variety of energy mixes are possible.

Still several patterns prevail. Progress or lack of moving toward a secure sustainable energy mix is largely related to government capacity, especially at the extremes as illustrated by Haiti with the least capacity and Costa Rica with the most. However, explaining progress in intermediate cases like Dominican Republic and Nicaragua requires additional insights. Are leftist regimes more concerned with the environment and energy security? Is there a political economy effect where democracies tend to place more emphasis on energy affordability?

Completing the picture

The country case studies suggest the four broad country groupings used to organize this volume provide an effective way of narrowing down the manner in which countries are likely to prioritize the elements in the energy trilemma. As a starting point these groupings show a gradual progression of increased concern over climate security as one progresses from carbon producers through carbon reduction countries (Table 1.1).

The first three columns represent each country's average score on each of the three main energy goals in the period 2013–2015. The second three columns represent the World Energy Council's assigned grade for the progress made in each area, while the final column identifies the two areas of priority (based on each country's rankings in the first three columns).

To see if it was possible to arrive at a clearer picture of countries likely to prioritize climate security, several other aspects were considered. For instance, do democracies predictably come to a different set of priorities than is usually the case with autocratic or authoritarian regimes? Do higher levels of per-capita income get translated into a greater effort at increasing climate security?

Rearranging our case countries into two groups, (a) those who prioritize energy sustainability either one or two in the energy trilemma and (b) those who place energy sustainability third (Table 1.2), it's clear that the average level of democracy is considerably higher (lower numbers represent a higher ranking) while per-capita incomes are a bit lower for those countries that prioritize energy sustainability. One cannot easily argue therefore that only the rich countries have the luxury of taking action to improve the environment.

The statistical analysis undertaken in Appendix A sheds additional light on the factors that might facilitate or accommodate increased emphasis on environmental sustainability. The empirical results suggest a key role for democracy in improved climate security efforts. Democracy enters in in two ways: affecting the country's preference for giving climate security a priority over either energy security or energy affordability. Once a country has opted for climate security increased levels of democracy are associated with improved levels of climate security.

These countries are likely to comply with global conferences that set specific targets for carbon reduction. The costs of doing this might come at the expense of energy security, but in an era of cheap oil and abundant natural gas, energy security might not be a strong trilemma-type constraint. This would be particularly the case if increased climate security was accomplished through the development of increasingly lower cost domestic green energy sources.

For countries that have not made a commitment to climate security, obtaining promises from them to voluntarily reduce carbon emissions may not be particularly effective in meeting global carbon reduction targets. However, it has been suggested that many of these countries have not given climate security a high priority, not because they don't want to, but because they simply do not have the financial capability in deploying green technologies on a large scale. For these countries a new international financial institution similar to the World Bank, but charged with lending for climate related investments might be an effective means towards improved climate security.

Democracy may also play a subtle role in those countries not giving climate security a high priority. Many countries, the United States, Canada, and South Africa being a good example, have difficulty in coming to a consensus over energy policies at the national level, especially

Table 1.2 Country groupings on environmental sustainability

Country groupings	*World Energy Council trilemma country rankings*			*Trilemma priorities*	*Democracy ranking*	*Per-capita income*
	Energy security	*Energy equity*	*Environmental sustainability*			
Group I						
UK	8.0	20.0	19.3	security, sustain	16	38,118
Brazil	33.0	83.3	17.7	sustain, security	51	15,111
Japan	64.3	18.7	41.0	equity, sustain	23	34,635
Pakistan	61.7	104.0	90.0	security, sustain	112	4,590
Jordan	114.7	61.3	103.7	equity, sustain	120	11,496
Turkey	66.0	77.0	72.7	security, sustain	97	18,869
France	42.0	9.7	10.7	equity, sustain	27	37,214
Denmark	3.7	43.0	10.7	security, sustain	5	42,757
Italy	67.3	44.0	22.7	sustain, equity	21	33,039
Dominican Republic	114.7	93.7	55.0	sustain, equity	60	12,652
Costa Rica	57.3	69.7	2.0	sustain, security	23	14,232
Nicaragua	103.7	97.7	71.0	sustain, equity	95	4,692
Average	*61.4*	*60.2*	*43.0*	*sustain, equity*	*54.2*	*22,283.8*
Group II						
Saudi Arabia	54.0	8.7	123.0	equity, security	160	49,537
Canada	1.0	2.0	62.3	security, equity	7	42,778
Russia	6.3	47.3	103.7	security, equity	132	23,292
Mexico	32.0	50.3	76.3	security, equity	66	16,284
South Africa	38.3	83.3	129.0	security, equity	37	12,446
China	19.3	87.3	127.3	security, equity	136	12,599
US	7.7	1.0	88.0	equity, security	20	52,118
Indonesia	17.0	78.7	95.0	security, equity	49	10,033

Country groupings	*World Energy Council trilemma country rankings*			*Trilemma priorities*	*Democracy ranking*	*Per-capita income*
	Energy security	*Energy equity*	*Environmental sustainability*			
Egypt	53.7	64.7	75.0	security, equity	134	10,046
Thailand	94.3	76.0	105.0	equity, security	98	15,012
India	68.3	106.3	122.0	security, equity	35	5,439
Germany	27.7	33.0	33.7	security, equity	13	43,602
Average	*35.0*	*53.2*	*95.0*	*security, equity*	*73.9*	*24432.2*

Note: The values are for 2014 and are in purchasing power parity, constant 2011 international dollars.

those policies relating to the environment and where there is the perception that an improved environment comes at the expense of jobs or higher energy prices. On the other hand, local or state governments that have seen the consequences of climate change in their areas, or simply have a genuine concern about the environment are often able to make significant strides in carbon reduction. Hopefully these can serve as a grass-roots model for eventually forcing improved climate security at the national level.

In the more authoritarian or autocratic countries the chance these movements will have a discernable impact is likely to be less, although the Thailand case holds out some hope. These movements are not encouraged and in countries like Algeria often suppressed. Leaders in these countries are usually more concerned with energy security and lower energy prices deemed necessary to keep popular discontent from creating instability. It is these countries in which global accords to combat carbon will likely find the most resistance and underperformance in meeting reduced global greenhouse gas emissions.

Notes

1 Cf. World Energy Council, Energy Trilemma Index for a complete description of the trilemma and estimates of its values for a variety of countries. www.worldenergy.org/data/.

2 In the German case, in 2015 energy equity and energy sustainability were ranked nearly identically after energy security (Table 1.1).

3 The 20–20–20 target is part of the EU's binding legislation to ensure the EU meets its climate and energy targets for the year 2020. The key targets in this package are: (a) 20% cut in greenhouse gas emissions from 1990 levels, (b) 20% of EU energy from renewables and (c) 20% improvement in energy efficiency.

2

Evolving factors affecting energy security

Marcus D. King

Introduction

The international energy system is in a state of constant change. However some global trends are discernible. These trends must be understood and taken into account as policymakers strive to balance considerations of energy security, climate security and economic competitiveness into policy frameworks that are, in most nations, increasingly designed to attain reductions in greenhouse gas emissions. Policies developed in response to intensified efforts to reduce greenhouse gases will have both positive and negative implications for global economies and populations.

Basic assumptions about energy supply and demand are shifting. Global markets are changing as North American oil and gas production has risen exponentially based on new discoveries and technologies. These factors as well as melting ice due to climate change have enabled the exploitation of vast hydrocarbon reserves in the Arctic if market forces are favorable. Population growth and rising consumer expectations are driving steady increases in energy demand in all sectors but the rate of this growth has been tempered by stagnant economies in several regions.

Trends in the geopolitics of energy are raising a number of concerns. Political turmoil in the oil supplying regions of the Middle East and Africa, Russian political threats to cut energy supplies to European markets, and growing instability around energy supply "chokepoints" are high on this list. These developments provide cause for many nations to reconsider their global alliances and strategies to insure energy security.

Physical threats to energy infrastructure, both natural and manmade, are another important factor. Vulnerabilities in energy production, trade and distribution systems are increasingly apparent. While the number of proven cyber-attacks against energy infrastructure remains low, growing adversarial capabilities present rising challenges for governments and corporations. In the natural world, slow-onset environmental changes such as water scarcity caused by droughts and the depletion of mineral resources may be overshadowed by the shock of extreme weather events but these threats are just as credible.

The development of clean energy technologies is an avenue that national governments can pursue in order to build economic competitiveness while advancing the other primary goals of energy and climate security. Rapid changes in clean energy technologies in such areas as nuclear energy, cleaner fuels and the possibility of climate geoengineering will no doubt alter the impacts of the other trends identified in this chapter. However, breakthroughs in these areas carry the potential to bring enormous benefits to mankind and insure energy security.

Energy security is an umbrella term that covers many concerns linking energy, economic growth and political power. Individual definitions vary widely based on whether security is perceived from an economic, social welfare, national security, sustainability or political viewpoint.[1]

Economists emphasize energy supply over other elements of a definition. A comprehensive economic definition is the availability of adequate, reliable, and affordable energy.[2] Winzer refers to this condition as "energy supply continuity."[3] Historically, the avoidance of oil supply disruptions and the resultant economic effects of price volatility are significant because oil price shocks frequently precede economic recessions.[4] Similarly, access to new energy reserves, the ability to develop new infrastructure, and stable investment regimes are the critical factors in insuring energy security for energy supply companies. At the microeconomic level, energy security is the ability of households and businesses to accommodate disruptions of supplies in energy markets.[5]

For national policymakers, energy security definitions focus on the risks of supply disruption and the security of infrastructure from terrorism, wars or natural disasters. Defense and security institutions offer more securitized definitions. For example, the Center for a New American Security, a U.S. think tank, deemphasizes price and affordability altogether, defining energy security as maintaining energy supplies that are "geopolitically reliable, and physically secure."[6] Others recognize that energy security contains a component of environmental sustainability or sustainable development.[7]

A more comprehensive definition of energy security takes human security into account. Sovacool observes that scholars can "no longer … envision and practice energy security as merely direct national control over energy supply, and it now necessitates careful cultivation of respect for human rights and the preservation of natural ecosystems along with keeping prices low and fuel supplies abundant."[8] Finally, Ladislaw and Nakano build on previous definitions by also taking political and social acceptability factors into consideration.[9] Trends that influence all of the dimensions of energy security identified herein will be discussed in the following sections.

Supply, demand and energy security

Global energy supply

The world hydrocarbon market has experienced a revolution in the availability of previously unexploited hydrocarbon resources. Historically, hydrocarbon resources were under the control of a few states centered in the Middle East. The geographic distribution of major energy resource has now shifted. The United States has become the world's leading producer of oil and natural gas combined,[10] having surpassed Russia in 2013. According to one scenario, North America became a net exporter of oil and gas in 2015, and will account for 66% of net global export growth from 2015–2035.[11] North America switches from importing 6% of its energy in 2013 to exporting 19% by 2035. Accordingly, oil accounts for over 60% of that reversal and the region becomes a net oil exporter in 2018.[12]

The current surge in hydrocarbon production is generally confined to North America and is due to a revolution in the accessibility of oil and natural gas from shale, sandstone, carbonate, and other tight geologic formations. U.S. and Canadian producers have taken advantage of efficient, cost-effective drilling and production techniques. These include horizontal drilling and fracking, a technique that uses the injection of high-pressure fluid to release gas and oil from rock formations.[13]

This surge in supply is expected to intensify because significant North American liquefied natural gas (LNG) exports are bottlenecked due to the fact that there is now only one operable natural gas export terminal located in Alaska.[14] This terminal is used primarily for exports to Japan. However, exports will increase with the construction of new terminals. The U.S. Federal Energy Regulatory Commission (FERC) has approved the construction of seven additional terminals and six are under construction.[15]

The increase in U.S. and Canadian production is a swing factor that has caused disruption in the global price of oil. The price until now has been largely determined by the Organization of Petroleum Exporting Countries (OPEC). OPEC's share of global oil output dropped from 42% in 2008 to 39% in 2014.[16] Therefore, the balance between OPEC and non-OPEC countries that governed the oil market for the last thirty years has been suspended, at least for now, and there is no guarantee that OPEC will regain its market share.

In general terms, increased U.S. natural gas supply will put downward pressure on gas prices in Europe and Asia in the years ahead.[17] Relatively inexpensive oil and gas supply has challenged Europe's green energy security policy agenda that promotes the development of more expensive renewable energy technologies over gas for energy production. However, although North American output has substantially increased, the unpredictability of energy suppliers in Russia, the Middle East and North Africa will continue to be a source of price instability in Europe in the shorter term, complicating the picture for energy planners.[18]

The rise in U.S. energy production is considerable and it has boosted energy security. Increased production is attributable in part to unique factors such as financiers with a tolerance for risk, a property rights regime that allowed for ownership of underground reserves and highly-developed network of delivery infrastructure. No country other than Canada enjoys such an industrial environment at this time.[19] In addition to a greater availability of fossil fuels, the diversity of U.S. energy supply has also increased. Solar electricity generation has increased 20-fold since 2008, and electricity generation from wind has more than tripled.[20] However, to maintain this energy security status U.S. policymakers will have to strike a balance between legitimate concerns over damage to the environment and safety risks associated with expanded energy production (such as those raised by opponents of hydraulic fracking) and securing the full economic benefits of the energy boom.

An expansion of oil and gas availability in the Arctic also signals a rise in the global supply of energy. The melting of the Arctic ice sheet, accelerated by climate change, is expected to increase access to hydrocarbon reserves. The Arctic contains about 22% of the total undiscovered, technically recoverable resources in the world. More specifically, the Arctic accounts for about 13% of the undiscovered oil, 30% of the undiscovered natural gas, and 20% of the undiscovered natural gas liquids.[21]

The path of Arctic energy extraction is difficult to predict. In 2015, after years of preparation, two major oil companies, Shell and Statoil announced plans to abandon drilling based on economic and regulatory concerns.[22] The profit margin of Arctic oil has diminished due to low world prices and reductions in demand.

Global energy demand

Based on current growth trends, world population will continue to grow until at least 2050. Projections indicate that the estimated world population of approximately 7.3 billion in 2016 could reach 9 billion by 2040.[23] Most of this population growth will occur in the developing world. Regional population growth signals an overall increase in global energy demand.

Differences in regional demand for energy and the type of energy that will be required for economic growth can be explained by variations in geographic distribution of population, income growth, and technological conditions.[24] While little population growth is expected in the global North, substantial population and income growth is expected in more southerly latitudes. Electricity demand for cooling would increase in these regions.[25]

According to scenarios developed by the International Energy Agency (IEA), growth in energy demand will be driven primarily by India, China, and countries in Africa, the Middle East and Southeast Asia. The IEA is linked to the Organization for Economic Cooperation and Development (OECD), a membership group consisting of developed and emerging economies. Under the IEA scenario, during the period from 2015–2040, non-OECD countries account together for all the increase in global energy use, as demographic and structural economic trends, allied with greater efficiency, reduce collective consumption in OECD countries. These declines are led by the European Union (-15% over the period to 2040), Japan (-12%) and the United States (-3%).[26]

Energy demand in the power sector is likely to correlate more closely with demographic change. India will be responsible for a large amount of increased demand for energy with population and incomes on the rise; an additional 315 million people are anticipated to be living in India's cities by 2040. India is expected to make three-quarters of its energy investments in the power sector, which needs to almost quadruple in size to keep up with projected electricity demand.[27] In contrast, the share of primary energy devoted to power generation in North America is expected to increase just slightly over a similar period.[28]

Liquid fuels have been an exception to the overall trend in demand growth. During the last two years, increased global supply of oil and gas has been accompanied by a decline in demand for these commodities. The recent downward demand shift for natural gas was driven by slowing global economic growth, specifically in key developing economies of China, India, and Brazil.[29] The demand drop in China is the most dramatic.[30] Other influential factors include greater energy efficiency in vehicles in response to previously high oil prices and reductions in the cost of some renewable energy technologies.[31]

The drop in global liquid fuel demand may be ephemeral, however. The faster that oil prices drop, the faster they might rebound and become an engine for economic growth. The U.S. Energy Information Administration (USEIA), estimates in their "reference case" projection scenario that world liquid fuels consumption will increase by more than one-third by 2040.[32]

Evolving energy geopolitics

In addition to increased energy security, the U.S. appears to be a geopolitical winner. The shale gas and tight oil revolution has done much to counter a global narrative that the U.S. is a geopolitical power in decline. The huge boom in U.S. oil and gas production, combined with the country's other enduring sources of military, economic and cultural strength should enhance U.S. power. Likewise, the geopolitical influences of some countries that have used their oil revenues in ways that are inimical to U.S. interests are likely to shrink. Iran, for example, has the highest fiscal breakeven price for oil at $150 per barrel. As of October 2015, the current price per barrel of oil stood at $45.25.[33]

While the revolution in shale oil and tight gas production began in North America, it is slowly spreading to other parts of the world.[34] Production increases in nations outside of North America are beginning to expand their geopolitical influence. Increases in production elsewhere in the Americas such as Brazil and Argentina, could help ensure increased availability of liquid fuels supplies for many years.[35] Likewise, there is potential for recent legislative changes in

Mexico will reverse that country's recent trend of slowly declining oil production.[36] Despite the existence of substantial reserves, timetables for production in Europe are longer. Experts estimate that if European nations were interested in exploiting domestic gas reserves, it could take 15 years to assess these reserves, build infrastructure and produce gas.[37]

Russia's position of influence appears stable, especially in Europe. Although an increase in global oil and gas supply will likely diminish its stature, with rich resources and new investment in exploration, Russia is likely to remain an important liquid fuels producer in the future.[38] At present, much of Russia's oil production comes from fields in the country's West Siberian Basin; however, interest is shifting toward undeveloped resources in East Siberia, the Russian Arctic, the northern Caspian Sea, and Sakhalin Island.[39] Previous Russian threats to cut off gas supplies have placed European economies dependent on Russian gas supply in a vulnerable position.[40]

Rising global political instability

Instability in developing nations can affect energy systems in a variety of ways as institutions become less functional. However, interruption of energy supply is the threat in which policy-makers and security organizations from more developed nations are most interested.[41]

Growing political instability often linked to internal violence, among major oil exporters such as Venezuela, Nigeria, and countries in the Middle East and North Africa (MENA) has wider geopolitical implications.[42] Within Africa, the situation in Nigeria is precarious. Since the 1990s, rebel groups in southern Nigeria such as the Movement for the Emancipation of the Niger Delta (MEND), where the majority of oil infrastructure is located, have reacted to political and income disparities by pirating oil, sabotaging oil equipment, and holding oil company employees hostage.[43] This insurgency demonstrates that energy systems can be attractive targets for attack when conflict ignites.

Other potential hot spots for supply disruptions are in areas adjacent to sensitive maritime chokepoints for oil transport. Protection of maritime transportation is essential to maintaining access to energy supplies. Approximately 84% of the world's maritime crude oil and petroleum trade flow through the top seven chokepoints: Strait of Hormuz, Strait of Malacca, Suez Canal and Suez-Mediterranean Pipeline (SUMED), Bab al-Mandeb, Danish Straits, Turkish Straits and the Panama Canal.[44]

Indonesia has experienced some political instability in the province of Aceh, home to an active insurgency for several decades. If the central government continues to prove unable to respond to disasters, separatists might renew piracy in the Straits.[45] Pirates based in Somalia have occasionally intercepted oil tankers in the Arabian Sea. From 2008–2012, actual and attempted robberies against ships in this region outnumbered those in the Straits of Malacca by 447 to 9.[46] The total amount of oil seized has been small and it is generally returned to the world market after the shippers have paid ransom.[47]

A gradual opening of the Arctic due to melting icepack and the potential exploitation of vast energy resources will also have geopolitical consequences. Between 2008 and 2011, a spate of major policy announcements and actions by nations with territorial claims focused on re-militarizing the region. These actions suggest the possibility of emerging interstate competition for control and access to the region's resources.[48]

Evolving environmental factors

As energy systems evolve, the way they are interacting with the natural world is changing. Environmental change and growing energy scarcity are the forces that will constrain energy resources.

Climate change

The direct physical impacts of climate change, such as increased frequency and severity of storms, heat waves, and droughts will impact energy security in a number of direct and consequential ways. Climate change is the "actor" that may 1) create second-order effects that may exacerbate social instability and disrupt energy systems; 2) directly impact energy supply and/or systems.[49] Climate change's direct impacts on energy systems and resources are likely to have more negative than positive impacts on the global supply of energy.[50] Accounting for regional variations, climate change will likely cause a net increase in global energy demand.[51]

Growing water scarcity

Several issues converging at the nexus of water and energy have gained substantial and growing attention. Climate change is one of several factors, including overwithdrawal contributing to global water scarcity. Maintaining adequate water supply in the face of climate change is a major emerging issue. In many regions, climate change is likely to reduce precipitation, increase surface water evaporation, and decrease river flows.[52]

The scale of water use for energy production is tremendous. The energy sector accounts for about 15% of world water withdrawal, a figure second only to agriculture. This percentage of total water withdrawals is projected to increase by 20% between 2010 and 2035.[53]

All power generation technologies use at least some water.[54] Nuclear reactors and fossil fuel electric generation plants use water for functions including cooling, steam generation, and waste disposal. In the U.S., for example, a nuclear plant was forced to shut down one of its reactors in 2012 because the ocean water used to cool emergency diesel generators and other safety-related equipment was too hot.[55]

Although it is perhaps not as intuitive as other effects, solar power generation also faces limitations due to water scarcity. Concentrated solar power is a promising technique involving the deployment of hundreds or thousands of revolving mirrors used to focus solar energy on a column of liquid, which in turn generates steam for electric turbines. These large-scale facilities require as much, or more, water for cooling as thermoelectric power plants.[56]

Growing water scarcity will certainly diminish hydroelectric generation capacity in many nations, posing the risk of plant shutdowns with associated impacts on local and regional economic activity.[57] Hydropower is by far the largest of renewable energy sources in the current global electricity mix.[58] The southeastern United States, Europe, eastern China, southern Africa, and southern Australia are particularly vulnerable.[59] Brazil, the world's second-largest producer of hydroelectricity after China, has been the victim of a prolonged and brutal drought.[60] Dwindling Himalayan glaciers may also decrease the potential for hydroelectric generation in China and in South and Southeast Asia.

Water shortages will put particular stress on China's energy sector where hydroelectric capacity has already been in decline for several years due to droughts.[61] Hydropower is China's second-largest energy source after coal. The country's installed hydropower capacity is set to rise to 350 gigawatts (GW) by 2020, up from 300 GW in 2015.[62] China is home to half the world's 80,000 dams, more than the United States, Brazil, and Canada combined.[63]

In the face of these environmental conditions, Chinese policymakers are likely to increase reliance on coal-fired power plants, the cheapest alternative to hydropower for energy generation. Existing carbon capture and storage (CCS) technologies can substantially decrease coal plant emissions. However, adding CCS technologies to coal plants would more than double their water consumption.[64]

Every energy sector requires water. In the transportation sector, biofuels and synthetic fuels production competes with other water uses, including agriculture and human consumption. Extractive technologies such as hydraulic fracturing are also a very water intensive process because the fluids for oil and gas recovery are largely comprised of water. Many of the world's unexploited oil and gas reserves are located in arid zones.[65] In sum, water scarcity is a key factor that will reduce options in nations working to increase energy security through the promotion of less carbon-intensive energy strategies.

Extreme weather

Taken in aggregate, a narrative over recent decades indicates a statistical trend toward more frequent and intense extreme weather events.[66] Rising frequency of heavy downpours is an expected consequence of a warming climate. Increased precipitation will have a slightly positive impact on hydroelectric potential in Asia with diverging patterns across other regions.[67] However, extreme heat waves experienced in some parts of the world decrease the efficiency of power plants during periods when electricity demand is highest, placing additional stresses on the electricity systems, including grids.[68]

Energy infrastructure is susceptible to disruption by weather conditions even in the most developed countries.[69] A blackout that crippled most of the U.S. northeast in 2003 occurred on a hot summer day when electricity demand was high and an overheated power line in a small Ohio town sagged and came into contact with a single tree.[70] This normally unremarkable incident interacted with several other power system failures to create a major regional blackout that affected 50 million people in the U.S. and Canada and caused financial losses between $4 and $10 billion in the United States.[71]

Physical and cyber threats

Although threats to energy infrastructure are global, power outages in the U.S. grid system caused by extreme weather serve as quotidian tests of the grid's resilience and regional coherence.[72] In this way, past natural events can offer insight into how the grid will react to cyber-attacks at similar energy chokepoints in the future. Because the U.S. grid is relatively regionally dispersed – it is made up of three regional systems with equipment administered by over 3,000 companies and municipalities – a high-impact cyber-attack is likely to have limited reach.[73] In 2014, the U.S. Department of Energy reported hundreds of outages caused by natural hazards and physical damage, but only three "suspected" cyber-attacks took place.[74]

At the same time, the potential for a cyber-attack to cause damage to energy equipment, with significant implications for customers, should not be understated. In the U.S., the complexity of the U.S. grid system means that there are many vulnerable crossover points.[75] The precedent set by the 2010 Stuxnet attack – the first deployment of cyber-physical weapons technology in which a computer worm caused Iranian nuclear centrifuges to spin out of control – and the increased frequency of cyber-attacks targeting energy companies in general are evidence of the changing threat landscape.[76]

The exact level of vulnerability can depend on grid architecture. The implementation of "smart grid" technology to increase economic and energy efficiency in response to climate change mitigation policy would connect homes to energy stations using Internet-based systems and could increase exposure to cyber-attack.[77] However, microgrids or localized grids can also disconnect from the traditional grid to operate autonomously and help mitigate grid disturbances to strengthen grid resilience, can play an important role in transforming the nation's electric grid and reduce the possibility of widespread damage resulting from a cyber-attack.[78]

According to Admiral Mike Rogers, director of the U.S. National Security Agency, major powers, such as China and Russia, already have the capability "to shut down, [and] forestall our ability to operate our basic infrastructure, whether it's generating power across this nation, [or] whether it's moving water and fuel."[79] Increasing robustness across the board to withstand high-risk, low-impact events is the first step to enhancing the ability to guard against anomalous events, such as cyber threats and extreme weather. Doing so in most nations will depend on the coordination of the federal government and private companies. According to Burke and Schneider, "no matter how much money a company spends on a cyber-security program, if aging equipment is not replaced, maintenance is inadequate, or there is inadequate power supply to meet consumer demand, then there is a significant vulnerability to an outage."[80] Ultimately, nations will have to learn to balance the process of infrastructure modernization, implementing systems such as the "smart grid" cyber security measures, and implementing the best strategies to maintain resilience against extreme weather.

Evolving climate change mitigation policies

Climate change mitigation is the "stabilization of greenhouse gas concentrations at a level that would prevent dangerous anthropogenic interference with the climate system."[81] In the coming decades, stronger policies to mitigate climate change will have a significant effect on societal norms, the cost and usage of energy, land use, and economic development strategies.

Heralded as a global achievement, on December 12, 2015, 195 countries convened the 21st Conference of the Parties (COP 21) under the aegis of the United Nations Framework Convention on Climate Change (UNFCCC). They reached a comprehensive voluntary agreement to reduce global greenhouse gas emissions with a goal of keeping temperature rise within 2 degrees Celsius.[82] Momentum leading up to the agreement reached in Paris was underscored by the presence of 150 presidents and prime ministers, the largest single day gathering of heads of state.[83] This agreement will be a key factor guiding national mitigation policies as well as international collective action to manage climate change.

The Paris Agreement articulates two long-term emissions goals. First, a peaking of emissions as soon as possible with recognition that it will take developing countries longer to reach the target and a goal of greenhouse gas neutrality (expressed as a "balance between anthropogenic emissions by sources and removal by sinks") in the second half of the century.[84]

During COP 21, national governments proposed unique greenhouse gas reduction targets known as Intended Nationally Determined Contributions (INDCs). The agreement creates a process whereby countries will submit INDCs every five years with clear expectations that they will represent a progression from previous years.

National adherence to the Paris Agreement will foreground some existing mitigation strategies and promote new approaches. A distinct feature of the Paris Agreement is that it clearly encourages voluntary contributions by developing countries. The creation of emissions targets in new countries moves countries away from the old North-South paradigm and into a paradigm of mutual dependency and creates a more complicated and dynamic climate change mitigation regime. Unlike previous agreements reached under the UNFCCC framework, the Paris Agreement includes provisions that encourage nations to take individual or collective action to utilize carbon markets.[85] Less developed nations may now choose to join carbon trading schemes, expanding those structures dramatically.

The Paris Agreement will also influence the trajectory of spending for research and development of clean energy technologies in many countries. Under the agreement, 20 nations have pledged to double their public research and development funding in clean energy under an initiative called Mission Innovation. A parallel private initiative by 28 of the world's richest private sector investors will support these public investments by providing capital for high-risk projects.[86] The agreement will open for signature in April 2016 and enter into force after 55 countries that account for at least 55% of global emissions have ratified it.[87] If states ratify the agreement quickly it could enter into force before 2020.[88]

An initiative to encourage non-state actors to enter pledges was also highly successful. Nearly 11,000 commitments from 2,250 cities, 150 regions, 2,025 companies, 424 investors and 235 civil society organizations were registered.[89] The emergence of some of these new actors will also change the nature and scope of existing climate change mitigation regimes.

Mitigation policies enacted to support the targets established by the Paris Agreement could also be strengthened by the implementation of a revised set of UN Sustainable Development Goals (SDGs). Earlier in 2015, 193 nations agreed on a revised set of goals including the achievement of universal global access to energy by 2030.[90] Specifically, the signatories agreed to: expand access to reliable and modern energy services; increase the share of renewables in the global energy mix and double the global rate of improvement in energy efficiency.

To meet the SDGs, developed nations must revise their international energy policy postures and development strategies. Like the Paris Accord, the SDGs promote increased cooperation in technology research, and development and investment in energy infrastructure in least developed countries.[91] Implementation of policies in support of the SDGs will benefit economies focused on manufacturing and diffusion of clean energy technologies.

Nuclear energy generation

Energy production and use account for around two-thirds of global GHG emissions today.[92] Considering the global emissions targets established by the Paris Agreement, nuclear energy is the only proven technology that can scale up to meet the world demand for carbon free sources,[93] but its share of global electricity generation has been declining since 1993.[94]

However, nuclear energy has great potential to improve energy security in the long run. It is expected to play an important role in the energy mix due to factors such as demand for electricity in the developing world, climate change concerns, security of energy supply and price volatility for other fuels.[95] By 2030, a low projection by IAEA indicates 17% growth in world total nuclear power capacity by 2030, while the high projection suggests as much as a 94% growth.[96] The following table demonstrates projected global growth in nuclear power. As Table 2.1 below illustrates, China and India are expected to lead the world in new reactor construction.

Nuclear power presents operational, financial, and political risks. The capital costs of new nuclear reactor construction can be astronomical. A study by the Massachusetts Institute of Technology estimates that a new nuclear plant in the U.S. would have an initial cost of $4 billion. Financing charges could add another $2 billion.[97] One company, Electricité de France estimates that the new Hinkley Point plant in the United Kingdom is expected to cost $36 billion.[98] A low price of natural gas and development of renewables would also hedge against nuclear growth.

Environmental risks are manifested in two ways. While there is widespread scientific agreement about how the waste disposal should be approached, the politics are complicated.[99] Globally, there is currently no permanent facility for the storage of high level nuclear waste.

Table 2.1 Nuclear reactors around the world

Country	*Reactors operable*	*Reactors under construction*	*Reactors planned*	*Reactors proposed*
USA	99	5	18	24
France	58	1	0	1
Japan	43	3	9	3
Russia	35	8	25	23
South Korea	25	3	8	0
China	30	24	42	136
India	21	6	24	36
Canada	19	0	2	3
UK	15	0	4	9
Ukraine	15	0	2	11
World Total	440	65	173	337

Source: World Nuclear Association (www.world-nuclear.org), as at March 2016.

Sites have been identified in Sweden and Finland, but these are projected to open only in the next decade or two.[100]

Environmental safety concerns surrounding nuclear power are considerable. The 2011 accident at the Fukushima reactor in Japan resulting from an earthquake and tsunami caused explosions and radiation release. It is the probable explanation for a significant, if possibly temporary, global slowdown of new power reactor installation.

Some civilian nuclear technologies can provide a basis for weapons development programs. Iran is one country located in a politically unstable region where attempts to build such a program have been successful.[101] These concerns are not confined to the Middle East. Every nation surrounding the volatile South China Sea that does not possess nuclear power – Vietnam, Malaysia, Indonesia, the Philippines and Singapore – is considering the acquisition of nuclear power reactors[102] presenting a regional security concern.

Advances in technology could facilitate growth in the nuclear sector. New nuclear fuel cycles and reactor technologies that address some of the concerns are under development creating hope that some of the gravest concerns can be allayed.[103] Some of the most promising technologies will be discussed later in this chapter.

Climate mitigation policy and energy security

Some long-run solutions to climate security will require higher prices for gasoline, electricity, and home heating oil[104] diminishing energy security for many consumers in the short term.[105]

Climate and energy security can be cross-compatible in the area of efficiency when these measures provide near-term cost reductions and serve to maintain or increase energy supply availability and reliability.[106] Whether energy consumption reductions can completely offset cost increases associated with more efficient technologies is situational and remains a point of contention.

There are many instances where policies designed to mitigate climate change and that promote energy security can be mutually reinforcing. In more advanced economies, energy conservation is described as a "no regrets" strategy for enhancing energy security while reducing climate change. In many cases, policies that reduce demand for energy – especially oil – through technology innovation (such as advanced alternative vehicle development) require greater energy efficiency.

One tension inherent in policies for climate change mitigation and energy security is that policies addressing each may require implementation on different timescales. Climate mitigation may phase in greenhouse gas emissions (GHG) reductions over time because climate risks evolve over decades and many of the solutions, including capital stock replacement, also require decades to implement. However, the risks associated with energy security affect national economies on daily to annual time scales.[107]

Climate policies can undermine energy security by limiting near term energy supply options. Consequently, Bordoff et al. suggest that greenhouse gas emissions reductions would be less disruptive to energy security if they were implemented only after key technological solutions, such as carbon capture and sequestration, become available for large-scale deployment.[108] This recommendation is at odds, however, with other analysts such as Yohe who argue that significant GHG emissions reductions must begin immediately to achieve any long-term climate stabilization goal at the minimum cost.[109]

Regulatory uncertainty surrounding long-term climate policies, particularly in major greenhouse gas emitter nations, has also had an indirect negative impact on energy security. In the United States, this uncertainty has caused power companies to delay capital investment decisions, such as building new natural gas, nuclear or renewable generation facilities that would lower carbon emissions and diversify the fuel mix.[110] Construction of new coal-fired plants is also on hold, causing generation capacities to lag demand growth in some cases while the economics of renewable and nuclear energy plant construction remains hazy, also based, in part, on regulatory factors.

Greenhouse gas mitigation policies can also carry consequences for human security. Twenty% of all greenhouse gas emissions come from deforestation and forest degradation.[111] Organizations such as the World Bank and the United Nations Environment Programme (UNEP) have implemented policies known as Reducing Emissions from Deforestation and Forest Degradation (REDD) to stimulate forest management and pay for ecosystem services. Disputes over land rights have an impact on the livelihoods of people who depend on the forests.[112]

Evolving disruptive energy technologies

Economic security through technology development

National governments are pursuing policies designed to promote the development of clean energy technology with the goal of achieving energy security, mitigating the worst impacts of climate change and increasing economic security. It is predictable that investments by governments' research institutions and corporations are increasing the likelihood of breakthroughs in clean energy technology across the spectrum of energy sectors including extraction, efficiency, storage, and generation.

However new energy technologies must traverse several stages within an innovation and development pipeline before they can be commercialized. Generally, these stages can be described as research, development, demonstration, and deployment. The transition between stages can be difficult because each stage has varying financial, intellectual, and facility requirements. For example, while start-up funding may be drawn from government sources, latter stages of clean energy technology development require large infusions of corporate and venture capital.

Despite these barriers to market entry, economic security is a key motivation for national governments to promote the development of innovative clean energy technologies.[113] Motivated by concerns about energy security and competitiveness, several European countries have established themselves as leaders in this area. Spain, Germany, the U.K., and Denmark are the leaders among them. Spain, home to some of the world's most successful renewable energy companies, has an installed renewable energy generation capacity of more than 30%.[114] Likewise, Germany has focused its large industrial sector on the manufacture of wind and solar technologies. More than a quarter of its electricity generation comes from wind, solar and other renewable sources.[115]

In addition to these national-level measures, the European Union has established an overarching framework of policies and institutions, such as the Europe-wide carbon trading market, that have encouraged the development of clean energy technology. Policies that encourage the adoption of renewable technology at the household level have successfully incentivized the deployment of clean energy technology across Europe.[116]

China has also positioned itself as a world leader in clean energy technology development and production. China is the world's largest manufacturer of wind turbines and solar panels.[117] Chinese success can be partially attributed to inexpensive manufacturing conditions coupled with protectionist policies which have helped to gain maximum market leverage from clean energy and build the Chinese manufacturing base to the detriment of its competitors.[118]

The United Arab Emirates (UAE) is a perhaps unexpected example of a country that sees the huge potential in investing in clean energy technology. Despite its status as an OPEC nation and possessing nearly 10% of the world's proven oil reserves, it has launched initiatives to position itself as a world leader in renewable energy technology.[119] Oil titan Saudi Arabia is also seeking to become a global solar power.[120]

The following section reviews key emerging technologies in each energy sector. The common characteristic of these technologies is their trending and relatively high capacity to spark breakthroughs in coordinated attempts to reduce global emissions and promote energy security through continued access to reliable energy supplies. Some of the benefits and barriers to development of these technologies are discussed in detail below.

Extractive technologies

Switching to natural gas as an energy generation choice is an option for greenhouse gas mitigation because it is less carbon-intensive than coal. Hydraulic fracturing (fracking) is the dominant technique that has allowed producers to tap into large tight oil, shale oil, and shale gas deposits.

An international debate about fracking's environmental consequences has emerged as the use of this technique spreads from North America to other regions. Concerns surfaced by this debate are a key factor preventing fracking from reaching its full technological potential. First, water contamination and the possibility of methane emissions are the main environmental concerns.[121] These risks to water resources include fear of contamination of surface and groundwater during site preparation, drilling, well completion, and operation.[122] Second, methane release during the fracking process is another major issue. In the most extreme scenario, the total amount of GHG emissions saved by extracting clean natural gas from fracking at a particular site would be offset by "fugitive" GHG emissions from methane released during drilling, completion, and operation of an unconventional well including flaring at the wellhead.[123]

While preventing methane release still remains a challenge, much of the current research is focused on better ways to address fracking's effects on water supplies including through the use of waste water or minewater, liquids other than water, compressed gasses, potentially including carbon dioxide.[124]

Energy generation technology

Renewable power

Solar and wind are the fastest growing sources of renewable energy by an order of magnitude.[125] More energy from the sun reaches earth in one hour than all of the energy consumed on our planet for one entire year.[126]

The emergence of solar and wind technologies is enabled by convergence of complementary rising technologies such as battery storage, big data and smart grids. They have the most disruptive potential.[127]

Costs of wind and solar energy have declined steadily since the 1980s but they have fallen the most dramatically in the last decade, causing a vast expansion in global deployment. This advantage marked a corresponding increase in global wind power with an expansion from 48GW in 2004 to 318GW by the end of 2013 and a growth in photovoltaic solar capacity from 2.6GW to 139GW in the same time period. In some areas, wind and solar are now cost competitive with coal and natural gas much of the time.

Innovations such as more efficient solar collection panels continue to reduce the cost and increase the attractiveness of this generation option. Wind power is a relatively mature technology. But incremental improvements in components including turbines, towers, blades and materials, and reduced construction and maintenance costs are also likely to increase wind power's economic competitiveness.[128]

Microgrids also enable the integration of growing deployments of renewable sources of energy such as solar and wind. In addition, the use of local sources of energy to serve local loads helps reduce energy losses in transmission and distribution, further increasing efficiency of the electric delivery system.[129]

Despite these cost advantages energy output is limited to when the sun is shining or the wind is blowing so these technologies are impeded by their intermittent nature. Improved energy storage technologies are needed to overcome these challenges.

Energy storage

Energy storage is perhaps the decisive enabling factor in determining the rate of development for renewable energy. Lithium-ion batteries are the most successful storage devices to be developed in the last 20 years.[130] This is the type of battery most often found in new electric and hybrid cars that are appearing on the market.

In 2014, the electric car company Tesla announced plans to build a $5 billion factory in Nevada, U.S.A., that will produce batteries that can hold up to 50 gigawatt hours of electricity by 2020. This figure represents more than the total amount of electricity produced globally in 2013.[131]

Lithium batteries have a tendency to overheat but gradual improvements in cost, storage density and manufacturing economies of scale mean it's still the best choice for many applications. A limiting factor is that lithium deposits are concentrated in the hands of a few countries. Other advanced automotive technologies require significant quantities of other rare earth minerals; at least 50% are concentrated in China which has exerted geopolitical leverage by threatening to cut off supplies swapping one dependency (foreign oil) for another (foreign rare earth minerals), with significant implications for the geopolitical landscape.[132]

Nuclear fusion and small modular nuclear reactors

The promise of fusion is very high because it can produce theoretically unlimited power without any carbon emissions and very little radioactive waste. Although there are some radioactive waste products from fusion they will become inert within a few hundred years, as opposed to the thousands of years that waste from fission reactors stays dangerous.[133] Further, fusion power plants cannot go critical, produce runaway reactions or a meltdown in the way that nuclear fission does.[134]

In 2014, an American company, Lockheed Martin, conducted the first fusion experiment that yielded more energy than was needed to start it. The company announced in 2014 that they could deliver a prototype fusion reactor in five years.[135] Other international consortia in the U.S., Europe, Russia and Japan have conducted experiments with fusion.[136]

There has been a substantial level of investment in this technology. In May 2015, ARPA-E, an American research agency affiliated with the Department of Energy, invested $30 million in entrepreneurial companies. There has been $450 million in investment in new private schemes according to the British Atomic Energy Authority.[137] In the past few years there has also been a sprinkling of entrepreneurs conducting experiments.

Fusion research has been conducted for 60 years and estimates of when hydrogen reactors will be available vary but some fall within the next few decades. ITER, formerly known as the International Thermonuclear Experimental Reactor, is a long standing fusion project in France directed by international consortia. According to some estimates, it will be ready for operation in 2027 at more than twice the original price tag.[138] If all the technical and design refinements in successive experimental reactors go according to plan, it is expected that the very first fusion power plants could be producing electricity for the grid by 2045–2050.[139]

Multiple studies have been conducted in recent years to assess the features and benefits of smaller-sized reactor designs suitable for global deployment. Small Modular Reactors (SMRs) are an emerging technology that has potential advantages over larger plants because they provide owners more flexibility in financing, siting, sizing, and end-use applications.

Nuclear power plants traditionally have a large capital cost and SMRs can reduce an owner's initial capital outlay or investment because of the lower plant capital cost. These reactors can be built in a controlled factory setting and installed module by module, reducing the initial investment costs.[140] This is especially important for developing economies or smaller markets which typically have limited availability of capital funds.[141]

While some countries can accommodate large plants (>1000 MWe), smaller sized SMRs can provide power where large plants are not needed or that may not have the necessary infrastructure to support large power generation units including smaller electrical markets, isolated areas, smaller grids, or restricted water or acreage sites.[142]

Small reactors are generally classified as producing 300MWe or less. Four reactors meeting these criteria are now operational in China, India, Pakistan and Russia but they are not necessarily modular in design.[143] To fully realize all of the noted benefits of SMRs, additional research will be necessary to resolve challenges introduced by differences in the designs, technologies, and operational characteristics relative to traditional reactors.

Climate change mitigation technologies

Carbon capture and storage (CCS)

Carbon capture and storage (CCS) is the integrated process of capturing carbon dioxide (CO_2) from power generation or industrial activities, then storing (sequestering) it to prevent its release into the atmosphere.[144]

There is a lack of widespread commercial deployment of CCS due in part to the inherent weakness in establishing plant sites where CO_2 storage is available and economical.[145] Lewis observes that proposals to inject CO_2 into the oceans or deep underground geological formations are problematic because oceans are already experiencing issues with large harmful concentrations of acid and the risk of leakage associated with the use of underground formations is high.[146]

Despite these challenges, CCS technology has undisputed potential to serve as a key component of a carbon mitigation portfolio for the electricity, petrochemical and other industries. Moreover, there are numerous studies that have concluded that in the long-term CCS can be a cost effective measure to reduce global CO_2 emissions.[147]

CCS projects are capital intensive so they are caught in a classic policy dilemma. While some governments view CCS as a low carbon option, without favorable government regulations and or strong financial incentives to significantly reduce CO_2 emissions there is little or no incentive for the private sector to develop and deploy CCS technology.[148]

Climate geoengineering

Broadly, geoengineering is an attempt to change the earth's climate to support human life. The U.S. House of Representatives Committee on Science and Technology defines geoengineering as "the deliberate and large-scale modification of the earth's climate systems for purposes of counteracting and mitigating anthropogenic climate change."[149] Mankind's scientific knowledge surrounding the possible effects of geo-engineering is small but this approach might become more desirable as the established mitigation options for keeping warming of the climate system within the range of two degrees seem unreachable over time. Current attempts to achieve geoengineering follow two basic approaches: solar radiation management (SRM) through blocking sunlight to lower global temperatures and/or removing carbon dioxide (CDR) from the earth's atmosphere to allow heat to pass back through and escape to outer space.[150]

Geoengineering carries significant ecological and political risks so widespread implementation of the various technologies under research and discussion may be a risky gambit. The negative consequences could be very similar to those of climate change itself, such as the creation of weather systems that benefit some nations at the expense of others.

Furthermore, a systematic assessment of potential security risks posed by climate engineering does not exist. Existing carbon removal approaches such as the United Nations REDD program involve planting and cultivating more trees and have the potential to cause conflicts over land-use and require water in large quantities. Thus climate engineering could redistribute climate security risks and add new kinds of risk.[151]

Therefore, a troubling aspect of geoengineering is that current deployment would take place in the absence of governance or regulatory structure. There is no treaty or international body with a sufficiently large mandate to regulate all proposed climate engineering measures.[152]

Transportation Sector

Today, transport-related emissions account for over 20% of global energy-related CO_2 emissions and are set to increase without the strong uptake of alternative fuel vehicles.[153] The vast majority of the world's energy for transportation still comes from fossil fuels and this trend is expected to continue to 2035 and beyond (Institute for Energy Research, "BP Energy Outlook to 2035," February 26, 2015).[154]

Electric vehicles

Electric vehicles are already commercially available but the cost of such vehicles is high. Designing smaller more powerful batteries is the key challenge. Recharging at a massive scale puts pressure on the energy grid. Electric vehicle production has a range of possible negative impacts on the environment largely emanating from the vehicle supply chain.[155]

The industry has set ambitious targets. One car company, Toyota, is aiming to sell nearly zero regular gasoline vehicles by 2050, only hybrid fuel cell-driven models. The automaker is pushing to reduce average emissions from cars by 90% by about 2050, compared with 2010 levels.[156]

Hydrogen fuel cells are a transportation technology that is commercially available but in small numbers and at a high cost. Since they're powered entirely by electricity, fuel cell vehicles are considered electric vehicles ("EVs"). Unlike other EVs, full cell vehicles' range and refueling processes are comparable to conventional cars and trucks. The weight and design of fuel tanks and the lack of a hydrogen fuel delivery infrastructure inhibits rapid development of this technology at this time.[157]

All electric vehicles face a number of challenges related to cost, refueling infrastructure and customer preferences.[158] Electric vehicles will continue to face competition from other vehicle technologies, including diesels, grid-independent gasoline-electric hybrids, flexible fuel vehicles and more efficient conventional gasoline vehicles, all of which are likely to become more fuel-efficient in the next 20 years.[159]

Fuel technologies

Overreliance on the single commodity of oil in the transportation sector has been an Achilles heel for many countries' national and energy security due to price volatility and uncertainty that can effect investment decisions.[160] In recognition of this, countries have moved toward two options. One is to move toward cleaner fuels within the suite of fossil based options and the other is to move toward non-fossil fuel based sources. Biofuels are a non-fossil option that have been produced and consumed for many years as additives and more recently as stand-alone options.

In 2008, biofuels accounted for less than two% of the world's transportation fuel but that amount is growing rapidly.[161] Biofuels generally emit much lower levels of CO_2 than fossil fuels. This relative advantage is especially pronounced in the case of cellulosic biofuel which is produced from wood, grasses, or the inedible parts of plants and advanced biofuels such as those derived from algae.[162]

The key problem with biofuels globally is the scarcity of potentially productive land that is not already dedicated to food or livestock cultivation. Therefore cellulose-based biofuel derived from grasses and crop residue affect food security. Algae-based biofuel is potential alternative technology although it is unproven at commercial scale. However, it would likewise not diminish food security. In some countries such as Brazil, clearing the forests to provide source material can result in the release of more greenhouse gases emissions than were abated through their use.

A desire to reduce reliance on foreign oil and take advantage of abundant coal reserves has led some countries to explore coal-to-liquid fuel conversion processes (CTL). Emissions from these fuels exceed those of fuels obtained from crude oil by a factor of two.[163] While the technology is proven and commercially operative, the costs of new CTL fuel F-T plants are very high though lower than source material in natural gas. However, the production of these fuels does not contribute to GHG emissions reduction, unless CO_2 sequestration or processing is possible.[164]

Simultaneous technological breakthroughs in the extraction, generation, mitigation and transportation sectors are extremely unlikely. However, a breakthrough in one sector will change the overall energy supply and demand equation in unpredictable ways. These changes will be, on balance, beneficial to both the environment and mankind if well-informed policies are established to exploit the most positive aspects of these changes.

Conclusion

In the coming decades, national policymakers will strive to address a security trilemma: how to maintain energy security, addresses concerns over climate sustainability, and balancce economic competitiveness.

This decision must be made in the context of global trends identified in this chapter including: growth in the supply of fossil fuels coinciding with a rising awareness of their harmful effect; reduced global energy dependence on OPEC and rising instability in oil producing regions; environmental impacts of climate change and extreme weather; physical impacts of infrastructure vulnerability and the likely advent of disruptive technologies.

In 2015, nearly every nation on the globe committed to the Paris Agreement on Climate Change and the UN Sustainable Development Goals. Collective adherence to these agreements would signal global transitions toward a carbon-free future. It is clear that some technologies are poised to enable this transition. In the energy generation sector, nuclear energy is the only proven technology that can scale up to meet the world demand for carbon free sources at this time. Incremental improvements in lithium-ion batteries hold the greatest promise to improve energy storage capacity that will in turn enable more rapid development of renewable energy. Geoengineering of the climate holds great promise in the area of climate mitigation but substantial technical and regulatory issues highlight potential unintended consequences of this option. Anticipated policy decisions supporting additional research, development and deployment of these and other technologies will increase the potential for "breakthroughs" that profoundly affect global energy use. Policies that manage and exploit these opportunities can create win-win-win solutions to the policy trilemma.

Globalization will insure that no country can remain unaffected by the evolving trends outlined in this chapter. It is equally inevitable that events we cannot identify or even envision will intervene to influence the challenging choices framed by the security trilemma.

Notes

1 Benjamin K. Sovacool, ed., "Defining, Measuring, and Exploring Energy Security," in *Routledge Handbook of Energy Security* (Abingdon: Routledge, 2010).

2 B. C. Staley et al., *Evaluating the Energy Security Implications of a Carbon-Constrained US Economy* (Washington, DC: World Resources Institute and Center for Strategic and International Studies, 2009).
3 C. Winzer, *Conceptualizing Energy Security* (Cambridge: Electricity Policy Research Group, University of Cambridge, 2011).
4 J. D. Hamilton, "Oil and the Macroeconomy since World War II," *Journal of Political Economy* 91, no. 2 (1983).
5 US Congressional Budget Office, *Energy Security in the United States* (Washington, DC: US Congressional Budget Office, 2012), www.cbo.gov/sites/default/files/cbofiles/attachments/05-09-EnergySecurity.pdf.
6 S. E. Burke and C. Parthemore, *A Strategy for American Power: Energy Climate and National Security*. Solarium Strategy Series, Center for New American Security, 2008, www.cnas.org/files/documents/publications/Burke_EnergyClimateNatlSecurity_June08.pdf.
7 R. S. Acepias, "The Sustainable Development Dimension of Energy Security," in *The Routledge Handbook of Energy Security*, ed. Benjamin K. Sovacool (Abingdon: Routledge, 2011), 96.
8 Sovacool, "Defining, Measuring," 9.
9 S. Ladislaw and J. Nakano, "Leader or Laggard on the Path to a Secure, Low-Carbon Energy Future?" Center for Strategic and International Studies (2008), https://csis.org/files/publication/110923_Ladislaw_ChinaLeaderLaggard_Web.pdf.
10 US Department of Energy, *The Quadrennial Energy Review 2015* (Washington, DC: US Department of Energy, 2015), http://energy.gov/epsa/downloads/quadrennial-energy-review-full-report.
11 BP, *BP Energy Outlook 2035: Focus on North America*, BP, 2015, www.bp.com/content/dam/bp/pdf/energy-economics/energy-outlook-2015/Energy-Outlook-2035-Focus-on-North-America.pdf, 24.
12 Ibid., 24.
13 R. Blackwill, and M. O'Sullivan, "America's Energy Edge: The Geopolitical Consequences of the Shale Oil Revolution," *Foreign Affairs* 93, no. 2 (2014), www.foreignaffairs.com/articles/united-states/2014-02-12/americas-energy-edge.
14 US Federal Energy Commission, *North American LNG Export Terminals Approved*, US Federal Energy Commission, 2016, www.ferc.gov/industries/gas/indus-act/lng/lng-approved.pdf.
15 Ibid.
16 "A New Era in the Oil Market," *IISS Strategic Survey 2015: The Annual Review of World Affairs* (September 2015): 37–48, www.iiss.org/en/publications/strategicsurvey/issues/strategic-survey-2015-5d6e/ss15-04-strategic-policy-issues-0ef3.
17 US Energy Information Administration (USEIA), *International Energy Outlook 2014*. Washington, DC: US Energy Information Administration, 2014. www.eia.gov/forecasts/ieo/pdf/0484(2014).pdf.
18 "A New Era in the Oil Market."
19 Blackwill and O'Sullivan, "America's Energy Edge."
20 "A New Era in the Oil Market," 45.
21 US Geological Survey, "90 Billion Barrels of Oil and 1,670 Trillion Cubic Feet of Natural Gas Assessed in the Arctic," *USGS Newsroom*, July 23, 2008, www.usgs.gov/newsroom/article.asp?ID=1980#.VjUUukKRZUQ.
22 Dan Joling, "Statoil Announces It Will Exit Alaska Offshore Exploration," Associated Press, November 17, 2015.
23 US Census Bureau, "International Programs," US Census Bureau, 2016, www.census.gov/population/international/.
24 D. J. Arent et al., "Key Economic Sectors and Services," in *Climate Change 2014: Impacts, Adaptation, and Vulnerability, Contribution of Working Group II to the Fifth Assessment Report of the Intergovernmental Panel on Climate Change*, ed. C. B. Field et al. (Cambridge: Cambridge University Press, 2014).
25 M. Klare, "Climate Change Blowback: The Threats to Energy Security," *SAIS Review of International Affairs* 35, no. 1 (2015).
26 US Energy Information Administration (USEIA), *International Energy Outlook 2015*. Washington, DC: USEIA, 2015. www.eia.gov/forecasts/aeo/pdf/0383%282015%29.pdf.
27 US Energy Information Administration (USEIA), *International Energy Outlook 2015*. Washington, DC: USEIA, 2015. www.eia.gov/forecasts/aeo/pdf/0383%282015%29.pdf.
28 BP, *BP Energy Outlook 2035*, 14.
29 C. K. Ebinger, "World Oil Demand: And Then There Was None," *Planet Policy* (blog), October 17, 2014, www.brookings.edu/blogs/planetpolicy/posts/2014/10/17-world-oil-demand-ebinger.

30 US Energy Information Administration (USEIA) *International Energy Outlook 2014*. Washington, DC: US Energy Information Administration, 2014. www.eia.gov/forecasts/ieo/pdf/0484(2014).pdf.
31 Ebinger, "World Oil Demand."
32 US Energy Information Administration (USEIA), *International Energy Outlook 2014*. Washington, DC: US Energy Information Administration, 2014. www.eia.gov/forecasts/ieo/pdf/0484(2014).pdf.
33 "Nasdaq – Crude Oil," October 22, 2015, www.nasdaq.com/markets/crude-oil.aspx.
34 US Energy Information Administration (USEIA), *International Energy Outlook 2014*. Washington, DC: US Energy Information Administration, 2014. www.eia.gov/forecasts/ieo/pdf/0484(2014).pdf.
35 Ibid.
36 Ibid.
37 S. Ladislaw, M. Leed and M. Walton, *New Energy, New Geopolitics: Geopolitical and National Security Impacts* (Washington, DC: Center for Strategic & International Studies/Rowman & Littlefield, 2014) http://csis.org/files/publication/140605_Ladislaw_NewEnergyNewGeopolitics_background2_Web.pdf.
38 US Energy Information Administration (USEIA), *International Energy Outlook 2014*. Washington, DC: US Energy Information Administration, 2014. www.eia.gov/forecasts/ieo/pdf/0484(2014).pdf.
39 Ibid., 12.
40 M. Frank et al., "Crossing the Natural Gas Bridge," Center for Strategic and International Studies, June 26, 2009, http://csis.org/publication/crossing-natural-gas-bridge; National Intelligence Council, *Global Trends 2030 Alternative Worlds* (Washington, DC: National Intelligence Council, 2012).
41 M. King, and J. Gulledge, "Climate and Energy Security: Evidence, Emerging Risks and a New Agenda," *Climatic Change* 123, no. 1 (2013).
42 "A New Era in the Oil Market."
43 M. Werz, and L. Conley, "Climate Change Migration and Conflict in Northwest Africa: Rising Dangers and Policy Options across the Arc of Tension," Center for American Progress, April 2012, www.americanprogress.org/issues/2012/04/climate_migration_nwafrica.html.
44 US Energy Information Administration (USEIA), *World Oil Transit Chokepoints*. Washington, DC: USEIA, 2014. www.eia.gov/beta/international/analysis_includes/special_topics/World_Oil_Transit_Chokepoints/wotc.pdf.
45 J. W. Busby, *Climate Change and National Security: An Agenda for Action*, report no. 32 (New York: Council on Foreign Relations, 2007).
46 ICC, International Maritime Bureau, *Piracy and Armed Robbery against Ships 2012 Annual Report* (London: ICC, International Maritime Bureau, 2013).
47 King and Gulledge, "Climate and Energy Security."
48 R. Huebert et al., *Climate Change and International Security: The Arctic as a Bellwether* (Arlington, VA: Center for Climate and Energy Solutions, 2012).
49 King and Gulledge, "Climate and Energy Security."
50 Klare, "Climate Change Blowback."
51 IPCC, 2014: *Climate Change 2014: Synthesis Report. Contribution of Working Groups I, II and III to the Fifth Assessment Report of the Intergovernmental Panel on Climate Change* [Core Writing Team, R.K. Pachauri and L.A. Meyer (eds)]. IPCC, Geneva, Switzerland.
52 "IPCC 2013: The Physical Science Basis Summary for Policy Makers," in *Climate Change 2013: The Physical Science Basis*, ed. T. F. Stocker et al. (New York: Cambridge University Press, 2013).
53 IAE, "World Energy Outlook," 2015. www.worldenergyoutlook.org/resources/water-energynexus/.
54 King and Gulledge, "Climate and Energy Security."
55 Hannah Northey and Hema Parmar, "Long, hot summer raised questions about how power plants might fare in warming world," *GreenWire,* September 14, 2012.
56 Klare, "Climate Change Blowback," 39.
57 Arent et al., "Key Economic Sectors and Services," 665.
58 Ibid.
59 Ibid.
60 Klare, "Climate Change Blowback," 33.
61 Patrick M. Cronin, and Robert D. Kaplan, "Cooperation from Strength: U.S. Strategy and the South China Sea," *CNAS*, January 2012. www.cnas.org/files/documents/publications/CNAS_CooperationFromStrength_Cronin_1.pdf.
62 International Hydropower Association, "Country Profile: China," 2015. www.hydropower.org/country-profiles/china.

63 Beth Walker, and Liu Qin, "The Hidden Costs of China's Shift to Hydropower," *The Diplomat*, July 29, 2015, http://thediplomat.com/2015/07/the-hidden-costs-of-chinas-shift-to-hydropower/.
64 Atlantic Council, *Energy for Water and Water for Energy: A Report on the Atlantic Council's Workshop How the Nexus Impacts Electric Power Production in the United States*, Atlantic Council, 2011, www.acus.org/publication/energy-water-and-water-energy.
65 International Energy Agency (IEA), *Resources to Reserves 2013: Oil, Gas and Coal Technologies for the Energy Markets of the Future*, IEA 2013. www.iea.org/publications/freepublications/publication/Resources2013.pdf.
66 D. Huber, and J. Gulledge, "Extreme Weather and Climate Change," Center for Climate and Energy Solutions, December 2011, www.c2es.org/publications/extreme-weather-and-climate-change.
67 Arent et al., "Key Economic Sectors and Services," 667.
68 M. Davis, and S. Clemmer, *Power Failure: How Climate Change Puts Our Electricity at Risk – and What We Can Do*, Union of Concerned Scientists, April 2014, www.ucsusa.org/sites/default/files/legacy/assets/documents/Power-Failure-How-Climate-Change-Puts-Our-Electricity-at-Risk-and-What-We-Can-Do.pdf.
69 CNA, *National Security and the Threat of Climate Change*, 2007, www.cna.org/CNA_files/pdf/National%20Security%20and%20the%20Threat%20of%20Climate%20Change.pdf.
70 King and Gulledge, "Climate and Energy Security."
71 CNA, *National Security and the Threat of Climate Change*, 2007, www.cna.org/CNA_files/pdf/National%20Security%20and%20the%20Threat%20of%20Climate%20Change.pdf.
72 S. E. Burke, and E. Schneider, "Enemy Number One for the Electric Grid: Mother Nature," *SAIS Review of International Affairs* 35, no. 1 (2015): 75.
73 Ibid.
74 Ibid., 77.
75 Ibid., 75.
76 Ibid., 77.
77 Ibid., 79.
78 US Department of Energy, Office of Electricity Delivery and Energy Reliability, "The Role of Microgrids in Helping to Advance the Nation's Energy System," accessed January 16, 2016, http://energy.gov/oe/services/technology-development/smart-grid/role-microgrids-helping-advance-nation-s-energy-system.
79 Burke and Schneider, "Enemy Number One for the Electric Grid," 78.
80 Ibid., 82.
81 United Nations Framework Convention on Climate Change. Convention on Climate Change. 1992. https://unfccc.int/resource/docs/convkp/conveng.pdf.
82 United Nations Framework on Convention for Climate Change (UNFCCC), "Historic Paris Agreement on Climate Change," press release, UNFCCC, December 12, 2015, http://newsroom.unfccc.int/unfccc-newsroom/finale-cop21/.
83 Center for Energy and Climate Solutions, *Outcomes of the UN Climate Change Conference in Paris* (Arlington, VA: Center for Energy and Climate Solutions, 2015).
84 Ibid.
85 B. Hulac, "Paris Talks Shifted Emissions Trading into High Gear," *ClimateWire*, December 16, 2015.
86 E. Lehmann, "White House Says Climate Pact Will Unleash Private Cash in Clean Energy," *ClimateWire*, December 17, 2015.
87 UNFCCC, "Historic Paris Agreement."
88 Center for Energy and Climate Solutions, *Outcomes of the UN Climate Change Conference*.
89 Ibid.
90 IEA, "World Energy Outlook Special Report: Energy and Climate Change," 2015. www.iea.org/publications/freepublications/publication/WEO2015SpecialReportonEnergyandClimateChange.pdf.
91 United Nations, "UN Sustainable Development Goals: 17 Goals to Transform our World," United Nations, 2015, www.un.org/sustainabledevelopment/energy/.
92 International Energy Agency (IEA), *World Energy Outlook Special Report 2015: Energy and Climate Change*, IEA, 2015, www.iea.org/publications/freepublications/publication/WEO2015SpecialReportonEnergyandClimateChange.pdf.

93 N. Lewis, “Powering the Planet,” presented at Symposium on Climate and Energy Imperatives for Future Naval Forces, Johns Hopkins University, Washington, DC, July 2010, www.jhuapl.edu/ClimateAndEnergy/Book/Author/Lewis,%20Nathan.pdf, 162.
94 O. R. Edenhofer et al., eds, *Climate Change 2014: Mitigation of Climate Change* (New York: Cambridge University Press, 2014), www.ipcc.ch/report/ar5/wg3/.
95 International Atomic Energy Agency, *Energy Electricity and Nuclear Power Estimates for the Period up to 2050*, Reference Data Series, no. 1 (Vienna: International Atomic Energy Agency, 2013).
96 Ibid.
97 C. D. Ferguson, *Nuclear Energy, What Everyone Needs to Know* (New York: Oxford University Press, 2011), 69.
98 Andersen, “Nuclear Power.”
99 USEIA, “International Energy Outlook 2011.”
100 Anderson, “Nuclear Power.”
101 Ferguson, *Nuclear Energy*, 129.
102 Center for New American Security, “Cooperation from Strength: The United States, China and the South China Sea,” www.cnas.org/files/documents/publications/CNAS_CooperationFromStrength_Cronin_1.pdf 2012.
103 Edenhofer et al., *Climate Change 2014*.
104 J. Bordoff et al., *An Economic Strategy to Address Climate Change and Promote Energy Security*. Hamilton Project Strategy Paper (Washington, DC: Brookings Institution, 2007), www.brookings.edu/papers/2007/10climatechange_furman.aspx.
105 UK MOD (Ministry of Defence) (2010), *Global strategic trends out 2040*. www.mod.uk/nr/rdonlyres/38651acb-d9a9-4494-98aa1c86433bb673/0/gst4_update9_feb10.pdf. Accessed February 20, 2012.
106 Bordoff et al., *An Economic Strategy to Address Climate Change*.
107 King, and Gulledge, “Climate and Energy Security.”
108 Bordoff et al., *An Economic Strategy to Address Climate Change*.
109 G. W. Yohe, “Addressing Climate Change through a Risk Management Lens,” in *Assessing the Benefits of Avoided Climate Change: Cost Benefit Analysis and beyond, Proceedings of the Workshop on Assessing the Benefits of Avoided Climate Change*, ed. J. Gulledge et al. (Arlington VA: Pew Center on Global Climate Change, 2009).
110 C. D. Ferguson, “Nuclear Power’s Uncertain Future,” *The National Interest*, March 15, 2012, http://nationalinterest.org/commentary/nuclear’s-uncertain-future-6643.
111 J. Steimel, “Backdraft: Minimizing Conflict in Climate Responses,” *New Security Beat* (blog), August 5, 2011, www.newsecuritybeat.org/2011/08/backdraft-minimizing-conflict-in-climate-change-responses/.
112 Ibid.
113 CNA, “Powering America’s Economy, Energy Innovation at the Crossroads of National Security Challenges,” Alexandria, VA, July 2010. www.cna.org/cna_files/pdf/MAB3book_8-3-2010.pdf.
114 Pew Charitable Trusts, *The Clean Energy Economy, Repowering Jobs, Businesses and Investments across America* (Philadelphia, PA: Pew Charitable Trusts, 2009).
115 J. Rankin, “Germany’s Planned Nuclear Switch-off Drives Energy Innovation,” *The Guardian*, November 2, 2015, www.theguardian.com/environment/2015/nov/02/germanys-planned-nuclear-switch-off-drives-energy-innovation.
116 CNA, “Powering America’s Economy.”
117 K. Bradsher, “China Leading Global Race to Make Clean Energy,” *New York Times*, January 31, 2010.
118 CNA, “Ensuring America’s Freedom of Movement: A National Security Imperative to Reduce US Oil Dependence,” October 2011, www.cna.org/EnsuringFreedomofMovement.
119 CNA, “Powering America’s Economy: Energy Innovation at the Crossroads of National Security Challenges,” July 2010. www.cna.org/cna_files/pdf/MAB3book_8-3-2010.pdf.
120 R. Manning, *Renewable Energy’s Coming of Age: A Disruptive Technology?* (Washington, DC: Atlantic Council, 2015), 2.
121 R. W. Howarth, R. Santoro, and A. Ingraffea, “Venting and Leakage of Methane from Shale Gas Development: Response to Cathles et al.,” *Climatic Change* 113, no. 2 (2012): 65.
122 National Intelligence Council, *Global Trends 2030 Alternative Worlds*, 37.
123 Ibid.
124 Ibid.
125 Manning, *Renewable Energy’s Coming of Age*.

126 Lewis, "Powering the Planet," 166.
127 Manning, *Renewable Energy's Coming of Age.*
128 Ibid., 6.
129 US Department of Energy, Office of Electricity Delivery and Energy Reliability, "The Role of Microgrids in Helping to Advance the Nation's Energy System," accessed January 16, 2016, http://energy.gov/oe/services/technology-development/smart-grid/role-microgrids-helping-advance-nation-s-energy-system.
130 "Batteries Included?" *The Economist*, February 2, 2013, www.economist.com/news/science-and-technology/21571117-search-better-ways-storing-electricity-hotting-up-batteries.
131 Manning, *Renewable Energy's Coming of Age.*
132 S. E. Burke, "Natural Security," Center for New American Security, 2009. www.cnas.org/node/2712; Christine Parthemore, "Elements of Security: Mitigating the Risks of U.S. Dependence on Critical Minerals," CNAS, 2011. www.cnas.org/files/documents/publications/CNAS_Minerals_Parthemore.pdf.
133 A. Jha, "When You Wish upon a Star: Nuclear Fusion and the Promise of a Brighter Tomorrow," *The Guardian*, January 25, 2015, www.theguardian.com/science/2015/jan/25/iter-nuclear-fusion-cadarache-international-thermonuclear-experimental-reactor-steven-cowley.
134 Ibid.
135 "Lockheed Martin Pursuing Compact Nuclear Fusion Reactor," press release, Lockheed Martin, October 15, 2014, www.lockheedmartin.com/us/news/press-releases/2014/october/141015ae_lockheed-martin-pursuing-compact-nuclear-fusion.html.
136 Ibid.
137 "Stellar Work," *The Economist*, October 24, 2015, www.economist.com/news/science-and-technology/21676752-research-fusion-has-gone-down-blind-alley-means-escape-may-now-be, 74.
138 "Stellar Work," *The Economist*, 74.
139 Jha, "When You Wish upon a Star."
140 Marcus King, LaVar Huntzinger and Thoi Nguyen, *The Feasibility of Small Modular Reactors on U.S. Military Installations* (Alexandria, VA: CNA Corporation, 2011).
141 Ibid.
142 Ibid.
143 World Nuclear Association, "Small Nuclear Power Reactors," December 2015, www.world-nuclear.org/info/nuclear-fuel-cycle/power-reactors/small-nuclear-power-reactors/.
144 John Banks, and T. Boersma, *Fostering Low Carbon Energy: Next Generation Policy to CCS in the United States* (Washington, DC: Brookings Institute, 2015).
145 IAE, "World Energy Outlook Special Report: Energy and Climate Change," 2015.
146 Lewis, "Powering the Planet."
147 Banks and Boersma, *Fostering Low Carbon Energy.*
148 Ibid.
149 Elizabeth L. Chalecki, *Environmental Security: A Guide to the Issues* (Santa Barbara, CA: Praeger, 2013).
150 Ibid.
151 A. Maas, and I. Comardicea, "Climate Gambit: Engineering Climate Security Risks?," in *Backdraft: The Conflict Potential of Climate Change Adaptation and Mitigation*, ed. Geoffrey D. Dabelko et al. Environmental Change and Security Program Report 14, no. 2 (Washington, DC: Woodrow Wilson International Center for Scholars, 2013), 44.
152 Maas and Comardicea, "Climate Gambit," 41.
153 IEA, "World Energy Outlook Special Report: Energy and Climate Change," 2015.
154 Manning, *Renewable Energy's Coming of Age.*
155 Troy R. Hawkins et al., "Comparative Environmental Life Cycle Assessment of Conventional and Electric Vehicles," *Journal of Industrial Ecology* 17, no. 1 (2013): 53–64, doi: 10.1111/j.1530-9290.2012.00532.x.
156 Yuri Kageyama, "Toyota Plans to Stop Selling Traditional Gasoline Cars by 2050," *Huffington Post*, October 14, 2015, www.huffingtonpost.com/entry/toyota-to-stop-selling-traditional-gasoline-cars-by-2050_561e5a7be4b0c5a1ce61380a.
157 Union of Concerned Scientists, "How Do Hydrogen Fuel Cells Work?," Union of Concerned Scientists, 2016, www.ucsusa.org/clean-vehicles/electric-vehicles/how-do-hydrogen-fuel-cells-work#.Vpvze_krKUk.

158 IEA, "World Energy Outlook Special Report: Energy and Climate Change," 2015.
159 US Energy Information Administration (USEIA), "Economics of Plug-in Hybrid Electric Vehicles," USEIA, 2009, www.eia.gov/oiaf/aeo/otheranalysis/aeo_2009analysispapers/ephev.html.
160 CNA, "Ensuring America's Freedom of Movement: A National Security Imperative to Reduce US Oil Dependence," October 2011, www.cna.org/EnsuringFreedomofMovement.
161 Ibid.
162 National Renewable Energy Laboratory, US Department of Energy, Office of Energy, "Research Advances, Cellulosic Ethanol," NREL: Golden, CO, March 2007.
163 CNA, "Ensuring America's Freedom of Movement."
164 Ibid., 28.

3

The climate and security imperative

Francesco Femia and Caitlin Werrell

Climate change, "a large-scale, long-term shift in the planet's weather patterns or average temperature,"[1] has significant implications for human, national, regional and international security. It does so by acting as a "threat multiplier," exacerbating existing risks to the security of communities, nations, and regions, and placing strains on the resource-security underpinnings of state legitimacy.[2] The threat comes not from climate change by itself, but rather, from how it interacts with the existing security landscape, and how it interfaces with the ability or inability of governments to effectively manage these conditions, and provide basic resources, and/or prosperity, to their respective publics. This suggests that climate change may present a serious challenge to state sovereignty in a number of places around the world. Given that international security rests on an international system of viable, sovereign states, it follows that climate change may present a significant threat not just to critical infrastructure, but to international security as well. However, the future does not need to look so dire. There are a range of practical and structural solutions for addressing climate change in a way that is commensurate to the likelihood and scale of the threat, and these solutions form the basis of an international "climate and security imperative."

Growing appreciation of existing climate risks

Increasingly, climate change is being assessed and addressed according to the risks it poses to security. This is in part due to better and more data on the links between climate change and other indicators of human and national security, and the ability to monitor and predict future trends. For example, in 2008, the U.S. Intelligence Community produced a National Intelligence Assessment on the National Security Implications of Global Climate Change to 2030,[3] and the UK Ministry of Defence produced a Global Strategic Trends report,[4] which included robust sections on climate risks and likely future scenarios.

In addition to assessing the future risks of climate change to security, there is an increasing amount of work evaluating the current risks. The U.S. Department of Defense, for example, in its 2014 Climate Change Adaptation Roadmap, noted that climate was "an immediate risk to national security."[5] A chapter on the observed intersection of climate change and human security was included in the Intergovernmental Panel on Climate Change (IPCC) 5th

Assessment Report.[6] This growing evidence of current human and national security risks associated with climate change is coupled with a gradual shift from addressing climate change as a "tolerable" risk primarily to infrastructure, to a possibly "strategically-significant risk" to national security.[7] This gradual up-scaling and incorporation of climate risks into broader risk assessments and policies is part of an ongoing need to better prepare for and mitigate the security risks amplified by a changing climate. While a considerable amount of progress has been made over the last several decades, there remains a gap between the risks nations and peoples face, and the capacity and political will to respond to these risks. This chapter explores both what has been done, and what remains to be addressed.

How security establishments view the climate change threat

Security establishments, including the U.S. military and intelligence community, and in over 70 countries around the world, have identified climate change as a national security threat.[8] This is reflected in a range of strategic documents, assessments, actions and statements made by national security, military and intelligence institutions across the globe – including important outputs from the United States such as the Quadrennial Defense Review and the National Intelligence Council's "Global Trends 2030: Alternative Worlds"[9] and from the UK Ministry of Defence[10] among others. The reasons for this appreciation are clear. Observations of existing climate change impacts on security have increased in the past decade, and projections regarding the influence of climate change on future security are consistently dire. Generally speaking, security practitioners have described climate change as a "threat multiplier."[11] This essentially means that it has the potential to exacerbate other drivers of insecurity. This includes factors such as water, food and energy insecurity, which are drivers that can contribute to significant infrastructure risk, state fragility, and even conflict.[12]

Infrastructure and geostrategic threats

In this context, security establishments in the United States, and globally, generally acknowledge two broad types of risks presented by the "threat multiplication" of a changing climate: "infrastructure" threats and "geostrategic" threats. Though this is an imperfect categorization – as there are indeed connections between the two, and nuances that the categories cannot fully capture – it is a helpful guide to understanding how security communities approach this threat.

Infrastructure threats

Climate change has direct security implications through its effect on the critical infrastructure underpinning a nation's security. This includes sea level rise, storm surge and other extreme weather risks to military installations and personnel that can both inhibit a nation's readiness to conduct critical military and humanitarian assistance operations, and severely damage essential civilian infrastructure, including the financial centers, agricultural hubs, and energy grids that undergird a nation's economic viability. Using the case of the United States as an illustrative example, this involves the multiplication of direct threats to military installations and capabilities.

The latest climate change projections forecast a future of more intense and frequent extreme weather events in North America, including slow-onset risks such as sea level rise (SLR) and droughts, and more quick-onset risks such as floods and storms. These events, either separately or occurring in a cascading manner, can devastate coastal zones, energy infrastructure and agricultural hubs in the United States whose viability rests on predictable rainfall patterns.[13] U.S.

domestic military installations are also at risk. For example, the U.S. Department of Defense (DoD) has devoted resources to assessing and preparing for climate change-driven phenomena, including the potential impact of more frequent and intense dust storms, droughts and rising temperatures, as well as extreme temperatures, on military installations in the Southwest United States.[14] The DoD also examines the impact of sea level rise on its numerous coastal military installations, including the low-lying Hampton Roads region, which according to the former Secretary of Defense Leon Panetta, is home to the "greatest concentration of military might in the world."[15] According to Matt Connolly, the critical military infrastructure at Hampton Roads is vulnerable in the following key ways:[16]

- Rising sea levels and climate disruptions are a "present security threat, not strictly a long-term risk."
- The Hampton Roads region, home to 1.7 million people and over two dozen military sites, is the second most vulnerable region in the U.S. to hurricanes, storms, and sea level rise.
- Some scenarios project that the sea level around Norfolk will rise by seven feet or more in the next 100 years.
- A U.S. Army Corps of Engineers risk assessment found that, by the second half of this century, 60 to 80% of Naval Station Norfolk could be flooded during storms the size of Hurricane Isabel in 2003.

U.S. military installations abroad are also at serious risk. For example, the U.S. Navy's Task Force Climate Change (TFCC)[17] and cross-agency efforts such as SERDP and ESTCP[18] conduct assessments of, and offer solutions to, the current and projected impact of sea level rise on naval installations worldwide.

The damage and degradation of military infrastructure driven by a changing climate can have implications for other military capabilities as well. Extreme weather events in "theater," for example (i.e. regions of the globe where the U.S. military conducts operations), can increase the vulnerability of armed forces by disconnecting them from reliable water and fuel supplies. The protection of military convoys transporting water and fuel is at times a very dangerous mission. For example, casualties resulting from attacks on vulnerable fuel convoys represented "one-third of U.S. Army casualties in Afghanistan in 2007."[19] This is one of the reasons why the DoD devotes resources to equipping its armed forces with portable water desalination and filtration devices, as well as mobile hybrid and renewable energy systems. The U.S. Army's "Energy to the Edge" program is one such effort to reduce the operational risks associated with protecting traditional fuel corridors.[20]

Geostrategic threats

On a broader and more diffuse level, climate change also presents a geostrategic threat by exacerbating stresses on the critical resources underpinning national security, including water, food and energy. These stresses, at their most acute, can degrade a nation's capacity to govern, which can have implications for national, regional and international security. Decreases in water, food and energy availability can devastate livelihoods, and contribute to a broad range of destabilizing trends, including mass population displacement, migration, and political unrest. These pressures in turn can contribute to state fragility, internal conflict, and potentially, state collapse.[21] Climate change can also indirectly change or disrupt existing international security dynamics in geostrategically-significant environments, such as the South China Sea and the Arctic.[22]

Cumulatively, these threats can place significant strains on a world order built on an international system of cooperating sovereign states. As state sovereignty is built on a legitimacy founded on a state's ability to both provide basic resources and/or prosperity – what one may call "output legitimacy" – and/or to offer a voice to its publics – "input legitimacy" – the effect of climate change, particularly on the output legitimacy of states, can have a profound influence on international security, and more broadly, world order.[23] Here is a look at four broad categories for how climate change interacts with and multiplies geostrategic threats:

- increased state fragility;
- heightened globalization of hazards;
- added risks to key geostrategic environments;
- direct threats to state sovereignty.

Increased state fragility

Climate change may increase the fragility of states. This impact is perhaps most acute on already fragile states. The populations of nations that are poorly-governed and resource-stressed are likely to be on the front lines of a changing climate. Fragile states such as Sudan, Ethiopia and the Central African Republic in Africa, Pakistan and Bangladesh in South Asia, and fragile states in the Middle East and North Africa such as Yemen and Syria,[24] are projected to experience some of the most dramatic effects of climate change in terms of rainfall variability and sea level rise.[25] According to the U.S. National Intelligence Council's 2015 Worldwide Threat Assessment, climate change will stress already stressed global food supplies, presenting a particularly acute problem in "Africa, the Middle East, and South Asia."[26] The following are two short case studies to further illustrate this point.

Libya

Libya is a clear case of this nexus between existing state fragility, climate change and water insecurity. Increases in drought days along the most populated coastal areas of Libya bordering the Mediterranean are, according to recent studies, likely to double from 101 days to 224 days a year.[27] Libya is also highly dependent on non-renewable groundwater from aquifers shared with neighboring states – water that is delivered to the coastline via the "Great Manmade River Project" constructed under the Ghaddafi regime. This means that in the not-too-distant future, Libya may run out of groundwater, and find itself more dependent than ever on rainfall in the winter – rainfall that is experiencing a sharp decline as a result of a changing climate.[28] Given that Libya is already experiencing a significant amount of political instability, and continues to be a destination for refugees and migrants throughout Africa, this added stress from climate change could contribute to chronic instability and/or state failure. An increase in water stress is a prime example of how climate change can act as a "threat multiplier."

Syria

Another example of how climate change can increase the fragility of a state is Syria. In 2010–2011, Syria was considered by many political analysts to be a relatively stable state compared to other countries experiencing the so-called "Arab Spring."[29] However, there were underlying climatic and natural resource stresses that made the country far more vulnerable than it seemed. In 2011, a study commissioned by the National Oceanic and Atmospheric Administration (NOAA) determined that climate change was very likely to be a major contributing factor in winter precipitation decline in and around the Mediterranean basin, including North Africa and the Middle East, since 1971. Syria, according to that study, was one of the hardest hit, experiencing a very significant decline from 1971–2010.[30] A subsequent study showed that the extreme drought Syria experienced from 2007–2010 (the worst in its history of records), was made 2–3 times more likely because of climate change.[31] As highlighted in a 2012 report by the Center for Climate and Security – building off research conducted by the United Nations[32] – this drought, coupled with significant natural resource mismanagement by the al-Assad regime, including the subsidization of water intensive crops such as cotton and wheat, and the widespread use of wasteful agricultural practices, such as flood irrigation, contributed to the decimation of a significant percentage of Syria's agricultural sector, and its pastoral rangeland. This directly contributed to a mass internal displacement of around two million farmers and herders during the period of the drought.[33] That displacement, which was reported on but largely missed by political analysts assessing Syria's fragility,[34] may have been a key factor in driving political unrest in the country, which has ultimately led to Syria being one of the most fragile countries in the world.[35]

In the absence of significant adaptation efforts, a slowing of the rate of climatic change, or significant improvements in natural resource governance, such nations are likely to become even more fragile than they already are, leading to the possibility of an increased incidence of state failures.

Heightened globalization of hazards

Climate change-related weather events also have the capacity to disrupt global markets that are critical for the resource security of states. Due to the global nature of these markets, climatic events in one part of the world can have dramatic impact on locations sometimes thousands of miles away. This phenomenon can be described as "the globalization of hazards."[36] Take, for example, the global wheat market and the case of Egypt. Egypt, like many of its neighbors, is one of the world's most highly dependent on the global wheat market.[37] In 2010, major drought and heat wave events in China and Russia – the latter explicitly connected to climate change by two separate studies,[38] devastated local wheat harvests, which drove China to make extraordinarily large purchases of wheat on the global food market.[39] This was a major factor in driving up the price of wheat in Egypt by about 300% in 2010–2011.[40] Egyptian "bread subsidy" policies were unable to bring the price down in many rural areas.[41] While urban protests were occurring in Cairo and other cities, the appeal of the revolutionary movement in Egypt broadened to the countryside, which saw at least three major food riots in 2011.[42] In this way, a climatic hazard in China and Russia contributed in a significant way to food insecurity in Egypt, due to the latter's dependence on the global food market. Such "globalized hazards" are likely to increase, as extreme weather events become more frequent, and more intense.

Added risks to key geostrategic environments

Climate change also places pressure on shared geostrategic, and often geographically ill-defined, environments, which may exacerbate existing international tensions, or create new ones. Two of the clearest examples of this nexus between climate change and key geostrategic environments are the South China Sea and the Arctic Ocean.

The South China Sea is a critical geostrategic choke point. According to a report from the Center for a New American Security (CNAS), ocean-going vessels carry $1.2 billion in U.S. trade annually through its waters. On top of this, sovereignty over parts of the Sea is bitterly contested by adjacent countries, and the U.S. and China have perennially competed over its control (with the U.S. viewing Chinese expansionism in the sea as a threat to international security, and the security of key allies and partners in the region).[43] On top of this dynamic, a warming ocean coupled with over-fishing is driving fish stocks northward into contested waters.[44] As nations bordering the sea, such as Vietnam, are heavily dependent on fish stocks in the South China Sea as a source of protein for their populations (30% of protein intake in Vietnam, for example), Vietnamese fishing fleets are likely to venture further north into waters that are subject to competing claims between China, its neighbors, and the United States.[45] These dynamics could lead to an increasing number of regional security disputes. Such disputes can quickly escalate into international security incidents when the United States becomes involved in support of a claimant.[46]

In the Arctic, by contrast, there has been extraordinary cooperation between nations in what might otherwise be a "Wild West" of ice and water. This may be due to the existence of successful cooperative forums, such as the Arctic Council, but there are other dynamics at play as well. Some scholars argue that as climate change contributes to increasing economic activity in the Arctic, greater cooperation between Arctic nations may become more common due to a perception that "stability is good for business."[47] However, it would be unwise to ignore the possible international security consequences of diplomatic tensions elsewhere in the world (such as disputes between Russia and the United States in Eastern Ukraine and Syria) on the security landscape in the Arctic.[48] Since the outbreak of violence in Ukraine, for example, the North Atlantic Treaty Organization (NATO) requested that all members suspend military cooperation with Russia (including in the Arctic), and Russia has withdrawn from a major forum for military cooperation in the Arctic, the Arctic Security Force Roundtable.[49] A future of an even more open Arctic, coupled with disagreements over security dynamics elsewhere in the world, could provide additional opportunities for tension, and even conflict, between major powers, and this has possible negative consequences for the maintenance of world order. While the current probability of conflict in the Arctic is low,[50] this will be an important space to watch in the future.[51]

Direct threats to state sovereignty

Climate change may also contribute to the disappearance of certain low-lying states through its contribution to sea level rise, as well as the loss of significant territory for other states. This includes small island states such as the Maldives,[52] and large swaths of countries, such as the low-lying coastal zones of Bangladesh.[53] Essentially, climate change presents an existential threat for some countries – a potentially total loss of sovereignty. The international community has no experience in managing the disappearance of nations as a result of environmental processes.[54] In fact, there are no international legal norms designed to account for such an eventuality, including no formal recognition of "climate refugees" or "environmental refugees."[55] The loss of entire states, or large zones within states, may contribute to a mass increase in stateless people in the international system, which could present both a humanitarian and international political and security crisis of the highest order.[56] The full nature of the consequences of such an event is not broadly understood, and that uncertainty presents a unique and unprecedented challenge to international security.

The 3 Ds of prevention and response

Addressing both the infrastructure and geostrategic risks of climate change means managing the unavoidable, and avoiding the unmanageable.[57] First, governments and societies must commit the resources necessary to manage climate change impacts on critical infrastructure, on food, water and energy security, and on geopolitical dynamics in strategically-significant parts of the world. Second, governments and societies must take measures to avoid worst-case climate change scenarios which may be very difficult for nations and international institutions to manage effectively. These solutions involve applications across the so-called "3 Ds" of security: defense, diplomacy and development.

Defense

Nearly all militaries are tasked with addressing both likely and unlikely threats to their respective nations, and developing plans and contingency plans for all such eventualities. This includes the high impact, high likelihood risk of climate change. While the previous section explored how climate change can impact military infrastructure, in particular, this section explores the broader "risk matrix" that militaries contend with, and how those risks affect the three distinct yet interrelated elements of military effectiveness: readiness, operations and strategy.

Readiness: Military "readiness" involves the capacity of a military's infrastructure to assist in the execution of operations in a timely and effective manner. This includes the maintenance of a secure, stable military infrastructure, including installations, supply lines and logistics, for the purposes of being able to carry out missions on short notice. Sea level rise and increased storm surge, for example, have the ability to degrade essential coastal military installations that are critical for operational effectiveness, including major military infrastructure such as the Hampton Roads region in the U.S. Mid-Atlantic, which is home to the largest number of military sites in the world.[58] As mentioned previously, droughts, wildfire and extreme temperatures can also have a significant effect on a military's readiness.

Operations: Outside of operational restrictions that flow from degraded military readiness, climate change can also have a direct influence on military operations, including war-fighting operations and humanitarian missions. The effects of climate change can significantly stress military supply chains, and challenge the logistical capacity of militaries in their operational environments. Mega droughts and flooding events in the war-fighting "theater," for example, can affect the water supplies of armed forces. Drought can also increase the probability of non-state actors, including terrorist organizations such as the Islamic State, seizing water resources to use as leverage against opponents and target populations.[59] The increase in the scale and frequency of natural disasters, and the possibility of an increased incidence of "cascading disasters" or "cascading consequences" where multiple disasters happen either simultaneously or in close proximity to one other, may also make it more difficult for militaries to provide humanitarian assistance and disaster relief (HADR).[60]

Strategy: Military strategy can also be affected by the increasing probability of destabilizing security conditions in strategically-significant regions of the world. In the Middle East and North Africa, climate stresses on water security, coupled with natural resource mismanagement, demographic change, and continued political tension, are very likely to increase the probability of persistent instability in the future. Melting ice in the Arctic region, in concert with ongoing tensions between Arctic nations, such as Russia and the United States, may increase the likelihood of conflict. Rainfall variability in the broader Asia-Pacific region, coupled with increasing urbanization, a growing coastal population, and a greater demand for more energy, will pose enormous risks to security in a part of the world that is of increasing importance to the international security community, including the U.S. military in the context of the "U.S. Asia-Pacific rebalance" strategy.[61] In the South China Sea, fish stocks migrating northward due to a warming ocean may pressure the fishing fleets of nations to move into disputed waters. As mentioned previously, this may precipitate increased tensions between China, its smaller neighbors, and the United States, who support the claims of smaller nations in the region, and these events can draw in military assets and actions.[62] Cumulatively, these pressures could increase the likelihood of militaries being called on to resolve conflicts, or provide post-conflict assistance. All of these dynamics will put stresses and strains on military strategies.[63] Glacial melt and flooding in Central Asia, a region that depends on Himalayan glaciers for most of its freshwater, coupled with existing security dynamics (such as the proliferation of nuclear materials and the persistence of international terrorist organizations), can prove to be a very volatile mix.

Diplomacy and Development

Climate change is also likely to strain diplomacy and development, though it also opens up opportunities in this space. These risks and opportunities exist in the context of both policy responses to climate-related threats to international security, as well as policy responses to climate change itself.

Diplomacy: As climate change challenges state fragility, exacerbates food, water and energy insecurity worldwide, and alters the landscape in which geopolitical dynamics play out, inter-governmental institutions, national governments and communities at the sub-national level will be forced to develop more innovative and sophisticated ways of addressing these complex crises. Disputes over addressing climate change can also spill over into other areas of international security cooperation, potentially fraying relationships between states and within intergovernmental institutions. However, given that climate change represents a threat to international security, responding to the threat also provides opportunities for increasing cooperation – on climate change and a broader array of issues. This includes strategic engagement opportunities. As an example, for diplomatic reasons, it is in the interest of the United States to support the climate resilience of its allies, partners and prospective allies in the Asia-Pacific region, as it advances its "Asia-Pacific rebalance" strategy. This could enhance its diplomatic leverage in the region, and increase its influence vis-à-vis regional actors with which the U.S. enjoys "coopetition," such as China.[64]

Development: The least developed nations are the most vulnerable to the effects of a changing climate. Developing countries are also most likely to experience instability and conflict as a result of dynamics exacerbated by a changing climate. In this context, government agencies and international institutions will need to ensure that assistance to developing nations is climate sensitive, as well as sensitive to the possible effects of climate change on instability and conflict. Making sure that climate policies and investments themselves are conflict-sensitive, is also of great importance, as maladaptation could do more harm than good, no matter the intentions.[65] Support for climate resilience in the developing world will also need to address fundamental power structures and inequities that drive instability and conflict in certain countries, in order to avoid reinforcing structures responsible for persistent poverty. As the G7-commissioned "A New Climate for Peace" notes:

> *Climate change will stress our economic, social, and political systems. Where institutions and governments are unable to manage the stress or absorb the shocks of a changing climate, the risks to the stability of states and societies will increase.*[66]

In the developing world, why those existing economic, social, and political systems exist will need to be better appreciated in order for investments in climate resilience to be most effective.

How does climate change compare to other security priorities?

Climate change is similar to other so-called "new security risks" or "transnational security risks," in that it is widely recognized as a high probability, high consequence risk.[67] This effectively means climate change is happening, and has potentially very significant, negative implications for international security. Despite this general consensus, the response to climate change from most governments, to date, has not yet been commensurate to the risk. The detonation of a weapon of mass destruction, for example, has been considered as a low probability, yet high consequence risk by experts.[68] This suggests that though the probability of such a weapon being detonated by a state or a non-state actor is low, such an event would be unacceptably catastrophic. Further, low probability events happen all the time. Given the legitimate "low tolerance" for such an eventuality, a regime of international laws, and significant state resources, have been marshaled and deployed to track and prevent the proliferation of weapons of mass destruction. Though significant intolerable risks related to climate change have been identified, an approach comparable to non-proliferation efforts has not fully materialized in the climate change sphere of policy-making. The scale and comprehensive nature of recent international agreements on climate change suggest that this may be changing. However, it is worth noting that even the most recent and ambitious international climate agreement in Paris is not legally-binding, unlike most major agreements regarding nuclear, biological and chemical weapons.[69]

Though comparisons between the probability and consequence of different transnational security risks are useful, placing climate change in a hierarchical "rank" of security risks can falsely separate it from other risks, and contribute to a fractured understanding of the risk landscape. Climate-driven water insecurity, for example, can increase the probability of state instability, which in turn could enhance the influence of non-state actors, who could help facilitate the proliferation of the fissile materials necessary for producing nuclear weapons,[70] interconnections make the case that it may be less important to rank climate change next to other security risks, and more important to treat climate change as part of a comprehensive "risk" or "threat" matrix.

Securitization vs. militarization

There exists an academic dialogue regarding the "securitization" of climate change which involves analysis of the "climate and security" discourse. This dialogue, however, tends to artificially conflate the recognition of climate change as a security risk, with the concept of "militarizing" responses to climate change.[71] It is important to address this mischaracterization, as "securitization" does not imply "militarization." The former suggests that climate change presents risks to security, while the latter implies that climate change is being primarily considered as a military problem, and that responses to it should therefore be led by military institutions.

In practice, evidence suggests that the treatment of climate change, and other transnational challenges, as a "security" problem has broadened the scope of responses to its risks, rather than narrowing that scope to military or security-oriented institutions. For example, intergovernmental security institutions and militaries, such as the UN Security Council and the U.S. Department of Defense, recognize the critical role of civilian agencies in combating "new security threats" such as poverty, health vulnerabilities, water and food insecurity, and a changing climate. The United Nations Security Council, for example, released a Presidential

Statement on Climate Change which identifies a non-security intergovernmental organization, the UN Framework Convention on Climate Change (UNFCCC), as a "key instrument for addressing climate change."[72] The U.S. Department of Defense's 2010 Quadrennial Defense Review (QDR) report includes a section on climate change which states that civilian agencies, such as the Department of State and the Department of Energy, should be leading actors in addressing the climate risk.[73] In other words, rather than narrowing the field of action, the climate and security dialogue broadens the scope and scale of action to include a greater number of agencies with different competencies. This helps ensure that few stones are left unturned in the effort to address the high probability, high consequence threat of a changing climate.

Better understanding climate and security risks

In order for climate change to be addressed in a manner that is commensurate to the threat to international security it poses, its "risk profile" must first be elevated among security establishments across the globe. This will require the improvement of analytical tools aimed at assessing climate risks to security, as well as improved channels for delivering prognoses of climate and security risks to senior government officials with the authority and resources to act robustly, and to act early. For example, evidence suggests that the predictive tools and indices governments and analysts use to assess the fragility of states are outdated, and may not adequately account for climate and natural resource-related risks to state fragility.[74] This has led analysts to miss vulnerabilities in seemingly stable states.

Brittle states

One of the reasons improved risk assessments are necessary is because the international security landscape plays host to a number of seemingly stable states that are nevertheless quite vulnerable under the surface, from a resource perspective. It is important to distinguish such "brittle states" from "fragile states," as in brittle states, the appearance of stability can lead analysts and policymakers to fail to anticipate fragilities, and to make poor political, economic and natural resource management choices. These states may score relatively high in "state fragility indices"[75] when compared against states that are more widely considered as "failed" or on "the brink of failure," such as Somalia, Sudan and the Central African Republic.[76] Often, measurements of the fragility of such states focus primarily on social, political and economic circumstances that suggest the likelihood of collapse to be low. However, such measurements often fail to fully take into account significant natural resource vulnerabilities and climatic stresses under the surface that may make them far more fragile than they seem.[77]

The problem is not so much the lack of data, as it is a lack of an appropriate use of existing data. Nations and analysts must better account for climate and natural resource stresses in the tools they use to determine the degree of a state's fragility. They must also ensure that information about environmental drivers of instability reaches senior decision-makers that are responsible for foreign, defense and national security policy. A failure to do so may result in further misdiagnoses of state fragility, and unpreparedness for the security breakdowns that can follow.

Underestimating and oversimplifying the connections between climate change and security

Popular commentary, often dependent on the drama of "two-sided" debate, tends to either oversimplify or underestimate climate change effects on security.[78] This dynamic can have a deleterious effect on public policy, and should therefore be corrected as much as possible. Syria, for example, is not the first "climate war," and climate change is not a proximate cause of the Syrian refugee crisis, despite the implications of some newspaper headlines to the contrary. If these kinds of simplistic claims are accepted at face value, commentators risk absolving governments, such as the al-Assad regime, of a responsibility to protect their publics. Similarly, the underestimation of climate change risks to security could lead governments and societies to miss critical vulnerabilities. In the case of Syria, there is evidence that climate change was a contributing factor to Syria's fragility.[79] Ignoring or underestimating this contribution could also risk leaving governments and populations unprepared.

A reasonable level of certainty

Though the future is not entirely predictable, evidence that climate change can contribute to state fragility, and other stresses on security, is growing. While additional research is critical for enhancing that certainty, policy-makers cannot wait for perfect certainty before making decisions, particularly in a complex and rapidly-changing environment, where delayed action could be implicated in state failure, conflict and humanitarian crisis.

Dr. Jay Gulledge notes that unlike academic scholars, governments are more concerned, by necessity, with false negatives than false positives.[80] If a government fails to detect a risk, and that risk materializes, that government will be held accountable for failing to protect its public. In terms of climate change impacts on security issues such as state fragility, there is a sufficient degree of certainty to justify comprehensive action to mitigate climate risks. Demanding that governments rest policy decisions on near-perfect certainty about climate change impacts before acting is not sensible, and inconsistent with how governments treat a range of other critical security risks. As General Gordon Sullivan, USA (ret.) stated:

> *People are saying they want to be perfectly convinced about climate science projections … But speaking as a soldier, we never have 100% certainty. If you wait until you have 100% certainty, something bad is going to happen on the battlefield.*[81]

In short, governments and publics must lower the "certainty threshold" for triggering preventive solutions to climate change, in order to avoid potentially catastrophic scenarios.

Existing risks and black swans

While the medium and long-term risks of climate change are projected to be destabilizing, climate change is affecting international security already. NASA notes that the globe is warming at a faster rate than it ever has before.[82] The U.S. Department of Defense's 2014 Climate Change Adaptation Roadmap highlights that climate change presents "immediate risks to national security."[83] The National Oceanic and Atmospheric Administration implicates climate change in the significant decline in winter precipitation across the Middle East and the Mediterranean from 1971–2010.[84] Arctic ice melt is fundamentally changing the geopolitical landscape of the High North. The IPCC 5th Assessment Report's Human Security chapter recognizes that climate change can indirectly increase the likelihood of violent conflict.[85]

Though climate impacts are already occurring, and likely to get worse, lower probability climate events, or "black swans," must also be planned for. Abrupt climatic changes, and gradual changes that precipitate abrupt stresses to food, water and energy security, have the potential to be destabilizing.[86] Changes in the jet stream, more rapid than expected sea level rise and glacial melt, and global disasters such as simultaneous shocks to major grain-producing nations, are a few potential abrupt consequences. While climate change models are strong in their predictive power, unknowns remain.

What to do: a few recommendations

Addressing climate change risks involves managing the unavoidable, and avoiding the unmanageable. Governments and international institutions must commit the necessary resources for managing climate change impacts on infrastructure, food, water and energy security, and geopolitical dynamics. These solutions involve applications across the so-called "3 Ds" of security: defense, diplomacy and development.

Defense

There are a range of actions militaries and security establishments can take, in concert with their civilian partners, to address this multi-tiered threat. These include:

- including a robust consideration of climate-related security threats in national defense strategies;
- elevating the "threat profile" of climate change within defense ministries;
- designating a senior defense official as a climate change lead;
- expanding traditional international security cooperation to encompass environmental and climate security matters;
- conducting annual assessments of the implications of climate change for operational missions;
- including attention to climate risks in security assistance programs, particularly as it relates to conflict avoidance and prevention;
- raising the profile of climate change at international defense summits and other security forums;
- developing cooperative military strategies for addressing climate impacts on military operations;
- incorporating climate concerns into military-military and civilian-military cooperation on disaster risk reduction; and
- addressing climate change at international and regional security institutions (NATO, UN Security Council, African Union, ASEAN Defense Ministers Forum, etc.).

Diplomacy

Foreign ministries, heads of state and diplomatic corps across the globe can take practical steps to address the broader security risks associated with a changing climate. These include:

- designating a senior diplomatic official with a "climate and security" mandate;
- elevating climate change as a priority within foreign ministries;
- addressing climate change concerns at multi-lateral institutions with security competence, such as the UN Security Council and the G7;
- creating new international institutions to address climate-related challenges;
- elevating climate and security issues as a strategic priority in all bi-lateral relations;
- responsibly integrating diplomatic and development efforts with military efforts to address the observed and projected security risks of climate change;

- promoting the appointment, by the UN Secretary General (UNSG), of a UN Special Representative for Climate and Security;
- advancing the creation, by the UNSG of a Joint Task Force on Climate and Security consisting of expert representatives from relevant UN institutions, to produce an assessment of the capacity of the UN as a whole to address the security implications of climate change; and
- developing the international legal and institutional structures to manage migration as a climate adaptation strategy.

Development

As most vulnerabilities associated with a changing climate will be related to the level of a nation's or a community's economic and political development, government agencies and international institutions concerned with development should:

- ensure that development assistance is sensitive to the possible effects of climate change on instability and conflict;
- advance climate policies and investments that are conflict-sensitive;
- integrate climate and security factors in development agency investments in conflict prevention and post-conflict reconstruction;
- ensure a robust consideration of climate change-related human and national security risks in periodic reviews of development assistance strategies and programs;
- improve understanding of how climate change pressures interact with state stability and state legitimacy; and
- commit significant resources to climate resilience in unstable parts of the world, and develop climate mitigation and adaptation strategies that are consistent with international security priorities.

Conclusion

Populations are urbanizing and growing, sea levels are rising, and extreme weather events are developing more frequently and intensely. Such dynamics will place additional strains on the food and water resources that underwrite the security of nations and populations around the world, and this may increase the scale of mass migration. Despite this, policies designed to comprehensively address the full picture of climate risks are lacking, despite recent advancements among security institutions, and in international climate negotiations. Waiting for security and humanitarian crises to hit the front pages before acting has often proven a great failure of the international community. Whether or not the international community learns from that failure will be the measure of its resilience.

In this context, improving, augmenting, and possibly even creating new international, regional, national and sub-national structures for addressing climate change, may be critical for a secure future. This is a step that will require far more than technical or technological fixes. Rather, it will require that nations, and international institutions, place climate change at the top of the international security agenda, and find ways of collaborating on reducing those risks that present the greatest challenge to a functioning world order. Such an international climate and security imperative is both necessary, and achievable.

Notes

1 Met Office, "What is Climate Change?" Met Office, 2010, www.metoffice.gov.uk/climate-guide/climate-change.
2 CNA Corporation, *National Security and the Threat of Climate Change* (Arlington, VA: CNA Corporation, 2007); US DoD, *2014, Climate Change Adaptation Roadmap* (Washington, DC: U.S. Department of Homeland Security, 2014).
3 T. Fingar, *National Intelligence Assessment on the National Security Implications of Global Climate Change to 2030*, Statement to the House Permanent Select Committee on Intelligence, House Select Committee on Energy Independence and Global Warming, 25 June 2008.
4 United Kingdom Ministry of Defence, *Global Strategic Trends – Out to 2045*. 5th ed., 2014. Available at: www.gov.uk/government/uploads/system/uploads/attachment_data/file/348164/20140821_DCDC_GST_5_Web_Secured.pdf.
5 U.S. Department of Defense, *2014 Climate Change Adaptation Roadmap* (Washington, DC: U.S. Department of Homeland Security, 2014).
6 Y. Hijioka et al., "Asia," in *Climate Change 2014: Impacts, Adaptation, and Vulnerability. Part A: Global and Sectoral Aspects. Contribution of Working Group II to the Fifth Assessment Report of the Intergovernmental Panel on Climate Change* (Cambridge: Cambridge University Press, 2014), 1327–1370.
7 U.S. Department of Homeland Security, *The 2014 Quadrennial Homeland Security Review* (Washington, DC: U.S. Department of Homeland Security, 2014).
8 American Security Project, *Climate Security Report*, American Security Project, 2012, Washington, DC, www.americansecurityproject.org/climate-security-report/.
9 US DoD, *2014 Climate Change Adaptation Roadmap* (Washington, DC: U.S. Department of Homeland Security); National Intelligence Council, *Global Trends 2030: Alternative Worlds* (Washington, DC: National Intelligence Council, 2012).
10 United Kingdom Ministry of Defence, *Global Strategic Trends – Out to 2045*.
11 CNA Corporation, *National Security and the Threat of Climate Change*; U.S. Department of Defense, *2014 Climate Change Adaptation Roadmap* (Washington, DC: U.S. Department of Homeland Security, 2014).
12 U.S. Department of Defense, *National Security Implications of Climate-Related Risks and a Changing Climate* (Washington, DC: U.S. Department of Defense, 2015).
13 K. Burks-Copes, *Risk Quantification for Sustaining Coastal Military Installation Assets and Mission Capabilities* (No. RC-1701) (Washington, DC: SERDP and ESTCP, U.S. Department of Defense, 2014).
14 C. A. Alaimo, "Military Taps UA Expertise to Cope with Impact of Climate Change," *Arizona Daily Star*, March 3, 2012, http://tucson.com/news/local/wildfire/military-taps-ua-expertise-to-cope-with-impact-of-climate/article_22e62435-a227-568d-9db9-3ba40f9ce8f5.html#ixzz1oGdBhrGm.
15 Leon Panetta, "Secretary of Defense Speech: Hampton Roads Chamber of Commerce," October 19, 2012, US Department of Defense, http://archive.defense.gov/Speeches/Speech.aspx?SpeechID=1729.
16 M. Connolly, *New BRIEFER: Hampton Roads, Virginia and the Military's Battle against Sea Level Rise* (Washington, DC: Center for Climate and Security, 2015).
17 U.S. Navy, "Climate Change: Task Force Climate Change (TFCC)," Green Fleet, accessed February 23, 2016, http://greenfleet.dodlive.mil/climate-change/.
18 SERDP, "Climate Change and Impacts of Sea Level Rise," SERDP and ESTCP, accessed January 23, 2016, www.serdp.org/Featured-Initiatives/Climate-Change-and-Impacts-of-Sea-Level-Rise.
19 Joshua Zaffos, "U.S. Military Forges Ahead with Plans to Combat Climate Change," *Scientific American*, April 2, 2012, www.scientificamerican.com/article/us-military-forges-ahead-with-plans-to-combat-climate-change/.
20 A. Z. Sanders, "Rapid Equipping Force Develops 'Energy to the Edge' Program," Fort Belvoir, VA,, August 5, 2011, United States Army, www.army.mil/article/62936/.
21 L. Rüttinger, Dan Smith, Gerald Stang et al., *A New Climate for Peace: Taking Action on Climate and Fragility Risks* (Washington, DC: Woodrow Wilson International Center for Scholars, European Union Institute for Security Studies, 2015).
22 F. Femia, and C. E. Werrell, "A Climate-Security Plan for the Asia-Pacific Rebalance: Lessons from the Marshall Plan," in *The U.S. Asia-Pacific Rebalance, National Security and Climate Change*, ed. C. E. Werrell, and F. Femia (Washington, DC: Center for Climate and Security, 2015).
23 V. A. Schmidt, "Democracy and Legitimacy in the European Union," in *The Oxford Handbook of the European Union*, (Oxford: Oxford University Press, 2012).

24 N. Haken et al., *Fragile States Index* (Washington, DC: Fund for Peace, 2015).
25 Y. Hijioka, E. Lin, and J. J. Pereira, *Climate Change 2014: Impacts, Adaptation, and Vulnerability. Asia.* IPCC 5th Assessment Report, 2014 (Cambridge and New York: Cambridge University Press).
26 J. R. Clapper, *Worldwide Threat Assessment of the US Intelligence Community*, Statement for the Record, Senate Armed Services Committee, James R. Clapper Director of National Intelligence, February 26, 2015, Washington, DC. www.dni.gov/files/documents/Unclassified_2015_ATA_SFR_-_SASC_FINAL.pdf.
27 J. W. Busby, K. L. White, and T. G. Smith, *Climate Change and Insecurity: Mapping Vulnerability in Africa*, Climate and Energy Paper Series (Washington, DC: The German Marshall Fund of the United States, 2010).
28 F. Femia, and C. E. Werrell, *A New Libya in a New Climate: Charting a Sustainable Course for the Post-Gaddafi Era* (Briefer No. 5) (Washington, DC: Center for Climate and Security, 2011).
29 C. E. Werrell, and F. Femia, eds, *The Arab Spring and Climate Change* (Washington, DC: Center for Climate and Security, Stimson Center, and Center for American Progress, 2013).
30 M. Hoerling, J. Eischeid, J. Perlwitz et al., "On the Increased Frequency of Mediterranean Drought," *J. Clim* 25 (2012): 2146–2161, doi: 10.1175/JCLI-D-11–00296.1.
31 C. P. Kelley, S. Mohtadi, M.A. Cane, et al., "Climate Change in the Fertile Crescent and Implications of the Recent Syrian Drought," *Proc. Natl. Acad. Sci.* 112 (2015): 3241–3246, doi: 10.1073/pnas.1421533112.
32 E. Wadid, B. Katlan, and O. Babah. *Global Assessment Report on Disaster Risk Reduction. Drought Vulnerability in the Arab Region Special Case Study: Syria.* Geneva: ISDR. 2010, www.preventionweb.net/english/hyogo/gar/2011/en/bgdocs/Erian_Katlan_&_Babah_2010.pdf.
33 F. Femia, and C. E. Werrell, *Syria: Climate Change, Drought and Social Unrest* (Washington, DC: Center for Climate and Security, 2012).
34 C. E. Werrell, F. Femia, and T. Sternberg, "Did We See It Coming?: State Fragility, Climate Vulnerability, and the Uprisings in Syria and Egypt," *SAIS Review of International Affairs* 35 (2015): 29–46, doi: 10.1353/sais.2015.0002.
35 Haken et al., *Fragile States Index*.
36 T. Sternberg, *Chinese Drought, Wheat, and the Egyptian Uprising: How a Localized Hazard Became Globalized, The Arab Spring and Climate Change* (Washington, DC: Center for Climate and Security, Stimson Center, and Center for American Progress, 2013).
37 Ibid.
38 S. Rahmstorf, and D. Coumou, "Increase of Extreme Events in a Warming World," *Proceedings of National Academy of Science* 108 (2011): 17905–17909. doi: 10.1073/pnas.1101766108; F. Otto, N. Massey, G.J. van Oldenburgh, et al., "Reconciling Two Approaches to Attribution of the 2010 Russian Heat Wave," *Geophysical Research Letters* 39 (2012): L04702, doi: 10.1029/2011GL050422.
39 T. Sternberg, "Chinese Drought, Bread and the Arab Spring," *Applied Geography* 34 (2012): 519–524, doi: 10.1016/j.apgeog.2012.02.004.
40 Sternberg, *Chinese Drought, Wheat, and the Egyptian Uprising.*
41 S. Johnstone, and J. Mazo, *Global Warming and the Arab Spring, The Arab Spring and Climate Change* (Washington, DC: Center for Climate and Security, Stimson Center, and Center for American Progress, 2013).
42 Sternberg, *Chinese Drought, Wheat, and the Egyptian Uprising.*
43 P. M. Cronin, ed., *Cooperation from Strength: The United States, China and the South China Sea* (Washington, DC: Center for New American Security, 2012).
44 M. D. King, "Climate Change and Vietnamese Fisheries: Opportunities for Conflict Prevention," in *The U.S. Asia-Pacific Rebalance, National Security and Climate Change?*, ed. C. E. Werrell, and F. Femia (Washington, DC: Center for Climate and Security, 2015).
45 Femia, and Werrell, "A Climate-Security Plan."
46 King, "Climate Change and Vietnamese Fisheries."
47 M. Bert, *A Strategy to Advance the Arctic Economy* (Memorandum no. 14) (Washington, DC: Council on Foreign Relations, 2012).
48 U. Friedman, "The Arctic: Where the U.S. and Russia Could Square off Next," *The Atlantic*, March 28, 2014.
49 A. J. Bailes, "A New Arctic Chill? Reactions in the North to New Tensions with Russia," Scottish Global Forum. 2015, www.scottishglobalforum.net/alyson-bailes-arctic-chill.html#_ftnref3.

50 T. C. Gallaudet, *Charting the Arctic: Security, Economic, and Resource Opportunities*, 2015, Committee on Foreign Affairs, Subcommittee on Europe, Eurasia, and Emerging Threats and Subcommittee on the Western Hemisphere, US House of Representatives, November 17, 2015, http://docs.house.gov/meetings/FA/FA14/20151117/104201/HHRG-114-FA14-Wstate-GallaudetT-20151117.pdf.
51 J. Stavridis, "Once Again, Europe Needs America," Politico, 2016, www.politico.eu/article/europes-security-disorder-demands-american-help-pressure-refugees-social-strain-open-borders/; E. Rosenberg, D. Titley, and A. Wiker, *Arctic 2015 and Beyond: A Strategy for U.S. Leadership in the High North* (Washington, DC: Center for New American Security, 2014).
52 R. McLean, L. A. Nurse, et al., "Small Islands," in *Climate Change 2014: Impacts, Adaptation, and Vulnerability. Part A: Global and Sectoral Aspects. Contribution of Working Group II to the Fifth Assessment Report of the Intergovernmental Panel on Climate Change* (Cambridge: Cambridge University Press, 2014).
53 Hijioka et al., "Asia."
54 J. G. Stoutenburg, *Disappearing Island States in International Law* (Leiden: Brill, 2015).
55 B. Glahn, "'Climate Refugees?' Addressing the International Legal Gaps," International Bar Association, June 11, 2009, www.ibanet.org/Article/Detail.aspx?ArticleUid=B51C02C1-3C27-4AE3-B4C4-7E350EB0F442.
56 L. Goff, and N. Samaranayake, "Climate Change, Migration and a Security Framework for the Asia-Pacific Rebalance," in *The U.S. Asia-Pacific Rebalance, National Security and Climate Change?* ed. C. Werrell, and F. Femia (Washington, DC: Center for Climate and Security, 2015).
57 R. M. Bierbaum, J. P. Holdren, M. MacCracken, et al., *Confronting Climate Change: Avoiding the Unmanageable and Managing the Unavoidable* (Washington, DC: United Nations Foundation and Sigma XI, 2013).
58 U.S. Department of Defense, *2014 Climate Change Adaptation Roadmap* (Washington, DC: U.S. Department of Homeland Security, 2014).
59 M. D. King, "The Weaponization of Water in Syria and Iraq," *Washington Quarterly* 38 (2016), 153–169.
60 National Homeland Security Consortium, "Protecting Americans in the 21st Century: Communicating Priorities for 2012 and Beyond," 2012, www.iafc.org/files/1DISASTERmgntHOMEsec/hs_2012NHSCWhitePaper.pdf.
61 Femia, and Werrell, "A Climate-Security Plan."
62 White House, *Fact Sheet: U.S. Response to Typhoon Haiyan* (Washington, DC: White House, 2013).
63 Center for Climate and Security, *Why Do Militaries Care about Climate Change?* Climate Survey 101 (Washington, DC: Center for Climate and Security, 2015).
64 Femia, and Werrell, "A Climate-Security Plan."
65 G. D. Dabelko, L. Herzer, S. Null, et al., eds, *Backdraft: The Conflict Potential of Climate Change and Adaptation* (Washington, DC: Environmental Change and Security Program, Woodrow Wilson Center, 2013).
66 Rüttinger et al., *A New Climate for Peace.*
67 World Economic Forum, *Global Risks 2014*. 9th edn. (Geneva: World Economic Forum, 2014).
68 R. G. Lugar, *The Lugar Survey on Proliferation Threats and Responses*, 2005, https://fas.org/irp/threat/lugar_survey.pdf.
69 United Nations Framework Convention on Climate Change, Adoption of the Paris Agreement, Proposal by the President, 2015.
70 C. Parthemore, *Climate Change & Nuclear Risks* (Washington, DC: Center for Climate and Security, 2016).
71 B. Hayes, and D. Deering, "The Secure and Dispossessed: Security for Whom?" *Open Democracy* 2016, www.opendemocracy.net/daniel-deering-and-ben-hayes/interview-with-dr-ben-hayes-on-his-recently-released-co-edited-book-ent.
72 Permanent Mission of Spain, Permanent Mission of Malaysia, Open Arria Formula Meeting on the Role of Climate Change as a Threat Multiplier for Global Security, 2015, www.spainun.org/wp-content/uploads/2015/06/Concept-Note_ClimateChange_20150630.pdf.
73 US Department of Defense, *Quadrennial Defense Review Report* (Washington, DC: U.S. Department of Defense, 2010).
74 Werrell, Femia, and Sternberg, "Did We See It Coming?"
75 Ibid.
76 Haken et al., *Fragile States Index.*
77 Werrell, Femia, and Sternberg, "Did We See It Coming?"

78 C. Werrell, and F. Femia, *On Syrian Refugees and Climate Change: The Risks of Oversimplifying and Underestimating the Connection* (Washington, DC: Center for Climate and Security, 2015).
79 Kelley et al., "Climate Change in the Fertile Crescent."
80 J. Gulledge, *Countries Should Assess Climate Risk the Way They Assess Other Security Risks* (Washington, DC: Center for Climate and Security, 2015).
81 CNA Corporation, *National Security and the Threat of Climate Change.*
82 H. Riebeek, "Global Warming: Feature Articles," National Aeronautics and Space Administration, 2010, http://earthobservatory.nasa.gov/Features/GlobalWarming/page3.php.
83 U.S. Department of Defense, *2014 Climate Change Adaptation Roadmap* (Washington, DC: U.S. Department of Homeland Security, 2014).
84 Hoerling et al., "On the Increased Frequency of Mediterranean Drought."
85 W. N. Adger, J. M. Pulhin, J. Barnett, et al., "Human Security," in: *Climate Change 2014: Impacts, Adaptation, and Vulnerability. Part A: Global and Sectoral Aspects. Contribution of Working Group II to the Fifth Assessment Report of the Intergovernmental Panel on Climate Change* (Cambridge: Cambridge University Press, 2014).
86 P. Schwartz, and D. Randall, *An Abrupt Climate Change Scenario and Its Implications for United States National Security*, 2003, California Institute of Technology Pasadena, Jet Propulsion Lab, http://oai.dtic.mil/oai/oai?verb=getRecord&metadataPrefix=html&identifier=ADA469325.

4

Climate change and energy security policies

Are they really two sides of the same coin?[1]

Peter R. Hartley

It has been claimed that energy security and climate policy should be considered "two sides of the same coin." For example, on a visit to the United States in October 2006, former UK Prime Minister Tony Blair said, "We must treat energy security and climate security as two sides of the same coin." Other leaders in Europe, members of the United States Congress and many commentators echoed Blair's statement.

Blair was addressing the claim that the Iraq War was at least partly motivated by concerns about the security of supply of crude oil. His point was that the industrialized world would not be so concerned about Middle East politics if it were not so dependent on Middle East oil supplies. For ensuring energy security, reducing oil consumption is a substitute for a military presence and military action in the Middle East.

At the same time, fossil fuel combustion adds carbon dioxide to the atmosphere. Since CO_2 is a greenhouse gas, increasing its concentration in the atmosphere should impede the outgoing transmission of infrared radiation. This could, in turn, have harmful effects by triggering changes in climate, especially a rise in global surface temperatures. Reducing these possibly harmful effects from climate change is the goal of climate policy.

Hence, reducing fossil fuel combustion should increase energy security while also reducing potentially harmful climate change. Although we have two policy goals, they should be treated as one, since one policy instrument can simultaneously further both goals.

To examine this argument more carefully we need to discuss what we mean by energy security and climate policy. We also need to investigate the range of policies that could address the two policy goals. Only then can we assess whether both goals are best addressed by the same policies, or whether policies that are best to further one goal might compromise attainment of the other goal.

Possible meanings of energy security

The above argument identifies energy security with national security. We can associate increased energy security with a reduced need to maintain influence in countries, such as those of the Middle East, that possess substantial energy resources but may be politically unstable or hostile to the West.

Another connection between national security and energy security is the fact that modern military forces require substantial refined oil products. The United States military, for example, consumes about 130 million barrels (about 20 billion liters) of oil products a year.

The notion of energy security also has economic dimensions. Sudden, large increases in energy prices have preceded most of the post-World War II recessions. Large energy price increases are thought to retard economic growth via a number of mechanisms. The need to spend more on energy commodities constrains household consumption of non-energy goods and services. Similarly, to save on non-energy related costs, firms reduce employment and investments in non-energy related capital. Productivity also declines as resources are reallocated in response to the energy price changes.

Higher energy prices also increase financial flows from net importers to net exporters of energy commodities. Many net exporters are smaller economies with limited capacity to absorb additional investment funds. The funds therefore have to be recycled by international capital markets and the resulting financial flows can disrupt exchange rates and international financial markets more generally.

Energy price variability also increases uncertainty about future energy prices, which in turn deters investments in competing types of energy or competing high cost locations such as the deep water Gulf of Mexico. Such investments typically are large and long-lived and thus made much more risky if energy prices are more uncertain. In so far as volatile energy prices reduce investments in domestic alternatives, they exacerbate the initial instabilities by concentrating production in less stable regions.

Several policies have been proposed to deal with energy insecurity. Strategic petroleum reserves can reduce the effects of supply shortages, and especially provide emergency supplies in case they are needed for military purposes. It is arguable whether such reserves increase security of supply on net, since they likely reduce the incentives for private firms to store energy commodities. Regardless of one's view on that issue, however, resource stockpiles are designed to contend with short term emergencies.

Longer term energy security for the world can be increased most effectively by diversifying energy sources and the range of regions from which they come, and especially by increasing supply from more stable countries. Increased substitutability among energy sources also increases resilience to supply disruptions. *National* energy security is also enhanced by a greater variety of *domestic* energy supply sources.

CO_2 emissions and climate policy

Almost all scientists who have investigated anthropogenic CO_2 emissions agree that they will change climates. However, far fewer scientists claim that the *overall* effects of such emissions under a so-called "business as usual" scenario, where no specific policies are taken to artificially constrain fossil fuel use, are likely to be significantly harmful. I will return to this proposition shortly. Before I do so, however, I want to discuss another issue. Even if we conclude that CO_2 emissions are likely to be significantly harmful on net, limiting CO_2 emissions is not the only possible policy response.

I will classify climate policy actions into five categories. The first is reducing the emissions of greenhouse gases, particularly CO_2. Other components of the atmosphere have stronger greenhouse effects than CO_2. In particular, water vapor absorbs a far wider range of outgoing radiation, and therefore has a more significant warming effect. However, the concentration of water vapor in the atmosphere is not thought to be *directly and significantly* influenced by human activity and therefore is not considered to be *directly* amenable to policy. Indeed, the main global computer climate models assume that a trend in CO_2 concentration in the atmosphere is the main determinant of trend changes in water vapor content so that controlling CO_2 also indirectly controls water vapor. Apart from water vapor, other greenhouse gases, including methane and hydrofluorocarbons, also have much stronger radiation trapping effects than CO_2. The main argument for focusing on CO_2 is that, since we release more of it, it could provide a more significant policy lever for reducing the total greenhouse effect.

A second category of responses to the threat of climate change from CO_2 emissions involves an offsetting sequestration of greenhouse gases, particularly CO_2. Already CO_2 is used for enhanced oil recovery. This involves extracting additional oil from a mature well by injecting CO_2 to increase the pressure and, with suitable additives, also alter the geochemistry. The permanence of such sequestration has been questioned, however, since some of the injected CO_2 may be produced along with the oil.

Another proposal under investigation involves injecting CO_2 into so-called methane hydrates. These are a form of solidified methane, or natural gas, found under great pressure and low temperatures on many continental shelves. Methane hydrates are thought to contain more energy than all other known remaining sources of fossil fuel combined. Laboratory experiments have shown that CO_2 can displace the methane in the hydrate structure both sequestering the CO_2 and liberating the methane to be used as a fuel.

There are also a few demonstration projects that bury CO_2 in underground reservoirs such as abandoned natural gas wells or salt domes, or in the deep ocean. Unlike the examples just discussed, these projects do not produce anything worthwhile in exchange. As a result, they are very expensive and unlikely to be used on a large scale without further technological innovations.

Another set of sequestration proposals relies on the fact that CO_2 is an input into photosynthesis, whereby plants use CO_2, water and the energy of sunlight to produce carbohydrates and then other organic compounds. Those organic compounds in turn, of course, sustain most of the rest of the life on earth, including us. When we oxidize the carbohydrates we not only get energy but also produce CO_2 (and water) that we return to the atmosphere in our breath. Planting forests, or reducing deforestation, sequesters some CO_2 until the wood decays. A related idea involves partially oxidizing plant material to produce charcoal, which is then buried. Since the charcoal is slow to decompose, this sequesters the carbon for a long time. It also can improve soil quality. It is unclear, however, whether it could be done on a large scale at a sufficiently low cost.

Other firms are working on using CO_2-enriched greenhouses to grow algae that have been genetically engineered to produce compounds that can be turned into synthetic liquid fuels as substitutes for oil-based fuels. Although those fuels also release CO_2 when they are burned, it is recycled CO_2 previously taken from the atmosphere by the algae. The same idea underlies producing ethanol and other biofuels from sugars produced by plants.

Proposals to seed the oceans with iron also aim to enhance the absorption of CO_2 by plants, in this case, phytoplankton. In the middle of the ocean, insufficient iron limits the growth of phytoplankton. Increasing the amount of iron would promote plankton growth and increase photosynthesis. Some of the additional carbon compounds so produced would fall to the ocean floor to be sequestered for a long time.

A third category of responses to the threat of climate change involves various geo-engineering projects aimed at offsetting the radiation trapping effect of greenhouse gases by increasing the direct reflection of incoming solar radiation. One way to do this is by increasing the amount of low cloud cover, which in turn reflects incoming solar radiation. A change in cloud cover of just a few% could completely offset the predicted warming effect from a doubling of CO_2 concentration in the atmosphere.

The above proposals try in some way to limit the ability of CO_2 emissions to change climate. The last two categories of response take a different tack altogether.

The fourth category of responses involves limiting the chance or magnitude of harmful consequences from future climate change. For example, levees or dykes can help reduce the costs of flooding, as can depopulating low lying areas. Improved building materials, and more stringent building codes, can lessen the damage and loss of life from strong winds. Better weather forecasts can help people get out of harm's way. Crops can be bred to be more resistant to droughts or wet weather. Farmers can be given better advice on which crops are best to plant under different seasonal weather forecasts.

The fifth category of responses involves taking better measures to deal with damaging weather events *after* they occur. Primarily, this would involve improving disaster relief measures. For example, very poor civil defense response, including the lack of effective co-operation between different levels of government, was a major cause of the loss of life from Hurricane Katrina in New Orleans in 2005, which was only the sixth strongest recorded in the United States. Other examples of policies that would fit into this category include having stockpiles of emergency medical and food supplies, faster ways of providing temporary housing, and better planning of evacuation routes, including contra-flow lanes.

When many policies could potentially address an issue, most people are likely to conclude that we should use all of them. The economic approach, however, argues that we should compare costs and benefits and implement first those policies with the lowest expected costs for a given level of expected benefits. In making these cost/benefit assessments we need to include in the costs and benefits the indirect effects as well as the direct ones. In particular, if a policy would also eliminate benefits that would otherwise have been obtained, those foregone benefits should be counted as additional costs of the policy.

Revisiting climate policy and energy security

Having discussed energy security and climate policy separately, we return to discuss the relationship between them. The two objectives will be related only if reducing fossil fuel use is part of the efficient response to energy insecurity or reducing CO_2 emissions is part of the efficient response to the threat of climate change. In both cases, the policies will be efficient only if their cost per unit of benefit delivered is lower than the cost/benefit ratio for all the competing policies. If other policies are less costly, restrictions on CO_2 emissions should be used only after further exploitation of those other actions has become just as costly as restricting emissions.

It has been suggested that many unexploited low cost options for increasing energy efficiency could be used to further both the energy security and climate goals. Energy efficiency is the primary energy or electricity input needed to perform a certain task, such as the amount of fuel needed to drive a vehicle a given distance, or the amount of electricity needed to run a light bulb of given luminosity.

If people are ignoring cost-effective measures to improve energy efficiency, however, it is reasonable to ask why. Part of the answer is that electricity prices often do not reflect the real cost of supply – especially the way those costs vary over time. People then do not have appropriate incentives to conserve. Another answer is that consumers may not be aware of new technologies that can more than repay a higher upfront cost through lower energy needs. Some consumers, especially those who face borrowing constraints, might also have a high discount rate and hence devalue future energy savings relative to the upfront capital costs. Landlords may also undervalue the benefits of increased energy efficiency for renters, although that raises the question of why the lower energy costs are not reflected in higher rental payments. A final explanation is that consumers may reject cost-effective energy efficient options because they have other drawbacks. For example, more energy efficient light bulbs may produce a worse quality of light, or more fuel efficient vehicles might be lighter and less safe. If so, implementing policies to force more energy-efficient choices might not be an economically efficient policy option.

Some have suggested that forcing, or using subsidies to encourage, an *immediate* shift to non-fossil fuel sources could be an efficient policy to further both climate and energy security goals. The main problem with this approach is that it would be enormously expensive.

Large amounts of capital invested in the current system for producing, delivering and using energy will have to be replaced if fossil fuels are eliminated. This would be very expensive to do over a short period. Since the investment funds could instead be used for other purposes, replacing otherwise productive capital would come at the cost of reduced prosperity and economic growth.

In addition, alternative sources of energy currently are much less efficient than fossil fuels, especially when one takes account of limitations such as their frequent unavailability and extreme short term variability, the inability to schedule their time of supply, their remoteness from markets, their low energy density, and their non-CO_2 related environmental costs. Some of these problems no doubt could be ameliorated through suitable R&D. For example, better energy storage technologies would greatly enhance the competitiveness of wind and solar energy and electric vehicles. Some subsidization of *basic* research into energy technologies could be justified as part of an efficient energy policy, but it will take time to solve the problems. Meanwhile, the very high cost of non-fossil sources of energy makes it unlikely that restricting the use of fossil fuels through taxes or direct controls is the most efficient response to the threat of climate change from CO_2 emissions.

Another fortunate turn of events is lessening the need to force the premature adoption of non-fossil sources of energy. The energy system is evolving toward much greater reliance on natural gas, which is the least carbon-intensive fossil fuel. Technological developments in the production of natural gas from shale have substantially increased estimated economically recoverable reserves. The consequent reduction in expected future natural gas prices is encouraging firms to invest in natural gas fired power plants. Efficiency gains in combined cycle gas turbine generating plants, and restrictions on emissions of sulphur dioxide, nitrous oxides, particulate matter and mercury from coal-fired plants, have increased this tendency. The result will be a slowing in the growth in CO_2 emissions in a way that does not raise energy costs and therefore is consistent with continued growth in economic prosperity.

Ultimately, the finite supply of fossil fuels means that we are not talking about unbridled emissions of CO_2 forever. Even in the absence of policies to restrict fossil fuels, they will continue to be burned only until their costs rise above the cost of alternatives, as someday they surely will. In addition, the physics underlying the greenhouse effect implies that the marginal effect of additional CO_2 emissions will decline as more accumulates. Adding paint to a pane of glass provides an analogy. As extra coats are added, each new coat reduces the transmitted light by a smaller amount.

More fundamentally, continuing CO_2 emissions until non-fossil energy sources take over will have some beneficial effects that offset costs from the climate change they produce. As noted above, CO_2 is not a pollutant in the sense that it directly harms people or other life on earth. On the contrary, CO_2 is food for plants and hence the foundation of the food chain. Experiments have shown that increased CO_2 in the air increases plant growth, makes plants more resistant to drought, disease, pollutants such as ozone and low light conditions, and increases yields of seeds or fruit. These beneficial effects have also been observed around natural CO_2

seeps. Their commercial value has been demonstrated by the addition of CO_2 to greenhouses to increase plant yields. The experimental evidence suggests that at least 10 percentage points of the increase in wheat and rice yields since 1750 is the result of the roughly 35% increase in CO_2 in the atmosphere that has occurred over the same period. The free fertilizer provided to farmers by continued CO_2 emissions over the next few decades could be essential for feeding the expected world population in 2050.

In addition, while increased CO_2 in the atmosphere will affect climates, the effects will vary quite a bit geographically and not all the changes will be harmful. In particular, while it has been claimed that a warmer climate will be associated with more extreme weather events, some scientists have made theoretical arguments and presented empirical evidence that warming is more likely to *reduce* the frequency of such events. As with changes in mean temperatures, however, the effects of CO_2 on weather variability also are likely to vary geographically.

With regard to direct impacts on human health, numerous studies in many countries, and therefore for a range of climates, have shown that abnormal cold snaps have more adverse direct effects on health than do abnormal heat waves. Consistent with this finding, significantly more people die on average in the winter than in the summer.

Climate models predict that the largest temperature increases from CO_2 will occur in the coldest air masses in the middle of winter. Since such air masses have temperatures far below zero degrees centigrade, an increase of even five degrees centigrade is likely to be more beneficial than harmful. Furthermore, insofar as high latitude air masses warm also in autumn and spring, high latitude grain producing areas in Canada, Northern Europe and Russia could be expected to benefit considerably from the longer growing seasons.

A recent article in the *American Economic Review* examined the effect of past weather fluctuations on United States agricultural output. It then asked what would happen to the value of US agricultural output if the climate changed from what it is now to what it is predicted to become according to several climate models. The answer was that, overall, the value likely would rise. Furthermore, this calculation did not take account of the fact that in a changed climate farmers would re-optimize the crops they plant to take better advantage of the new conditions. The calculation also did not take into account the increased agricultural productivity from the direct fertilizer effects of CO_2.

Throughout European history at least, more rapid economic and social progress has tended to be associated more with warm climatic phases, such as the Minoan, Roman and Medieval warm periods, than with cold periods, such as the Dark Ages. A possible reason is that agriculture in Europe was more productive in warmer periods. The period of maximum average global temperatures since the last ice age used to be called the "Holocene climate *optimum*" until it became unfashionable to think of warmth as being associated with "good times."

The latter discussion raises another point. Climate is always changing. Controlling the concentration of greenhouse gases in the atmosphere at best does something about just *one* source of climate change. Climate will still change regardless of what happens to CO_2. This reduces the certainty with which we can claim benefits from controlling CO_2. For example, we might incur costs to reduce temperature increases only to discover that we are heading into a natural cooling cycle that could make temperature rises more beneficial. Uncertainty about the benefits of controlling CO_2 in turn raises the risk and therefore reduces the value of investing in technologies aimed at reducing CO_2 emissions.

The ample historical and natural evidence of past natural climate changes raises another issue. How much of the recent temperature change is natural, how much is attributable to anthropogenic sources other than CO_2, such as land clearing, large scale irrigation, and urbanization, and how much results from the accumulation of CO_2? For example, cooling forces evidently have been offsetting sources of warming, including CO_2, for at least the last decade.[2] As another example, half of the recovery in average temperatures from the Little Ice Age cold period, which occurred from the mid-1600s to the early 1800s, took place before 1930, and therefore could not have been due to CO_2 accumulation in the atmosphere. More generally, there have been many previous periods of warming and cooling of natural origin and comparable magnitude to late twentieth century warming.

The larger the non-CO_2 components of climate change, and the more variable the effects geographically, the stronger the case for limiting the chance or magnitude of harmful consequences from climate change, or taking measures that improve recovery from damaging weather events. These types of measures will help protect against climate change *regardless* of its source, while limiting CO_2 addresses only one source. In addition, such measures would enable us to retain the benefits of increased CO_2 in the atmosphere, such as the stimulus to plant growth and any induced climate changes that are beneficial, while limiting the costs associated with climate changes that are harmful. Furthermore, many policies for handling damaging weather events would also be useful for events that have nothing to do with climate change, such as earthquakes or terrorist attacks. This further reduces their cost/benefit ratio compared to constraining fossil fuel use.

Even if we decided that restricting fossil fuel use is part of the best climate policy for the world as a whole, we have to ask how effective it would be to limit fossil fuel use in only part of the world. Developing countries are not going to slow their economic growth by avoiding low cost fossil fuels. As those with large populations, such as China, India, Brazil and Indonesia, increase their standard of living, their increased CO_2 emissions already are swamping reductions made elsewhere. For example, CO_2 emissions from energy consumption in China increased by more than 167% over the decade 1999–2009, while the absolute *increase* in India over the same period was around two and a half times the *decrease* in the United States. Only at later stages of development does energy use per unit of GDP begin to decline as economies shift into services and use more energy efficient, but also more expensive, production technologies.

With continuing growth in CO_2 emissions from developing countries, any reduction in emissions in the *developed* world will have lower marginal benefits. Restricting fossil fuel use in the developed world alone might even be counterproductive from the perspective of reducing global CO_2 emissions. Raising the cost of energy in developed economies alone will drive industry to the developing countries, more than likely resulting in a net increase, not a decrease, in world energy use and world CO_2 emissions. There are two reasons for this. Energy efficiency tends to be lower in developing economies than in developed ones, while industry relocation probably would raise energy used for transportation of manufactured goods.

Finally, we need to consider that the *cost*, including the energy security cost, of forgoing the use of fossil fuels is likely to vary geographically. For example, the United States and Canada together have very large deposits of coal, tar sands, oil shale, shale oil, shale gas and methane hydrates. The World Energy Council estimates that the United States has around 30% of the world's known coal resources, which is more than any other country. They also estimate that Canada has more than 70% of the world's known bituminous oil, and the United States has more than 70% of the world's known oil shale resources. While oil imports are currently an energy security issue for the United States, absent concern about CO_2 emissions the United States and Canada could together produce, at costs competitive with the current price of crude oil, all the petroleum products they need until alternative energy technologies become competitive. However, producing liquid fuels from unconventional oil resources, or from coal, releases additional CO_2 before the fuels are burned. For the United States and Canada, therefore, energy security and restrictions on CO_2 emissions are in conflict. For western Europe and

Japan, where there are far fewer remaining indigenous supplies of fossil fuels, restricting the use of fossil fuels is of much lesser consequence.

While Australia is not as well endowed with fossil fuel resources as is North America, it has very large per capita endowments of coal and natural gas. As a result, cheap energy is a major source of comparative advantage for the Australian economy and energy is a relatively large part of its exports either directly or embedded in other products. Furthermore, econometric studies, such as those done by the Center for Global Trade Analysis at Purdue University, suggest that countries in Asia and the Middle East that would not impose taxes or controls on fossil energy would be close substitute locations for many energy-intensive industries currently located in Australia. Given the high costs to Australia of restricting fossil fuel use, it is quite reasonable for Australia to insist on *very* strong evidence that the sacrifice will yield comparably high benefits before it agrees to incur the costs.

When I first started writing on energy economics issues in Australia in the early 1980s I wrote a paper titled "Cheap Resources into Expensive Energy." At the time, I was writing about how inefficient state-owned monopolies in the electricity industry were turning Australia's abundant, and often non-tradable and therefore cheap, coal into expensive electricity. This was hampering Australia from obtaining maximum benefit from its resource endowment. From a purely domestic perspective, the proposition that Australia should tax or otherwise constrain the use of its fossil fuels could be seen as "Cheap Resources into Expensive Energy Revisited." For Australia, as for the United States and Canada, energy security and climate change are not "two sides of the same coin."

Of particular relevance for current policy discussions, it would be an expensive but futile exercise for Australia to restrict CO_2 emissions *unilaterally*. Since Australia currently produces less than 1.4% of total anthropogenic CO_2 emissions, if the Australian economy were *completely shut down*, the resulting effect on global surface temperatures would not be measurable even if the climate models predicting the largest effect from CO_2 accumulation turn out to be accurate.

Nor is it likely that substantial reductions in emissions from Australia would have any effect on policies adopted by others. Even in the extremely unlikely event that all OECD nations implemented controls on CO_2 emissions to the level specified in the Kyoto Protocol it would amount to little in the face of continuing strong growth in emissions from high population developing countries. As a result, the marginal effects of such aggregate OECD reductions on surface temperatures would be tiny. The *proponents* always acknowledged that the controls in the Kyoto Protocol were only a *first step* toward limiting climate change from CO_2 accumulation.

In summary, the costs of reducing CO_2 emissions are currently so high, especially for developing countries, that it is most unlikely that meaningful controls on CO_2 emissions could be instituted at the global level before alternatives to fossil energy become competitive. At that time, policies to force reduced fossil energy consumption would be unnecessary. In the interim, policies aimed at encouraging basic research to lower the cost of new energy technologies, limiting the harmful consequences from climate change, or contending better with damaging weather events would yield far greater expected benefits for a comparable level of expected costs.

Notes

1 A previous version of this paper was presented as the 2011 Reid Oration at the University of Western Australia. It has been lightly edited mainly to convert a paper that was initially delivered orally into written form.

2 Recall that this was written in 2011. What is now known as "the pause" has continued for another four years.

5

Renewable energy in the MENA

Why did the Desertec approach fail?

Luigi Carafa and Gonzalo Escribano

Introduction

A Desertec Industrial Initiative (Dii) was created in 2009 with an ambitious mission to source 15% of Europe's power needs from North African desert solar power by 2050. As in other developing countries, deploying solar energy in the Middle East and North Africa (MENA) implies significant investment risks which are higher than in developed countries. Higher risks for investors are linked to uncertainties in a number of areas such as weak regulations for the promotion of electricity from renewable energy sources (RES-E), lack of qualified staff, construction delays, and so forth (see section 3). Higher investment risks translate into higher overall project finance costs.

The Desertec approach was aimed to reduce the above uncertainties and create a pipeline of projects across the region, but failed to do so. In October 2014, the export-based Desertec vision collapsed.[1]

At the time of writing, engineers are completing Ouarzazate Noor I Concentrated Solar Power (CSP) plant in Morocco – the first large-scale solar project which most resembles the initial Desertec concept. 500,000 parabolic mirrors placed in 800 rows follow the sun as it moves throughout the day. Each solar mirror is 12 metres high and focused on a steel pipeline carrying a heat transfer solution. This solution is warmed during the process, getting into a heat engine where it is mixed with water to create steam that turns power generating turbines. This 160 MW plant will become operational by the end of 2015, providing electricity to 275,000 homes. It will be followed by Noor 2 and 3 plants later in 2017.[2]

Why did political and industrial initiatives built around the Desertec vision fail in the end? This chapter seeks to answer this question. To do so, Section 2 takes stock of the state of development of renewable energy in the MENA. Section 3 tracks the trajectory of industrial initiatives and political cooperation such as Dii and the Mediterranean Solar Plan, and it delves deep into the causes of failure of the Desertec vision. Section 4 analyses a number of misperceptions regarding energy security that negatively affected the realization of the Desertec vision. Section 5 elaborates on the need to re-think regional and industrial cooperation around a more inclusive narrative of sustainable energy development. For the sake of clarity, this chapter restricts its focus to Europe's MENA neighbouring countries, excluding Turkey, Yemen, Sudan, Saudi Arabia, UAE, Iraq, Kuwait, Qatar, and Bahrain.

This chapter makes three main contributions. Firstly, it investigates why the Desertec approach failed to boost a broad regional energy transition in the MENA – arguing that the lack of public acceptance of a vision based on electricity export towards Europe coupled with the changing political situation in several MENA countries resulted in the lack of a clear project pipeline across the region. Secondly, it contributes to the growing body of literature on the economics of renewable energy in the MENA. Thirdly, this study makes a contribution to the debate on the future of regional energy cooperation in the MENA.

Renewable energy in the MENA: the state of development

Electricity is a key concern for the countries of the Middle East and North Africa. Power generation takes up the largest share of primary energy, i.e. mostly fossil fuels. The electricity sector equally represents the largest source of carbon emissions in the region – with 943 million tonnes of CO_2 per year ($MtCO_2$/year) accounting for 42% of total emissions in 2011.[3]

MENA countries account for almost half of total pre-tax global energy subsidies, with expenditure reaching USD 236 billion in 2011 compared to a global total of USD 481 billion.[4] Coupled with this, booming population and economic development make electricity consumption grow twice as fast as the world average.[5]

Power generation capacities will have to be quadrupled until 2050. In other words, MENA countries will need infrastructure investments of at least USD 676 billion by 2050.[6] Under a business-as-usual scenario, growing electricity demand will continue to rely on fossil fuels mostly. This implies that net exporting countries will experience a reduction in their hydrocarbon export capacity; net importing countries will experience an increase in their fossil fuel import bill and its associated fiscal costs. As a result, power-related carbon emissions are set to steadily increase rather than decrease.

To escape business-as-usual and reduce carbon emissions, MENA countries need to shift towards electricity from renewable energy sources (mostly solar and wind) as well as increase energy efficiency. In order to better understand the state of progress and the challenges that lie ahead in the MENA, three important factors have to be taken into account: technology development, finance and policy.

From a technology development perspective, photovoltaic (PV) and wind technologies registered dramatic progress over the last decade and became fully competitive for power generation vis-à-vis fossil fuel power generation. Solar thermal technologies, also known as concentrated solar power, hold a great potential in the MENA but still need to advance along its cost curve. Electricity from CSP can be supplied 24 hours-a-day, 7 days-a-week, 365 days-a-year with no need for a fossil fuel back-up. Given the high upfront costs, CSP developments have been limited to hybrid plants or pilot projects such as the Ouarzazate Noor I plant – which is expected to be operational by the end of 2015. The pipeline of the CSP project is slowly growing as technology development progresses along the cost curve. Chiefly, further technology development of heat storage is needed to bring costs down. For ongoing CSP projects to be profitable in the MENA, concessional finance has been of primary importance. The case of the Noor CSP complex in Morocco is a telling example.[7]

From a finance perspective, a transition to RES-E poses remarkable challenges to MENA countries. Firstly, conventional power generation investments have to be shifted from high-carbon to low-carbon technologies. Secondly, investments need to be further topped up by USD 739 billion by 2050[8] – with fossil fuels still dominating the electricity fuel mix, and with a large development of non-hydro renewables (mainly solar and wind). Solar and wind power plants require higher upfront costs than fossil fuel-based power plants. In addition to that, fossil fuel subsidies also have a downside distortive effect on upfront costs. Meanwhile, solar and wind power plants require smaller maintenance costs than fossil fuel-based power plants.

The total renewable energy finance attracted by MENA countries substantively increased from $0.6 billion in 2004 to $12.6 billion in 2014.[9] Renewable energy investment towards the MENA increased every year by 36% on average, registering the world's second-highest mean annual growth rate only following China. Over the last decade, the largest shares within the whole region went to Jordan, Egypt, and Morocco – the latter tops the list with $635 million for the 160MW Ouarzazate Noor I CSP project, (see Table 5.1).

These data show a positive upward trend, registering a dramatic increase in investments towards the MENA between 2012 and 2014. This reflects the high-quality of solar and wind energy resources as well as the vast untapped potential which the MENA is endowed with. However, the region as a whole is still lagging behind at a global scale. The MENA only accounted for around 1–1.5% of the global total renewable energy investment between 2004 and 2011. In 2014, the MENA only accounted for 4.7% of the global total renewable energy investment – a long way after China, Europe, Asia and the United States (in this order).

From a policy perspective, MENA countries are still lagging behind. The Arab Future Energy Index registers the state of play of MENA governments in this respect.[10] With the only exceptions of Morocco and Jordan, all the other MENA countries score very low in terms of policy frameworks, institutional capacity, market structure and financing capacity. Egypt, Tunisia, Palestine, and Algeria score below 40%. Lebanon stops at 26%. While most of the MENA countries identified clear renewable energy targets, only Morocco and Jordan started to develop consistent national policies and have been able to attract significant public and private investment to date going beyond $500 million investment per project.[11]

Cooperation has been reinforced and further developed at regional, multilateral, and industrial levels. At regional level, the European Union (EU) and its MENA neighbours are engaged in intergovernmental cooperation within the Mediterranean Solar Plan. A number of regional technical platforms were established. At multilateral level, IFIs are implementing a USD 750 million MENA CSP Investment Plan endorsed by the Clean Technology Fund.[12] At industrial level, three private industrial consortiums were established, namely the Desertec Industrial Initiative, Medgrid, and Res4Med.

To this point, the landscape for solar and wind deployment has changed favourably as compared to 2004 – a time when developments were extremely limited. Most positive developments occurred at the national level, where there is an increasing awareness by governments of the urgent need to act as well as of the opportunities unleashed by renewables. There are big yet surmountable challenges for MENA governments which call for an increasing reform of their electricity sectors. Beyond the national level, International Financial Institutions (IFIs) played a key role in facilitating concessional finance towards reference projects such as Ouarzazate Noor I CSP plant in Morocco. By contrast, industrial and political approaches have failed or achieved little change so far.

Why did Desertec and the Mediterranean Solar Plan fail to boost a regional transition to renewables in the MENA?

This section explores the reasons why industrial and political approaches built around the Desertec vision failed to boost a sustained transition to renewables in the MENA region.

Table 5.1 Renewable energy investment in the MENA, $ billion (2004–2014)

Unit ($bn per year)	*2004*	*2005*	*2006*	*2007*	*2008*	*2009*	*2010*	*2011*	*2012*	*2013*	*2014*
MENA	0.6	0.8	1.1	2.4	2.3	1.7	4.2	2.9	10.4	8.7	12.6
China	3.0	8.2	11.1	16.6	25.7	39.5	38.7	49.1	62.8	62.6	83.3
Europe	23.6	33.6	46.7	66.4	81.6	81.2	111.1	120.7	89.6	57.3	57.5
Asia (without China and India)	7.2	9.2	10.0	12.5	13.6	13.7	19.3	24.1	30.5	44.7	48.7
United States	5.4	11.6	29.1	33.0	35.1	24.3	35.1	50.0	38.2	36.0	38.3
World total	45.1	72.9	112.1	153.9	181.8	178.5	237.2	278.8	256.4	231.8	270.2
MENA share of world total	1.3%	1.1%	1%	1.5%	1.3%	0.9%	1.8%	1%	4%	3.7%	4.7%

Own calculations based on data from BNEF, *Global Trends*.

An important premise is needed at this point. From a project financing perspective, developing solar and wind power plants in the MENA implies significant investment risks which are higher than in developed countries.[13] Higher risks are linked to uncertainties in a number of areas such as sub-optimal regulations for the promotion of RES-E, lack of qualified staff, construction delays, weak government institutions, corruption, and so forth.[14] When compared to a similar project in developed countries, higher investment risks translate into an extra cost in the overall project finance costs. Industrial and political approaches built around the Desertec vision aimed to reduce the above uncertainties and create a pipeline of projects across the region.

Behind the so-called Desertec concept there was a partnership between the Club of Rome, the Hamburg Climate Protection Foundation and the National Energy Research Centre of Jordan founded in 2003 under the label Trans-Mediterranean Renewable Energy Cooperation Network (TREC). In collaboration with scientists of the German Aerospace Centre (DLR), TREC put forward a vision of an EU-MENA community of shared clean energy and water interests.[15] The Desertec concept gained support from both politics and industry in Germany, and later at European level. A White Book for Desertec was presented at the European Parliament in November 2007. The Desertec concept received further support at legislative level with Directive 2009/28/EC on the promotion of the use of energy from renewable sources. Article 9 of this directive would virtually allow EU member states to import clean electricity from third countries.

In July 2009 a group of mostly German companies established the Desertec Industrial Initiative (Dii) with the aim of turning the Desertec concept into a reality by 2050. In July 2010 a French initiative called Medgrid was created, later followed by an Italian industrial initiative called Renewable energy solutions for the Mediterranean (Res4Med).

Initially, Dii started out with the aim of developing large-scale projects across the MENA. With the global downturn and the Arab uprisings, however, this business model became challenging. Dii had to face the defection of key German shareholders Bosch, Siemens, E.ON and associated partner Bilfinger – which changed their business strategies, in some cases (i.e. Siemens) even closing down their CSP departments. The lack of public acceptance of the Desertec vision (i.e. electricity export towards Europe) coupled with big uncertainties about the changing political situation in several MENA countries resulted in the lack of a clear project pipeline across the region. This paved the way for a fundamental transformation of Dii in October 2014.[16]

At political level, intergovernmental cooperation was framed under the Mediterranean Solar Plan (MSP) umbrella. The Union for the Mediterranean (UfM) initiated a multi-stakeholder process, in an attempt to create a regional strategy for the deployment of 20 GW of installed renewable energy capacity by 2020 along with the necessary transmission capacity and cross-border interconnections.[17] In addition to this, the European Commission created in 2010 a technical assistance project named 'Paving the Way for the Mediterranean Solar Plan' (PWMSP). With a budget of EUR 4.6 million, this project provided technical analysis of issues as varied as regulatory frameworks, support to investment, infrastructure requirements and transfer of knowledge.[18]

Such a multi-stakeholder process managed to achieve a low-common-denominator strategy in Jordan in May 2013.[19] This regional strategy, called 'Master Plan', was supposed to be submitted to the UfM Energy Ministerial Conference for political endorsement in December of the same year, but did not make it to the final political stage. While objections in the field of energy cooperation traditionally arose from partners in the South of the Mediterranean, this time things went differently. Spain had little option other than to oppose the submission of this document to the Energy Ministers. Madrid did not support a German-Moroccan proposal of statistical transfer (rather than physical transfer) of electricity. Likewise, Madrid expressed its concerns about clean electricity trade from Morocco to Europe when there is an overcapacity in Spain.[20]

Spain currently exports electricity to Morocco, covering as much as 10% of the domestic electricity consumption. However, more sustained electricity cooperation between Spain and Morocco could be mutually beneficial in the long run – unleashing possibilities to better integrate renewables into the energy mix in both countries. Better connection of this larger grid with the European grid would offer export avenues for overcapacity via France.

For this to happen, France and Spain should put an end to an old dispute on the reinforcement of their cross-border electricity interconnections. Paris maintains the Iberian Peninsula in the condition of an 'electricity island' with an interconnection capacity as low as 3%. In October 2014, the European Council asked the European Commission and EU countries to take urgent measures to reach a 10% minimum electricity interconnection capacity target by 2020. The European Council also added a non-binding target of 15% interconnectivity by 2030.[21] This requires real commitment of all national governments (France included) in order to escape politics and focus on serious policy.

Debunking the myths of RES-E in the MENA

The deployment of RES in the EU's Southern Neighbourhood has been hampered by European misperceptions regarding energy security. This continues to be one of the main arguments put forward by its critics. It has been argued that contrary to domestic RES generation, which improves energy security through the reduction of energy imports and energy dependency, importing RES electricity from the Southern shore of the Mediterranean would increase European energy dependency and therefore its supply insecurity. This section takes the opposite view, showing that under certain conditions RES imports do not necessarily harm European energy security and can even improve it.

In fact, vested interests and RES protectionism in the European energy sector may well be behind these rather unspecified energy security concerns. The Desertec timing (as happened with the Mediterranean Solar Plan) was especially unfortunate because the Arab Spring caught its promoters by surprise. The increased perception of geopolitical risks associated with the project was not adequately factored in, nor were its long-term economic development spillovers, as will be seen in the next section. Because of this, Desertec was unable to capitalize the role of RES in fostering economic development and therefore long-term political stability. Instead of constructing a discourse for sustainable and inclusive development aligned with the demands expressed by the Arab Spring, Desertec got entrenched in an obsolete Eurocentric and elite-driven regional approach unable to articulate a new narrative addressing the contribution of RES to mitigate geopolitical risks.

The energy security implications for Europe of importing RES from Mediterranean Partner Countries (MPCs) have been analysed applying portfolio choice theory and its risk-cost trade-off. Assuming that solar electricity generation from North Africa is included in the EU's energy portfolio under the same risk conditions prevailing in the EU (same return/cost variability), both PV) and thermo-solar (CSP) technologies yield higher returns in MPCs due to higher insolation levels, pushing the efficient risk/cost frontier upwards. Thus, the equilibrium shifts towards a more efficient portfolio with lower cost levels and similar risks. In turn, if lower costs come with higher risk levels, the equilibrium may not achieve any significant improvement in the cost-risk trade-off. In the worst-case scenario the integration of MPCs RES into the EU could outpace the above-mentioned cost reductions with even higher risk levels.[22] It has been suggested that electricity policies in MPCs are path-dependent, and that technical and financial assistance should target areas with a particular de-risking potential: namely capacity-building on project assessment, project finance, and grid management.[23]

The usual claim is that RES imports from MPCs imply the same geopolitical risks associated with energy dependence in fossil fuels: dependency upon foreign resources. However, this argument is not supported by a careful economic analysis of international RES and risks.[24] In the first place, access to RES generation from abroad entails the diversification of geographical origins, energy sources, or both. In that case, the European vulnerability could actually decrease for a given energy dependency ratio.[25] New RES resources and technologies from new exporting countries increase European energy diversification, reducing vulnerability even if they come from the same countries that traditionally supply the EU with fossil fuels. For instance, importing solar electricity from Algeria into the EU does increase both diversification and dependence for both the EU (supply) and Algeria (demand). From a portfolio perspective such a diversification of sources and suppliers reduces vulnerability, compensating for the deterioration of energy dependence indicators.

Second, the technical characteristics of RES also limit their use as a political weapon. Electricity generation from RES cannot be stored or re-routed as easily as hydrocarbons. Electricity storage costs will remain high in the medium term, while re-directing electricity exports is not possible in the absence of expensive and lengthy to build infrastructures, like high tension lines and interconnectors. Therefore, a Mediterranean Partner Country could not just cut its electricity supply to the EU without simply wasting the resource, at least in the short term. In the longer run, redirecting green electricity supply towards domestic markets would require transmission investments and could increase its cost above either politically or economically acceptable levels in countries already afflicted by the dual problem of poor household difficulties with energy access and the associated (and fiscally unsustainable) cost of energy subsidies to mitigate energy poverty.

Regulation and geopolitics constitute the main risks faced by RES investments in MPCs. Starting with regulatory risks, it is worth mentioning that they are not the monopoly of MPCs, as shown by the recent German and Spanish regulatory changes in their respective RES support schemes. As stated above, the upward shift in the risk/cost efficient frontier can only occur under similar levels of regulatory risk and property rights protection that prevails in the importing market. In this regard, the risk to European consumers of RES imports from Southern Mediterranean producers depends upon the regulatory and institutional convergence of these countries towards the energy-related *acquis communautaire*. While such a convergence can be initially limited to renewable electricity, it may incrementally spill over the whole electricity sector, given its physical and normative inter-connectedness.

To reduce regulatory risk, the EU followed a dual track approach. On the European side, the Commission delivered an institutional framework in Directive 2009/28 on renewable energy, introducing the possibility of developing joint projects and support mechanisms, and eventually extending flexibility mechanisms to neighbouring countries. On the other hand, MPCs should converge towards the EU energy-related *acquis communautaire*, at least in the field of renewable energy. This convergence entails highly technical issues, like the interoperability of electricity systems and the related harmonization of certifications and standards. But it also touches on more sensitive issues that may alter national political economy balances, like support mechanisms and its control, access to the grid, public procurement rules or authorizations.

Both paths failed in several respects, and not only due to unfortunate timing. The mechanisms established by the Commission were thought to reduce the regulatory risks of exporting RES electricity to the EU. Nobody anticipated (in a 2009 Directive) that the financial crisis was to pulverize all electricity demand projections, and therefore that no RES electricity imports will be needed to fulfil the 2020 national RES targets (20%). Moreover, electricity demand stagnation in the EU left European utilities afflicted by excess capacity. As of today, if there were infrastructures and regulation frameworks in place, it would be European utilities exporting electricity to MPCs, to be sure, mainly non-renewable electricity. In fact, the only functioning synchronized electricity interconnection between the EU and the MPCs is used to export electricity from Spain to Morocco (as much as 10% of Moroccan domestic electricity consumption). However, it should be taken into account that the existing interconnections with EU neighbours are based on regulated exchanges, which has necessitated to date extraordinary non-market figures.[26]

Finally, RES support fatigue in the midst of a fiscal crisis erased the appetite for new (and even worse) transnational support mechanisms in countries. This is why Spain blocked the Mediterranean Solar Plan on the grounds that before interconnecting the EU with its neighbours it would be wise to interconnect the EU itself. This was the death certificate for the only proposal to date that has tried to explore a differentiated regulatory framework limited to renewable energies outside the EU. So, the regulation enacted by the EU became obsolete in market terms: the demand was simply not there and European utilities' appetite for RES disappeared within the EU, not to mention those Mediterranean countries where revolutions were taking place. The risks the European Commission wanted to address became irrelevant because of new market conditions, but also due to the emergence of geopolitical risks. Compared to geopolitics, regulatory risks are much easier to manage (or should be), so when geopolitical instability enters the scene, concerns on regulation tend to fade away.

The irruption of the geopolitical logic also led us to the failure of normative convergence from the Mediterranean side. With the only exception of Morocco, by 2011 the countries targeted by Desertec (mainly Egypt and Tunisia) entered difficult and uncertain political times. The Middle East dimension (the Gulf Cooperation countries) was closed due to the Syrian and then Iraqi wars. In this context, the deployment of renewables was the last priority for these countries. In North Africa, only Algeria and Morocco remained stable enough to advance in reforming its energy sector. And only Morocco has made significant in-roads in converging, however slowly and in a fragmented manner, towards EU-like, or at least EU-compatible RES rules. Furthermore, in Morocco it has led to a process of normative diffusion through policy networks with potential spillovers on the whole energy sector. Egypt and Tunisia are trying to catch up, but it will take time for regulatory improvements to be more visible than political instability or security concerns. Adaptation and mitigation policy incentives are equally lagging well behind those enacted by the EU.[27]

Only in this regard, RES geopolitics is not that different from hydrocarbons. While there are clear limitations on the capacity to project power through RES, access to renewable resources involves geopolitical risks to the same extent that accessing oil or gas fields, but no more. The point here is not that RES do not imply such kinds of risk, but rather that this is not a risk specific to RES: certainly it is going to be very difficult to develop RES investments in Libya in coming years, but this is exactly the case with oil and gas investments. In the same vein but from the regulatory perspective, RES investments in Algeria face the same regulatory and fiscal burdens that have almost paralysed investment in the hydrocarbon sector. But this does not mean that RES cannot be developed in politically stable countries involved in a (slow) reform process of their energy policies, like Morocco; or that it should not be conceived as a tool to foster energy development and sustainability (environmental and economic) in countries like Tunisia, Jordan or Egypt.

There is no great regulatory design that can counteract current geopolitical drivers, be it a Euromed Energy Community or a Mediterranean Platform for renewable energies as proposed by the European Commission. However, a more differentiated approach that takes account of the different geopolitical and regulatory contexts and delivers tangible results in the short term would be needed to deal with diverging MPCs preferences. After all, differentiation entails geopolitics by other (normative) means. On a country level: taking stock of Moroccan stability, proximity, existing infrastructures and gradual energy reforms; and offering Tunisia the incentives to develop its solar and wind resources. On the deployment strategy: favour institutional innovation, reduce scale and forget about RES exports in the short term. On the timing: leave scaling up for the medium term with technologies' cost reduction and focus on supplying domestic demand. Regional designs, including imports and exports in a homogeneous Mediterranean energy space, are urgent, but for the long term.

Re-framing regional cooperation around a narrative of sustainable energy development

Geopolitics and lack of regulatory convergence were not the only external obstacles to the Desertec initiative. As shown, the institutional framework was not sufficiently propitious for its de-risking strategies to deliver real incentives to investors. Moreover, the European narrative focused on narrow economic and technical issues, without developing a coherent and inclusive discourse that could make a difference when compared with traditional energy relations. On the geopolitical arena it did not address the fundamental question that RES could improve not only European, but also North African energy security, especially in fossil fuel importers like Morocco, Tunisia, Jordan and Egypt.

Concerning regulatory convergence, the 'Europeanization' narrative turned out to be counter-productive, as happened with other 'external governance' European initiatives in the neighbourhood. However 'normative' they were labelled, they were received as geopolitics by other (normative) means. For instance, the Neighbourhood Policy has been criticized for proposing an empire by example or imposing regional normative hegemony,[28] or for neglecting that its soft normative approach has hard consequences for the people living in the European neighbourhood.[29] At any rate, external governance and Europeanization cannot constitute a sufficiently attractive narrative for sustainable energy cooperation in the Mediterranean. The most descriptive definition of such a narrative came from the Algerian government when it labelled the Neighbourhood Policy '*gouvernance par télécommande*', and few moves have been made from the European side to change that perception in MPCs.

The European narrative has failed to deliver a credible framework partly because it has not addressed MPCs' preferences properly. For sure, these countries were interested in profiting from structural comparative advantages like high insolation or wind load factors, as well as abundant land and labour. But they were, and still are, more interested in building dynamic advantages like industrial clusters, innovative regulation, technological transfers, training and technical skills. This section identifies the lack of a European narrative on RES cooperation being a driver for economic development in MPCs as one of the main weaknesses in the European approach. Instead, at best it was perceived in the Southern shore of the Mediterranean as a eurocentric project to serve European environmental preferences. Less indulgent observers only saw a strategy to promote European utilities, industries and engineering firms, and something that could be termed renewable imperialism in the sense that it replicates previous (and to some extent existing) European patterns of resource extraction.[30]

Perhaps more importantly, the European discourse was completely disconnected from the more constructive narrative that the Union for the Mediterranean inherited from the Barcelona Process: achieving a shared space of peace and prosperity in the Euro-Mediterranean region. The EU missed the opportunity to elaborate such a narrative for the first time in energy issues, portraying RES as an instrument of sustainable energy development in the Mediterranean. Energy development would imply RES supplying part of the increase in its electricity demand; contributing to modernize MPCs energy services and alleviate energy poverty; and using their renewable energy resources to dynamize the economy and create jobs. Their demands to the EU were the provision of technical cooperation, training and technology transfers to fully reap the overall benefits and externalities of RES deployment.

The literature on the impact of RES deployment confirms the potential for significant economic gains in MPCs. De Arce et al. conclude that RES deployment will push Moroccan GDP between 1.17% and 1.9% at the end of the period (2040) according to different scenarios.[31] Always depending on the scenario, between 267,000 and 482,000 jobs could be created. Other simulation exercises also confirm the economic viability of renewable energies in an eventual Euro-Mediterranean power system.[32] For instance, the optimization of RES goals could actually decrease overall energy system costs in the region.[33] However, some authors highlight that institutional lags remain in place. Even in Morocco, by far the most advanced MPC in this field, the literature identifies obstacles related to uncertainty and informality, concluding that in the CSP sector any significant increase in foreign investment requires removing several financial and legal barriers.[34] Other studies show that, by 2050, RES shares close to 100% are possible for North Africa by 2050, with RES-E exports gaining importance only after 2030.[35]

It is true that renewable energy deployment cannot be a 'saviour-for-all', and that it requires fundamental domestic reforms, like existing energy market pricing mechanisms or providing different fiscal and regulatory incentives.[36] Another frequent criticism is the top-down approach followed by both the EU and MPCs. Cambini and Franzi find that this is the preferred option by MPC regulators to deliver policy diffusion, and therefore rule change.[37] However, this approach has proved to date unable to offer an inclusive model that deviates from the usual elite-driven processes that pervade Euro-Mediterranean relations, especially in energy matters. RES were narrowly conceived as a strategy to achieve European environmental objectives and promoting its renewable industries, utilities and engineering firms, neglecting the dynamic benefits expected from MPCs. But it also suffered from a 'more of the same' syndrome that tends to affect the reception of EU external governance mechanisms in its neighbourhood.

That was a missed opportunity, because few projects can reassemble features like sustainability, modernization, social and economic development, decentralization and rent-avoidance. All these elements should have been considered to develop a soft power narrative able to attract reform-minded actors and offering incentives to local populations: RES electricity access, jobs and ownership of their natural resources. Whether the lessons from the failure of Desertec or the Mediterranean Solar Plan were learned remain to be seen. The European Commission pretends to have adopted a more bottom-up approach in its new Mediterranean Renewable Energy Platform, but the inclusive and sustainable development narrative is still absent and there are no significant incentives in sight for RES to really become a game changer or a catalyst for further energy integration in the Euro-Mediterranean space.

Conclusion

The electricity sector represents the largest source of carbon emissions in the MENA region. Booming population and economic development makes electricity consumption grow twice as fast as the world average. The renewable energy finance attracted by MENA countries has substantively increased between 2012 and 2014. However, the region as a whole is still lagging behind at a global scale, accounting in 2014 only for 4.7% of global renewable energy investment.

This chapter has shown that, from a project financing perspective, developing renewables in the MENA implies higher investment risks than in developed countries due to the uncertainties in areas such as sub-optimal tools to promote renewable electricity, lack of qualified manpower, construction delays, weak government institutions, and the like. Industrial and political initiatives built around the Desertec vision aimed to reduce the above uncertainties and create a pipeline of projects across the region. However, the lack of public acceptance of the Desertec vision (i.e. renewable electricity exports towards Europe) coupled with big uncertainties about the changing political situation in several MENA countries resulted in the lack of a clear project pipeline across the region.

This chapter has sought to show that there is no great regulatory design that can compensate for current geopolitical drivers, whether a Euromed Energy Community or a Mediterranean Platform for renewable energies as proposed by the Commission. However, a more differentiated approach that takes account of the different geopolitical and regulatory contexts would be needed in order to accommodate diverging MPCs' preferences. In this regard, differentiation entails geopolitics by other (normative) means. At a country level, this chapter's conclusions point to the need to take stock of Moroccan stability, geographical position, infrastructures and incremental energy reforms. It also suggests offering Tunisia the incentives to develop its solar and wind resources. Regarding the deployment strategy, our conclusion highlights the need to foster institutional innovation, reducing the scale of deployment, and being pragmatic by not insisting on the issue of RES exports, at least in the short term. Scaling up should wait for further cost reduction in RES technologies, and focus on supplying domestic demand. Regional designs, including imports and exports in a homogeneous Mediterranean energy space, are a long-term endeavour.

Finally, the Desertec top-down approach has proved unable to offer an inclusive model that deviates from elite-driven processes. RES were narrowly conceived as a strategy to achieve European environmental and economic objectives, neglecting the dynamic benefits expected by MPCs and distorting their reception of EU external governance mechanisms in the domain of RES-E. This was a missed opportunity, because Desertec or the Mediterranean Solar Plan could have developed a much needed Euro-Mediterranean energy narrative based on ideational drivers like sustainability, modernization, social and economic development, decentralization and rent-avoidance. Such a soft power European energy narrative could have attracted reform-minded actors, but also local populations by offering them improved electricity access, jobs and ownership of their natural resources. While the European Commission has adopted a more bottom-up approach for its brand-new Mediterranean Renewable Energy Platform, the narratives of inclusive and sustainable development are still absent for RES to become a driver for energy integration in the Euro-Mediterranean energy space.

Notes

1 Desertec Industrial Initiative (Dii), *Dii is Entering a New Phase: Dii Adapts its Business Model to Provide Concrete Services for Renewable Energy Projects*, press release (Rome: Dii, 2014).
2 "Morocco Poised to Become a Solar Superpower with Launch of Desert Mega-project," *The Guardian*, 26 October 2015.
3 S. Devarajan, L. Mottaghi, F. Iqbal, et al., *MENA Economic Monitor: Corrosive Subsidies* (Washington, DC: World Bank, 2014), http://documents.worldbank.org/curated/en/2014/10/20272046/mena

-economic-monitor-corrosive-subsidies; World Bank, *World Development Indicators: CO_2 Emissions from Electricity and Heat Production, Total (Million Metric Tons)* (Washington, DC: World Bank, 2014).
4 C. Sdralevich, R. Sab, Y. Zouhar, et al., *Subsidy Reform in the Middle East and North Africa: Recent Progress and Challenges Ahead* (Washington, DC: International Monetary Fund, Middle East and Central Asia Department, 2014).
5 Observatoire Méditerranéen de l'Energie, *Mediterranean Energy Perspectives 2011* (Paris: Observatoire Méditerranéen de l'Energie, 2011).
6 World Energy Council (WEC), *World Energy Scenarios: Composing Energy Futures to 2050* (London: World Energy Council, 2013).
7 L. Carafa, G. Frisari, and G. Vidican, "Electricity Transition in the Middle East and North Africa: A De-risking Governance Approach," *Journal of Cleaner Production* (forthcoming), http://dx.doi.org/10.1016/j.jclepro.2015.07.012.
8 WEC, *World Energy Scenarios.*
9 Bloomberg New Energy Finance (BNEF), *Global Trends in Renewable Energy Investment 2015* (Paris: UNEP-Division of Technology, Industry and Economics, 2015).
10 Regional Center for Renewable Energy and Energy Efficiency (RCREEE), *Arab Future Energy Index: Renewable Energy 2015* (Cairo: RCREEE, 2015).
11 RCREEE, *Arab Future Energy Index 2015*; Regional Center for Renewable Energy and Energy Efficiency (RCREEE), *Arab Future Energy Index: Renewable Energy 2013* (Cairo: RCREEE, 2013).
12 Clean Technology Fund, 'Clean Technology Fund, Concept Note for a Concentrated Solar Power Scale-up Program in the Middle East and North Africa Region' (CTF/TFC.3/7) (prepared for the Meeting of the CTF Trust Fund Committee, Washington, DC, 2009).
13 Carafa, Frisari, and Vidican, 'Electricity Transition'; N. Komendantova, A. G. Patt, L. Barras, et al., 'Perception of Risks in Renewable Energy Projects: Power in North Africa', *Energy Policy* 40 (2012): 103–109.
14 MIGA, *World Investment and Political Risk 2013* (Washington, DC: World Bank, 2013); O. Waissbein, Y. Glemarec, H. Bayraktar, and T. S. Schmidt, *Derisking Renewable Energy Investment. A Framework to Support Policymakers in Selecting Public Instruments to Promote Renewable Energy Investment in Developing Countries* (New York: United Nations Development Programme, 2013).
15 Desertec Foundation, *White Book – Clean Power from Deserts: The Desertec Concept for Energy, Water and Climate Security* (Hamburg: Desertec Foundation, 2008).
16 Dii, *Dii Is Entering a New Phase.*
17 Resources and Logistics, *Identification Mission for the Mediterranean Solar Plan* (Lot 4-N° 2008/168828) (Brussels: ENPI Information Centre, 2010); European Investment Bank, *FEMIP Study on the Financing of Renewable Energy Investment in the Southern and Eastern Mediterranean Region* (European Investment Bank, 2010), www.eib.org/attachments/country/study_msp_en.pdf; L. Carafa, *The Mediterranean Solar Plan through the Prism of External Governance* (Barcelona: Euro-Mediterranean Study Commission, 2011).
18 Paving the Way for the Mediterranean Solar Plan, *Roadmap Morocco* (Brussels: Consortium MVVdecon/ENEA/RTE-I/Terna/Sonelgaz, 2013).
19 K. Petrick, S. Erdle, M. Strauss, et al., *The Mediterranean Solar Plan – Master Plan* (Barcelona: Union for the Mediterranean, 2013).
20 Interviews EU, 2014.
21 European Council, *Conclusions on 2030 Climate and Energy Policy Framework* (SN 79/14) (Brussels: European Council. 2014).
22 Gonzalo Escribano, José M. Marín, and Enrique San Martín, "RES and Risk: Renewable Energy's Contribution to Energy Security. A Portfolio-based Approach," *Renewable and Sustainable Energy Reviews* 26 (2013): 549–559.
23 Carafa, Frisari, and Vidican, 'Electricity Transition'.
24 Gonzalo Escribano, 'RES in the 'Hood and the Shrinking Mediterranean Solar Plan', in *A Guide to EU Renewable Policy*, ed. I. Solorio et al. (Cheltenham: Edward Elgar, forthcoming).
25 A. Aslani, E. Antila, and K. V. Wong, 'Comparative Analysis of Energy Security in the Nordic Countries: The Role of Renewable Energy Resources in Diversification', *Journal of Renewable and Sustainable Energy* 4, no. 6 (2012): 062701.
26 Alessandro Rubino, and Michael Cuomo, 'A Regulatory Assessment of the Electricity Merchant Transmission Investment in EU', *Energy Policy* 85 (2015): 464–474.

27 A. Katsaris, 'Europeanization through Policy Networks in the Southern Neighbourhood: Advancing Renewable Energy Rules in Morocco and Algeria', *Journal of Common Market Studies* (January 2015), doi: 10.1111/jcms.12320.

28 J. Zielonka, 'Europe as a Global Actor: Empire by Example?', *International Affairs* 84, no. 3 (2008): 471–484; H. Haukkala, *The EU-Russia Strategic Partnership: The Limits of Postsovereignty in International Relations* (New York: Routledge: 2010).

29 N. Tocci, ed., *Who is a Normative Foreign Policy Actor? The EU and its Global Partners* (Brussels: Center for European Policy Studies Paperbacks, 2008).

30 Escribano, 'RES in the 'Hood.'

31 Rafael de Arce, R. Mahía, E. Medina, and G. Escribano, 'A Simulation of the Economic Impact of Renewable Energy Development in Morocco', *Energy Policy* 46 (2012): 335–345.

32 A. Calzadilla, M. Wiebelt, J. Blohmke, et al., 'Desert Power 2050: Regional and Sectoral Impacts of Renewable Electricity Production in Europe, the Middle East and North Africa' (working paper no. 1891, Kiel Institute, Kiel Germany, 2014).

33 B. Brand, and J. Zingerle, 'The Renewable Energy Targets of the Maghreb Countries: Impact on Electricity Supply and Conventional Power Markets', *Energy Policy* 39, no. 8 (2011): 4411–4419.

34 Eva Medina, Rafael de Arce, and Ramón Mahía, 'Barriers to the Investment in the Concentrated Solar Power Sector in Morocco: A Foresight Approach Using the Cross Impact Analysis for a Large Number of Events', *Futures* 71 (August 2015): 36–56.

35 I. Boie, C. Kost, S. Bohn, et al., 'Opportunities and Challenges of High Renewable Energy Deployment and Electricity Exchange for North Africa and Europe e Scenarios for Power Sector and Transmission Infrastructure in 2030 and 2050', *Renewable Energy* 87 (2016): 130e144.

36 Laura El-Katiri, *A Roadmap for Renewable Energy in the Middle East and North Africa* (MEP paper no. 6) (Oxford: Oxford Institute for Energy Studies, 2014), www.res4med.org/uploads/studies/1402069246Oxford.pdf, 2.

37 C. Cambini, and D. Franzi, 'Assessing the EU Pressure for Rules Change: The Perceptions of Southern Mediterranean Energy Regulators', *Mediterranean Politics* 19, no. 1 (2014): 59–81.

6

Frameworks for regional co-operation

The EU

Benjamin Görlach, Matthias Duwe and Nick Evans

The European Union

Current status of the European Union

Almost 60 years after its inception as the European Economic Community, the European Union of today is a peculiar construction. It shares common markets for goods and services, capital and labor, freedom of movement for its 500 million citizens, common institutions and an extensive body of common legislation. Several of its Members share the euro as a common currency as well as open borders within the Schengen Area. In these regards, the EU is far more than just an organization to enhance regional cooperation.

At the same time, faced with the rise of nationalist and populist policies and an anti-EU sentiment in several Member States, the question of what the EU is or what it should become is more pressing than ever. The answers range from those that see the EU primarily as a common market (with some complementary policies to improve the functioning of the common market) to those that want to see the "ever closer union among the peoples of Europe" achieved through a full political union, or even a European confederation.

In addition, the EU faces a number of internal and external challenges, which also cast doubt on the prospects for further European integration. The financial and economic crisis beginning in 2008, and the ensuing sovereign debt crisis, have raised questions about the future of the common currency, the euro, and highlighted the need for a more coordinated economic policy among the eurozone members. Rising tensions with Russia since 2013 over the Ukraine conflict and the annexation of Crimea have shed light on differences among the EU Member States, testing the limits of the EU's Common Foreign and Security Policy. And finally, the recent unprecedented levels of migration to Europe have exposed the EU's difficulties in reaching a coordinated response, and continue to challenge the policy of open borders within the Schengen area. Thus, some of the landmark achievements of economic and political integration in Europe – the common currency, open borders under the Schengen agreement and the Common Foreign and Security Policy – currently face an uncertain future.

Yet at the same time, the day-to-day processes of policy making and policy implementation continue to function well. In the area of climate and energy policy, the EU continues to exert a significant influence, both in terms of domestic EU policies – where many of the national policies are shaped by EU-level policy frameworks – but also at the international level. Regarding domestic efforts, the EU has set itself a number of fairly ambitious climate and energy targets for the short, medium and long term (2020, 2030 and 2050). It is currently well on track towards achieving the 2020 targets (or has already achieved them) and with additional efforts may also achieve the more demanding 2030 targets.[1] The current set of climate and energy targets is complemented by a suite of policy instruments and governance mechanisms tailored to achieve the EU's emissions reduction targets, including tools for greenhouse gas (GHG) emissions monitoring, improving energy efficiency, promoting renewable energy sources and emissions trading.[2] At the international level, the EU successfully employed its soft power and its esteemed climate leadership to make an agreement possible at the 21st Conference of the Parties to the UN Framework Convention on Climate Change in Paris (COP21). While there are of course plenty of fathers and mothers to the Paris Agreement, it is fair to say that the EU played an important and positive role in forming the coalitions and forging the compromises that made the agreement possible.[3]

The historical roots of EU energy and climate policy

Energy has historically been at the heart of EU policy making. Two of the founding treaties of what is today the European Union were meant to promote cooperation and mutual control in energy-related sectors: the 1951 Treaty of Paris establishing the European Coal and Steel Community (ECSC) and the 1957 Treaty establishing the European Atomic Energy Community (Euratom).[4] Following the adoption of the Paris Treaty in 1951, a mere six years after the end of World War II, the formal entry into force of the ECSC in 1952 marked the very first step towards European integration. The ECSC was set up to promote cooperation between the former adversaries, France, Germany and Italy as well as Belgium, the Netherlands and Luxemburg, by creating common markets for coal and steel and thus establishing mutual control over two sectors that had been critical for the war effort. While the ECSC formally existed until 2002, its institutions and functions were eventually merged with the European Economic Community, which was formed under the Treaty of Rome in 1957.

The year 1957 also saw the establishment of another European Community, the European Atomic Energy Community (Euratom), which was tasked with promoting peaceful uses for nuclear energy. While the executive functions of the Euratom treaty were taken over by the European Economic Community in 1965, the Euratom Treaty formally continues to exist until today. Given the controversial and divided views on nuclear energy among the current EU Member States, however, the treaty has little practical impact, aside from providing a source of funding for nuclear fusion research.

Despite the fact that two of the founding treaties of the current EU were energy-related, the EU as such for a long time lacked the competence to become active in the area of energy policy.[5] Shared competence between the EU and its Member States on energy policy matters was only established with the entry into force of the Lisbon Treaty in 2009. However, while the Lisbon Treaty established shared competence for energy, Article 194 also enshrined the right of each Member State to "determine the conditions for exploiting its energy resources, its choice between different energy sources and the general structure of its energy supply."[6]

The historical roots of EU climate policy are much younger and appear only in the late 1980s, corresponding to growing international awareness of the challenges and threats posed by human interference with the climate. Indeed, when charting the emergence of European climate policy, it is helpful to think of the EU and international climate measures as developing in "tandem," i.e., as a co-evolution in policy-making.[7]

Moreover, from the very beginning, there are many ways in which the EU sought to influence the UN climate regime but was also influenced itself by international developments in turn. The global nature of the climate problem lent itself to the EU taking it on as an issue on which the EU could engage in "regime building"[8] – one specific example being the EU carbon/energy tax proposed in May 1992, at the time of the adoption of the UN Framework Convention on Climate Change (UNFCCC) and in anticipation of the Rio Earth Summit. This formative period in EU climate policy was also key in identifying the three major fields of action that remain the pillars of European climate policy today – emissions reduction, renewable energy promotion and energy efficiency.

The 1990s are generally seen as an era of consolidation and strengthening of European Environmental Policy, following the adoption of the Single European Act of 1987, which provided more formal competencies for environmental protection to the European Commission.[9] Since the Single European Act, the European Council and the European Parliament can determine climate policy, as part of environmental policy, on the basis of a qualified majority, a factor that has strongly aided the evolution of a common EU-wide climate policy.[10] And yet, initially little progress was made on individual climate policy instruments, with the tax proposal failing due to opposition from a group of Member States led by the UK, which deemed it overreach on the part of the EU into national fiscal policy.[11] As a result, the further development of specific policies and measures was largely put on hold, until the Kyoto Protocol in 1997 spurred the EU into a renewed phase of dynamic policy-making. In 2000, the European Climate Change Programme was launched, as a process to identify policies that would enable the EU to meet its goal of reducing greenhouse gas emissions 8% below 1990 levels. Being negotiated in parallel to the European Climate Change Programme, the Renewable Energy Directive (2001/77/EC) was adopted, which saw the introduction of indicative quantitative targets for share of energy from renewable sources in each Member State.

As a replacement for the failed tax proposal, the focus had shifted to discussions of a cap and trade system for large CO_2 point sources – which could be agreed under the rules for environmental, not tax policies. With the Kyoto obligation in mind and the concept of "tradeable emission permits" enshrined under the Kyoto system, the EU Emissions Trading Scheme (EU ETS) was finally adopted in 2003, as a central pillar of European greenhouse gas reduction policies.[12]

Over the next half a decade these fledgling EU instruments would lay the groundwork and provide the necessary learning opportunities for future developments in climate policy. All the while, shared competencies between the EU and its constituent Member States in regards to climate issues were largely unchanged until the adoption of the Lisbon Treaty in 2009, which placed a clear emphasis on the supranational level in implementing international climate agreements.[13]

Traditional and new objectives in EU climate and energy policy

Climate and energy policy in the EU is guided by a multitude of goals, objectives and associated targets, which differ in their legal nature – ranging from mere political declarations to objectives enshrined in primary EU law, to internationally binding obligations. Likewise, while some of the policy aims can be clearly identified as chiefly energy or climate-related, most of them are relevant in both fields.

In terms of primary EU law, Article 194 of the 2009 Lisbon Treaty, establishing the shared competence for energy, defines the four core objectives of EU energy policy as (a) ensuring the functioning of the energy market, (b) ensuring security of energy supply in the Union, (c) promoting energy efficiency and energy saving and the development of new and renewable forms of energy and (d) promoting the interconnection of energy networks.

Building on these core objectives, the different successive frameworks, strategies and roadmaps that have come to define EU energy policy generally incorporate a triad of objectives, combining – with varying emphases – (a) the security of supply, (b) affordability for households and competitiveness of energy-using industries and (c) environmental sustainability, in particular through low-carbon or climate-friendly types of energy. These three objectives are often complemented with more procedural objectives, such as the completion of the single market for energy through the removal of technical and regulatory barriers.

Since 2007, the third, environmental dimension has been formally structured around three separate but interdependent climate targets – building on the three pillars of EU climate policy since the 1990s.[14] The current targets for 2020 consist of 1) at least a 20% reduction of greenhouse gas emissions below 1990 levels – a target which the EU in fact achieved five years ahead of time,[15] 2) a 20% share of renewable energy in EU gross final energy consumption and 3) a reduction of primary energy consumption by 20% below the projected levels for 2020.[16] For the latter two targets, a continuation of current trends will be sufficient to achieve the 2020 targets with a comfortable margin.[17]

The 2020 climate target triad, known collectively as the "20–20–20" strategy, was mainly implemented via legislative measures at the EU level through the Climate and Energy Package, which was formally adopted in the spring of 2009. The package marked a novel turn towards the integration of climate and energy policy and was seen as a beacon of EU climate ambition in the lead up to the UN Climate Conference in Copenhagen.[18]

While the target triad adopted in 2007 referred to 2020, they were embedded in a broader long-term strategy, with the objective of much more substantive emission reductions by 2050. This objective, specified by the EU Heads of State in 2009 as emission reductions of 80–95% below 1990 levels by 2050, is in line with, and was explicitly linked to, IPCC emissions reduction corridors required to meet a 2°C target and thereby avoid the worst effects of human-induced climate change.[19] The objective was reinforced in 2011 with the publication of a forward-looking Roadmap, describing the path towards a low-carbon economy in Europe.[20] Long-term EU aims have been reiterated multiple times in subsequent papers and Council Conclusions, and adopted by EU Heads of State and Government, indicating ample political support for decarbonization in the EU.[21]

While the first ten years of the new millennium thus saw rapid developments in EU climate policy, culminating in the robust 2008/09 package, the dynamic slowed down considerably in the aftermath of the economic crisis since 2008, and the lacklustre outcome from the UN Climate Summit in Copenhagen.[22] Part of the slowdown can also be attributed to the longer-term effects of the EU enlargement of 2004: several of the new Member States rely heavily on coal in their national energy mix, and public as well as political support for ambitious climate policies are by and large weaker in most of the new Member States.[23]

The 2008/9 package was and continues to be implemented, and some provisions have actually been strengthened or added, such as a Directive for energy efficiency (2012/27/EU) or regulations for transport vehicles. However, further reinforcement of the central pillars of the package – fostering renewable energy, the EU ETS and effort-sharing – has failed to find much legislative footing in recent years and instead most political attention was taken up with short-term measures to address the oversupply of allowances in a floundering EU ETS. Moreover, although it became evident that the EU was set to overachieve its 2020 target by a considerable margin, efforts to increase the emissions reduction ambition set in 2007 for the remainder of the current period (until 2020), as foreseen in 2007 in the case of a global climate change agreement, were not successful in part due to significant opposition from Central and Eastern European Member States, concerned about impacts to their largely coal-based energy structures.[24]

In the light of the prolonged economic crisis in since 2008, and the heightening Russia-Ukraine conflict, the emphasis of EU energy policy has shifted away from climate ambition and to the other two objectives that form the traditional triad of energy policy objectives, i.e., security of supply and competitiveness. This led to a substantially different political and economic atmosphere when it came time to decide on a post-2020 EU climate and energy framework.[25] As a result, the political process was contentious and highlighted critical gaps in ambition, objectives and goals between EU member states, which, as some observers argued, led to indistinct wording in the final conclusions.[26] Adopted in October 2014, the 2030 climate and energy framework continues to follow the familiar three-pronged approach of the "20–20–20" targets – increasing ambition in emissions reduction (40% by 2030 over 1990 levels), renewable energy shares (27% by 2030) and energy efficiency (27% by 2030 compared to projections of future energy consumption). But unlike its predecessor for 2020, the 2030 renewable energy target is binding only at the EU level – a sign of the slightly less favorable political climate. And as an echo of the concerns over energy security, the three-target structure was expanded with a fourth quantitative target on interconnection in the energy system. The 2030 framework represented the EU position towards the UN Climate Conference in Paris in December 2015, even though its legislative implementation – in the form of a new and revised climate and energy package – is still being deliberated, with many proposals expected in 2016 and discussions likely to take 1–2 years to finalize after that. See Table 6.1 for an overview of EU targets over time.

Trends and challenges in EU climate and energy policy

Going forward, there are a number of dynamics at work in EU climate and energy policy, which are not necessarily aligned. Thus, the institutional and political challenges the EU currently faces, and the debate about the desirability of further European integration, also affect the EU's climate and energy policies. This debate is also fuelled by tension between unified EU targets on the one hand, and starkly different views of EU Member States about their future energy supply on the other hand. Furthermore, the integration of energy and climate policies in the EU remains a challenge, closely linked to the question of whether these policies should become more or less centralized. More recently, the renewed emphasis on energy security in the wake of the Russia-Ukraine crisis has begun to change the dynamics of EU energy and climate policy, and will continue to do so. And finally, it remains to be seen how the climate agreement adopted in Paris in December 2015 will add new momentum to EU climate and energy policy, and if it will possibly swing the pendulum back to place greater emphasis on climate objectives. The following sections discuss some of these trends and dynamics in greater detail.

Centralization vs. decentralization of climate and energy policies

The struggle between centralized and decentralized EU governance is as old as the EU itself. Centralized governance, understood as the concentration of legal and administrative competencies at the EU level, offers the promise of greater efficiency in implementation and a

Table 6.1 Overview of EU greenhouse gas emission targets over time

Target time horizon	*2000*	*2010 (2008–12)*	*2020*	*2030*
Time the target was set	1990	1997	2007	2014
EU target ambition	Stabilize at (return to) 1990 levels	−8% from 1990 levels (original proposal 15%)	At least −20% from 1990 (unilaterally), −30% if others join as part of a global deal	At least −40% domestic from 1990 levels
Enshrined at UN level	The general ambition was included in the UNFCCC (Article 4) in text format (not as a quantified target)	−8% from 1990 levels	−20% from 1990 levels	At least −40% domestic from 1990 levels
Break down of targets to the national level	All MS (EU15) subscribed to UNFCCC goals	Internal EU15 burden sharing (from +28% to −27%)	Internal EU28 effort sharing for non-ETS emissions only (based on 2005 levels)	Internal differentiation into national binding targets for non-ETS sectors (based on 2005 levels)

"levelling of the playing field" all across the EU, thus minimizing distortions of competitiveness and the resulting economic inefficiency. Decentralized governance, understood as the delegation of legal and administrative competencies to national, regional and local levels, is seen to offer the advantage of a more flexible, responsive and accountable way of setting and implementing rules, which is then more easily adapted to different local circumstances or political preferences.

In the field of climate and energy policy, different factors are at work that influence the level of centralization of EU governance:[27]

- The degree to which a policy initiative is linked to the single market (including the single energy market). The single market is a central pillar of EU policies and a strong driving force for centralization, in order to eliminate barriers to competitiveness and level the playing field between producers across Europe. One important way in which single market rules affect climate policy is through the European state aid rules, which have been applied inter alia to certain design choices of support schemes for renewables or nuclear power.
- The applicable decision rules and associated voting thresholds: for instance, issues related to taxation, requiring unanimous support at the EU level, are not strongly centralized, since unanimity is nearly impossible to achieve.[28]
- Whether the policy issue is framed primarily as a matter of environmental policy (where EU competence is well established and accepted) or whether it is framed as a matter of, e.g., energy or transport policy (where EU competence is more contested).[29]
- The transboundary dimension of a policy and the benefits of cross-border cooperation in its implementation, e.g., where the effects of policy choices regarding the national energy mix are transmitted through shared gas or electricity grids.

- The existence of international obligations of the EU (with more comprehensive obligations supporting a higher degree of centralization).
- The role of contentious technological choices, e.g., related to the role of nuclear power or the use of coal. In light of the vastly different attitudes towards certain technologies, Member States have been adamant about their right to determine their energy supply.

The degree of centralization differs between climate and energy policy. Climate policy in the EU is largely centralized through a number of factors. To begin with, there is a common, internationally binding emission reduction target at the EU level (8% below 1990 levels in the 1st commitment period of the Kyoto Protocol, 20% in the 2nd commitment period, and 40% by 2030 as per the EU's Nationally Determined Contribution (NDC) under the Paris Agreement). This common EU obligation is shared by all Member States through the so-called "bubble" arrangement (allowed under Article 4 of the Kyoto Protocol) under which the target is broken down to different national shares.[30] Second, the European Council and the European Parliament can determine climate policy, as part of environmental policy, on the basis of a qualified majority rather than requiring unanimity. And third, there are a number of policy instruments and governance mechanisms in the area of climate policy, connected with the climate targets, which provide for a greater harmonization of approaches across the EU. The degree of harmonization ranges from energy efficiency and renewables policies – where the EU sets the targets and increasingly the frameworks for policy design but where implementation is still predominantly national – to the EU ETS, where both the rule-making and the implementation are increasingly centralized at the EU level.

By contrast, energy policy in the EU continues to be predominantly determined at the national level.[31] There are some driving forces for a greater centralization of EU energy policy: the historical roots of the current EU (Euratom, ECSC), in which energy-related questions played a central role; the internal market for energy among the Member States, which remains part reality and part aspiration; a body of shared regulation for the energy sector (e.g. through state aid rules or regarding nuclear safety) as well as EU-funded infrastructure for greater interconnection (the Trans-European Networks for Energy, TEN-E). And yet, the shared competence of the EU and its Member States is still a fairly recent evolution in European energy policy, having been introduced through the 2009 Lisbon Treaty.[32]

Diverging views among the Member States

At the same time, there are also very strong drivers to keep control over energy policies at the national level. Historically, the EU Member States have made very different technology choices for their energy supply. These choices reflect strongly diverging views on the risks and the benefits of different technologies, different weightings of the priorities of energy policy (environmental protection and decarbonization vs. security of supply and reduced import dependence vs. affordability and competitiveness), but also the different domestic endowments of natural resources. As a result, the structure of energy supply differs considerably between EU Member States – and hence decarbonization of energy supply entails different payoffs for each Member State. Some Member States are already well advanced, with renewables accounting for half of the energy used in heating and cooling in Finland and Latvia or two-thirds of the power generation in Austria and Sweden.[33] Nevertheless, for other countries, the challenge is considerably greater: in Estonia and Poland, coal accounts for half of the domestic energy consumption and still more than a third in Bulgaria and the Czech Republic.[34]

The discrepancy between EU Member States is also evident when it comes to the most contentious technologies, i.e., nuclear power. Half of the 28 EU Member States have at least one nuclear power plant in operation. Moreover, while some of the countries that operate nuclear power plants have started to phase them out (most notably Germany), others have plans to increase their nuclear capacity (such as the UK). The vast discrepancies in resource endowments and general attitudes are one main reason why Member States have always defended their prerogative to choose their own energy sources and determine the general structure of their domestic energy supply, which is enshrined in Article 194 of the EU Treaty.

In recent years, some observers have pointed to a re-nationalization of energy policies in recent years.[35] Most fundamentally, this re-nationalization can be explained through disagreements about the priority afforded to the different objectives of energy policy – most notably the German "Energiewende," which is driven by the desire to phase out nuclear and transition to renewable energy as opposed to concerns about import dependence on Russian gas taking central stage in Central European EU Member States.[36] The fundamental disagreement about political priorities is mirrored by a lack of coordination at the operational level: in a shared electricity grid, national energy policies are bound to be felt in neighboring countries, through transboundary electricity flows and the resulting effect on electricity prices.[37] Since regional cooperation on the management of such transboundary effects is only just beginning to emerge, some Member States have effectively taken steps to limit their integration into the European electricity grid. A further factor contributing to re-nationalization is the absence of European solutions. This is apparent in the case of electricity market design and capacity mechanisms, where several Member States are pursuing different options that are not necessarily compatible.[38] Another case in point is the failure of the EU ETS to generate a meaningful carbon price, which has led several EU Member States to implement complementary national policies – such as the UK carbon price floor, or Germany's decision to mothball 2.7 GW of lignite-fired power generation by transferring eight units into a "capacity reserve." To some extent, the EU initiative of creating an Energy Union (discussed in greater detail below) can also be seen as an attempt to reverse the re-nationalization of energy policy.

Diverging perspectives on energy policy among Members States also have critical implications for EU climate policy. The last few years can be characterized by a "trend towards polarization" in discussions surrounding EU climate and energy legislation, with most notably Germany and Denmark pushing for heightened ambition and the Visegrad countries, headed by Poland, strongly resisting.[39] Moreover, backed by a coalition of Central and Eastern European countries, Poland threatened to veto the 2008 climate and energy package, arguing that it did not adequately address the varying energy circumstances in each MS.[40] Finally in 2011 and 2012, Poland made good on this threat, effectively blocking the Commissions Roadmap to a Low Carbon Economy for 2050.[41] Seemingly averse to any mention of long-term emissions cuts or "decarbonization" in EU legislation, Poland's stance is that European action on climate change should not surpass international ambition. As has been argued, Poland's inflexibility in EU climate policy-making is due to a mismatch between EU objectives, current domestic policies and the country's "energy-economic situation" and correspondingly the strength of incumbent energy actors.[42]

Integration of climate and energy policies

Over the years, the fields of climate and energy policy have become increasingly intertwined, to the point where they are effectively inseparable. To achieve the EU's long-term climate objective of an 80–95% emission reduction by mid-century requires a decarbonization of the

European economy, i.e., an almost complete phase-out of fossil fuels.[43] While this entails challenges for most parts of the economy, it requires a revolution of the energy sector in particular, which needs to be transformed nearly completely to zero-carbon energy sources. Within the energy sector, electricity generation has a pivotal role to play: first, because of the emissions already associated with electricity generation in today's energy mix and second because of the role of electricity for the decarbonization of other sectors, such as the electrification of transport and space heating. Which, of course, will only be compatible with the decarbonization objective if the electricity used for transport and space heating comes from zero-carbon sources.

The interaction between climate and energy policies is increasingly also felt in the operation of the energy system: renewables have evolved from a niche energy source to a significant share of energy supply, now accounting for a sixth of gross final energy consumption and more than a quarter of all energy generation in the EU.[44] This is being felt in different ways: first, the electricity grid has to accommodate a growing share of intermittent generation from renewables that fluctuates in time and space. This requires greater flexibility on the demand side (implementing solutions to adjust the demand to fluctuations in generation, and corresponding mechanisms to reward such flexibility), but also on the supply side (maintaining sufficient flexible, dispatchable generation capacity that can be ramped up and down quickly to make up for shortfalls in renewables). And finally, energy storage technologies may have a crucial role to play in aligning supply and demand, provided that they become economically feasible.[45] All of these challenges are linked through electricity markets: as such, rewarding flexibility on the side of electricity consumers or generators, providing incentives for investment in renewables and coordinating the different efforts, should be genuine tasks for the market, and correspondingly guided by the electricity price.

Unfortunately, the way in which traditional electricity markets operate is not well suited to accommodating a high share of renewables; as they do not have to pay fuel costs, electricity from renewables will enter the market at a marginal cost of nearly zero. In an electricity market based on marginal cost pricing, this suppresses wholesale prices, via a mechanism known as the merit order effect.[46] However, in a market where non-dispatchable renewables such as wind and solar dominate power generation, an electricity price purely based on marginal cost pricing would fall to nearly zero for most of the year. This means that operators of conventional power plants will find it increasingly difficult to recover the capital cost of their investment – or would only be able to do so if the price is allowed to rise to very high levels in the periods where renewable generation capacity is not available. Part of the solution to this problem could be the establishment of some type of capacity mechanism, which rewards electricity generators for maintaining dispatchable backup capacity.

For the EU, the rise of renewable energy has several implications. First, the example goes to show how a policy – promoting renewable energy sources – originally considered as primarily climate-related is now affecting the operation and design of the electricity market – both issues that would have normally fallen under the auspices of energy policy. Second, the need for a new market design adds greater complexity to the already complex task of integrating electricity markets across the EU. Efforts to enhance energy integration are already trailing behind in reaching their stated objectives. A lack of physical interconnection capacity, as well as regulatory and economic barriers such as regulated tariffs or market dominance by incumbents, hinder the formation of an EU-wide integrated market.[47] As there is, as of now, no agreement on an EU-wide harmonized capacity mechanism, France and the UK have recently adopted their own, national solutions; other countries including Germany are contemplating different options.[48] However, since these solutions operate in physically and economically connected electricity markets diverging national decisions on how to solve the problem risk adding further barriers to the full integration of power markets.

The overlap between climate and energy policies is reflected in a growing number of joint policy developments at EU level, both in terms of substantive policies, but also in terms of institutional arrangements. In substantive terms, the 2008/09 Climate and Energy Package marked a step towards greater integration, defining climate and energy targets for 2020 as well as (most of) the corresponding legislation for both policy areas in the frame of one coordinated package. This approach was subsequently repeated in the EU Climate and Energy Framework for 2030, adopted in 2014, and the development of the "Energy Union" concept (see more below) – as the traditional climate related energy policy aspects were specifically complemented with energy market integration elements (such as the interconnection target). In institutional terms, with the establishment of the new Commission in 2014, the Commission President Jean-Claude Juncker decided to restructure the institutional responsibilities by having a single Commissioner for Climate Action and Energy (in the person of Miguel Arias Cañete), overseeing the two Directorates-General of Climate Action and Energy. The jury is still out whether this institutional integration has resulted in a strengthening or a weakening of the two policy areas – experts point both to the potential for a more integrated, streamlined approach, but also to the risk of sidelining climate policy ambition in cases where climate and energy objectives are in conflict.[49]

Going forward with a joint EU climate and energy policy, one of several challenges that remain is the different legal bases for energy and climate policies in the EU governance system – and, as a result, different voting rules and different degrees of centralization (see above). As this situation is unlikely to change significantly in the near future, the particular challenge of formulating policy solutions that can straddle the divide between decentralized energy policies and more centralized climate policies remains an important art to master for effective and sustainable EU policy-making in the area.

Renewed attention to security of supply

Security of supply did not always feature prominently in EU energy policy. In the 1990s, in a time of generally low energy prices and geopolitical optimism, energy relations in Europe were seen as a normal part of trade relations. Import dependence and security of supply were not seen as major concerns – despite the fact that Europe's dependence on Russian gas imports was much higher in 1990 (with 55% of gas imports from Russia) than in 2013 (39%).[50]

At a time when security of supply was less of a concern, climate concerns had a stronger effect on EU climate and energy policy. The Climate and Energy Package of 2008/9, adopted in the run-up to the Copenhagen climate summit, was very much a climate-driven package.[51] In the following years, enthusiasm for ambitious climate policies cooled down considerably, as explained in section 2 above.

As attention to climate policy declined, the security of energy supply and dependence on fuel imports from politically unstable regions received growing attention. In 2012, the EU depended on imports for more than half of its energy supply; it imported 86% of the oil and two-thirds of the gas consumed in the EU. All these shares have been increasing over recent decades and are expected to increase further.[52] At the same time, imports are concentrated on a small number of supplier countries. In particular, for natural gas, Russia, Norway and Algeria alone accounted for more than 80% of EU imports. Several EU Member States – especially smaller ones in Northern and Eastern Europe – depend entirely on a single supplier of natural gas, and often on one supply route.[53]

Import dependence makes a country vulnerable, certainly if it is exacerbated by the physical dependence on a single connection to a single supplier. Yet whether this import dependence also translates into a threat to the security of supply depends on the perceived (political) reliability of the supplier country in question, and whether it is conceivable that the supplier country will exploit the vulnerability to leverage power,[54] in the sense that depending on Norwegian gas imports entails a different risk than depending on Russia. In Europe, the risks of import dependence became a concern in the winters of 2006, 2008 and 2009, when Russia disrupted natural gas supplies to Ukraine and thereby also to the EU. In the light of rising tensions between the EU and Russia since 2013 over the Russia-Ukraine conflict and the annexation of Crimea, concerns about the security of supply in Europe grew further, raising the issue on the political agenda. The renewed political uncertainty in the MENA region since the Arab Spring further added to these concerns.

As an immediate response to the heightened concerns about energy security, the EU adopted the European Energy Security Strategy in May 2014.[55] The strategy represents a mix of urgent and immediate measures to reduce the exposure of the most vulnerable Member States, as well as longer-term elements to reduce Europe's dependence on energy imports. The immediate actions included preparedness for possible disruptions of (Russian) gas supply to Europe, to take effect already in the winter of 2014/2015, as well as strengthened emergency mechanisms to improve the coordination among Member States in the case of supply disruptions and shortages. Longer-term elements include the reduction of energy demand, diversification of supply, increasing domestic energy production and completion of the single market for energy. The 2014 strategy thereby intensified prior efforts – already in 2011, Member States had decided to establish a mechanism to exchange information on existing intergovernmental agreements on energy supply, concluded between EU Member States and supplier countries.

In parallel, the role of energy security has also grown in the EU's Common Foreign and Security Policy. Since 2011, the EU Commission has been seeking to develop its activities in the field of energy diplomacy, with a view to developing an External Energy Policy for the EU. Part of these efforts was to deepen the dialogue with supplier countries: such energy dialogues had been established with Russia (since 2000), OPEC (since 2005), Norway, Ukraine, the Gulf Cooperation Council, the Caspian Sea Region, and several other countries or regional bodies. These dialogues are institutionalized to different degrees, conducted at different intensity, and are often combined with other dialogue and exchange processes. In addition, energy security has become an established element of the European Neighbourhood Policy, the principal tool through which the EU seeks to enhance economic, political and regulatory cooperation with its Eastern and Southern periphery. The European Neighbourhood Policy includes both some of the most important suppliers of oil and gas to Europe (such as Russia and Algeria), as well as pivotal transit countries (such as Ukraine and Turkey).

And yet, in substance, achieving greater coordination and cooperation remains a challenge. Thus, progress remains limited on the completion of a single market for energy, a measure that is supposed to make Europe's energy system more integrated, more unified and thereby more resilient against external shocks. Progress is uneven in terms of infrastructure measures: while the Member States have acted swiftly to enable reverse flow in gas pipelines (i.e. from West to East) in an effort to hedge against disruptions of supply from Russia, progress is more limited when it comes to increasing interconnection capacity between European countries (such as between France and Spain), or connecting energy islands to the European grids (such as the Baltic countries).[56] By contrast, the major pipeline projects connecting the EU to Russia and the Caspian Region – the North Stream connecting Russia and Germany through the Baltic Sea, and the failed Nabucco and South Stream pipeline projects that were supposed to connect

Russia/Turkey with Southeastern Europe – turned out to be highly controversial and divisive projects, revealing the difficulties for EU Member States of agreeing to a common position. Thus, one of the ultimate goals of a European External Energy Policy – an EU that speaks with one voice vis-à-vis third parties – remains elusive for the time being.

3.5 The "Energy Union" as the solution?

Faced with tendencies towards a re-nationalization of energy policies, and in light of the renewed attention to energy security in the wake of the Russian-Ukrainian conflict, the EU in 2015 launched the Energy Union as a new framework for its energy policy.[57] The concept of the Energy Union is not entirely new: over recent years, different political actors put forward several proposals on how to strengthen, better coordinate and re-politicize EU energy policy. In 2010, the then-president of the European Parliament Jerzey Buzek and the former EU Commission President Jacques Delors had proposed the creation of a "European Energy Community," with a strong emphasis on a system that would guarantee security of supply and inter-European solidarity. Yet, while their proposal received some attention, it failed to gain political traction.[58] This changed in 2014, when the then Polish Prime Minister Donald Tusk, in an opinion piece in the *Financial Times*, proposed the creation of an "Energy Union" as a direct response to the Russian-Ukrainian conflict, with the aim of establishing a unified EU as a counterpart to Russian dominance. Elements of his proposal included a mechanism for jointly negotiating energy contracts with Russia, solidarity mechanisms between Member States in case of supply disruptions, EU support for the diversification of gas infrastructure – but also making full use of the fossil fuels available within the EU.[59] Thus, Tusk's original proposal for the Energy Union was motivated predominantly by the Russia-Ukraine crisis, focusing exclusively on security of supply, with particular emphasis on gas supplies.[60] With this push, Tusk also sought to rectify a perceived imbalance in EU energy policy, in which – in the views of some Central European Member States – climate and environmental considerations had taken undue precedence over matters important to them, above all security of supply and affordability.[61] In his article, Tusk made clear where he thought priorities ought to lie: "We need to fight for a cleaner planet but we must have safe access to energy resources and jobs to finance it."

The break-through for the Tusk proposal came when the incoming President of the European Commission, Jean-Claude Juncker, took up Tusk's proposal and made the Energy Union one of the core elements of his political agenda, and installed with Maroš Šefčovič a separate Commission Vice-President for the Energy Union. But for the concept to be politically acceptable to all Member States, and for it to succeed as an overarching framework for energy and climate policy, the Energy Union had to incorporate all key areas of climate and energy policy, and align the objectives in a more balanced way. Thus, the Strategy for a "Resilient Energy Union with a Forward-Looking Climate Change Policy" which the European Commission proposed in 2015 follows a much broader approach than the original Tusk proposal. This strategy is structured around five dimensions:[62]

1 Energy security, solidarity and trust (including the issues of diversification of supply, solidarity between MS, a stronger role for the EU in global energy markets, and more transparency on gas supply).
2 A fully integrated energy market (including the issues of interconnections, market rules and state aid guidelines, regional cooperation, energy poverty, and empowering consumers).
3 Energy efficiency to moderate demand (with a focus on the buildings sector and on transport).

4 Decarbonizing the economy (including an ambitious EU climate policy, and the target of becoming the world leader in renewable energy).
5 Research, innovation and competitiveness.

As a result, the new, now much broader focus of the Energy Union fully encompasses the existing targets laid out in the EU's 2030 climate and energy framework (reducing greenhouse gas emissions, expanding renewable energy sources, improving energy efficiency), as well as the target for increasing interconnections adopted by the October 2014 European Council.[63] Thus, in an effort to bring everyone on board and to ensure compatibility with other strands of EU energy and climate policy, the European Commission has greatly broadened the scope of the Energy Union and expanded the number of targets it is expected to achieve. But this greater acceptance comes at a cost of greater ambiguity on what the Energy Union stands for, especially since the underlying (perceived) conflicts between climate ambition and security of supply have not been resolved.

One novel element of the Energy Union is that it comes with a new means of monitoring and measuring progress – and of interaction between the Commission and the Member States. The new governance structure is supposed to help integrate information from different policy areas and place them in the context of individual, nationally specific energy policy – and in doing so, help streamline existing processes and reduce administrative efforts. As per the process currently foreseen, so-called "National Climate and Energy Plans" would form the backbone, largely replacing existing, separate planning strands for renewables, efficiency, etc. These plans are expected to form a central part of an EU energy governance system, in which the Commission supports Member States in defining national contributions and measuring progress through a set of key indicators.[64] Starting in November 2015, the European Commission itself issues an annual "State of the Energy Union" report.[65] Thus, the governance mechanism on the one hand would keep Member States in charge of defining national climate and energy policies and measures in the context of their respective individual circumstances, while also keeping them accountable for the fulfilment of the common objectives set at the EU level.

Going forward, it thus remains to be seen how much concrete impact the Energy Union will have. The European Commission is firmly committed to the Energy Union, and politically invested in its success. The EU Member States are overall supportive of the Energy Union – but this may also be explained through the ambiguity of the concept, which allows every Member State to emphasize the aspects that are most in line with domestic priorities and interests. It is an open bet whether, in the current political climate, EU Member States would maintain their support for the Energy Union if that involves giving up some of the domestic priorities and interests in exchange for more coordination and solidarity. In any case, it is quite possible that the greater impact of the Energy Union will not come from the targets it embodies – which represent, by and large, a summary of existing targets – but rather through the new governance structure it implements.

The Paris effect?

With the adoption of the Paris Agreement in December 2015, there is a new impulse that has the potential to influence EU climate and energy policy significantly. Its full impact cannot yet be properly assessed at the time of writing, as it will depend to a large extent on a) the broader political narrative about "what Paris means" that becomes dominant in the EU and globally and b) the actual decisions on remaining technical details that will be the subject of follow up negotiations (including on the stringency of the monitoring system, etc.) and their subsequent implementation in practice. This section, is therefore, largely a collection of substantiated possibilities.

The European Union was heavily engaged and invested in the preparations for Paris and the global diplomatic effort aimed at brokering a compromise. The EU was not in a comfortable spot for Paris. Its 2030 emission target compromise had been hard won internally, and many Member States remained sceptical about the degree and timing of more ambitious climate action. On top of that, it did not have much to bring to the table in terms of explicit financial commitments, which could have won the EU more support from developing countries. Still, the Paris Agreement is also, to some extent, a success of EU climate diplomacy, which had an important and positive role in forming the coalitions and forging the compromises that made the agreement possible.[00]

So what can be said at this point in time about the potential effect of the Paris Agreement on EU climate policy? Remembering the negative impact of the 2009 Copenhagen summit on the further development of EU climate policy, there can be no doubt that the successful conclusion of the Paris talks will provide a supportive impulse for the completion of the broad and complex legislative process that is going to take place in the EU in 2016–2018 – as it attempts to put into legislation the detailed measures through which the 2030 Framework and its targets shall be achieved.

Paris may lend more weight to the climate change concerns in the broader climate and energy framework/Energy Union approach and may further strengthen the superior role of the greenhouse gas reduction target within the 2030 framework. This could in turn, mean more focus on the EU ETS and the non-ETS target sharing among Member States (maybe provide a basis for advancing stronger sectoral climate policy, on transport or agriculture), but less focus and political backing from Paris for renewable energy and efficiency as "secondary" targets. In this way, Paris could re-emphasize the tendency of a prioritization of targets that is visible in how the 2030 framework compares to 2020.

There is also the chance that after Paris a debate will start on reopening the GHG target. The formulation of "at least" 40% domestic emission cuts chosen in October 2014 could be interpreted to mean that now is the time to talk about a higher target. The fact that the Paris Agreement sets the long-term target for a temperature increase at "well below" 2°, with an aspiration to even aim for 1.5°, whereas the current EU long-term target was defined on the basis of the 2° target, could further support the case for a more ambitious EU 2030 target.

Certainly, the request for five-year updates on national contributions in the Paris Agreement represents a hook to potentially revise the target upwards in the not too distant future. A dedicated target review clause of such frequency does not exist in EU climate legislation and would thus need to be integrated (with several options as to how that could be achieved). Such an innovation, prompted by Paris, could, through the new governance tool of National Climate and Energy Plans for 2030, be extended from GHGs only to broader energy policy goals also, specifically renewables and efficiency.

Another potentially significant impulse could arise from the combination of explicit long-term goals and the request to develop "long-term low greenhouse gas emission development strategies" towards 2050 (and submit them by 2020). While the UNFCCC regime had talked about such strategies before, this explicit reference in the new Agreement could serve to revive this process and have it be integrated into the new governance structure under the Energy Union (as, for example, a 2050 dimension of the National Plans). An organized EU process to formulate a long-term decarbonization strategy would imply an explicit discussion of the implications for all sectors of the economy – with the likely greatest effect for the energy sector. In this way, Paris could strengthen the case for an integrated European energy policy aimed at decarbonization.

But it is also clear that, once the euphoria following the successful conclusion of the Paris conference has subsided, the underlying political realities and conflicts in the EU will still be around. The lasting strength of any Paris effects is thus to be seen only over time.

Outlook: energy and climate policies as a lighthouse project in troubled times?

For the future of Europe's energy and climate policy, one thing is sure: there is no lack of ambitious goals. While inhibiting clarity, the multitude of goals has created enough wriggle-room to accommodate the diverse interests of the EU Member States – with their diverging energy mixes, diverging resource endowments, diverging political priorities, and path dependencies in terms of investment, technologies and infrastructure. The multitude of goals has also made it possible to flag up different subsets of goals, responding to shifts in the political agenda – be they caused by changing political fashions or by external developments, or both.

And yet the multitude of goals can mask the fact that, when push comes to shove, Member States still have very different ideas about the relative importance of the different goals, the best ways of achieving them, and who should bear responsibility for the process. In a Europe of 28 separate energy systems, Member States still enjoyed great freedom to pursue their own energy policies. But as energy systems in Europe become more and more interconnected both physically and economically, and as external pressures remind Europe of the need for greater integration and better coordination, the limits of national approaches become evident. And with increasing integration, the conflict lines become clearer.

The use of coal is a case in point here. On the one hand, especially those Member States with significant domestic coal resources consider coal as essential to guarantee their energy independence and a secure energy supply. On the other hand, if the EU takes the decarbonization of its energy supply seriously, the Member States that rely on coal for a significant part of their energy mix will either need to invest massively into the (technologically and economically) highly uncertain prospect of carbon capture and storage, or will have to start preparing for a phase-out of coal in the coming decades.

In the current political climate, with multiple political crises rocking the EU, key achievements of European integration under fire, and the rise of populist anti-EU platforms in several Member States, the question is whether the EU is unified enough to assume a stronger role in energy policy, both internally and externally. To be sure, a foreign energy policy could be a field where the benefit of a unified EU, negotiating with one voice, would seem obvious in the current geopolitical climate – yet this is also true of the common currency or an EU without internal border controls, and in neither of these cases do the obvious benefits of a common approach trump the strong attachment to national sovereignty over European solutions.

To frame the matter differently, it could be asked whether energy policy could (once more) be the issue that re-invigorates European integration, 60 years after the European Coal and Steel Community, in the face of a common threat of supply disruptions from Russia, and the common challenge of decarbonizing Europe's energy system. After all, one facet of the debate that remains woefully under-explored is the link between decarbonization and energy security. Of course, the relevant European strategies and roadmaps dutifully mention the (near-obvious) fact that energy efficiency and the use of renewables will help to reduce the dependence on imported fossil fuels. But the full economic and political implications of connecting the two targets remain insufficiently explored. Thus, for instance, while, overall, decarbonizing its economy will reduce the EU's dependency on fossil fuel imports, this process proceeds at different speeds for different fuel types. Whereas oil imports are expected to go down soon and substantially, the EU expects gas to continue playing a role for a longer time still. And as fossil fuel imports decline, there will be increasing imports of low-carbon energy – in the form of (solid, liquid or gaseous) biofuels, or as electricity produced from renewable sources. Since the latter depends on the availability of infrastructure to transmit electricity into the EU, import dependence is expected to become more regional than global, and will increasingly be defined by the supply grid.[67] Yet what this means in terms of changing roles for old and new energy suppliers, and the EU's role in managing this transition, is only beginning to emerge.

Finally, it remains to be seen to what extent the Paris Agreement will be able to re-invigorate the climate and energy debate in Europe. In the short term, the Paris effect may not be strong enough to reopen a hard won compromise on the EU's 2030 target. And yet over time, the Agreement might help to turn "decarbonization" from a lofty, long term aspiration into a concrete strategy. The combined impact of the frequent target review and the need to develop a long-term strategy could thus move the debate, and provide space to talk about the practical implications of going low carbon in the space of 30 years in a more serious fashion – including the need for a more integrated European energy system.

Notes

1 European Environment Agency, *Trends and Projections in Europe 2015: Tracking Progress towards Europe's Climate and Energy Targets* (EEA Report 4/2015) (Copenhagen: European Environment Agency, 2015).
2 Andrew Jordan, and Tim Rayner, "The Evolution of Climate Policy in the European Union: A Historical Overview," in *Climate Change Policy in the European Union: Confronting the Dilemmas of Mitigation and Adaptation?*, ed. Andrew Jordan et al. (Cambridge: Cambridge University Press, 2011).
3 "Foie Gras, Oysters and a Climate Deal: How the Paris Pact Was Won," *Climate Change News*, December 14, 2015, www.climatechangenews.com/2015/12/14/foie-gras-oysters-and-a-climate-deal-how-the-paris-pact-was-won/.
4 Benjamin Görlach, and Nils Meyer-Ohlendorf, *Energy Policy in the Constitutional Treaty: Future Options for a European Energy Policy and Implications for the Environment* (Berlin: Ecologic Institute for International and European Environmental Policy, 2003), 6.
5 Ibid., 13.
6 European Union, "Treaty of Lisbon Amending the Treaty on European Union and the Treaty Establishing the European Community" (OJ C 306), 2007, http://eur-lex.europa.eu/legal-content/EN/TXT/PDF/?uri=OJ:C:2007:306:FULL&from=EN.
7 Sebastian Oberthür, and Marc Pallemaerts, "The EU's Internal and External Climate Policies: An Historical Overview," in *The New Climate Policies of the European Union: International Legislation and Climate Diplomacy* (Institute for European Studies 15) (Brussels: VUB Press and Brussels University Press, 2010), 27.
8 Christian Hey, "EU Environmental Policies: A Short History of the Policy Strategies," in *European Union Environmental Policy Handbook: A Critical Analysis of EU Environmental Legislation*, ed. Stefan Scheuer (New York: International Books, 2006).
9 Pamela M. Barnes, "The Role of the Commission of the European Union: Creating External Coherence from Internal Diversity," in *The European Union as a Leader in International Climate Change Politics*, ed. K. Rüdiger, W. Wurzel, and James Connelly, Routledge/UACES Contemporary European Studies 15 (London: Routledge, 2011).
10 Jordan, and Rayner, "The Evolution of Climate Policy," 53; Jos Delbeke, and Peter Vis, eds, "EU Climate Leadership in a Rapidly Changing World," in *EU Climate Policy Explained* (New York: Routledge, 2015), 12.
11 Oberthür, and Pallemaerts, "The EU's Internal and External Climate Policies," 27.
12 Frank Convery, "Origins and Development of the EU ETS," *Environmental and Resource Economics* 43, no. 3 (2009): 391–412; Benjamin Görlach, "Emissions Trading in the Climate Policy Mix: Understanding and Managing Interactions with Other Policy Instruments," *Energy & Environment* 25, no. 3 (2014): 733–750, doi: 10.1260/0958-305X.25.3-4.733.
13 Barnes, "The Role of the Commission," 44.

14 At their spring meeting of that year, EU Heads of State and Government agreed on a specific formula of three targets of 20% by 2020, which had been presented by the European Commission at the beginning of the same year. European Commission, *Limiting Global Climate Change to 2 Degrees Celsius – The Way Ahead for 2020 and Beyond* (COM(2007)2) (Brussels: European Commission, 2007). This move was intended to send an early signal about the EU's resolve and ambition also to non-European partners in the international negotiations, which were moving towards discussing a framework for the time after 2012, when the original commitment period of the Kyoto Protocol ended. At the time, European leaders made an explicit link to the adoption of a broad and ambitious global climate deal and put forth the option of raising their greenhouse gas target to 30% in case such an agreement was concluded. European Council, *Presidency Conclusions 8/9 March 2007*, 2007, http://register.consilium.europa.eu/pdf/en/07/st07/st07224-re01.en07.pdf.
15 European Environment Agency, *Trends and Projections in Europe 2015*, 17.
16 European Council, *Presidency Conclusions 8/9 March 2007*; Barbara Schlomann, and Wolfgang Eichhammer, "Interaction between Climate, Emissions Trading and Energy Efficiency Targets," *Energy & Environment* 25, no. 3–4 (2014): 709–32, doi: 10.1260/0958–305X.25.3–4.709.
17 European Environment Agency, *Trends and Projections in Europe 2015*.
18 Claire Dupont, and Sebastian Oberthür, "Decarbonization in the EU: Setting the Scene," in *Decarbonization in the European Union: Internal Policies and External Strategies* (Basingstoke: Palgrave Macmillan, 2015), 4.
19 Intergovernmental Panel on Climate Change [IPCC], "IPCC Fourth Assessment Report: Climate Change 2007: Working Group II: Impacts, Adaptation and Vulnerability (Glossary A–D)," IPCC, 2007, www.ipcc.ch/publications_and_data/ar4/wg2/en/annexessglossary-a-d.html.
20 European Commission, *A Roadmap for Moving to a Competitive Low Carbon Economy in 2050*, 2011, http://eur-lex.europa.eu/LexUriServ/LexUriServ.do?uri=COM:2011:0112:FIN:EN:PDF.
21 European Council, *Presidency Conclusions 15265/1/09*, 2009; Dupont, and Oberthür, "Decarbonization in the EU"; Severin Fischer, and Oliver Geden, *Updating the EU's Energy and Climate Policy New Targets for the Post-2020 Period* (Berlin: Friedrich-Ebert-Stiftung, International Policy Analysis, 2013).
22 Severin Fischer, "The EU's New Energy and Climate Policy Framework for 2030: Implications for the German Energy Transition," *SWP Comments* 55 (Berlin: Stiftung Wissenschaft und Politik, 2014).
23 European Commission, *Climate Change* (Special Eurobarometer Report 409) (Brussels: European Commission, 2014).
24 Fischer, and Geden, *Updating the EU's Energy and Climate Policy*.
25 Nils Meyer-Ohlendorf, Matthias Duwe, Katharina Umpfenbach, et al., *The Next EU Climate and Energy Package: EU Climate Policies after 2020* (Berlin: Ecologic Institute, 2014).
26 Fischer, "The EU's New Energy and Climate Policy Framework," 3.
27 Camilla Bausch, Benjamin Görlach, and Michael Mehling, "Ambitious Climate Policy through Centralization? Evidence from the European Union," *Climate Policy* (forthcoming).
28 Jordan, and Rayner, "The Evolution of Climate Policy," 60; Delbeke, and Vis, "EU Climate Leadership in a Rapidly Changing World," 12; Convery, "Origins and Development of the EU ETS," 393.
29 Andrew Jordan, Dave Huitema, Tim Rayner, and Harro van Asselt, "Governing the European Union: Policy Choices and Governance Dilemmas," in *Climate Change Policy in the European Union: Confronting the Dilemmas of Mitigation and Adaptation?* (Cambridge: Cambridge University Press, 2010), 37.
30 Jordan, and Rayner, "The Evolution of Climate Policy," 65.
31 Jos Delbeke, Ger Klaassen, and Stefaan Vergote, "Climate-Related Energy Policies," in *EU Climate Policy Explained*, ed. Jos Delbeke, and Peter Vis (New York: Routledge, 2015), 62.
32 Ibid., 61.
33 European Environment Agency, *Trends and Projections in Europe 2015*, 98.
34 Eurostat, *EU Energy in Figures: Statistical Pocketbook 2015* (Luxembourg: Publications Office of the European Union, 2015), 23.
35 Camilla Bausch, Ennid Roberts, Lena Donat, et al., *European Governance and the Low-Carbon Pathway: Analysis of Challenges and Opportunities Arising from Overlaps between Climate and Energy Policy as Well as from Centralisation of Climate Policies* (CECILIA 2050 Project Deliverable D 4.2) (Berlin: Ecologic Institute, 2015), 62; Severin Fischer, and Oliver Geden, "Limits of an 'Energy Union': Only Pragmatic Progress on EU Energy Market Regulation Expected in the Coming Months," *SWP Comments 28* (Berlin: Stiftung Wissenschaft und Politik, 2015).
36 Fischer, "The EU's New Energy and Climate Policy Framework."

37 Jakob Schlandt, "Germany's Energy Transition in the European Context," *Clean Energy Wire*, June 25, 2015, www.cleanenergywire.org/dossiers/germanys-energy-transition-european-context.
38 Ibid.
39 Fischer, "The EU's New Energy and Climate Policy Framework," 2.
40 Jon Birger Skjærseth, *Implementing EU Climate and Energy Policies in Poland: From Europeanization to Polonization?* (8/2014) (Lysaker, Norway: Fridtjof Nansen Institute, 2014).
41 European Commission, *A Roadmap for Moving.*
42 Skjærseth, *Implementing EU Climate and Energy Policies*, 5.
43 Dupont, and Oberthür, "Decarbonization in the EU," 2.
44 European Environment Agency, *Trends and Projections in Europe 2015*, 43.
45 Stefan Lechtenböhmer, and Sascha Samadi, "The Power Sector: Pioneer and Workhorse of Decarbonization," in *Decarbonization in the European Union: Internal Policies and External Strategies*, ed. Claire Dupont, and Sebastian Oberthür (New York: Macmillan Press, 2015), 61.
46 Frank Sensfuß, Mario Ragwitz, and Massimo Genoese, "The Merit-Order Effect: A Detailed Analysis of the Price Effect of Renewable Electricity Generation on Spot Market Prices in Germany" (working paper, Sustainability and Innovation No. S 7/2007, Karlsruhe, 2007).
47 Delbeke, Klaassen, and Vergote, "Climate-Related Energy Policies," 63.
48 Schlandt, "Germany's Energy Transition."
49 Bausch et al., *European Governance and the Low-Carbon Pathway*, 31.
50 Tom Casier, "The Geopolitics of the EU's Decarbonization Strategy," in *Decarbonization in the European Union: Internal Policies and External Strategies*, ed. Claire Dupont, and Sebastian Oberthür (New York: Macmillan Press, 2015), 161; European Commission, *European Energy Security Strategy* (COM (2014) 330) (Brussels: European Commission, 2014), 2.
51 Jordan, and Rayner, "The Evolution of Climate Policy," 76; Dupont, and Oberthür, "Decarbonization in the EU," 4.
52 Delbeke, Klaassen, and Vergote, "Climate-Related Energy Policies," 62.
53 European Commission, *In-depth Study of European Energy Security* (SWD(2014) 330 final/3) (Brussels: European Commission, 2014), 44.
54 Casier, "The Geopolitics of the EU's Decarbonization Strategy," 163.
55 European Commission, *European Energy Security Strategy.*
56 European Commission, *State of the Energy Union 2015* (COM(2015) 572) (Brussels: European Commission, 2015).
57 European Commission, *Energy Union Package: A Framework Strategy for a Resilient Energy Union with a Forward-Looking Climate Change Policy* (COM(2015) 80) (Brussels: European Commission, 2015).
58 Fischer, and Geden, "Limits of an 'Energy Union,'" 2.
59 Donald Tusk, "A United Europe Can End Russia's Energy Stranglehold," *Financial Times*, April 21, 2014, www.ft.com/cms/s/0/91508464-c661-11e3-ba0e-00144feabdc0.html#axzz3xnVfjM8P.
60 Katharina Umpfenbach, *Streamlining Planning and Reporting Requirements in the EU Energy Union Framework: An Opportunity for Building Consistent and Transparent Strategies* (Berlin: Ecologic Institute, 2015), 8.
61 Fischer, and Geden, "Limits of an 'Energy Union,'" 2.
62 European Commission, *Energy Union Package.*
63 Umpfenbach, *Streamlining Planning and Reporting Requirements*, 8.
64 Ibid.
65 European Commission, *State of the Energy Union 2015.*
66 "Foie Gras, Oysters and a Climate Deal," *Climate Change News.*
67 Casier, "The Geopolitics of the EU's Decarbonization Strategy," 167.

7

Regional coordination in energy systems and its impact on energy security

Julia Nesheiwat

Regional coordination in energy systems is not a zero-sum game, but can benefit the national interest of all parties. There are challenges and risks involved, but given the mutual problems of climate change and energy security, there is good reason for countries to cooperate. The integration of regional energy markets increases diversification of supply and delivery, cost-savings, and energy efficiency. Working together, countries can reduce climate change effects while promoting sustainable energy solutions and global energy security for the future.

Regional integration of electricity markets and renewables

Renewable energy is a key component of plans to reduce carbon emissions. However, renewable sources such as wind and solar must supply large geographic areas to be cost effective due to variable weather conditions and the difficulties of electricity storage. In the coming years, the issue of energy storage may play a less significant factor due to emerging powerwall technologies.[1] Until that time, though, such constraints must be taken into account. Electricity markets in some regions are stressed, and integration can improve reliability and costs by facilitating demand aggregation and a broader range of power plants. A complementary mix of generation capacity and differing peak usage times in neighboring countries promotes the utilization of the cheapest and cleanest generating sources whenever available, while ensuring an uninterrupted supply.[2]

Sharing of energy knowledge and technologies

With particular emphasis on renewables, energy efficiency, and clean technologies, the sharing of such knowledge amongst countries is needed to help promote responsible energy policies. Given that a state's greenhouse gas emissions can negatively affect the climate shared by neighbors, it is beneficial for all to share the knowledge that can help reduce pollutants.[3]

Challenges to regional integration

Energy security risks make some countries reluctant to embrace the import and cross-border trade of electricity. Dysfunction between system operators or geopolitical changes could result in blackouts. Thus, some governments still prefer to generate power locally rather than import at cheaper costs. A lack of transmission lines and inefficiencies in current infrastructure are a further constraint. There is need for a standardized regulatory framework and government commitment. Moreover, some countries that benefit from a resource endowment of cheap energy may be reluctant to integrate their markets due to the potential price disruptions.[4]

Regional coordination in Africa

In 2011 the UN began the Sustainable Energy for All initiative (SE4ALL). Africa had 42 of its countries opt into the initiative, which emphasized expansion of energy access to universal levels, increasing energy efficiency, and doubling the global energy mix's portion of renewables.[5] These objectives are especially critical for the continent, which has 600 million inhabitants who have no access to electricity. Even of those that do, 700 million must rely on solid fuels for basic electricity needs.[6] Such numbers are even more shocking when the potential of the continent is considered, as it has a largely untapped renewable energy capability. Due to the huge benefits involved for Africa and its overwhelming participation in the initiative, the African Development Bank hosted in May 2013 the SE4ALL Africa Hub, to manage the program's implementation throughout the continent. The Hub is supported by the Sustainable Energy Fund for Africa (SEFA) and the Pilot Africa Climate Technology and Finance Center.[7] SE4ALL is not the only energy initiative for Africa. Many have come up recently, such as the African Renewable Energy Initiative. While such a surge in efforts appears to be a positive step, it underscores the need for regional coordination even more. The African Development Bank's Kurt Lonsway has tried to stress this importance by stating, "There is a need for a coordination framework that is able to ensure coherence and avoid duplication of effort, and maximize reach and impact across the continent."[8] To facilitate this regional coordination, the Africa-EU Energy Partnership has begun mapping initiatives to maximize the effectiveness of all programs.

Regional coordination in South America

Energy resources in South America are vast and vary greatly, yet many countries struggle to maintain an adequate energy supply. Though it would greatly benefit from regional coordination, there has been reticence due to political differences as well as a mismatched vision of the energy future within the region.[9]

Countries with massive fossil fuel deposits, such as Venezuela, appear content with their current energy mix, which is heavily reliant on major CO_2 emitters. Bolivia, Ecuador, and Peru cumulatively invested $40 million in fossil fuel subsidies. Others, such as Uruguay, envision a greener future and have invested much of their energy sector in renewables. It is expected that demand for energy in the region will double by 2030, and a coordinated effort will be the best chance it has to meet consumer needs. For years, it has been suggested that the best approach to regional coordination would be a South American Energy Treaty. The South American Union South American Energy Council (UNASUR) has made several attempts to draft a treaty, but issues, such as the energy mix, have been too divisive for them to make progress.[10]

Regional coordination in Europe

To achieve an integrated, interconnected, resilient, secure, and sustainable energy market system, the European Union has plans for an "Energy Union."[11] In 2006, the European Regulators Group for Electricity and Gas (ERGEG) introduced seven regional electricity markets to start the electricity market integration process. By 2014 the power grids in southern and northwestern Europe were also linked, accounting for an annual consumption of roughly 2400 TWh to cover approximately 70% of European customers. Slovakia, the Czech Republic, Hungary, and Romania likewise plan to link to the rest of Europe. An underwater line to Sweden also connects Poland with the northwestern region of Europe. Leaders of the EU expect energy trading through electricity connectors to increase 10% by 2020. Its renewables and energy market integration cost €94 billion for the upgrades and new power lines. The EU found 100 power bottlenecks that presented a challenge to integrating national grids, with 80% relating to renewables such as wind and solar. Interconnection goals for each country equate to 10% of generating capacity to achieve the trans-EU electricity infrastructure. By 2020, the common EU goal is a 20% reduction in greenhouse gas emissions, a 20% renewables share in energy consumption, and a 20% improvement in EU energy efficiency.

Connected electricity and energy markets reduce costs, emissions, and improve supply and interdependency. Ukraine, Georgia, and Belarus are important transit countries in the region, and their stability is paramount to energy security. However, Russia's annexation of Crimea has made it apparent that new transit countries are needed.

Turkey: coordinating multiple regions

Crimea is located in Ukraine. However, it was annexed by Russia after Ukraine underwent a change in government due to mass riots on March 18, 2014.[12] This move was met with sanctions and condemnations from the international community. It also damaged Ukrainian–Russian relations. The Russian annexation of Crimea and subsequent political ramifications have led to the exploration of Turkey as a possible new transit hub. While Turkey lacks abundant fossil fuel reserves, its position in Europe's southern corridor and Asia gives it the potential to draw from many oil producers, while bypassing Russia.[13]

At the time of invasion, Russia supplied 31% of Europe's oil and Europe accounted for 71% of its crude oil exports.[14] While some European countries are less dependent on oil, such as France, others, such as Germany, are highly dependent on it to meet energy needs. Oil to Western Europe from Russia was mostly transferred via pipelines in Ukraine. With the new developments, though, the stability of that oil supply was called into question. Further, it gave pause to many countries who are dependent on Russian oil but disagree with Russian policy. This left an opportunity in the market to supply oil to Europe.

Turkey has been steadily rising as an energy transporter to Europe, largely due to Europe's shift away from Russia. Russia supplies 3% of the oil that Turkey imports. However, that is down from 12% in 2010. The crisis in Syria, which has created tension between the two countries, could be seen as a motivating factor for this change. Iran used to supply most of Turkey's oil. Its share, though, has been reduced from 51% to 26%.[15] This is because Iran and Turkey have an increasingly unfriendly relationship. Fearing what would happen if the Kurdistan Workers Party (PKK), a Marxist Kurdish group within Turkey that is designated as a terrorist organization, stops fighting in Turkey and the Kurdish struggle were to focus on the Kurds in Iran, Iran has given weapons and encouragement to the PKK. Regionally, Turkey and Iran have been on opposing sides on important regional issues like the Syrian Civil War. Much of this has to do with the increasing importance of the Sunni-Shia divide in the Middle East, as Iran is Shia and Turkey is Sunni and it causes them to pursue opposing policy. They are also both competing to be major regional powers. Further, due to sanctions, Iranian oil is more difficult to export.[16] In order to reduce dependence on Iran and Russia, Turkey has looked to other oil suppliers in recent years, which is in line with Europe's diversification goals.

Two major pipelines have contributed to Turkey's ability to move away from Russian and Iranian oil. First, the Baku-Tbilisi-Ceyhan (BTC) pipeline has allowed Turkey to import oil from Azerbaijan and Kazakhstan. While it only accounts for a small portion of Turkey's oil imports, it is still significant. Next is Turkey's pipeline with Iraq. It has two branches. Its first branch is the Kirkuk-Ceyhan pipeline, which is administered by the Iraqi government. The second line is connected to the Kirkuk-Ceyhan pipeline, but begins in Taq Taq oilfields and is overseen by the Kurdish Regional government. This pipeline significantly reduced the need for Iranian oil and helped Iraq overtake Iran as the main exporter of oil to Turkey. These instances of regional energy coordination though, have not been without their complications.[17]

First, there are some obstacles to the success of the Kurdish pipeline. Disputes with the Iraqi government over profit sharing and the legality of the pipeline itself caused it to have a slow initial start after it became active in 2013. However, it was a major contributor to Turkish oil imports this past year, mostly due to a deal struck in December 2014 between the KRG and Iraqi government.[18] Much of that, though, was born of necessity.

Two important developments have occurred recently: the rise of Daesh and the escalation of hostilities between the Turkish government and the PKK. Daesh's seizure of portions of Iraqi territory has greatly disrupted the flow of oil from the main Iraqi pipeline since 2014. Turkey's pipeline with Iraqi Kurdistan was able to offset that disruption. However, rising antagonism between the PKK and Turkey has led the PKK to sabotage the Iraqi–Kurdish pipeline. It also puts the safety of other pipelines in jeopardy, as many of Turkey's pipelines flow through its Kurdish areas, leaving them vulnerable to PKK attacks.[19]

These issues are notable, but not insurmountable. They do, however, highlight a major challenge of regional energy coordination, which is its interaction with regional politics, especially in areas that suffer from instability. Even so, opportunities present themselves in such situations that further reinforce the need for regional coordination. Turkey's newfound stature as an alternative transit hub incentivizes it to make political decisions that maintain or advance its position. One of the decisions it may be forced to grapple with is how to deal with its eastern question, possibly pushing it to seek a peaceful resolution with the PKK. Though no recent behavior has indicated that will be the case, the opportunity is there, and it would be an instance of how regional energy coordination can affect security issues outside of the energy sector.

Oil is not the only resource Turkey can enable Europe to diversify on. Natural gas is another fossil fuel that Turkey could potentially act as a transit hub for through increased regional energy coordination. Currently, Europe remains reliant on Russia for its natural gas needs. Turkey's geography opens the possibility of an influx of natural gas from the Middle East and Central Asia instead.

In 2009, under the title "Southern Corridor-New Silk Road," the European Commission (EC) launched the Southern Gas Corridor (SGC) policy initiative to diversify its energy away from dependence on Russia.[20] In March 2015, construction began on a key portion of the SGC, which is the Trans-Anatolian Pipeline (TANAP).[21] By the time of completion, the pipeline is expected to move 16 billion cubic meters (bcm) of gas per year. The gas from the pipeline is projected to reach Turkey in 2018 and Europe by 2019, after the completion of the Trans Adriatic Pipeline (TAP).[22] Currently, Azerbaijan is the sole supplier of natural gas through the pipeline, but the opportunity for others throughout the Middle East and Central Asia to contribute is open.[23] If more suppliers are not added and the project is not expanded beyond the projected 16 bcm of gas, the SGC will do little to diversify Europe's gas consumption. Improved regional cooperation would bring on board more suppliers and allow the SGC to reach its diversification goal. Further, it would be a mutually beneficial situation for all parties. Turkey would be an important transit hub, Middle Eastern and Central Asian countries would be able to expand their export market, and Europe would have more autonomy over its affairs vis-à-vis Russia. Expansion of suppliers is easier said than done, however. Some countries, like Turkmenistan, can be thrown into the mix more immediately. Others, such as Kazakhstan and Iran, have extenuating political conditions surrounding their export of natural resources. Emerging sources of natural gas, such as Israel, Cyprus, and Iraq, still need to be established and have the added hurdle of complicated political problems.[24] Again, the great benefits of regional energy coordination can be seen with the SGC, as can the barriers for progress associated with the pursuit of widespread collaborations.

Regional coordination in Southeast Asia: prospects of the Trans-ASEAN Gas Pipeline (TAGP)

As Southeast Asian economies and populations grow, regional energy demand will likely increase 80% by 2035 from 2011 levels.[25] TAGP is a regional effort to respond to this energy demand by interconnecting electricity and natural gas networks in ASEAN countries, with the end goal of increasing regional energy security and sustainability. Thus far multilateral pipelines have not been constructed, but 13 bilateral links spanning 3,631 kilometers have been built. The rising interest in LNG in recent years has put the TAGP's relevance into question. Following a re-evaluation in 2012, the project now includes LNG regasification terminals (RGTs).[26]

Benefits of regional cooperation on TAGP include greater energy security, diversification, and less reliance on coal. Working together as a region, individual countries will be less vulnerable to volatile markets. Further they will receive the benefits of solidarity as a larger energy consumer and will be able to pool resources for the exploration and development of untapped resources.

Challenges and drawbacks to the project include the enormous infrastructure and development costs and inherent difficulty of harmonizing regulations, standards, and legal frameworks.[27] Moreover the wide diversity of regime types and market structures among ASEAN countries will hinder coordination and add the risks of a potentially unstable geopolitical situation.

Regional coordination in the global context: COP21

Regional coordination is extremely important, not just for the immediate benefits it provides to countries, but because it is the foundation of international progress. Climate change is a global issue, and to be tackled properly everyone must be involved. It is easiest to approach when countries are already involved in some kind of external framework to limit moving parts.

In December 2015, the 21st session of the Congress of Parties (COP21) unanimously agreed to prevent global temperatures from rising up from pre-industrial times more than 2 degrees Celsius. Further, they seek to limit it to 1.5 degrees, if possible. A long-term goal of emitting no more GHG than can be absorbed naturally by trees, oceans and oil has been set to be accomplished between 2050 and 2100. Every five years a review will take place to evaluate whether or not countries hold true to their pledge. Additionally, developed countries will provide monetary aid to developing countries to enable them to cope with climate change. Criticisms of the deal exist.[28] Many take issue with the fact that it is not legally binding. Some view the goal of limiting climate change to 2 degrees Celsius while endeavoring to cap it at 1.5 degrees Celsius as not ambitious enough. Developing nations have voiced concern that the monetary aid being pledged is insufficient, while some developed countries were reluctant to provide even that amount.[29] Even with these concerns, a global climate change initiative that was unanimously passed, involving 195 countries, is a major step in global reform. It is also a larger scale reminder of what regional coordination is key. In order to meet the objectives laid out by COP21, a lot of regional coordination will have to be implemented so that countries are given the tools to meet their pledges.

One example of regional coordination that will most likely occur on the heels of COP 21 is the re-emergence of Japan as an energy leader to enable emerging economies to implement their pledges. In its submitted pledge, Japan affirmed its commitment to its Joint Crediting Mechanism (JCM). Japan's JCM enables Japan to develop greener energy technologies and distribute them to emerging economies. This could enable countries throughout the region, and beyond, to enhance their energy security while satisfying their role in the international framework.

Concluding remarks

Regional energy coordination is an integral part of efforts to mitigate climate change, increase the global energy supply, and increase energy efficiency while reducing costs. It will not be easy. There are risks associated with undertaking such an endeavor, and they should be acknowledged. However, the payoffs of regional cooperation will only increase in the future, and outweigh the associated concerns. Thereby, regional coordination should be encouraged and supported whenever feasible.

Notes

1 Stratfor, "Climate Agreement Will Only Hasten Transition beyond Oil," Stratfor, December 13, 2015, www.stratfor.com/analysis/climate-agreement-will-only-hasten-transition-beyond-oil.
2 Manuel Baritaud, and Dennis Volk, *Seamless Power Markets: Regional Integration of Electricity Markets in IEA Member Countries* (International Energy Agency, 2014), www.iea.org/publications/freepublications/publication/seamless-power-markets.html, 8–9, 14–17.
3 Nigel Lucas, "Energy Security in Asia: Prospects for Regional Cooperation" (ADB Economics working paper no. 407) (Asian Development Bank, 2014), www.adb.org/publications/energy-security-asia-prospects-regional-cooperation, 19–20.
4 Ibid., 23.
5 African Development Bank Group, "Sustainable Energy for All," African Development Bank Group, www.afdb.org/en/topics-and-sectors/initiatives-partnerships/sustainable-energy-for-all-se4all/.
6 African Development Bank Group, "Lighting Up Africa Requires Coordination in Energy Initiatives," African Development Bank Group, October 12, 2015, www.afdb.org/en/news-and-events/article/lighting-up-africa-requires-coordination-in-energy-initiatives-15196/.
7 African Development Bank Group, "Sustainable Energy for All."
8 African Development Bank Group, "Lighting Up Africa."
9 Adilson De Oliveira, "Energy Security in South America: The Role of Brazil" (International Institute for Sustainable Development, 2010), www.iisd.org/library/energy-security-south-america-role-brazil, i.
10 Katell Abivan "Latin America Divided between Green Energy and Oil," *Phys.Org*, April 1, 2015, http://phys.org/news/2015-04-latin-america-oil-green-energy.html.

11 European Commission, "Cornerstones of the New EU Energy Union," European Commission, March 13, 2015, http://ec.europa.eu/commission/2014-2019/sefcovic/announcements/cornerstones-new-eu-energy-union_en.
12 Chris Arnold, "How Russia's Annexation of Crimea Could Hurt its Economy," *NPR Blog*, March 26, 2014, www.npr.org/blogs/parallels/2014/03/26/294877200/how-russias-annexation-of-crimea-could-hurt-its-economy.
13 Emil Suleimonov, and Josef Kraus, "Turkey: An Important East-West Energy Hub," Middle East Policy Council, April 14, 2014, http://mepc.org/journal/middle-east-policy-archives/turkey-important-east-west-energy-hub.
14 Guy Chazan, and Ed Crooks, "Europe's Dangerous Addiction to Russian Gas Needs Radical Cure," *Financial Times*, April 3, 2014, www.ft.com/cms/s/0/dacfda08-ba64-11e3-8b15-00144feabdc0.html#slide0.
15 US Energy Information Administration (USEIA), "Turkey," US Energy Information Administration, August 6, 2015, www.eia.gov/beta/international/analysis.cfm?iso=TUR.
16 Orhan Coskun, "Iraq Minister Sees Deal Soon with Kurds on Oil Exports," Reuters, April 9, 2014, www.reuters.com/article/2014/04/09/iraq-turkey-oil-idUSL6N0N11Z820140409.
17 USEIA, "Turkey."
18 US Energy Information Administration (USEIA), "Iraq," USEIA, www.eia.gov.
19 Ibid.
20 Friedbert Pfluger, "The Southern Gas Corridor Finally Becomes Reality," *Caspian Report* (Fall 2013), www.hazar.org.
21 Robert M. Cutler, "The Role of the Southern Gas Corridor in Prospects for European Energy Sector," *Caspian Report* (Winter 2014), www.hazar.org.
22 Gareth M. Winrow, "Final Investment Decision for Shah Deniz II Boosts Prospects for Southern Gas Corridor," *Caspian Report* (Winter 2014), www.hazar.org.
23 Mubariz Kasanov, "Some Remarks on Economic Benefits of Tanap for Turkey," *Caspian Report* (Spring 2014), www.hazar.org.
24 Aura Sabadus, "Southern Gas Corridor and the Potential for Genuine Diversification," *Caspian Report* (Spring 2014), www.hazar.org.
25 International Energy Agency, "A Pipeline Alternative to Asian LNG," *IEA Journal* 7, no. 4 (2014), www.iea.org/ieaenergy/issue7/a-pipeline-alternative-to-asian-lng.html.
26 ASEAN Council on Petroleum, "Trans ASEAN Gas Pipeline Project (TAGP)," ASEAN Council on Petroleum, www.ascope.org/projects.html.
27 Tilak K. Doshi, "ASEAN Energy Integration: Interconnected Power and Gas Pipeline Grids," in *Enhancing ASEAN's Connectivity*, ed. Sanchita Basu Das (Singapore: Institute of Southeast Asian Studies, 2012), 144–145.
28 Helen Briggs, "Global Climate Deal: In Summary," *BBC World News*, December 12, 2015, www.bbc.com/news/science-environment-35073297.
29 Andrew Restuccia, "The One Word that Almost Sunk the Climate Talks," *Politico*, December 13, 2015, www.politico.eu/article/one-word-almost-sunk-climate-talks-legally-binding-cop21-deal-global-warming/.

Part II

Energy transitions in the carbon producing countries

8

In the furnace

Saudi Arabia and the dynamics of global climate change

James A. Russell

Saudi Arabia sits the middle of the world's climate furnace – there are few hotter, drier places on the planet. It's only going to get even hotter and drier throughout Saudi Arabia and the Middle East over the rest of the century as the world continues to dump carbon into the atmosphere. Since 1995, the world's atmosphere has seen carbon amounts increase from 360 parts per million to an estimated 400 parts per million by 2015. The world's atmosphere has never had more carbon in it.[1] Some researchers estimate an increase in temperature of 3 degrees Celsius throughout the Middle East by 2050. According to climate change researchers, the Arabian Peninsula eventually will become too hot for people to remain outdoors for more than six hours at one time. Writing in the journal *Nature Climate Change*, Jeremy Pal and Elfatih Elfatir categorically state "... by the end of the century certain population centres in the same region are likely to experience temperature levels that are intolerable to humans owing to the consequences of increasing concentrations of anthropogenic greenhouse gases (GHGs)."[2]

As if on cue, other "canary in the coal mine" indicators have emerged pointing to inexorable climate change trends in Saudi Arabia and the wider Middle East. As noted by *New York Times* columnist Tom Friedman, a heat index temperature of 163 degrees Fahrenheit was reported in the Iranian city of Bandhar Mahshar on July 31, 2015, described by a weatherman at the time as "one of the most extreme readings ever in the world."[3] Meanwhile, in the midst of its war with the Islamic State, Iraqi citizens in Baghdad rose up in spontaneous protest at the inability of its government to deliver enough electricity to keep the city's air conditioning humming to deal with the intolerable heat. As noted in a poignant report describing every-day life in Baghdad during the summer of 2015, "the lucky ones drive around in their cars with the air conditioning on, visit shopping malls, or wait for the air coolers to switch on and huddle around them in a single room. Those without that wherewithal find cool where they can, sometimes swimming in dirty, sewage-tainted pools and canals."[4]

Elsewhere in September 2015, Israel experienced its worst sandstorm since records started being kept – a storm that was almost certainly made worse by abandoned farmland in Syria as a result of an ongoing drought and debilitating civil war.[5] During the storm, air pollution in Jerusalem reached 173 times the national average and power usage in the country broke all national records as Israelis tried to keep cool in the searing heat.[6] The same storm produced high winds and torrential rains in Mecca, Saudi Arabia and undoubtedly played a role in the collapse of a crane that killed 107 and injured 238.[7] Elsewhere in the region, Iran remained in the grip of a seven-year drought as reservoirs throughout the country sank to new lows.[8]

It would be easy to think of these extreme cases as scenes created for Hollywood movies, but in fact they will become the future norm, regardless of the December 2015 Paris Agreement. These summer 2015 snapshots provide a view into the challenges that await Saudi Arabia and the wider Middle East as the world inevitably gets hotter and its ecosystems become ever more stressed. The tidal wave of looming environmental stresses adds yet another systemic factor to a region already being torn asunder by four civil wars, failed states, dangerous regional balance of power politics, militant Islamic extremism, massive displaced refugee populations, military interventions by outside states, and oppressive governments seeking to rein in their citizens' demands for different forms of governance. All in all, it's a scenario for the perfect storm of a wider and destabilizing long-term regional crisis. Indeed, the world may be seeing only the opening phase of that crisis that promises to get progressively worse as the political and environmental stresses converge and multiply over the rest of the century. The storm will indeed break in Saudi Arabia and the Middle East if it has not already.

This chapter examines the case of Saudi Arabia as the world attempts to systemically address the issue of climate change and the associated challenges of environmental stress as it seeks to control the release of carbon into the world's atmosphere. As noted at the outset, while Saudi Arabia sits in one of the hottest pieces of real estate in the world, it similarly sits in the hot seat of the global debate over how to control the release of carbon into the world's atmosphere. There can be no effective global accord to limit carbon emissions without the agreement of Saudi Arabia (and its Gulf State neighbors) – an agreement that the Saudis understandably show little enthusiasm for. The chapter will address the dimensions of the environmental changes unfolding on the Arabian Peninsula and the wider Middle East, Saudi Arabia's strategic dilemma in responding to these environmental stresses, the politics of climate change and energy markets, and the implications of these issues for regional security and stability.

The gathering storm

To state the obvious, Saudi Arabia is located in the Middle East and is subjected to the same environmental stresses that are common to the region. Indeed it is hard to think of many environmental stresses that aren't present there. As poignantly noted by researchers writing in 2009: "the Arab countries are in many ways among the most vulnerable in the world to the potential impacts of climate change, the most significant of which are increased average temperatures, less and more erratic precipitation, and sea level rise (SLR), in a region which already suffers from aridity, recurrent drought and water scarcity."[9] These are systemic, cross-regional problems that will affect life throughout Saudi Arabia and the Middle East.

The region is already getting hotter. The year 2010 was the hottest year since data started being collected in the late 19th century, and five Arab countries set new high temperature readings in 2011.[10] It is not just going to get a lot hotter in the Middle East; it's going to become even drier in a region today that already contains less than 1% of the world's fresh water resources. The summers will get hotter and winters shorter and drier, with annual rainfall forecast to decrease by as much as 20% over the rest of the century. Some estimates suggest that annual rainfall may decline by as much as 50%.[11] All but six of the Arab countries of the Middle East suffer today from water scarcity, defined by the World Bank's minimum requirement of individuals having access to 1,000 cubic meters per year of fresh water. What little fresh water there is above ground will decline significantly by the end of the century. Higher temperatures will see a reduction of water runoff of 10% by 2050 at the same time that demand for fresh water is forecast to increase by 60% over the same period.[12] Two of the region's major aboveground fresh water sources, the Jordan and Euphrates rivers, are forecast to see dramatically decreased water flows by the end of the century. Those aboveground water sources that remain will be severely stressed by pollution and decrepit public works infrastructures. Researchers estimate that several cities in the Middle East lose as much as 40% of their fresh water due to leaky pipes.[13]

The story of the region's underground sources of fresh water isn't any better. Underground aquifers are being sucked dry by growing, fresh-water hungry societies. In war torn Yemen, for example, the capital of Sanaa and its projected population of over 4 million may exhaust its ground water supplies by 2025. Some sources estimate that as much as 50% of Yemen's population of 24 million already does not have access to safe drinking water.[14] Iran is in the midst of what some describe as an epic water crisis as a result of persistent drought and disastrous water management practices. It has exhausted 70% of its groundwater supplies over the last 50 years.[15] Jordan and Saudi Arabia are sucking an estimated 9 billion cubic meters a year from Disi/Saq aquifers – water that has been underground for tens of thousands of years that has also been found to contain potentially dangerous levels of naturally produced radiation.[16] Today, an estimated 50% of Saudi Arabia's freshwater comes out of the ground. In short, the region's already scarce natural supplies of water through rainfall and aquifers are drying up.

The region is running out of fresh water at the same time as it will have to start coping with the threat of SLR, that will further stress fresh water supplies due to salt water intrusion. According to the Intergovernmental Panel on Climate Change (IPCC), the average rate of sea level increase has doubled since 1993 to an annual rate of 3.2 millimeters a year. Researchers forecast annual increases in sea level of 16 millimeters a year by 2081. By the end of the 21st century, the world's oceans may rise a total of 1 meter as areas in the Arctic and Antarctic inexorably melt.[17] Egypt is the regional state most vulnerable to SLR, although the Gulf States also will not be immune. An estimated 6% of Egypt's GDP is at risk with an SLR of 1 meter. Qatar, Kuwait, and the United Arab Emirates are also particularly vulnerable to SLR over the rest of the century. Qatar's land area could be reduced by anywhere from 2.6–13% depending on the level of SLR.[18] Saudi Arabia is somewhat less vulnerable than its GCC neighbors to this phenomenon, although it too will see its coastlines change dramatically with the rising waters.

Two systemic stresses must also be added to the cauldron of environmental problems: population growth and urbanization. For centuries, the Middle East and North Africa population hovered around 30 million, reaching an estimated 60 million early in the 20th century. Since, then the region has experienced one of the most rapid population growth rates in the world. By 1950, the region's population reached 100 million.[19] The region's population today is estimated to total between 340 and 350 million and is projected to increase to 588 million by 2050.[20] The region's population growth rates are slowing from their rates of the mid-20th century, but the bow wave of population growth will break over the region during the next quarter century.

Like the rest of the world, the Middle East is in the midst of a systemic change in the distribution of its population between rural and urban areas. Populations around the planet are leaving the countryside and moving to cities – a process over the last 60 years that has occurred at a dizzying and unprecedented pace.[21] In 2007, more people lived in cities than in the countryside for the first time in history. By 2050, the United Nations estimates that world will see one-third of its population live in rural environments, two-thirds in cities – reversing the disposition of the world's population as recorded in middle of the 20th century.[22] The movement of these populations will introduce a variety of political, economic and social stresses in societies everywhere, but particularly in the developing world. Historically, the process of urbanization is associated with sweeping political, economic, cultural, and social transformation in the affected societies. These transformations are not always peaceful.[23]

In the Middle East, an estimated 56% of the region's population currently lives in urban centers. That percentage is expected to rise to 75% by the middle of the century. Persistent drought and hotter temperatures will help speed the growth of the region's cities, as agrarian subsistence farming becomes untenable. The emptying of Syria's countryside due to the extreme drought between 2006 and 2009 is a window in the future of what awaits other societies in the region.[24] The move to the cities from the country will also see a decrease in the production of food, making regional societies ever more dependent on food imports. The shift of the region's populations from rural environments to cities will hardly be an orderly process. Indeed, newly arriving urban inhabitants will find themselves in cities with inadequate, dilapidated and aging infrastructures, overcrowded housing, and governments already struggling to provide water, electricity and basic social services. In Cairo, for example, the city's population is expected to double by the middle of the century, reaching 40 million inhabitants.[25] Life promises to be difficult in the region's urban, concrete heat islands where a minority of well off citizens will sit in air conditioned buildings while the less well off will literally bake outside.

According to the United Nations, Saudi Arabia's population is expected to grow from approximately 30 million in 2015 to 40 million by the middle of the century. Over 90% of Saudi Arabia's inhabitants will live in urban areas by 2050.[26] The populations of these urban areas are and will be overwhelmingly youthful. The World Economic Forum estimates that 60% of the Middle East and North Africa population is under the age of 25, with unemployment rates in some states for this group as high as 40%.[27] In Saudi Arabia, 37% of the population is below the age of 14 and 51% is under the age of 25. Youth unemployment is variously estimated at somewhere between 20 and 30%.[28] As a demographic, this group is generally well educated, having passed through the Kingdom's extensive educational system, with as many as 100,000 graduates annually entering the job market. By 2030, Saudi Arabia's labor market will have to accommodate 4 million additional workers. Education received in the Saudi system, however, remains overwhelmingly focused on religion and rote memorization – subject areas of little use in a diversified, private sector driven job market. That job market remains concentrated in the state sector, with what few private sector jobs there are going to foreign nationals. The International Monetary Fund estimates that 60% of Saudi Arabia's labor force totaling 3.3 million people are employed in the public sector.[29] According to the Saudi Arabia Monetary Agency, only 10% of the Saudi labor force works in the private sector.[30]

The Middle East remains systemically vulnerable to the social, political and economic forces unleashed by urbanization. The region's political systems remain controlled by security sector and familial elites with little space for more widespread popular participation and the development of civil societies. To substitute for the lack of private sector development, regional states have created bloated and inefficient government bureaucracies to provide jobs, which, in the petroleum states like Saudi Arabia, are used to distribute wealth to the citizens in exchange for their acquiescence at being ruled by the elites. For example, in Kuwait it is estimated that 90% of the work force is employed by the government. Perhaps most important, the region has not opened up opportunities for women to participate in the economic, social, political, and economic development of their respective states. In a particularly religious and tradition-bound country like Saudi Arabia, it means that the state is receiving almost no economically beneficial contribution from a significant portion of its population – an intolerable and unsustainable situation for a country entering the modern era.

Saudi Arabia's strategic dilemmas and opportunities

Saudi Arabia confronts the region's troubled strategic landscape and faces both opportunities and dilemmas as it attempts to transition to a new economic model and reduce its reliance on carbon. The central strategic problem facing the Kingdom is simply this: it depends on pumping ever increasing amounts of oil out of the ground to keep the ship of state afloat now and for the foreseeable future at a time when the world is attempting to implement a system created under the December 2015 Paris Agreement to limit carbon release that comes from burning fossil fuels. There is an inherent contradiction between trying to slow the increase of the world's temperature and Saudi Arabia's requirement to keep on pumping oil to fund its state and feed the world's growing thirst for oil. At stake is the viability of Saudi Arabia's rentier state, which is built on the premise of distributing wealth from energy sales revenues to its citizens in exchange for their acquiescence in being ruled by the House of Saud without meaningful political representation.[31] The ways in which the Kingdom addresses this contradiction will almost certainly determine the survival of the state as currently constituted and its ruling royal family.

The Kingdom is in a race against time as it simultaneously addresses a series of strategic, political, economic, and environmental challenges:

1 Build and pay for an infrastructure to accommodate its burgeoning population that includes the environmental mitigation and adaptation program.
2 Make the necessary investments and political commitment to diversify its economy away from domination by the public sector. Saudi Arabia needs the private sector to become the engine of economic growth that will provide jobs for an estimated 4 million Saudis that will enter the work force over the next 20 years. One recent study estimates the price tag of these investments could total as high as $4 trillion.[32]
3 Transition from a carbon-based economy to one in which the Kingdom's energy needs are increasingly met by renewable and cleaner energy sources to help enable it to meet its carbon reduction commitments under the Paris accord.
4 Continue to manage the politics of energy markets while simultaneously meeting its commitments to reduce carbon output in such a way that the Kingdom will maintain its relative power and influence around the world.
5 Preserve the shape and identity of the state and the ruling family's position as it manages all the above.

It is difficult to understate the magnitude of these simultaneous challenges, bringing potentially wrenching and even revolutionary changes in a society that is arguably only just beginning to enter the modern era.

It's hard to understate where Saudi Arabia would be today without oil sales revenues. Indeed, it's hard to conceive of Saudi Arabia as anything other than the quintessential petro state. Oil sales revenue today accounts for approximately 85–90% of the government budget, 90% of all export revenues, and 45% of GDP. In 2015, some estimates indicate that oil revenues generated $160 billion of government revenue in a total budget of $223 billion, with $38 billion of that total supported by deficit financing.[33] Saudi revenues are distributed to its citizens in many forms, such as government jobs, generous unemployment benefits, subsidized gasoline, water and housing, and free education and healthcare. The Kingdom spent an estimated $36 billion in 2013 on subsidies for gas, water and electricity.[34] Saudi consumers paid 45 cents a gallon for gasoline in 2015 and received electricity at a fraction of the cost paid by consumers around the world.

During 2015, Saudi Arabia's oil production totaled between 9.75 and 10.3 million barrels per day, an increase of 6% over 2014 levels, reaching a 30-year high in production levels.[35] The Kingdom is thought to have a production capacity of approximately 12.3 million barrels, 2.6 million barrels of which constitutes spare production capacity. No other oil producer has this kind of production flexibility. Saudi Arabia has historically used this capacity to smooth supply disruptions on world markets.[36] The world is expected to continue needing more oil – despite the commitments to limit carbon release. Indeed, the world's continued economic growth depends on access to reasonably priced supplies of energy. According to the US Energy Information Administration, if left unchecked, world demand for oil may reach 120 million barrels a day by 2040,[37] an increase from daily production of 95 million barrels per day in 2015.[38] If this scenario is realized, much of the additional oil needed to keep pace with global demand will come from Gulf State producers, including Saudi Arabia, where oil is much cheaper to get out of the ground than in other oil producing states.

The contradictions between the need for more oil and the need to mitigate climate change are addressed with great clarity in the scientific community. In addition to modeling forecasts about increased temperatures and SLR, recent research conclusively demonstrates that much of the world's vast fossil reserves (coal, natural gas and oil) must be kept in the ground indefinitely if the world is to have any chance of limiting the world's temperature increase to 2 degrees Celsius over the rest of the century. Figures published in the journal *Nature* indicate that as much as 260 billion barrels of oil in the Middle East would have to remain in the ground under such a scenario – roughly the size of Saudi Arabia's estimated oil reserves.[39] Such an outcome would certainly reorder the world's geostrategic map that places Gulf State oil producers at the epicenter of global energy markets, to say nothing of the associated political and economic challenges that would be created in each of the producing states if significant parts of their oil reserves remain in the ground. In the past, Saudi Arabia has demanded that it be compensated for any oil left in the ground as part of an agreement to limit global carbon release.

The strategic dilemma confronting Saudi Arabia is significant, but, importantly, it is better positioned than many of its poorer regional neighbors to withstand the previously outlined systemic pressures. Therein lies the potential for opportunity to sensibly manage the Kingdom's transition from a carbon-based economy to one with reduced fossil fuel income and output. Over the decade of 2003–2013, oil prices skyrocketed from $36 a barrel to $110 a barrel by 2011. The decade saw an unprecedented period of oil-fueled economic growth in Saudi Arabia during which time the Kingdom doubled its GDP to an estimated $750 billion (somewhat larger than Sweden and Switzerland) and became the world's 19th largest economy. GDP grew at an annual rate of 6% over the decade – one of the most rapid rates of growth in the world. Monthly household income increased by 75% from $2,100 to $3,600 and 1.7 million jobs were created (1 million of which were in the public sector) as money poured into the Kingdom's coffers. Moreover, women started to enter the Saudi work force in numbers over the period, reaching an estimated 1.8 million or 18% of the working age female population. Saudi Arabia is today regarded as a high income country with an estimated 2014 per capita GDP of $24,000.[40]

While the House of Saud has a deserved reputation for corruption and lavish lifestyles within its extended family, it must also be pointed out that the family responsibly invested oil sales revenues in the Kingdom's infrastructure during the period. Just as important, the regime also bought down the national debt and put money in the bank. It did not gamble with its money. Unlike much of the rest of the world, Saudi investments remained in conservative instruments during the boom in equities markets in the first half of the decade. As a result, during the world's 2008–2009 financial meltdown, Saudi Arabia remained in a strong financial position, buying down its debt at a time when many other countries underwent grave macroeconomic crises and had to borrow money to avert economic meltdowns. Over the decade it virtually eliminated its public debt and compiled financial reserves totaling nearly 100% of GDP. A staggering $1.6 trillion was invested in the Saudi economy during the decade, including $300 billion in foreign direct investment (mostly in the petroleum sector). McKinsey & Co. estimated that by 2015 the state had compiled $1.4 trillion in financial and other assets. Approximately $450 billion was spent in national development projects over the decade on transportation infrastructure, healthcare, the education system, and social welfare support. Electrical generation capacity was increased by 32%. Eighty-one new hospitals and 20 new universities were created during the period.[41] Mega construction projects like the Kingdom Tower in Jeddah were also launched, which, when completed, will reach 3,300 feet into the sky, making it the tallest building in the world.

In addition to infrastructure investment, the flexibility provided to the House of Saud by the stockpile of cash was dramatically illustrated during the uprisings around the region during the Arab Spring in 2011. To forestall potential unrest, the regime pumped another $130 billion into the country's economy virtually overnight in the form of higher salaries and bonuses for government workers, increased unemployment benefits, additional housing, and more mosques for the country's powerful religious establishment. An estimated $70 billion alone was spent on 500,000 low-income housing units.[42] The windfall has also provided the Al-Saud with political leverage over its neighbors. In a reversal of fortune from the bitter disputes of the 1960s Nasser era, Saudi Arabia has pumped an estimated $6.5 billion into the teetering Egyptian economy following the toppling of the Muslim Brotherhood government by the security services in 2013.

Environmental mitigation and adaptation investments figured prominently in the spending spree over the decade. Indeed, it is no exaggeration that the oil boom is funding one of the longest running and largest environmental mitigation programs ever attempted in history. At the top of the list is Saudi Arabia's investment in desalinated water production. The Kingdom consumes nearly 7 billion cubic meters a day, 50–60% of which is desalinated. Annual demand for water is growing at 8% a year. The Saline Water Conversion Corporation operates 36 plants that produce an estimated 3 trillion cubic meters of water a day, which provides 50% of the Kingdom's water and 70% of all water consumed in cities.[43] In 2014, the Kingdom brought the $7.2 billion Ras al-Khair plant into production – the world's largest desalinization plant. When completed, the plant, located northwest of Jubail, will produce 264 million gallons of fresh water per day (1.025 million cubic meters) as well as 2,600 megawatts of electricity. In 2014, the Kingdom invested an estimated $4.4 billion in desalinization projects.[44] Saudi Arabia will need an additional $53 billion in investment and an additional 20–30 new desalination plants over the next 15 years to meet projected demand for water.[45]

Unsurprisingly, Saudi Arabia will need more electricity in the years ahead. According to the Energy Information Administration, the Kingdom in 2013 began bumping up against its power generation capacity, estimated at 58.4 gigawatts. Air conditioning accounts for 70% of electricity demand, which is forecast to steadily increase, as the region gets hotter.[46] Saudi Arabia's inhabitants consume an estimated nine times more electricity per capita than their surrounding neighbors.[47] Saudi Arabia will need to generate an additional 35 gigawatts in electricity capacity over the next 15 years, requiring an investment up to $120 billion both to add to and modernize inefficient and aging turbine systems. Some estimates suggest that the Kingdom will have to double its power generating capacity over the next 20 years to 120 GW.[48] Up until now, Saudi Arabia has used oil and natural gas to power its electricity generating system, helping to fuel a steep increase in domestic energy consumption. Saudi Arabia today has one of the highest per capita energy consumption rates in the world. It currently consumes 25% of all the oil it produces, with internal consumption projected to grow at 7% annually. Saudi Arabia is expected to triple its domestic oil consumption over the next 15 years. If current trends continue, one study estimates that the Kingdom could become a net oil importer by the middle of the century. Depriving the Kingdom of its principal source of income could have potentially cataclysmic consequences.[49]

Diversification of energy sources figures into the Saudi approach. To help lead the effort, the Kingdom established the King Abdullah University of Science and Technology located on the Red Sea North of Jeddah. In May 2012, the Kingdom announced an ambitious 20-year $109 billion plan to generate 41 gigawatts of electricity through a new solar program, which would meet between 25 and 30% of its needs. In parallel, it sought to generate another 21 gigawatts in geothermal and wind power. There is virtually no human or industrial infrastructure in the Kingdom to execute and/or support a program of such magnitude. Perhaps unsurprisingly, the projects have been delayed. In January 2015, Hashim Yamani, president of the King Abdullah City for Atomic and Renewable Energy, the royal agency established to oversee renewable energy policy, delayed the projects by 8 years, stating: "We have revised the outlook to focus on 2040 as the major milestone for long-term energy planning in Saudi Arabia."[50] Under even the most optimistic of scenarios, renewable energy sources will provide no immediate relief from the Kingdom's overwhelming dependence on oil and natural gas as sources for power.

The Kingdom also has taken steps to increase energy efficiency. In December 2014, it announced the imposition of minimum fuel/mileage standards for new and used vehicles and light trucks and automobiles imported into the Kingdom. The standards, based on US Corporate Average Fuel, are to be phased in between 2016 and 2020. As part of the National Energy Phase II program launched in partnership with the UN Development Programme, the Kingdom launched a campaign to have energy efficiency ratings on all new air conditioners and replace aging and inefficient units still in use.[51] As many as 50,000 non-compliant air conditioning units have been seized and destroyed by the government under the joint government-United Nations program.[52] Last but not least, Saudi Arabia has adopted international building codes standards that make thermal insulation mandatory in all new building construction across the Kingdom's 23 cities.[53]

The politics of energy and climate change

Saudi Arabia has been seen around the world as one of the biggest impediments to a global deal to limit carbon output. During negotiations in Copenhagen in November 2014, one observer estimated that the Saudis were responsible for 40% of all the objections in the sessions.[54] Most recently, the Kingdom led a group of 22 nations that successfully lobbied nations to prevent the adoption of more aggressive climate change objectives. According to many observers, Saudi Arabia worked hard to water down the text throughout the final COP 21 negotiations in Paris in December 2015. It has also resisted pressures to contribute to a fund to help poorer developing countries shoulder the costs for climate mitigation efforts. Nonetheless, Saudi Arabia has signed on to the December 2015 Paris agreement of 195 nations.

Under terms of commitments made to the treaty as its Intended Nationally Determined Contribution, or INDC, submitted in November 2015, Saudi Arabia seeks to reduce carbon emissions by 130 tons annually by 2030. The plan offered no specifics on how its carbon reduction commitments will be met and offered no numeric benchmarks through which to judge its progress or lack thereof in meeting its targets. In 2012, Saudi Arabia was rated as the 14th largest emitter of greenhouse gases in the world, totaling an annual release of 572 metric tons, or 1.2% of the world's total.[55] The Saudi INDC submission stated: "The Kingdom will engage in actions and plans in pursuit of economic diversification that have co-benefits in the form of greenhouse gas (GHG) emission avoidances and adaptation to the impacts of climate change, as well as reducing the impacts of response measures."[56] The Saudi submission also vaguely warned that its emission targets would be adjusted between 2016 and 2020 if the Paris agreement creates an "abnormal" burden on the Saudi economy.[57] The plan acknowledges the link between climate mitigation and adaptation efforts and high levels of oil exports. In addition to pursuing economic efficiencies and diversified economic growth, the plan calls for Saudi Arabia to develop carbon capture and utilization technology. The plan calls for the Kingdom to build the world's largest capture and use plant, capable of purifying 1,500 tons of carbon per day that will be recycled back into the country's growing petro-chemical industry. Another initiative calls for the insertion of carbon into underground oil reservoirs. Using a technology called Carbon Dioxide-Enhanced Oil Recovery in a demonstration/pilot project, 40 million cubic feet of carbon dioxide will be captured, processed and injected into the Othmaniya oil reservoir as part of pilot project to determine the feasibility of the idea.

By recent standards, the Saudi submission to the Paris accord is significant, since it has been seen as one of the major roadblocks to any attempt to limit carbon release. It was accused by some observers of trying to wreck the whole deal, staunchly refusing to support more stringent actions to limit temperature increase targets to 1.5 degrees Celsius, baulking at the idea of trying to decarbonize the world's economy by the middle of the century, and demanding compensation for potential lost revenues from oil income.[58] The German organization Climate Action Tracker remained unimpressed by the Kingdom's INDC submission, referring to it as "inadequate," and among the worst plans in the world. The organization caustically stated that: "If all countries adopted this level of ambition, global warming would be likely to exceed 3–4 degrees Celsius this century."[59] The organization particularly criticized Saudi Arabia for refusing to establish any baseline levels for carbon emissions through which to measure progress and/or the lack thereof.

The politics of Saudi Arabia's approach to the Paris negotiations remain inextricably intertwined with the politics of international energy markets. As negotiators gathered in Paris to negotiate limits on carbon release, Saudi Arabia's oil production and exports reached an all-time high. In the arena of energy markets, Saudi Arabia has been playing a different sort of politics. After lining its coffers with oil averaging $110 a barrel between 2011 and 2014, prices tumbled to $50 in 2015, cutting Saudi oil revenues and those of other producing states by more than half. Breaking with its past practice of cutting production to restore price stability, the House of Saud instead opened the oil spigots, repeatedly ignoring the entreaties of its OPEC oil-producing partners. While on the one hand tumbling oil prices hurt the Kingdom's short-term revenues, lower prices serve a number of broader strategic interests. Conventional wisdom suggests that Saudi Arabia's refusal to moderate supply is a way to politically pressure oil dependent geopolitical rivals in Tehran and Moscow and to drive more expensive producers from the market to preserve Saudi market share.

Oil prices north of $100 a barrel helped stimulate a glut in world oil production in places like Canada and the United States, where it became economical to produce shale oil in ever increasing quantities. These additional oil supplies also importantly threatened Saudi Arabia's share of world oil markets, particularly in the world's fastest growing markets in Asia. By March 2015, for example, the US production of shale oil reached 5.3 million barrels per day, helping the dramatic overall increase in US oil production that reached a 40-year high of 9.4 million barrels per day in May 2015. There is some evidence that US shale oil producers are feeling the pinch, with production slated to decline in early 2016. Some analysis suggests that the Saudis are successfully preserving their market share around the world with their refusal to cut production.[60]

Pressuring high cost producers, however, represented only one objective, according to some analysts. Various analysts argued that a deeper and more profound fear drove Saudi actions – a leveling off of the global demand for oil.[61] According to this point of view, higher oil prices not only brought the more expensive oil onto the market, they also reduced demand for oil and further encouraged the move to renewable sources of energy. The slowing demand for imported oil in China was particularly concerning to Saudi officials, some of whom believe that depressing oil prices actually buys Saudi Arabia and its OPEC partners more time in averting what would be a disastrous leveling off of global demand in a higher priced market.[62]

Another line of analysis suggests that Saudi Arabia is attempting to maximize income from a commodity that will gradually recede in importance as the world transitions to a non-carbon based future. This argument suggests that the Saudis recognize that the inexorable move towards limits on carbon emissions will mean that some portion of their most valuable commodity – oil – will have to remain in the ground.[63] The essential logic of the argument is that any barrel of oil sold at a profit, however small, is more valuable than one that is not. Oil that remains in the ground eventually will cease to have value – a disastrous outcome for the Kingdom. Preserving the primacy of their market share would thus allow the Saudis to maximize the value of their commodity.

Strategic and policy implications

As noted at the outset of this chapter, Saudi Arabia sits in the middle of an already unstable region. Political and environmental challenges promise to continue multiplying throughout the Middle East over the rest of the century. Saudi Arabia will not be immune from the troubles of its neighbors whose societies are already fractured by sectarian politics and civil war. The Saudis will feel the political and strategic heat as the region's climate continues its inexorable climb up the temperature scale.

Whatever the Saudi motivations in their strategies to simultaneously manage the politics of energy markets and climate change, it is indeed a delicate balancing act with profoundly important strategic implications for the future of the Kingdom and the future of the planet. As noted at the outset of this chapter, the survival of the Saudi rentier state and the future of the house of Saud may depend on their ability to successfully perform this balancing act. The Kingdom is in a race against time in its attempt to preserve an economically and politically viable state that can survive the transition to a post-carbon future. The assessment in this chapter is that the rentier model cannot be indefinitely sustained as limits on carbon emissions become a global reality, although it is uncertain whether and/or how fast the House of Saud will take steps in this direction.

The economic pressures on the Kingdom are clear. It needs to continue pumping oil for at least the next couple of decades to generate the revenue needed for investment in infrastructure to accommodate its growing population. There is no substitute for oil income over the near term. The Kingdom needs more of everything and must figure out how to pay for it: electricity generating capacity, freshwater desalination plants, housing, public transportation and road systems, health care, and education – to name but a few of the claimants on Saudi resources. As previously noted, the Kingdom's strong financial position is a good start to meet the daunting economic challenges over the next quarter century.

Perhaps more difficult is the parallel political challenge that must also be addressed if the state is to move from the rentier model to one in which the private sector drives economic growth.

The domestic political challenges are complex and varied. If the political leadership attempts to develop a viable and vibrant private sector, it will be asking more of its citizens in a state where subsidies for energy, education, healthcare and public sector jobs are slowly but surely reduced. The work force will no longer be able to live off handouts from the regime and rely on public sector jobs, and receive religiously focused education.

A new and better-educated Saudi work force comprised of men and women will be needed if a viable private sector is to be developed to drive the Kingdom's economic growth. Restrictions on the participation of women in the political, economic, and cultural life of the Kingdom will have to be relaxed in this future. Opening up opportunities for women and changing the country's education system to give Saudi workers the skills they need will require the House of al-Saud to take on domestic political stakeholders like the religious establishment, which has been empowered and funded in the oil era.

Managing this transition as the world slowly but surely clamps down on carbon production and consumption adds another layer of complexity to an already difficult situation. Saudi Arabia has done a good job in slowing down the world's progress towards a climate accord, but the commitments made in Paris represent the opening attempt to limit emissions in the world's carbon producing countries. Saudi Arabia can survive for a time through oil revenues, but its days as a profligate welfare-spending state are slowly but surely coming to an end.

For much of the 20th century, Saudi Arabia and the Middle East served as a strategic epicenter for the West's own economic and strategic security. Cheap and readily available Saudi and Gulf State oil helped support a century of relative political stability and sustained economic growth. Saudi Arabia was the "prize" in the regional constellation, a role embraced by the House of Saud as it fashioned its security partnership with the United States to protect it from its outside enemies.[64] Outsourcing its external protection to the US suited the al-Saud as it focused on building a peaceful and stable internal political order. In some senses, the al-Saud always correctly foresaw that the most significant challenges to the state came not from external enemies but from internal ones. It remains unclear just how relevant that security system will be in the future as the challenges to the internal stability of states throughout the Middle East multiply in the years ahead. The wisdom of the House of Saud's prescient choices all those many decades ago will certainly be on trial over the next quarter century as the state confronts nothing less than a transformation of the Kingdom's society if it is to survive into the next, and hotter, century.

Notes

1 As noted by Juraj Mesik, "Arab World Vulnerable to Global Warming," *Arab News*, August 13, 2015.
2 Jeremy S. Pal, and Elfatih A. B. Elfatir, "Future Temperature in Southwest Asia Projected to Exceed a Threshold for Human Habitability," *Nature Climate Change* 6 (2016):197–200.
3 Thomas L. Friedman, "The World's Hotspot," *The New York Times*, August 19, 2015.
4 Anne Barnard, "120 Degrees and No Relief? ISIS Takes a Back Seat for Iraqis," *New York Times*, August 3, 2015.
5 Paul Rivlin, "Climate Change Has Happened: The Middle East's Climate Crisis," *Middle East Economy* (Tel Aviv University) 5, no. 10 (2015): 1–7.
6 "Sandstorm Plaguing Israel is the Worst in Country's History," *The Times of Israel*, September 11, 2015.
7 "Crane Collapse Kills at Least 107 in Mecca Grand Mosque," *Al Jazeera*, September 11, 2015.
8 Thomas Erdbrink, "Scarred Riverbeds and Scarred Pistachio Trees in a Parched Iran," *New York Times*, December 18, 2015.
9 Mostafa K. Talba, and Najib W. Saab, eds, *Arab Environment Climate Change: Impact of Climate Change on Arab Countries* (Beirut: Lebanon, 2009), 7.

10 Reported in Dorte Verner, ed., *Adaptation to Changing Climate in the Arab Countries*, MENA Knowledge and Learning: Quick Note Series no. 79 (Washington, DC: World Bank, 2013).

11 Cyprus Institute, *Climate Change and Impacts in the Eastern Mediterranean and the Middle East* (Cyprus: Cyprus Institute, 2008).

12 Verner, *Adaptation to Changing Climate in the Arab Countries*, 2.

13 Ibid.

14 Adam Heffez, "How Yemen Chewed Itself Dry," *Foreign Affairs*, July 23, 2013.

15 Erdbrink, "Scarred Riverbeds"; also see Jason Rezaian, "Iran's Water Crisis is the Product of Decades of Bad Planning," *Washington Post*, July 2, 2014.

16 Markus Becker, "Contaminated Acquifers: Radioactive Waters Threatens Middle East," *Der Spiegel*, November 5, 2012.

17 Nicholas Bakalar, "3.2 Millimeters: A Troubling Rise in Sea Level," *New York Times*, November 30, 2015.

18 Mahmoud Medany, "Impact of Climate Change on Arab Countries," 131, in Mostafa K. Tolba and Najib W. Saab, eds, *Arab Environment: Climate Change – Impact of Climate Change on Arab Countries* (Arab Forum for Environment and Development, 2009), 127–136.

19 United Nations data summarized in Population Reference Bureau, *Population Trends and Challenges in the Middle East and North Africa* (Washington, DC: Population Reference Bureau, 2001).

20 Pew Research Center, *The Future of World Religions: Population Growth Projections* (Washington, DC: Pew Research Center, 2015).

21 United Nations, *World Urbanization Prospects* (New York: United Nations, 2014).

22 Ibid., 7.

23 Donatella della Porta, and Mario Diani, *Social Movements: An Introduction*, 2nd edn. (Malvern, MA: Blackwell, 2006).

24 Henry Fountain, "Researchers Link Syrian Conflict to Drought Made Worse by Climate Change," *New York Times*, March 2, 2015.

25 Further background in Patrick Kingsley, "A New Cairo," *The Guardian*, March 16, 2015.

26 Saudi Arabia Country Profile, United Nations, Department of Economic and Social Affairs, Population Division, *World Urbanization Prospects: The 2014 Revision*.

27 Omar Kutayba Alghanim, "Solving the Problem of Youth Unemployment in the MENA Region," World Economic Forum, May 27, 2015.

28 Figures drawn from the World Bank and the United Nations cited in Stasa Salacanin, "GCC Countries and Their Diversification Efforts," *Business in Qatar and Beyond Magazine*, July 1, 2014. Also see Caryle Murphy, *Saudi Arabia's Youth and the Kingdom's Future*, Occasional Paper Series 2 (Washington, DC: Woodrow Wilson International Center for Scholars, 2011); Julia Glum, "Saudi Arabia's Youth Unemployment Problems among King Salman's Many New Challenges after Abdullah's Death," *International Business Times*, January 23, 2014.

29 International Monetary Fund (IMF), *Saudi Arabia: Selected Issues* (IMF country report 13/230) (Washington, DC: IMF, 2013). Also see figures from Jadwa Investment, *Saudi Labor Market Report – October 2015* (Riyadh, Saudi Arabia: Saudi-US Information Service, 2015).

30 2009 figures cited in John Sfakianakis, "Saudi Youth Struggle to Find Work Raises Urgency for Reform," Arab News, February 17, 2011.

31 Hazem Beblawi, "The Rentier State in the Arab World," in *The Arab State*, ed. Giacomo Luciani (Berkeley: University of California Press, 1990), 85–98; Steffen Hertog, *Princes, Brokers, and Bureaucrats: Oil and the State in Saudi Arabia* (Ithaca, NY: Cornell University Press, 2010).

32 McKinsey Global Institute, *Saudi Arabia beyond Oil: The Investment and Productivity Transformation* (London, San Francisco and Shanghai: McKinsey & Co., 2015).

33 Andrew Torchia, "Saudi 2015 Budget Based on Oil Price around $60 – Analysts," *Reuters*, December 28, 2014.

34 Figures cited by Saudi economist Turki al Haqeel as quoted in Ubaid Al Suhaimy, "Saudi Arabia: The Desalination Nation," *Asharq al-Awsat*, July 2, 2013.

35 YCharts, "Saudi Arabia Crude Oil Production Chart," YCharts, https://ycharts.com/indicators/saudi_arabia_crude_oil_production/chart/.

36 See figures in Grant Smith, "Saudi Arabia Set to Pump Maximum Crude as Battle for Market Share Heats Up," *Bloomberg News*, June 19, 2015.

37 Table A4, US Energy Information Administration (EIA), *International Energy Outlook 2014* (Washington DC: US EIA, 2014).

38 International Energy Agency (IEA), *Oil Market Report* (Paris: IEA, 2015).
39 Christopher McGlade, and Paul Ekins, "The Geographical Distribution of Fossil Fuels Unused When Limiting Global Warming to 2 Degrees Celsius," *Nature* 517 (January 2015): 187–190. Also summarized in Damian Carrington, "Leave Fossil Fuels in the Ground to Prevent Climate Change," *The Guardian*, January 7, 2015.
40 Figures drawn from a variety of sources and summarized in McKinsey Global Institute, *Saudi Arabia beyond Oil*, 19–44. The report draws data from the World Bank, International Monetary Fund, Saudi Ministry of Economy and Planning, Saudi Ministry of Labor, Saudi Ministry of Health, and the Saudi Arabia Monetary Agency, Saudi Ministry of Finance, Central Department of Statistics and Information, US Energy Information Administration, among others.
41 Ibid.
42 Neil MacFarquar, "In Saudi Arabia, Royals Buy Peace for Now," *New York Times*, June 8, 2011.
43 Al Suhaimy, "Saudi Arabia: The Desalination Nation."
44 Deema Almashabi, "Saudis Start Production at World's Biggest Desalination Plant," *Bloomberg Business*, April 23, 2014. Also see Jeffrey Ball, "Why the Saudis are Going Solar," *The Atlantic* (July/August 2015).
45 Y. Alyousef, and M. Abu-ebid, "Energy Efficiency Initiatives for Saudi Arabia on Supply and Demand," in *Energy Efficiency – a Bridge to Low Carbon Economy*, Zoran Morvaj, ed. (InTech, March 16, 2012), 279–308. Available online at www.intechopen.com/books/energy-efficiency-a-bridge-to-low-carbon-economy/energy-efficiency-initiatives-for-saudi-arabia-on-supply-and-demand-sides.
46 International Energy Agency (IEA), "Summer Demand Taxes Saudi Power Sector, but Kingdom is Working on Solutions," IEA, August 6, 2014, available online at www.iea.org/newsroomandevents/news/2014/august/summer-demand-taxes-saudi-power-sector-but-kingdom-is-working-on-solutions.html. This piece is an abridged version of a longer article that appears in *IEA Energy – The Journal of the International Energy Agency*, 7: 40–42, available online at www.iea.org/media/ieajournal/Issue7_WEB.pdf.
47 "Saudis Use Nine Times More Electricity Than Fellow Arabs," *Arab News*, February 19, 2014.
48 Abdulrahman Al Ghabban, *Saudi Arabia's Renewable Energy Strategy and Solar Energy Deployment Roadmap*, King Abdullah City for Atomic and Renewable Energy, 2013, powerpoint briefing available online at www.irena.org/DocumentDownloads/masdar/Abdulrahman%20Al%20Ghabban%20Presentation.pdf.
49 Glada Lahn, and Paul Stevens, *Burning Oil to Keep Cool: The Hidden Energy Crisis in Saudi Arabia* (London: Chatham House, 2011).
50 As quoted in Anthony DiPaola, "Saudi Arabia Delays $109 Billion Solar Plant by 8 Years," *Bloomberg News*, January 19, 2015.
51 Details of plan at United Nations Development Programme (UNDP), "National Energy Efficiency Program Phase II," UNDP, www.sa.undp.org/content/saudi_arabia/en/home/operations/projects/environment_and_energy/national-energy-efficiency-programme–phase-2-.html.
52 United Nations Development Programme (UNDP), "Saudi Arabia: Government Join Forces to Implement Energy Efficiency Labels," UNDP, www.sa.undp.org/content/saudi_arabia/en/home/ourwork/environmentandenergy/successstories/ee_implementation.html.
53 Irfan Mohammed, "All New Buildings Must Have Thermal Insulation," *Arab News*, March 24, 2014.
54 Antonio Juhasz, "Suicidal Tendencies: How Saudi Arabia Could Kill the COP 21 Negotiations in Paris," *Newsweek*, September 12, 2015. The piece quotes Alden Mayer, from the Union of Concerned Scientists, "most of them [40% of the objections] attempts to try and water down the language, to not express certainty and urgency, and the need to move to lower temperature ranges."
55 Alex Pashley, "World Carbon Emissions in One Handy Graphic," World Resources Institute, June 7, 2015, www.climatechangenews.com/2015/07/06/world-carbon-emissions-in-one-handy-graphic/.
56 *The Intended National Determined Contribution of the Kingdom of Saudi Arabia under the UNFCC*, Riyadh, November 2015. Posted by the United Nations at http://www4.unfccc.int/submissions/INDC/Published%20Documents/Saudi%20Arabia/1/KSA-INDCs%20English.pdf.
57 Ibid.
58 Suzanne Goldenberg, "Saudi Arabia Accused of Trying to Wreck Climate Deal," *The Guardian*, December 8, 2015.
59 Climate Action Tracker, Saudi Arabia, November 24, 2015. Available online at www.climatechangenews.com/2015/07/06/world-carbon-emissions-in-one-handy-graphic/.
60 Gaurav Aginhotri, "Decoding Saudi Arabia's Strategy in its Oil Price War," *Oilprice*, October 7, 2015; Anjil Ravai and Simeon Kerr, "Saudi Arabia Oil: No Pain without Gain," *Financial Times*, October 11, 2015.

61 Views summarized in Peter Waldman, "Saudi Arabia's Plan to Extend the Age of Oil," *Bloomberg Business*, April 12, 2015.

62 Waldman, "Saudi Arabia's Plan." The piece cites Mohammed al-Sabban, an economist who served as adviser to the Saudi Petroleum Minister Ali Al-Naimi from 1998–2013. The piece also quotes from WikiLeaks cables from the US Embassy reporting on the Saudi fears of collapsing demand for oil.

63 Elias Hinckley, "Everything has Changed: Oil and the End of OPEC," *Energy Trends Insider*, January 20, 2015.

64 A term coined in Daniel Yergin, *The Prize: The Epic Quest for Oil, Power and Money* (New York: Simon & Schuster, 2008).

9

Energy, climate and economic security, and Canada's road from oil exporter to deep decarbonization

Chris Bataille

Introduction

As of late 2014, international political consensus was shifting to acceptance that deep emission reductions were necessary to avoid dangerous climate change, exemplified by a 2015 G7 communiqué where all members committed to limiting the anthropogenic increase in global mean surface temperature to less than 2 degrees Celsius (°C) by 2050 relative to 19th century global temperatures, and full decarbonization by 2100.[1] According to IPCC, to ensure a better than even chance of remaining below a 2°C average surface temperature rise global annual greenhouse gas (GHG) emissions will need to be reduced by 42–57% by 2050 relative to 2010, and 73–107% by 2100 in order to stay within a global carbon budget of 960–1,430 $GtCO_2e$.[2] Achieving this target requires steep declines to near zero carbon intensity in all sectors.

Being a signatory to the G7 communiqué was a major visible shift of public policy for the Canadian federal government. Since withdrawing from the Kyoto Protocol to the United Nations Framework Convention on Climate Change (UNFCCC) at Durban in December 2011, and with the rapid growth of the oil sands since 2000, Canada has gained alongside Australia a reputation as a "bad boy" in the climate policy world. As of writing, however, the Keystone XL pipeline has been rejected, Canada has a new Liberal majority government that has renamed the Ministry of Environment the Ministry of Environment and Climate Change, and there is much hope that Canada will "return to the climate fold."

Reality is, as always, a bit more complicated. Canada is a geographically large federation with distinct powers assigned to the federal and provincial levels of government. While energy policy is normally a provincial power, critical to climate policy, the federal government becomes involved when energy commodities cross provincial or international borders. The federal government is also responsible for signing international treaties, and has overriding powers of national interest based on "peace, order and good government."[3] The test for using this power is very strong; however, versions of it have only been used during civil insurrection and wartime, and it is unlikely to be used to enforce climate policy in the foreseeable future. Its existence, however, is an overarching influence on federal provincial relations.

In terms of climate policy, while the Canadian federal government has withdrawn from the Kyoto Protocol, the federal government has imposed restrictions on coal generation (no new generation can be more emitting than an average combined cycle natural gas turbine), personal vehicle efficiency and GHG intensity regulations are equivalent to US regulations and on the same trend as European regulations,[4] and Canada has joined the US in announcing heavy vehicle regulations. Also, in 2009, Canada signed the Copenhagen Accord to the UNFCCC, which, unlike the Kyoto Protocol, is a non-binding agreement; Canada agreed to reduce its GHG emissions by 17% from its 2005 levels by 2020, which translates to a reduction of 124 Mt. This target is highly unlikely to be met, but Canada is an active participant in the "Durban Process" by which all participants, not just Annex I, agree to participate and offer some level of emissions reduction for the 2025–2030 period, or Intended Nationally Determined Contributions (INDCs). At time of writing this new voluntary process is arriving at its first milestone, the Convention of the Parties (COP) 21 of the UNFCCC in Paris, where all the voluntary pledges being made during 2015 will be registered, and a process for their renewal will be discussed.

Canada announced its INDC in May 2015, pledging a 30% reduction from 2005 levels by 2030, which translates into 524 Mt in 2030 from a forecast of 798 Mt (including land-use GHGs). Canada's INDC is deep by any measure given current emissions trends, and is likely to be dependent on a suite of aggressive provincial policies and new federal policies. Canada's INDC is on one of several possible emissions reduction pathways consistent with a 2°C objective (Figure 9.1).[5] With the INDC 2030 target achieved, it would then be another policy and technology stretch to reduce emissions from a forecast level of 16 tonnes per capita in 2050 to the DDPP goal of 1.7 tonnes per capita in 2050.

Perhaps more importantly in terms of climate policy, however, several provinces, which have strong powers related to resources, energy and especially electricity generation and use, have some of the strongest climate policies in the world. Ontario's target is equivalent to 37% below 1990 levels by 2030, which it intends to achieve with a suite of policies, including an existing

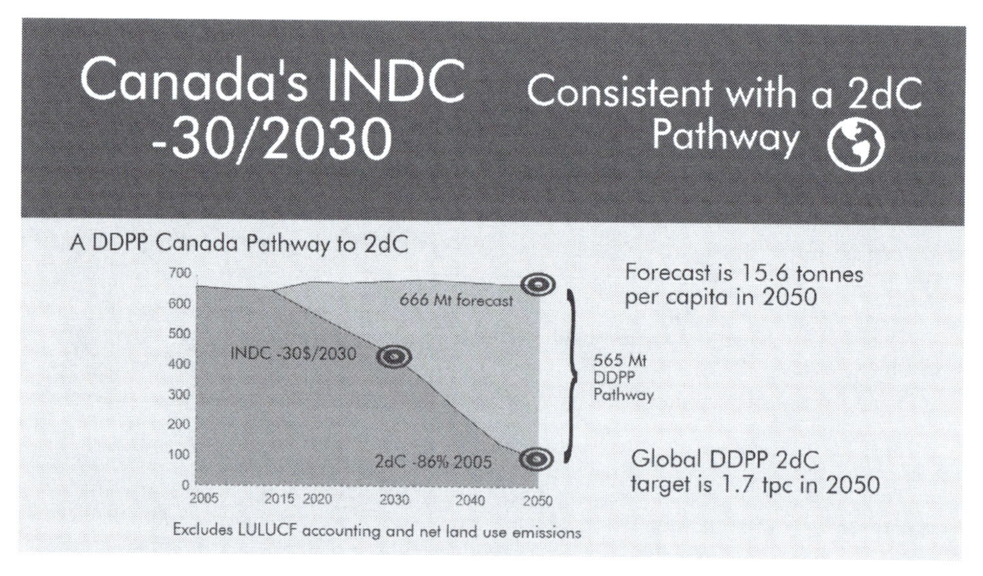

Figure 9.1 Canada's INDC and the DDPC 2°C pathway
Source: Bataille, Sawyer, and Melton, *Pathways to Deep Decarbonization in Canada*; originally in Bataille, and Sawyer, "Canada's INDC and its Two (Thirteen?) Speed Climate Policy."

coal phase out and joining the Western Climate Initiative (WCI) cap-and-trade system. Québec's target is 20% below 1990 levels by 2020, again the centerpiece being the WCI cap-and-trade system among a suite of other policies. B.C.'s target is 33% below 2007 levels by 2020, via system-wide carbon tax ($30/tonne CO_2e) and virtually 100% clean requirement for electricity. Alberta's target is 50 Mt below business as usual, via its own Specified Gas Emitters Regulation (SGER) intensity-based cap-and-trade system and CCS program (see the epilogue for an update on Alberta's climate policy).

Another key factor is that unlike many developed countries, Canada's population is growing quickly, and its economy with it (Table 9.1). Demand for energy services will increase, and how one perceives Canada's capacity to decarbonize depends on how one perceives the capacity to decarbonize all newly installed energy end uses.

The central theme of this book, of which this chapter on Canada is a part, is the policy trilemma currently facing governments, specifically the difficulty of simultaneously attaining *energy security*, sustainable climate (*climate security*), and economic competitiveness, or *economic security*. Canada's federal, provincial and municipal governments all face this challenge, but its nature is different because unlike most other developed countries Canada, and more specifically certain regions, are significant fossil fuel exporters. For example, by 2014, prior to the oil price collapse in 2015, annual crude oil and natural gas exports had risen to more than $92 billion and $16 billion respectively of total exports of $529 billion (all in 2014 CA dollars),[6] the largest single component of exports, having edged out motor vehicles and parts, the longstanding previous largest component ($75 billion). Almost all of these exports were originally sourced from British Columbia, Saskatchewan, Newfoundland and especially Alberta. Imports of energy products, about half crude oil and half refined petroleum products, were $43.6 billion. Crude oil is only imported in bulk to Ontario, Québec and New Brunswick; we will return to this important regional differentiation later in the chapter.

In this chapter, I will take maintenance of the 2°C target as a "sustainable climate," and define climate security as having been met if the 2°C target is maintained. Redefining sustainable climate as climate security in this context is useful, because it allows us to compare the three security priorities using a similar definition, and allows meaningful comparison. When economic security, energy security and climate security point in the same direction, they are mutually reinforcing and climate policy is likely to advance. When one differs with the others, specifically climate security because it is a global issue with focused local mitigation costs, then decision makers prioritize, usually at the cost of significant climate policy.

In this essay, I will make a sequence of three arguments based on evidence from the Canadian federation and the Canadian Deep Decarbonization Pathways Project (DDPP) report:[7]

1 *Energy supply security is a prerequisite for economic security, but this does not require fossil fuels.* I will make the case that it is actually easier to decarbonize in Canada than other countries, due to its large hydropower and other potential renewables resources compared to its population, as well as a large geological potential for carbon capture and storage.

Table 9.1 Development indicators and energy service demand drivers

	2010	*2020*	*2030*	*2040*	*2050*
Population [millions]	33.7	37.6	41.4	44.8	48.3
GDP per capita [$/capita, 2010]	37,288	49,787	57,754	67,500	78,882

2 Economic security is fundamentally different for energy importers and exporters, and *climate security will typically enhance economic security for fossil fuel importers, but reduce it for exporters*. To exemplify this difference, I will contrast the cases of Alberta and Québec in the Canadian DDPP deep decarbonization scenario. To better articulate this argument, I will walk through three scenarios of oil prices (long run $40, $80 and $114 per barrel in 2015 USD) in the Canadian DDPP project, and show what different oil prices do to Canadian economic structure.

3 Climate security, to the extent damages are not immediate, will face ambivalence and resistance from fossil fuel exporters, who will produce as long as there is a business case for production. If one producer shuts down in face of political pressure, another less pressured one will take its place; the only sure way to cease fossil fuel emissions to the atmosphere is to ruin the business case for producing them. Based on this, I will argue *a necessary component of climate security is to encourage alternatives that reduce demand*. I will return to the oil price scenarios described earlier to exemplify the role of demand (in this case from the United States) in setting prices and therefore fossil fuel production.

Before proceeding, we need to clarify that the term "energy" is not sufficiently accurate for this conversation. Firms and consumers consume electricity, refined petroleum products (gasoline and diesel), natural gas, and coal. They are all forms of energy, but that is where the similarity ends. The supply chains and markets for these energy forms are all very different, in terms of location, cost and carbon intensity. Their substitutability depends on the end use, e.g. electricity, natural gas, heating oil, and coal can all be substituted for home heating, but not for lighting or running appliances. In transport, bioliquids can be used in place of petroleum products, and technology is slowly making electricity more substitutable for gasoline and diesel (i.e. in the form of hybrid and battery powered vehicles). In general, electricity is the most substitutable for most end uses, but it is also the most expensive per unit energy, and difficult to store in volume or for any length of time. As we proceed, we will describe how energy, economic and climate security are affected by the differing characteristics of these energy forms.

Argument 1: Energy supply security is a prerequisite for economic security, but does not require fossil fuels.

There are two main approaches to considering energy security in relationship with economic and climate security. One mainstream approach, represented by a review of the literature on energy security in Winzer finds that while the term is used in many different ways depending on the time, context and place, the most durable and common vision defines energy security as "the continuity of energy supplies relative to demand."[8] Put more simply, a modern economy cannot run without an uninterrupted supply of electricity, heat and transport energy, and without them we would return to a late 1700s level of technology. The other stream, represented by Sovacool and Brown,[9] considers the environmental impacts of energy systems a "dimension" of energy security, and that the topics of energy security, sustainability, and economic efficiency are interlinked and overlapping. There are elements of truth in both approaches, and which is applied depends on the problem one wishes to study and the intellectual tools one wishes to bring to bear.

The International Energy Agency (IEA)'s working definition of energy security as "the uninterrupted availability of energy sources at an affordable price" further distinguishes between short and long term security, specific to short term supply chain events, and longer term maintenance of supply infrastructure.[10] Kruyt, van Vuuren, de Vries, and Groenenberg,[11] in a widely referenced study, further define Winzer's narrower definition along four dimensions:

availability (i.e. physical availability somewhere), accessibility (i.e. availability for purchase), affordability (i.e. reasonable prices as a portion of income and revenue) and acceptability (i.e. based on social and environmental considerations).[12] Sustainability concerns, specifically those associated with climate change, impinge on all four of these criteria. Using solar photovoltaics (PV) as an example, given the efficiency of PV cells we can reasonably make, is there enough land, building surfaces and solar incidence that we could physically generate enough to meet our needs? Can we purchase the necessary equipment from a manufacturer? Are enough materials available? How much of our income will it consume? What other necessities might it crowd out? And finally, will people complain about seeing solar panels everywhere?

Mathy, Criqui, Hillebrandt, Fischedick, and Samadi,[13] while they are probably not the original source, used a broader definition than that typically seen in the energy security literature, that a secure energy system is one that is robust (i.e. suited to very different economic or technological environments, at the domestic and international level) and resilient (i.e. that swiftly recovers its balance in the event of crises, accidents or acute instability).

Jewell, Cherp, and Riahi come perhaps closest to a modern, usable definition of energy security with "low vulnerability of vital energy systems, delineated along geographic and sectoral boundaries."[14] Cherp and Jewell further link it to the broader security literature by arguing that any discussion of energy security should address the questions of "Security for whom, for which values, and from what threats?"[15] As was laid out in the beginning and will be discussed earlier, these are very salient questions because "security for whom" is very different for energy importers and exporters.

None of the above definitions of energy security require, however, that energy be derived solely or mostly from fossil fuels, to which Jewell, Cherp, and Riahi expressly address themselves.[16] Most energy end uses (e.g. heating, cooling) can be electrified, and electricity can be made carbon free from hydropower, wind, solar, biomass, nuclear, and fossil fuels where the emissions are captured and stored underground. There are some questions about intermittency of wind and solar, but wider spread and more transmission coupled with some storage can allow a very high share of intermittent generation (50%–100% if the excess electricity is used to produce syngas or hydrogen). Decarbonization of heavy industry, freight transport, and aviation is more difficult. Carbon capture and storage (for heavy industry), hydrogen or biofuels could address these end uses, but they need significant research, development, prototyping and commercialization. All these options are currently more expensive than fossil fuels, however, and should be accompanied by a strong push for across the board energy efficiency to reduce primary energy requirements. Please see our 2015 report, *Pathways to Deep Decarbonization in Canada*,[17] for details or go to www.deepdecarbonization.org.

How might this be achieved in Canada from a policy perspective? In the Canadian deep decarbonization pathways report, we offered the following policy package based on three key physical concepts: *efficiency*, *decarbonization of energy carriers* (e.g., electricity, biofuels and hydrogen), and *fuel switching* to these clean carriers. All suggested policies are technology agnostic and performance orientated, i.e. regulations are based on energy and GHG intensity, not adoption of specific technologies:

1 *Best-in-class regulations* require the use of zero or near-zero emission technologies in the buildings, transport and electricity sectors, applied to all new installations and retrofits. The requirements are as follows:

   - *Mandatory energy and GHG intensity regulations for buildings, vehicles and appliances.* These follow current federal and provincial regulations to the early mid-2020s, and then start dropping to a 90–99% reduction in GHG intensity by 2045.

- *Buildings.* Regulations would trend down to require net-zero-energy residential buildings after 2025, and commercial buildings after 2035. This would be enabled by highly efficient building shells, electric space and water heaters with heat pumps for continuous load devices, solar hot water heaters and eventually solar PV as costs fall. Community heating opportunities identified thorough energy mapping are also an option.
- *Personal and freight transport.* Personal vehicles and light freight regulations, because these sectors have several options including efficiency, electrification, bio fuels, hydrogen and mode shifting, would be on a rolling 5-year renewal and tightening schedule, with the announced long-run goal being for all new vehicles to decarbonize in the early 2030s. Heavy freight vehicles that have more limited options (including some rail-based mode shifting, efficiency, biofuels and hydrogen – batteries are not sufficiently power dense for freight) would be on a schedule to decarbonize by 2040.

2 *Mandatory 99% controls for all landfill and industrial methane sources (landfill, pipelines, etc.).* Any remaining emissions would be charged as per the following policy.

3 *A hybrid carbon-pricing policy, differentiated by heavy industry and the rest of the economy:*

- *A tradable GHG performance standard for "downstream" point source heavy industry* (including electricity), evolving from −25% from 2005 in 2020 to −90% before 2050, using output-based allocations to address competitiveness concerns. This system has the advantage that it produces an incentive for early "lumpy" emissions reduction projects, such as carbon capture and storage in electricity with consequent innovation effects, the excess permits of which can be sold to other emitters. If desired, an absolute cap and trade system could be implemented instead with mostly similar effects.
- *A flexible carbon price, either a carbon tax or an upstream cap and trade,*[18] *covering the rest of the economy*, rising to CDN $50 by 2020 and then in $10 annual increments to 2050.[19] The funds are recycled half to lower personal income taxes and half to lower corporate income taxes. The charge would be flexible based on progress – the above charge was required to meet the DDPP target, but technological advancement driven by carbon pricing and complementary innovation policy would likely dramatically reduce the necessary price, as happened with the US SOx cap-and-trade system.

A land-use policy package that values the net carbon flows of large parcels of land. The policy would provide standardized valuation and accounting for net carbon flows on agricultural, forested, brownfield and wild private lands. Government lands would be managed including net carbon flows in the mandate.

Based on our modelling, Figure 9.2 provides an overview of the emissions trajectory and abatement drivers from the 2050 forecast to the 2050 DDPC target. Canada's +2°C budget works out to be about 75 Mt in 2050, or a 90% reduction below today's level. Per capita emissions transition from 20 tonnes per capita today to 1.7 tonnes by 2050 in the DDPC pathway. In terms of costs, deep decarbonization requires a significant restructuring of investment, but not a large overall increase. Overall investment per year increases only $13.2 billion (+8% over historical levels), but hides sectoral differences. Consumers spend $3.0 billion less each year on durable goods like refrigerators, cars, appliances and houses, while firms must

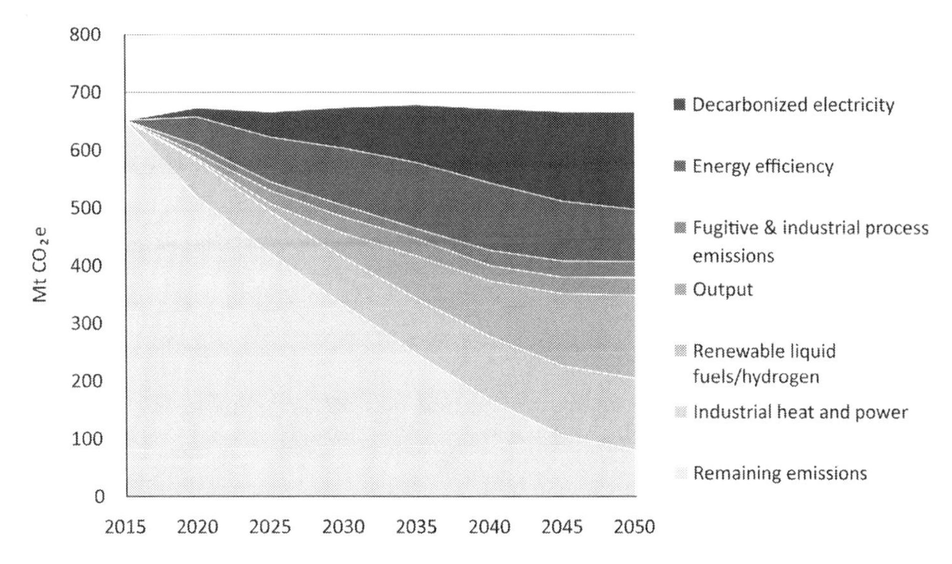

Figure 9.2 Canada's emissions and abatement drivers under the DDPP policy package
Source: Bataille, Sawyer, and Melton, *Pathways to Deep Decarbonization in Canada.*

spend \$16.2 billion more. \$13.5 billion of this is in the electricity sector (+89% over historical levels), by far the most important shift, and \$2.9 billion in the fossil fuel extraction sector (+6% over historical levels).

Beyond the "nuts and bolts" policy package above, a national and regional discussion is required on various pathways options to a deeply decarbonized future. The DDPP recipe for deep decarbonization, grounded on efficiency, decarbonization of energy carriers like electricity, and fuel switching from fossil fuels to these carriers, has many options within it, some more or less appropriate for a given region. Electrification was central to most country decarbonization plans in the DDPP, and was central in our Canadian pathways, but every region chose a different electricity generation mix, based on local resources, politics and social preferences. British Columbia, Québec, Manitoba, Newfoundland and the Maritime provinces largely adopted hydropower because of the available resources; Alberta and Saskatchewan went with a mix of fossil fuels and CCS with renewables with an emphasis on the former, while Ontario largely went with renewables with a smattering of everything else. Every region and country will choose a different generation mix based on local resources and politics, but they share decarbonization of electricity generation, and an increase in electricity use, typically around +40–50%.

In terms of dynamic pathway policy effectiveness and efficiency,[20] some actions and policies, such as efficiency, must be pursued early if they are to have substantial effect because of equipment stock turnover effects (i.e. replacement of high emission building, equipment and vehicles with low emission ones as they wear out), and in turn, efficiency provides insurance that less primary energy will be necessary if any one of the decarbonization energy carriers doesn't work out (e.g. if renewables, nuclear, or carbon capture and storage prove intractable).

Jewell, Cherp, and Riahi in their assessment of energy security under multiple decarbonization scenarios generated through the MESSAGE integrated assessment model, find that in most cases, because of the prominent place played by decarbonized electricity generation and fuel switching to electricity in most scenarios, decarbonization actually increases many measures of energy security, especially the proportion of primary energy sourced from outside a country, or the "net import dependence."[21]

Besides the energy and climate security benefits, there is some evidence that low-carbon technology options can contribute to enhanced economic security as a co-benefit via reduced exposure to fossil fuel market price volatility.[22] McCollum et al. also argue for increased well being from reduced air pollution induced by climate change mitigation, and show that costs for climate change mitigation could be substantially compensated for by the corresponding reductions in air pollution control and associated health costs.[23]

In sum, while the argument that energy supply security as defined by Winzer and other authors is necessary for economic security is fairly obvious, I argue that a secure energy supply need not be based on a system where fossil fuels are combusted in the open atmosphere, enhancing climate security for all.

Argument 2: Climate security can enhance economic security for fossil fuel importers, but reduces it for exporters

For energy importing regions, energy imports require that a certain amount of income leaves the region to purchase energy. In the long run and all other things being equal, this requires exporting something else to acquire foreign currency, has negative effects on the purchasing power of the domestic currency relative to other currencies, and less income is available for consumption, savings and hence investment. Normally, for highly developed nations this is not a problem – Japan imports almost 80% of its energy, and easily covers the cost of this from its GDP (energy imports only represent 2.6% of GDP).[24] Yet energy security is still a concern for Japan, and there are official policies to diversify both types of energy source and supply sources within each type.

For fossil energy exporting regions the exact opposite is true. Exports increase potential consumption, savings and investment, have positive effects on the domestic currency, and make more foreign currency available for imports. Economic security is bound up with fossil fuel exports.

Where this becomes interesting is when a country contains regions that are fossil fuel importers and exporters. The Canadian regions of British Columbia, Alberta, Saskatchewan and Newfoundland are all exporters of oil; when the price of oil goes up they are positively affected because, despite some increase in gasoline and diesel prices, government, corporate and labor revenues rise. The regions of Ontario, Québec, New Brunswick, Prince Edward Island and Nova Scotia are all energy importers; they are negatively affected by an increase in the price of gasoline and diesel with no positive revenue effects.

Ontario and Québec are particularly negatively affected because of their export orientated manufacturing base. When the price of oil rises, the demand for Canadian dollars to buy Canadian oil rises, and the Canadian dollar rises, making Ontario and Québec's manufactured products more expensive. In extreme cases this leads to what has been termed the "Dutch disease," first described in Ellman[25] in the case of Holland's natural gas boom of the 1960s, where dominance of exports in a given resource makes the currency too expensive, chilling exports of other products.

To provide an example of the relative effects of oil prices on Canadian regions, we ran three different long run oil price scenarios for the DDPP project:

- Deep decarbonization with a high oil price, based on the Canadian National Energy Board 2013 Reference scenario prices and output projections out to 2030. For 2031–2050, we carried out the 2030 price assumption while letting our models determine production based on the price. This scenario mirrors some of the International Energy Agency's 2°C scenarios that maintain high prices and relatively high demand. **HIDDPP** has oil prices rising to $114 in 2035 and then stabilizing to 2050; oil output rises 1.5 million barrels per day (mbpd) above our no-decarbonization reference case by 2050, topping out at 7.5 mbpd (Figure 9.3).

- Deep decarbonization with a medium oil price, based on the NEB 2013 low pathway where oil stabilizes at $80 per barrel in the long term, and the same methodology as HIDDPP past 2030. In **MIDDPP** output falls roughly 1.5 million barrels per day by 2050 compared to our no-decarbonization reference case.
- Deep decarbonization with a low oil price, based on conversations with global DDPP modellers (i.e. their view of possible oil prices in a deep decarbonized world) and using initial analysis from the 16 DDPP countries. The combined consensus view is that oil demand will likely fall about 50% in a deeply decarbonized world (Figure 9.4). Combined with existing resources producing based on variable costs without searching for new reserves, reduced demand coupled with ongoing supply stabilizes the long-term price at $40/barrel in today's USD in 2050. In **LODDPP** the price of oil evolves from 2015 in even steps down to average $40/barrel by 2050, with Canadian production falling to about 850,000 barrels per day. The remnant production largely represents the continuing operations of oil sands and shale facilities that have amortized their capital costs and operate until the resource is exhausted or the market price falls below variable unit production costs.

All three deep decarbonization scenarios include advanced low emissions oil sands technologies such as solvent extraction and direct contact steam generation, as well as large-scale implementation of CCS.

Source: Bataille, C., D. Sawyer, and N. Melton. (2015). Pathways to Deep Decarbonization in Canada.Table 9.2 provides a comparison of regional and national GDP under the scenarios. As can be seen, GDP at least doubles in all scenarios. Our major observation is that the uncontrolled combustion of fossil fuels is a high economic growth pathway, but is not necessary for continued rates of high GDP growth. The impact varies between provinces. Alberta's

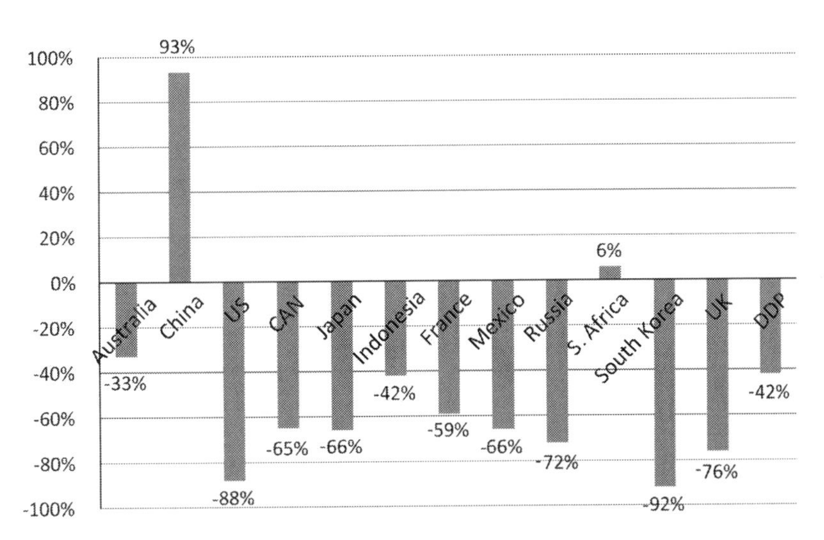

Figure 9.3 Oil demand compared to today in a deeply decarbonized world
Source: Bataille, Sawyer, and Melton, *Pathways to Deep Decarbonization in Canada.*

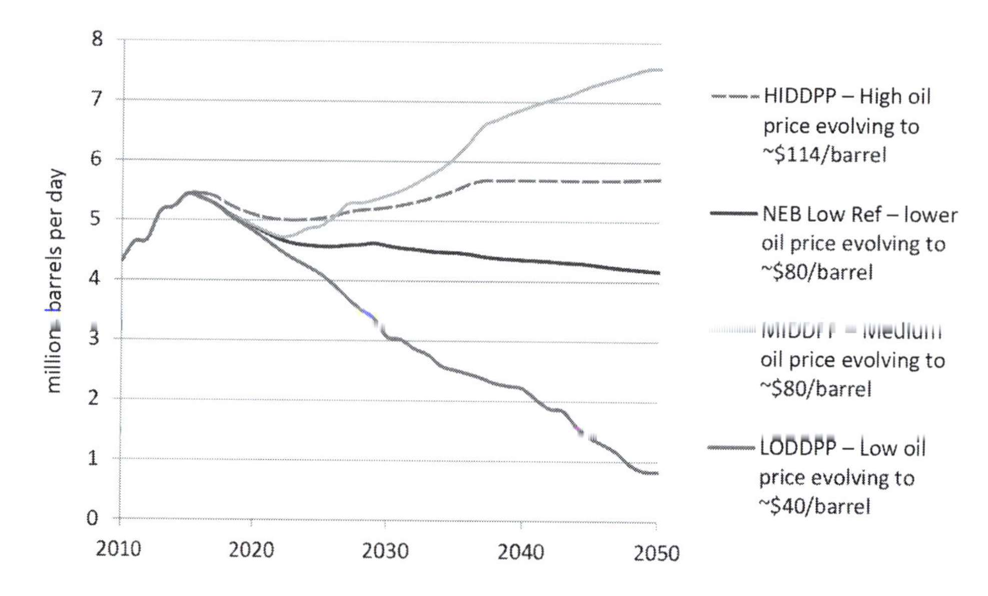

Figure 9.4 Canadian oil output under varying oil price (USD 2014) assumptions

Table 9.2 Changes from 2015 in regional GDP in 2050 (relative to 2015=1)

	2050 REF ($80/barrel)	*MIDDPP ($80/barrel)*	*HIDDPP ($114/barrel)*	*LODDPP ($40/barrel)*
AB	1.71	1.34	1.71	1.20
BC	2.42	2.14	2.10	2.17
SK	2.67	2.23	2.20	2.16
MB	2.71	2.50	2.47	2.51
ON	2.09	1.89	1.87	1.90
QC	2.29	2.47	2.40	2.57
AT	2.18	1.97	1.95	2.01
Canada	2.15	1.98	2.01	1.99

Source: C. Bataille, D. Sawyer, and N. Melton. (2015), *Pathways to Deep Decarbonization in Canada.*

economy is substantially smaller with decarbonization because of obvious effects on the fossil fuel industry, but still increases between 20 and 70% by 2050 in all cases relative to 2015. Québec's economy actually grows with decarbonization compared to our reference case because of its plentiful low-cost hydroelectricity, and benefits yet again from lower transport costs associated with lower oil prices in LODDPP.

Figure 9.5 shows the impact of the high (HIDDPP) and low (LODDPP) price scenarios relative to the DDPC reference case (MIDDPP). Alberta and Québec are again the most highly affected, with large gains to Alberta in the high oil pathway (HIDDPP) accompanied by some losses to Québec and Ontario, while the low oil price negatively affects Alberta and benefits Ontario and Québec.

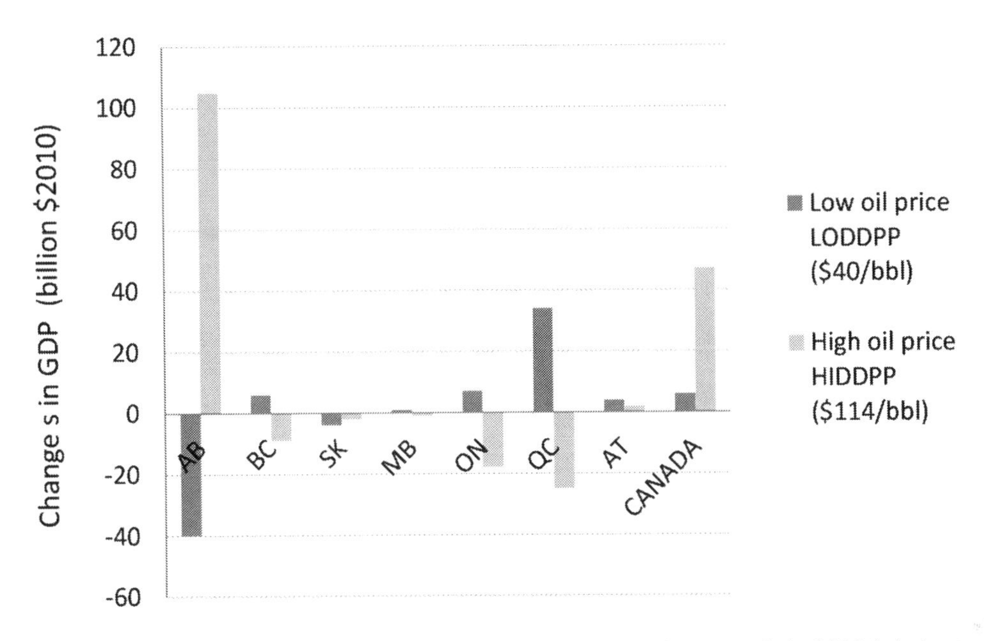

Figure 9.5 Impact of changes in oil price on GDP in decarbonization scenario in 2050 (relative to the MIDDPP scenario)

Source: Bataille, C., D. Sawyer, and N. Melton. (2015). *Pathways to Deep Decarbonization in Canada.*

Table 9.3 Changes from 2015 in sectoral GDP in 2050 (2015=1)

	2050 REF ($80/barrel)	*MIDDPP ($80/barrel)*	*HIDDPP ($114/barrel)*	*LODDPP ($40/barrel)*
Conv. oil	0.2	0.2	0.2	0.1
Oil sands	1.7	1.1	3.2	0.1
Refining	1.4	1.1	0.7	1.0
Natural gas	0.9	0.7	0.8	0.7
Electricity	1.2	5.9	5.9	6.0
Cement-lime	2.5	2.7	2.4	2.4
Pulp & paper	3.2	2.7	2.4	3.0
Iron & steel	1.7	1.3	1.2	1.3
Aluminum	2.7	2.8	2.6	2.8
Coal mining	2.0	2.9	2.7	3.0
Other industry	2.7	2.5	2.2	2.8
Services	2.3	1.9	1.9	2.0
Manufacturing	1.9	1.6	1.4	1.6
Trade	2.4	2.0	2.0	2.0
Transport	2.2	2.1	2.1	2.2
Government	2.5	2.2	2.2	2.2
Total	*2.2*	*2.0*	*2.0*	*2.0*

Now we look at the effect of the oil price pathways on economic structure, as defined by the output of firms. Table 9.3 gives detailed GDP by sector and scenario, showing changes due to the oil price.

Figure 9.6 shows the biggest sector impacts of decarbonization are on the electricity and services industries, as well as on resource rents collected by government. Electricity GDP grows strongly because output and prices increase, while services decrease (while still more than doubling from 2002) due to the overall drag associated with the decarbonization policies. Industry, manufacturing and transport have an inverse relationship with oil prices due to transport costs (low prices raises industry GDP and vice versa).

In sum, oil prices, and not national decarbonization policy, are the key determinant of Canadian oil production and therefore, to a certain extent, Canada's regional economic structure. Overall GDP is relatively unaffected, but with strong regional effects. Domestic deep decarbonization is feasible in all cases.

In terms of energy security, the analysis indicates high oil prices, and their high oil production and lower climate security, are directly related to higher economic security for fossil fuel exporting regions of Canada (e.g. Alberta), but not for Canada as a whole, and are directly detrimental to economic security for fossil fuel importing regions (e.g. Ontario and Québec).

Argument 3: A necessary component of climate security is the undoing of the business case for producing fossil fuels by providing alternatives that reduce demand.

We now arrive at the third part of this essay, a discussion of the necessary conditions for climate security from the perspective of a fossil fuel energy exporter. As discussed at the beginning, when climate security adds to energy or economic security, it is not likely to face any more

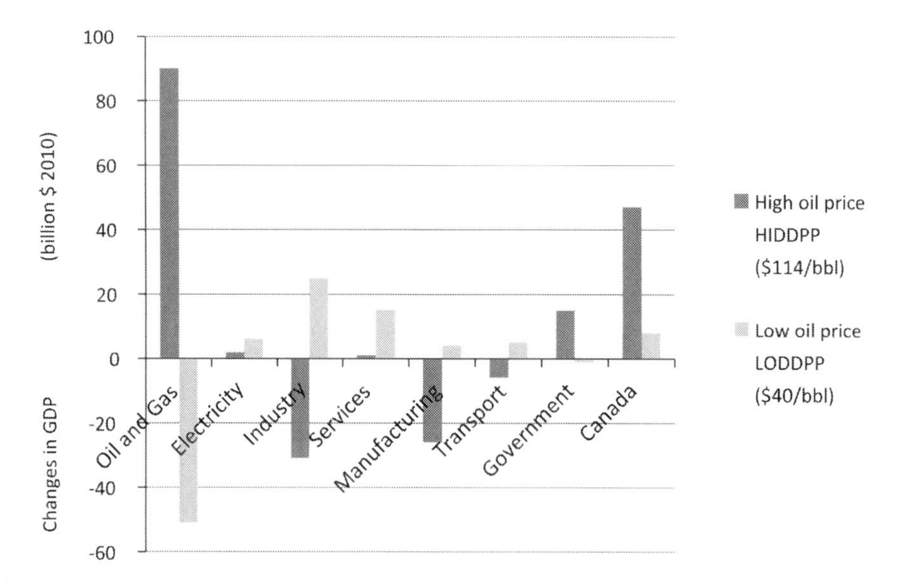

Figure 9.6 Impact of changes in oil price on sector GDP with DDPC policies in 2050 (relative to reference case)

Source: C. Bataille, D. Sawyer, and N. Melton. (2015). *Pathways to Deep Decarbonization in Canada.*

than usual political headwind or intransigence. When it works against economic security, as it does for current fossil fuel exporters, strong, even policy-fatal headwinds can be expected.

As we have already seen, basic economics would suggest that energy exporting regions like Alberta will continue to produce fossil fuel commodities as long as the expected market price is higher than the cost of production, including the cost of border penalties, carbon pricing or other regulations, for the life of the producing investment. On the other hand, if the market price for fossil fuel commodities falls below the all-inclusive cost of production, producers will gradually reduce and eventually cease production of their own accord, as per the full life-cycle capital amortization and variable operating costs of their individual production stock. This is explicitly seen in the LODDPP scenario, where the equilibrium price of oil falls to $40/barrel in today's dollars.

So, to enhance climate security, how does one go about bringing down fossil fuel demand, or the market price received by producers?

(A) *Direct supply side constriction regulations.* These raise the market costs, inducing demand reductions, but they also have a weakness. If they are not imposed globally, some other supplier will take the place of the constrained supplier (e.g. fracked oil replacing oil sands crude oil).

(B) *Demand side technology specific regulations.* These enforce a specific technology. While achieving a specific GHG intensity objective, they lock out cheaper or more appropriate options that may be available, and do not encourage technology or process innovation.

(C) *Demand side technology performance regulations.* Performance regulations, while they may be based on a given technology, do not dictate a specific technology but instead a performance level, like grams of CO_2 per kilometer driven. They allow the market place to find the cheapest way of meeting the target, but do not push technology innovation beyond the current regulation unless they explicitly get tighter through time.

(D) *Carbon pricing through carbon tax or cap and trade.* These operate by adding to the effective costs for consumers of fossil fuels, thereby directly reducing demand, and also by increasing the cost base for fossil fuel producers, effectively reducing their profit per unit of production. These are the standard first choice instruments of climate policy, and directly incentivize technology innovation in a carbon reducing way on both the demand and supply sides of the market. If, however, consumers of fossil fuels have no perceived or real alternatives in the near term, political resistance will build quickly, eventually to policy fatal levels. As seen in Bataille, Sawyer, and Melton[26] and many other sources (e.g. see the literature review in Pye and Bataille[27]), by itself, the carbon price necessary for 2°C compliance will be very high, on the order of several hundred dollars per tonne, and may not survive the political process by itself.

(E) *Encourage replacement substitutes for fossil fuel end use consumption*, e.g. subsidize and support development of more efficient, biofuel, syngas or electric cars in place of standard gasoline cars; encourage electrification of all building heating and cooling end uses; and encourage electrification or the use of hydrogen in heavy industry. While not economically efficient in the short term, subsidies and development support for low carbon end use alternatives provide implicit policy support for technology performance or pricing regulations. They show firms and consumers that options are available, or soon will become available. Once these low carbon alternatives are successful in capturing market share, their costs will fall with economies of scale, they will gradually lower fossil fuel demand, and eventually fossil fuel prices will fall according to supply and demand dynamics. There will be some "rebound" back to fossil fuels, but eventually the long run price for fossil fuels will fall

> below its subsistence level, and the business case for fossil fuel production will expire. One key weakness of direct subsidies, however, is that the scale required to transform overall global fossil fuel consumption is simply unaffordable. They can, however, be used to help incentivize and start technology innovation and market transformation.

I argue that some combination of C, D and E, *technology performance regulations, carbon pricing and encouragement of replacement substitutes*, is necessary to secure climate security. Given that subsidies are expensive, especially for a large scale transformation, and carbon pricing of any form is politically difficult, I argue that effective climate security will come from a combination of technology performance regulation and carbon pricing sticks, and development assistance and direct subsidization of end use alternative carrots. This game of sticks and carrots takes time, however, sufficient time for research, development, prototyping and commercialization of alternatives. This gives current fossil fuel exporters time to plan for falling production, and the associated economic structural adjustment.

Except for Norway and Statoil, no major fossil fuel producer has explicitly acknowledged that the time of uncontrolled release of fossil fuel combustion will come to an end. We still live firmly in the era of fossil fuel combustion; the world is not yet in a position to stop emitting GHGs and still meet human needs, especially for basic energy services in the developing world. The mortal danger to major oil, gas and coal producers is not from policy today or even targets in 2030; it is from inappropriate, unamortized production capital in the future. Adjustment to deep decarbonization for fossil fuel producers is a matter of investing according to something like the following algorithm:

- Maximize efficiency and minimize carbon intensity of existing facilities. Be especially careful with methane emissions from wells, pipelines and facilities.
- Consider the lifetimes of new production equipment and facilities. Is production of a given commodity (coal, oil, or natural gas) likely to be required over that period if climate targets are mandated?
- Consider stepping down the carbon intensity ladder for production and internal consumption: coal to oil, oil to gas, gas to syngas, and so on down to electricity, hydrogen, or biofuels.
- Consider reallocating new capital to lower and zero carbon sources.

If deep decarbonization comes to pass, economic security for fossil fuel producing regions like Alberta, will require a re-visioning and logistical planning process. What will be the working and non-working population to plan for? What associated socioeconomic infrastructure will be needed (roads, hospitals, schools, etc.)? What will be the potential tax base? Will the population need structural adjustment help, or will the transition be paced to new investment, and will all new workers and investment find their way to lower GHG intensity employment? If deep decarbonization occurs it will take a generation or more, but reorganizing and shaping the economies of fossil fuel exporters in the least painful way possible will also take as long.

Conclusion

In this chapter I have made three arguments based on the Canadian federation and the Canadian chapter of the Deep Decarbonization Pathways Project. I have argued that energy security is essential to economic security but does not necessarily require fossil fuels, that climate security normally improves energy and economic security for energy importers, but reduces economic security for exporters and therefore their likelihood of compliance, and finally, that climate security will require a combination of demand and supply side sticks and carrots.

There is a common conception that direct pressure on fossil fuel suppliers like the Canadian oil sands will increase climate security (e.g. by blocking expansion of pipelines like Keystone XL). While important politically, climate security cannot be achieved solely by convincing fossil fuel exporting regions to stop producing. If pressure on one supplier is successful, as it seems to have been for the oil sands, there is a long list of potential suppliers waiting to take up the slack, starting with North American tight oil through fracking. The key to real climate security is to create real and perceived alternatives to fossil fuel consumption, and apply the necessary carrots (subsidies, development support) and sticks (technology performance regulations, carbon pricing) to encourage the great majority of firms and consumers to transition to low carbon options. This will shrink the demand side of the market, and undo the business case for production by bringing the fossil fuel market price below where it is economic to produce for any producer.

Epilogue

Two major relevant events occurred after this chapter was written. The first was COP 21 in Paris in December 2015,[28] resulting in what has come to be known as "The Paris Agreement." Based on the voluntary INDC process established at Durban, the Paris Agreement is generally recognized as being a breakthrough success in international climate policy negotiations, in that all major emitters have committed to emissions reductions for the first time, with the commitments reflecting domestic circumstances. It is also generally recognized, however, that the sum of the committed INDCs commits the world to a temperature increase of somewhere between 2.7 and 3.5°C, depending on estimates. This is well over the 2°C increase agreed to at Durban and upon which the DDPP analysis was based, much less the aspirational 1.5°C target mentioned in the Paris Agreement. The agreement is also purposefully non-binding, specifically so that United States Senate approval is not required, where the agreement would almost certainly be rejected. The US administration negotiated from the point of view that the US 2025–30 INDC could be achieved through natural developments in the US electricity industry (i.e. switching from coal to natural gas) combined with administrative command and control regulation imposed by the executive branch through the US Department of Energy on the electricity and transportation sectors. While the Paris Agreement is a big step forward, a great increase in domestic and global policy ambition is required if the 2°C limit is not to be breached.

The second major event from the perspective of this article occurred in Alberta, home of Canada's oil sands. In the spring of 2015 a newly elected provincial government ordered a review of the province's climate policies. The review was set up as a three month stakeholder engagement process, based on written submissions, internet engagement, and community open houses. 25,000 online responses and 535 formal submissions were received, including the Canadian DDPP report, submitted by Carbon Management Canada. At the same time, the Keystone XL pipeline, a key conduit for new oil sands production, was cancelled due to US perception of the oil sands in relation to its climate policy goals and domestic energy agenda; this increased pressure on the Alberta domestic discussion to increase ambition.

On November 20, 2015, the outcome of the Alberta review was announced.[29] The report and announced policies recognize the importance of the fossil fuel industry to Alberta's current economic well-being while preparing the province for an eventual global low carbon economy, with a focus on economic diversification, decoupling energy use from economic growth and reducing the carbon intensity of energy. As recommended in the Canadian DDPP report, from which the recommended policy package was quoted prominently in the panel report, Alberta will be implementing a general carbon tax rising to $30 in 2018, a cap-and-trade system for large emitters like the oil sands using output based allocations, electricity decarbonization regulations, an aggressive methane control program, and energy efficiency policies for buildings and communities. The policy package is designed for coordination with trading partners and eventual increased ambition. It puts in place most of the main tools to eventually decarbonize the Alberta economy, and is a model of its kind for an energy exporter. Even more significant is the broad political consensus behind the policies, from households, environmental NGOs to oil sands firms, established using a first class stakeholder engagement process.

In sum, in late 2015 there was a significant public and political shift in favor of some level of global climate policy to ensure climate security, but it cannot be said that this shift is as yet strong enough to ensure climate security defined as staying below an average 2°C temperature rise.

Notes

1 *Leaders' Declaration, G7 Summit, 7–8 June 2015*, 2015, https://sustainabledevelopment.un.org/content/documents/7320LEADERS%20STATEMENT_FINAL_CLEAN.pdf.
2 O. Edenhofer, R. Pichs-Madruga, Y. Sokona, et al., eds, *Change 2014: Mitigation of Climate Change. Contribution of Working Group III to the Fifth Assessment Report of the Intergovernmental Panel on Climate Change* (Cambridge: Cambridge University Press, 2014).
3 British North America Act (1867), §91.
4 Chris Bataille, David Sawyer, and Noel Melton, *Pathways to Deep Decarbonization in Canada*. Canadian section authors: Bataille, C., D. Sawyer, and N. Melton. *Deep Decarbonization Pathways Project*, eds. H. Waismann, L. Segafredo, C. Bataille, R. Pierfederici. Sustainable Development Solutions Network (SDSN), and Institute for Sustainable Development and International Relations (IDDRI). New York/Paris. Downloaded from www.deepdecarbonization.org December 12, 2015.
5 Chris Bataille, and David Sawyer, "Canada's INDC and its Two (Thirteen?) Speed Climate Policy," (blog), June 26, 2015, www.blog-iddri.org/2015/06/26/canadas-indc-and-its-two-thirteen-speed-climate-policy/.
6 Statistics Canada, "Canadian International Merchandise Trade: Annual Review, 2014," *The Daily*, April 2, 2015, www.statcan.gc.ca/daily-quotidien/150402/dq150402b-eng.htm.
7 Bataille, Sawyer, and Melton, *Pathways to Deep Decarbonization in Canada*.
8 C. Winzer, "Conceptualizing Energy Security," *Energy Policy* 46 (2012): 36–48, doi: 10.1016/j.enpol.2012.02.067.
9 B. Sovacool, and M. Brown, "Competing Dimensions of Energy Security: An International Perspective," *Annual Review of Environment and Resources* 35, no. 1 (2010): 77–108, http://dx.doi.org/10.1146/annurev-environ-042509-143035.
10 International Energy Agency (IEA), *Energy Security and Climate Change: Assessing Interactions* (Paris: IEA, 2007); International Energy Agency (IEA), *World Energy Outlook 2007: China and India Insights* (Paris: IEA, 2007); International Energy Agency (IEA), "Energy Security," International Energy Agency, 2014, www.iea.org/topics/energysecurity/.
11 B. Kruyt, D. P. van Vuuren, H. J. M. de Vries, et al., "Indicators for Energy Security," *Energy Policy* 37, no. 6 (2009): 2166–2181, doi: 10.1016/j.enpol.2009.02.006.
12 Winzer, "Conceptualizing Energy Security."
13 S. Mathy, P. Criqui, K. Hillebrandt, et al., "Uncertainty Management and the Dynamic Adjustment of Deep Decarbonization Pathways," *Climate Policy* (2016).
14 J. Jewell, A. Cherp, and K. Riahi, "Energy Security under De-carbonization Scenarios: An Assessment Framework and Evaluation under Different Technology and Policy Choices," *Energy Policy* 65 (2014): 743–760, doi: 10.1016/j.enpol.2013.10.051.
15 A. Cherp, and J. Jewell, "The Concept of Energy Security: Beyond the Four As," *Energy Policy* 75 (2014): 415–421.
16 Jewell, Cherp, and Riahi, "Energy Security under De-carbonization Scenarios."

17 Bataille, Sawyer, and Melton, *Pathways to Deep Decarbonization in Canada.*
18 If the latter, the heavy industry cap and trade system would not be required, but the signal to adopt CCS (and consequent innovation) in electricity would be missing until the mid-2020s.
19 All prices are in constant 2015 Canadian dollars unless otherwise denoted.
20 Mathy et al., "Uncertainty Management."
21 Jewell, Cherp, and Riahi, "Energy Security under De-carbonization Scenarios."
22 Edenhofer et al., eds, *Change 2014.*
23 D. McCollum, V. Krey, K. Riahi, et al., "Climate Policies Can Help Resolve Energy Security and Air Pollution Challenges," *Climatic Change* 119, no. 2 (2013): 479–494, doi: 10.1007/s10584-013-0710-y.
24 K. Oshiro, M. Kainuma, and T. Masui, "Analysis of Decarbonization Pathways and Their Implication for Energy Security Policies in Japan," *Climate Policy* (2016) (forthcoming).
25 Michael Ellman, "Natural Gas, Restructuring and Re-industrialisation: The Dutch Experience of Industrial Policy," *Oil or Industry* (1981): 149–166.
26 Ibid.
27 S. Pye, and Chris Bataille, "Improving Deep Decarbonization Modelling Capacity for Developed and Developing Country Contexts," *Climate Policy* (2016) (forthcoming).
28 United National Framework Convention on Climate Change (UNFCCC), "The Paris Agreement (Draft decision /CP/2015/L)," UNFCCC, 2015, http://unfccc.int/resource/docs/2015/cop21/eng/l09r01.pdf.
29 Alberta Government Climate Change Climate Advisory Panel, *Leadership Report to Minister*, Alberta Government Climate Change Climate Advisory Panel, 2015, alberta.ca/documents/climate/climate-leadership-report-to-minister.pdf.

10

Energy transitions in carbon-producing countries

Russia

Jack D. Sharples

Introduction

In the context of the study of energy transitions in carbon-producing countries, Russia presents a case that is both highly significant and highly interesting, due to its specificity. Russia's significance as a carbon-producing country derives from its status as a world-leading energy producer: in 2014, Russia ranked second for oil and gas production, sixth for coal production, third for nuclear power production, fourth for hydroelectricity generation, and fourth for electricity generation overall.[1] Not only does Russia produce fuel for its own consumption, it facilitates the fuel consumption (and, therefore, CO_2 emissions) of other countries through its hydrocarbon exports.

As well as being the world's third largest energy consumer, Russia is also the fourth largest emitter of CO_2 in the world, after China, the United States, and India. Indeed, in 2012, Russia's CO_2 emissions were equal to those of the entire Middle Eastern region, and approximately one and a half times those of the non-OECD Americas and one and a half times those of the whole of Africa.[2] Clearly, trends in Russia's energy consumption and CO_2 emissions are highly significant factors in the global challenge of reducing CO_2 emissions.

This chapter examines how Russia's energy consumption patterns, and by extension, CO_2 emissions, have changed over the past decade. In doing so, it identifies trends that are likely to continue into the medium term (to 2020) and the causes of those trends. In addressing these questions, it is possible to draw conclusions as to whether Russia, as one of the world's most significant carbon producers, is undergoing an "energy transition," or whether Russia's energy consumption and CO_2 emissions will continue along the lines of "business as usual" for the medium term, and why this is the case.

In order to answer these questions, this chapter begins by establishing trends in Russia's levels of energy consumption and related CO_2 emissions (total and per capita), placing them into the international context. The second part of this chapter then seeks to explain these trends by examining a range of factors, including population growth, economic growth, levels of energy efficiency (measured in terms of energy consumption per unit of GDP produced). The third part of this chapter analyses trends in Russia's energy consumption by fuel and by sector, and in doing so, identifies areas in which Russia could make an "energy transition." Finally, this chapter examines the participation of the Russian government in international climate change politics, to draw inferences about the orientation of the Russian government, and the level of its commitment to managing Russia's fuel consumption and related CO_2 emissions.

Russia's energy consumption 2004–2014: identifying trends

Russia's total energy consumption has risen steadily over the past decade. In 2004, Russia's energy consumption was 649.5 million tonnes of oil equivalent (mtoe). By 2014, that figure had risen by 5% to 681.9 mtoe.[3] To put the gradual increase in Russia's energy consumption into context, Table 10.1 clearly illustrates that energy consumption in the most developed economies (United States, Japan, and EU-28) fell between 2004 and 2014. By contrast, the world's largest countries by population, China and India, exhibited dramatic increases in energy consumption. Mexico and Brazil demonstrated substantial increases in energy consumption, in line with their regional neighbours. In terms of current trends in total national energy consumption, Russia's closest comparative countries are Canada, South Africa, and Australia.

However, a slightly different picture emerges when trends in energy consumption per capita, measured in kilograms of oil equivalent, are considered. Per capita energy consumption declined, or at least stabilized, in the most advanced economies (United States, Japan, EU-28, Canada, Australia) and increased to the greatest extent in the world's largest countries by population, China and India, albeit from a low base. In terms of trends in energy consumption per capita between 2004 and 2012, Russia's closest comparative countries are South Korea, Brazil, Venezuela, and Belarus. However, while levels of energy consumption per capita in 2012 in Russia and South Korea are rather similar, per capita energy consumption levels in Brazil, Venezuela, and Belarus are considerably lower.

Russia's CO_2 emissions 2005–2014

According to data from the United Nations (UN), Russia's CO_2 emissions were 2,510 million tonnes in the baseline year of 1990. Thereafter, Russia's emissions declined every year to reach a trough of 1,441 million tonnes in 1998, before rising every year (except 2009) to reach 1,659 million tonnes in 2012.[4] This means that by 2012, Russia's CO_2 emissions had declined by 34% against the baseline of 1990, but had been also on an upward trend for more than a decade.

Data from the World Bank is not available for the base year of 1990, but states Russian CO_2 emissions of 2,082 million tonnes in 1992, falling to 1,498 million tonnes in 1998, and rising to 1,603 million tonnes in 2004 and 1,808 million tonnes in 2011. Therefore, according to the World Bank, Russia's CO_2 emissions grew by 12.8% between 2004 and 2011. Although the data from the UN is more complete, data from the World Bank has been used in Figure 10.1 below, due to the availability of comparable data from the same source, on per capita CO_2 emissions.

For comparison, World Bank data shows that between 2004 and 2011, US CO_2 emissions fell from 5,510 million tonnes to 5,306 million tonnes, EU CO_2 emissions fell from 4,052 to 3,574 million tonnes, China's CO_2 emissions rose from 5,288 million tonnes to 9,020 million tonnes, and India's CO_2 emissions rose from 1,349 million tonnes to 2,074 million tonnes. The current trend for CO_2 emissions to decline in developed countries and rise in developing and middle-income countries is borne out by the fact that between 2004 and 2011, OECD CO_2 emissions declined by 5.8% from 13,133 million tonnes to 12,377 million tonnes, while non-OECD CO_2 emissions rose dramatically by 44.5% from 15,411 million tonnes to 22,273 million tonnes.[5]

Table 10.1 Total Primary Energy Consumption (TPES) (million tonnes of oil equivalent – mtoe) and Energy Consumption per Capita (kilograms of oil equivalent)

Country	*TPES (2004–2014) and energy consumption per capita (2004–2012)*					
	TPES (mtoe)			*Energy consumption per capita (kg oil equivalent)*		
	2004	*2014*	*Change*	*2004*	*2012*	*Change*
Russia	*649.5*	*681.9*	*+5.0%*	*4,494*	*5,283*	*+17.6%*
Australia	115.3	122.9	+6.6%	5,598	5,644	+0.8%
Belarus	24.9	28.6	+14.9%	2,753	[illegible]	+17.1%
Brazil	201.3	296.0	+47.0%	1,129	1,392	+23.3%
Canada	315.6	332.7	+5.4%	8,365	7,226	−13.6%
China	1,573.1	2,972.1	+88.9%	1,265	2,143	+69.4%
Estonia	–	–	–	3,983	4,174	+7.2%
EU-28*	1,818.2	1,666.3	−8.4%	3,606	3,254	−9.8%
India	345.1	637.8	+84.8%	460	624	+35.7%
Iran	166.3	252.0	+51.5%	2,244	2,883	+28.5%
Japan	525.1	456.1	−13.1%	4,090	3,546	−13.3%
Kazakhstan	43.0	54.3	+26.3%	3,383	4,458	+31.8%
Latvia	–	–	–	1,961	2,171	+10.7%
Lithuania	9.1	5.4	−40.7%	2,779	2,469	−11.2%
Mexico	158.9	191.4	+20.5%	1,486	1,543	+3.8%
Norway	39.9	46.7	+17.0%	5,756	5,817	+1.1%
Saudi Arabia	174.4	239.5	+37.3%	4,980	6,789	+36.3%
South Africa	117.0	126.7	+8.3%	2,757	2,675	−3.0%
South Korea	213.8	273.2	+27.8%	4,337	5,268	+21.5%
Turkmenistan	18.6	31.3	+68.3%	3,919	4,943	+26.1%
Ukraine	137.7	100.1	−27.3%	3,031	2,690	−11.3%
United States	2,349.1	2,298.7	−2.2%	7,882	6,815	−13.5%
Venezuela	67.5	84.3	+24.9%	2,148	2,558	+19.1%

Sources: BP, *Statistical Review of World Energy*, 40; Eurostat, "Simplified Energy Balances: Annual Data," Eurostat, last modified February 2, 2016, http://ec.europa.eu/eurostat/product?code=nrg_100a&mode=view; World Bank, "World Data Bank: Energy Use (kg of Oil Equivalent per Capita)," World Bank, 2015, http://data.worldbank.org/indicator/EG.USE.PCAP.KG.OE/countries/1W-RU-ZA-AU-CA?display=default.

* Statistics for EU-28 for TPES are given for 2013, not 2014.

In terms of per capita emissions, the world average in 2011 was 4.9 metric tonnes per capita. Only 26 geographical entities, from a list of 218 given by the World Bank, had per capita CO_2 emissions above 10 metric tonnes. Of these 26, eight were geographic entities with very small populations (Aruba, Brunei, Cayman Islands, Faroe Islands, Greenland, Luxembourg, New Caledonia, and Palau) and six were small states with significant oil and/or gas exports (UAE, Bahrain, Kuwait, Oman, Qatar, Trinidad and Tobago). This leaves a group of 12 states with significant populations and high levels of CO_2 emissions per capita: Australia, Canada, Czech Republic, Estonia, Finland, Kazakhstan, South Korea, Netherlands, Russia, Saudi Arabia, Turkmenistan, and the United States. Of this group, only six (Russia, Kazakhstan,

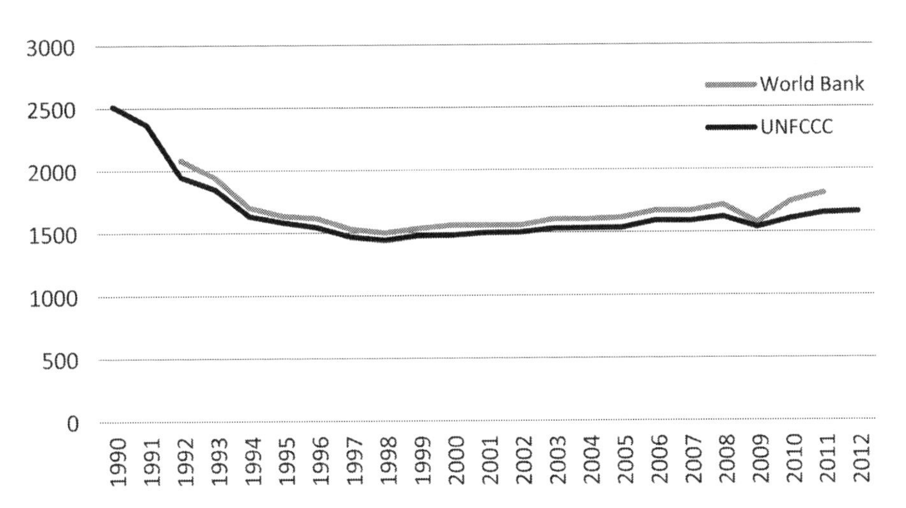

Figure 10.1 Russia's CO_2 emissions, 1990–2012
Source: UNFCCC, "Greenhouse Gas Inventory Data"; World Bank, "CO_2 Emissions (kt)," World Bank, 2015, http://data.worldbank.org/indicator/EN.ATM.CO2E.KT/countries.

Turkmenistan, Estonia, South Korea, and Saudi Arabia) recorded an increase in their per capita emissions between 2004 and 2011 (the latest year for which per capita data is available from the World Bank). By contrast, the advanced economies showed declines of between 3% (Australia) and 20% (Finland).[6] The growth rate of Russia's CO_2 emissions per capita during this period (13.7%) was broadly similar to those of Estonia (11.2%) and South Korea (17.9%), but far below the growth rates of Kazakhstan (37.9%) and Turkmenistan (32.3%). A further three states with emissions of between 5 and 10 metric tonnes per capita in 2011 exhibited growth rates of 11–12%, similar to those of Russia: Belarus, Bulgaria, and Venezuela.[7]

The trends in total and per capita energy consumption and CO_2 emissions discussed above suggest that Russia is following a different trend to the advanced economies of North America, Western Europe, and Japan, where levels of per capita energy consumption and CO_2 emissions are falling, and a different trend to the rapidly developing countries of China and India, whose emissions grew 50–60% during the 2005–2012 period, on the basis of a combination of economic and population growth. Clearly, Russia does not fit into either the group of advanced economies or the group of rapidly developing economies with large populations. To explain why Russia is following its own trends in energy consumption and CO_2 emissions, it is now necessary to examine possible causes of these trends, in order to draw conclusions about the likelihood of their continuation.

Causes of trends in Russia's energy consumption and related CO_2 emissions

Explaining trends in Russian energy consumption: population, economic growth (GNI per capita), and energy consumption per unit of GDP

In its report on world energy resources, the World Energy Council (WEC), noted, "Population growth has always been and will remain one of the key drivers of energy demand, along with economic and social development."[8] In the case of Russia, demographic data suggests that the latter factors are far more influential than the former.

Table 10.2 Trends in CO_2 emissions, total and per capita, 2004–2014

Country	*CO_2 emissions, total and per capita, 2004–2012*					
	CO_2 emissions (total-million tonnes)			*CO_2 emissions (tonnes per capita)*		
	2004	*2011*	*Change*	*2004*	*2011*	*Change*
Russia	*1,603*	*1,808*	*+12.8%*	*11.1*	*12.6*	*+13.7%*
Australia	343	369	+7.7%	17.0	16.5	-3.0%
Belarus	58	63	+9.1%	6.0	6.7	+12.0%
Brazil	338	439	+30.1%	1.8	2.2	+20.7%
Canada	332	405	[illegible]	[illegible]	14.1	−18.1%
China	5,288	9,020	+70.6%	4.1	6.7	+64.5%
Estonia	17	19	+8.3%	12.6	14.0	+11.2%
EU-28	1,052	3,571	−11.8%	[illegible]	7.1	−15.4%
India	1,349	2,074	+53.8%	1.2	1.7	+38.9%
Iran	447	587	+31.1%	6.5	7.8	+20.9%
Japan	1,259	1,188	−5.7%	9.9	9.3	−5.7%
Kazakhstan	172	262	+52.0%	11.5	15.8	+37.9%
Latvia	7	8	+9.4%	3.2	3.8	+20.2%
Lithuania	13	14	+3.1%	3.9	4.5	+15.0%
Mexico	412	467	+13.3%	3.8	3.9	+1.9%
Norway	43	46	+6.8%	9.3	9.1	−1.0%
Saudi Arabia	396	520	+31.4%	16.5	18.1	+9.8%
South Africa	425	477	+12.4%	9.1	9.3	+1.8%
South Korea	482	589	+22.2%	10.0	11.8	+17.9%
Turkmenistan	43	62	+43.9%	9.2	12.2	+32.3%
Ukraine	343	286	−16.6%	7.2	6.3	−13.4%
United States	5,763	5,306	−7.9%	19.7	17.0	−13.5%
Venezuela	152	189	+24.4%	5.8	6.4	+11.3%

Sources: World Bank, "CO_2 Emissions (kt)," World Bank, 2015. http://data.worldbank.org/indicator/EN.ATM.CO2E.KT/countries; World Bank, "CO_2 Emissions: Metric Tonnes Per Capita." 2015, http://data.worldbank.org/indicator/EN.ATM.CO2E.PC/countries.

Russia's population peaked in 1992 at 148.7m, and fell every year thereafter to 142.7m in 2008. Since then it has recovered to 143.8m in 2014. However, Russia's annual population growth rate remains low, at 0.0–0.2% per year, against a global average of 1.2% per year.[9] If Russia's population continues to grow at its 2009–2014 rate of 1m every five years, Russia will not surpass its 1995 population level until 2040. Therefore, any significant changes in Russia's total energy consumption and CO_2 emissions will occur on the basis of changes in consumption and emission per capita under relatively stable population conditions, rather than under conditions of rapid population growth. This puts Russia in stark contrast with the other major CO_2 emitters (United States, China, and India), but in a similar situation to Japan and its post-Soviet neighbors (Ukraine, Belarus, Estonia, Latvia, Lithuania), with the latter experiencing population stagnation or decline over the past decade.

With economic and social development highlighted by the World Energy Council as key drivers in energy demand, it is worth examining the relationship between socio-economic

Table 10.3 Population and GNI per capita (adjusted for purchasing power parity – PPP), 2004–2014

Country	*Population and GNI per capita (PPP), 2004–2014*					
	Population (million)			*GNI per Capita (USD, PPP)*		
	2004	*2014*	*Change*	*2004*	*2014*	*Change*
Russia	144.1	143.8	−0.2%	10,010	24,710	+146.9%
Australia	20.1	23.5	+16.7%	30,420	42,880	+41.0%
Belarus	9.7	9.5	−2.7%	8,510	17,610	+106.9%
Brazil	186.1	206.1	+10.7%	9,950	15,590	+56.7%
Canada	32.0	35.5	+11.1%	32,860	43,400	+32.1%
China	1,296.1	1,364.3	+5.3%	4,410	13,130	+197.7%
Estonia	1.4	1.3	−3.6%	13,950	25,690	+84.2%
EU-28	494.3	508.3	+2.8%	26,489	36,275	+36.9%
India	1,126.4	1,295.3	+15.0%	2,560	5,640	+120.3%
Iran	69.3	78.1	+12.7%	11,850	16,140	+36.2%
Japan	127.8	127.1	−0.5%	29,920	37,920	+26.7%
Kazakhstan	15.0	17.3	+15.2%	11,570	21,580	+86.5%
Latvia	2.3	2.0	−12.1%	13,130	23,150	+76.3%
Lithuania	3.4	2.9	−13.3%	12,810	25,390	+98.2%
Mexico	108.3	125.4	+15.8%	11,050	16,500	+49.3%
Norway	4.6	5.1	+11.9%	43,280	65,970	+52.4%
Saudi Arabia	24.1	30.9	+28.4%	32,260	(2013) 51,320	+59.1%
South Africa	46.7	54.0	+15.6%	9,070	12,700	+40.0%
South Korea	48.0	50.4	+5.0%	22,920	34,620	+51.0%
Turkmenistan	4.7	5.3	+13.0%	4,940	14,520	+193.9%
Ukraine	47.5	45.4	-4.4%	5,980	8,560	+43.1%
United States	292.8	318.9	+8.9%	42,260	55,860	+32.2%
Venezuela	26.3	30.7	+16.6%	11,510	17,230	+49.7%

Source: World Bank, "World Data Bank: Population," World Bank, 2015, http://databank.worldbank.org/data//reports.aspx?source=2&country=RUS&series=&period=#selectedDimension_WDI_Ctry; World Bank, "World Data Bank: GNI per Capita, PPP (Current International $)," World Bank, 2015, http://data.worldbank.org/indicator/NY.GNP.PCAP.PP.CD/countries.

development (measured as gross national income [GNI] per capita, adjusted for purchasing power parity – PPP) and per capita energy consumption. Data from the World Bank highlights Russia's impressive economic growth between 2004 and 2014. Russia is the only country in the world to have had a GNI per capita of above $10,000 in 2004, and still experience a growth in GNI per capita of over 100% during the decade that followed. Russia aside, the only countries to have recorded such growth rates between 2004 and 2014 were those that began their economic growth from a much lower base, China and India. Conversely, all countries with a GNI per capita (PPP) of over $10,000 in 2004, and a population above 2 million, recorded growth rates between 5 and 95%, with the exception of Libya, which recorded negative growth.[10]

However, while economic growth and per capita energy consumption may be linked in Russia's case, this is not true of all countries. To find the link, one must be more specific. The advanced economies of the United States, Canada, Japan, and the EU-28 all recorded GNI per capita (PPP) growth of approximately 27–37% between 2004 and 2014, whilst simultaneously reducing their per capita energy consumption. Conversely, several states that recorded the largest increases in energy consumption per capita (Brazil, India, and China) also recorded large increases in GNI per capita (PPP) from a low base. Russia has yet to follow the developed economies in breaking the link between economic growth and energy consumption, yet it also has a significantly higher GNI per capita (PPP) than less-developed countries that rely more strongly on increased energy consumption for economic growth (Brazil, China, India, Iran, Kazakhstan, Turkmenistan, and Venezuela). This is also illustrated in the table below, which highlights the fact that while Russia's energy efficiency in GNI creation is improving, it remains at a low level.

It is notable that by 2012/2014, Russia's GNI per capita (PPP) was in the upper middle-income range: significantly below the advanced economies (EU-28, United States, Canada, Australia, Japan, South Korea, and Norway), significantly above the lower middle-income countries (Brazil, India, China, South Africa, Mexico, Iran, Venezuela, Ukraine and Belarus), and broadly similar to several of its post-Soviet neighbors (Estonia, Latvia, Lithuania,

Table 10.4 Units of gross domestic product (GDP) produced per unit of energy consumed, 2004–2012

Country	*GDP per unit of energy consumed*		
	2004	*2012*	*Change*
Russia	3.78	4.41	+16.8%
Australia	6.81	7.53	+10.6%
Belarus	3.58	5.25	+46.6%
Brazil	10.63	10.74	+1.0%
Canada	4.71	5.90	22.9%
China	4.05	5.14	+26.9%
Estonia	5.05	5.64	+17.4%
EU-28	8.97	10.56	+17.8%
India	6.50	7.79	+20.0%
Iran	6.23	5.59	−10.2%
Japan	8.19	9.87	+20.5%
Kazakhstan	4.25	4.82	+13.6%
Latvia	7.87	9.49	+20.6%
Lithuania	6.86 (2005)	9.60	+40.1%
Mexico	10.00	10.47	+4.7%
Norway	10.83	10.94	+0.9%
Saudi Arabia	7.47	7.19	−3.8%
South Africa	3.89	4.63	+19.0%
South Korea	5.68	6.06	+6.6%
Turkmenistan	1.48	2.52	+69.8%
Ukraine	2.32	3.09	+33.5%
United States	6.17	7.42	+20.3%
Venezuela	6.42	6.92	+7.8%

Source: World Bank, "World Data Bank: GDP (PPP) per Unit of Energy Use," World Bank, 2015, http://data.worldbank.org/indicator/EG.GDP.PUSE.KO.PP.KD.

Kazakhstan). Yet to achieve this level of economic development, Russia's energy consumption per unit of GDP was broadly similar to that of South Africa and Kazakhstan, and slightly lower than that of Turkmenistan, and Ukraine.

This leads us to consider how Russia was able to achieve exceptional economic growth, in terms of GNI per capita (PPP), between 2004 and 2014, with relatively modest increases in energy consumption. Looking forward into the medium-term future, it is also worth considering the slowing of Russia's economic growth between 2013 and 2015, and the potential for several years of economic stagnation between 2015 and 2020, and the potential impact of this on Russia's energy consumption and carbon emissions.

Oil prices and exports: the link between energy and economic growth in Russia

For the past decade, Russia has been the world's second-largest exporter of crude oil and refined petroleum products and the world's largest natural gas exporter.[11] According to the Russian Ministry of Finance (MinFin), tax on oil and gas production accounted for 19.0% of federal budget revenues in 2012, while export duties (on all products, not just minerals such as oil and gas) accounted for 38.6%. MinFin also states that in 2012 the federal budget was equal to 20.7% of Russia's GDP.[12] The Russian state statistical service, RosStat, states that between 2005 and 2012, the share of "mineral products" in the total value of Russia's exports grew from 65% to 72%, up from 54% in the year 2000.[13] Therefore, if oil and gas accounted for 72% of Russia's export duties, the combined share of tax on oil and gas production and export duties levied on oil and gas exports would be 46.8% of federal budget revenues and 9.7% of Russia's GDP. However, the International Energy Agency (IEA) estimates the share of oil and gas in Russia's GDP at approximately 20%, and suggests that it could even be higher.[14]

It is interesting to note that, although such tax revenues are included in calculations of Russia's GNI and, therefore, in Russia's GNI per capita,[15] the definition of "Energy Use" by the World Bank does not class exports as consumption.[16] Statistics from the Energy Information Administration (EIA) show that between January 2004 and July 2008, the spot price of Brent crude oil rose dramatically, from $30 per barrel to $144 per barrel. Despite the slump in 2008–09, the price of Brent crude remained above $100 a barrel from February 2011 to August 2014.[17] Despite the fact that Russia's oil and gas exports increased by just 5 and 9% respectively between 2004 and 2014, Russia benefitted from a dramatic increase in oil and gas sector tax revenues (due to rising prices), which contributed to GNI growth.

The fact that Russia's oil and gas exports contributed significantly to Russia's GNI growth between 2004 and 2014, and yet were not classed as energy consumption, helps explain how Russia was able to achieve rapid economic growth without a dramatic increase in domestic energy consumption. The link between energy exports and nominal economic growth in Russia has been proven by research on the close correlation between international oil prices and Russia's annual GNI growth figures.[18] Given that a significant proportion of Russia's GNI growth between 2004 and 2014 was based on improved terms of trade for fuel exports, the dramatic increase in GNI, which rose faster than total energy consumption, may have created the illusion of an improvement in energy efficiency, as measured in GDP per unit of energy consumed. Conversely, a dramatic decline in export revenues (caused by declining prices), may create the illusion of a decline in energy efficiency, as measured in GDP per unit of energy consumed. Furthermore, the fluctuations of international oil and gas prices may exaggerate Russia's nominal GNI rates of growth/decline

Therefore, a continuation of the current depression of international oil prices (and, by extension, a depression in the value of Russia's oil exports and the rate of Russia's nominal economic growth) will not necessarily result in a decline in energy consumption and related carbon emissions in Russia. Given that population growth, economic growth, and the nominal energy efficiency of Russia's economy are insufficient explanatory factors for explaining trends in Russia's energy consumption and CO_2 emissions, it is necessary to analyse Russia's domestic energy consumption by fuel and sector.

Trends in Russia's energy consumption

Total Primary Energy Supply (TPES)

Russia's energy consumption is overwhelmingly dominated by hydrocarbons (see Tables 10.5, 10.6, 10.7 and 10.8). Together, oil, natural gas, and solid fuels accounted for 88.2% of TPES in 2012, a slight decline from 88.8% in 2004. As can be seen from the tables below, the most significant trends during this period were the increases in oil consumption and nuclear power generation, and an absolute decline (in addition to a relative decline) in the consumption of solid fuels. To understand why these shifts took place, it is necessary to consider how energy is consumed in Russia, by examining trends in Russia's final energy consumption.

Final energy consumption – heat and electricity generation and consumption

According to the latest data available from the IEA, the most significant use of energy in 2012 was for the generation of heat and electricity, accounting for 52.5% of total primary energy consumption. This includes 54.7% of total coal consumption, 62.3% of natural gas consumption, 100% of nuclear and renewable energy consumption, and just 6.1% of refined oil products consumption.[19]

The production of heat and electricity takes place at electricity plants, heat plants, and combined heat and power (CHP) plants. Nuclear and hydroelectric power stations already account for 94.6% of fuel inputs for electricity-only power generation stations, but just 0.2% of fuel inputs at CHP plants and are not used at all in heat-only plants.[20] Therefore, Russia's electricity-only power plants are already essentially carbon-free.

CHP plants account for 67.1% of Russia's electricity generation and 46.7% of Russia's heat production. Given that natural gas accounts for 71.0% of fuel used in Russia's CHP plants,[21] it

Table 10.5 Changes in Russian fuel consumption 2004–2014

Fuel	*Consumption of fuel (million tonnes of oil equivalent – mtoe)*					
	2004	*2008*	*2012*	*2014*	*Change 2004–14*	
Oil	126.2	133.9	145.7	148.1	+21.9 mtoe	+17.4%
Natural gas	350.4	374.4	374.6	368.3	+17.9 mtoe	+5.1%
Solid fuel	99.9	100.7	98.4	85.2	−14.6 mtoe	−14.7%
Nuclear	32.7	36.9	40.2	40.9	+8.2 mtoe	+25.1%
Hydro	40.2	37.7	37.3	39.3	−0.9 mtoe	−2.2%
Other renewables	0.1	0.1	0.1	0.1	0.0 mtoe	0.0%
Total	649.5	683.7	696.3	681.9	+32.4	+5.0%

Source: BP, *Statistical Review of World Energy*, 10, 22, 32, 35, 36, 40.

Table 10.6 Changes in Russian total primary energy supply balance, 2004–2014

Fuel	*Share in Russia's total primary energy supply (% of total)*				
	2004	*2008*	*2012*	*2014*	*Change 2004–14*
Oil	19.4	19.6	20.9	21.7	+2.3%
Natural gas	54.0	54.8	53.8	54.0	0.0%
Solid fuel	15.4	14.7	14.1	12.5	−2.9%
Nuclear	5.0	5.4	5.8	6.0	+1.0%
Hydro	6.2	5.5	5.4	5.8	−0.4%
Other renewables	0.0	0.0	0.0	0.0	0.0%
Total	100.0	100.0	100.0	100.0	

Source: BP, *Statistical Review of World Energy*, 10, 22, 32, 35, 36, 40.

Table 10.7 Electricity generation by fuel source in Russia, 2004–2012

Fuel	*Share in electricity generation (% of total)*			
	2004	*2008*	*2012*	*Change 2004–12*
Natural gas	45.3	47.6	49.1	+2.3%
Solid fuel	17.3	18.9	15.8	−1.5%
Oil	2.6	1.5	2.6	0.0%
Hydro	19.1	16.0	15.6	−4.1%
Nuclear	15.5	15.7	16.6	+1.1%
Waste and biofuels	0.2	0.2	0.2	0.0%
Non-hydro renewables	0.0	0.1	0.1	+0.1%
Total	100.0	100.0	100.0	

Source: IEA, "Russian Federation: Energy Indicators."

Table 10.8 Heat generation by fuel source in Russia, 2004–2012

Fuel	*Share in heat generation (% of total)*			
	2004	*2008*	*2012*	*Change 2004–12*
Natural gas	66.5	65.8	67.6	+1.1%
Solid fuel	24.4	20.9	19.8	−4.6%
Oil	7.0	5.6	5.4	−1.6%
Hydro	0.0	0.0	0.0	0.0%
Nuclear	0.2	0.2	0.2	0.0%
Waste and biofuels	1.9	1.9	1.8	−0.1%
Non-hydro renewables	0.0	0.0	0.0	0.0%
Other (not specified)	0.0	5.6	5.2	+5.2%
Total	100.0	100.0	100.0	

Source: IEA, "Russian Federation: Energy Indicators."

is not a viable option for Russia to further reduce its coal consumption by switching from coal to natural gas in its CHP plants. In Russia's heat-only plants, natural gas already provides 74.1% of the source fuel, while biomass provides 2.7%. Coal provides 15.3%, and oil products the remaining 7.9%.[22] As with Russia's CHP plants, the existing dominance of natural gas in the fuel supply mix for Russia's heat plants makes the "quick win" substitution of gas for coal unlikely.

If Russia is going to continue using CHP plants, the only non-coal alternatives to natural gas are either a significant increase in the use of biomass/waste in CHP plants, or the introduction of nuclear power to Russia's district heating system along the lines of the proposed (later abandoned) Loviisa-3 project in Helsinki.[23] The latter approach is an area with significant potential. For example, the Leningrad II nuclear power plant in St Petersburg is currently under construction, with plans to supply district heating as well as electricity.[24] However, given Russia's large fuel demand for its CHP plants (238 mtoe in 2012) and the small size of Russia's biomass/waste production (7.1 mtoe in 2012) relative to Russia's coal consumption in CHP plants (58.8 mtoe in 2012),[25] the large-scale replacement of coal with biomass in the medium term is highly unlikely.

SRegarding Russia's final electricity consumption by sector, industry accounted for 45.7% in 2012, followed by commercial/public services (20.9%), residential (17.9%), and transportation (12.4%). In terms of heat consumption by sector, residential accounted for 46.3%, followed by industry (38.4%), commercial/public services (11.2%), and agriculture/forestry/fisheries (4.2%).[26] Given these statistics, it is clear that patterns of energy consumption, particularly heat and electricity, in the industrial and residential sectors have the greatest potential to affect Russia's overall energy consumption and related CO_2 emissions.

Final energy consumption – the industrial and residential sectors

Of Russia's 461 mtoe of final energy consumption in 2012, 31.2% was consumed by the industrial sector, 23.8% by the residential sector, 20.3% by the transportation sector, 14.5% as "non energy use" (predominantly the petrochemicals sector), 7.8% by the commercial/public services sector, and 2.4% by the agriculture/forestry/fishing sector.[27] Given that they account for more than half of Russia's energy consumption, an examination of the industrial and residential sectors is key to understanding whether Russia is undergoing an "energy transition."

Russia's industrial sector is not only a major consumer of heat and electricity, it is also a significant final consumer of coal and natural gas, accounting for 83.3% of final coal consumption and 26.6% of final natural gas consumption. When the use of coal for the generation of heat and power is discounted, 79.5% of Russia's coal supply is used for the creation of coke, with the use of bituminous coal and lignite consumption in the industrial, residential and commercial/public service sectors accounting for the remainder.[28] The use of coking coal in blast furnaces for steel production is highly significant for Russia. The IEA notes that Russia was the fifth-largest steel producer in the world in 2012, while "The steel industry is the main consumer of coking coal by far."[29]

According to the Intergovernmental Panel on Climate Change,[30] the production of steel, cement, ammonia, aluminum, and paper are the world's most energy-intensive industries. In addition to its role as a major steel producer, Russia was the world's seventh-largest cement producer in 2014.[31] The IEA reports Russia's cement industry as being largely gas-fuelled, with just five of Russia's 50 cement plants running on coal.[32] The IPCC also states that in 2012, Russia was the world's third-largest producer of ammonia (for which natural gas is an essential feedstock),[33] and the second-largest producer of aluminum (a process requiring large amounts of electricity). Finally, Russia was the fourteenth-largest paper producer in the world in 2013.[34]

While Russia's annual nominal GNI growth or decline may be "exaggerated" by the impact of international oil prices, "real" economic growth/decline will continue to influence the activities of these energy-intensive industries and, therefore, levels of energy consumption in Russia's industrial sector. Given that these industries will continue to form an important part of Russia's economy, they will also continue to contribute to Russia's energy consumption, particularly with regard to coal, natural gas, and electricity, for at least the medium term future.

In the residential sector, heat accounted for 52% of energy consumption in 2012, along with natural gas (29.3%), electricity (10.3%), refined oil products (5.7%), coal (1.5%), and biofuel/waste (1.1%).[35] Russia's district heating system is key to understanding energy consumption in the country's residential sector. According to the IEA, approximately 70% of Russia's population is connected to the district heating system.[36] However, supply chain losses (during the processes of heat generation, transmission, distribution, and end use) average 30–40%, and can be as high as 60%.[37] Clearly, there are significant opportunities to increase the efficiency of the production, transmission, and consumption of heat in Russia's residential sector, although a complete shift away from the district heating system towards boilers in individual apartments is unlikely.

The structure of Russia's energy consumption in the industrial and residential sectors highlights two areas in which Russia could reduce its per capita energy consumption and CO_2 emissions. Firstly, there is great potential for increases in the efficiency of Russia's heat and electricity generation. Secondly, although Russia's energy-intensive industries will remain significant consumers of heat and electricity for the foreseeable future, there is great potential for energy saving in the residential sector, by reducing losses in the district heating system and by lowering domestic energy consumption through better insulation of Russia's housing stock and the encouragement of more efficient use of electricity through a combination of metering, higher domestic prices, and higher energy efficiency standards for electrical consumer goods.

The importance of heat and electricity for Russia's energy consumption, particularly in the industrial and residential sectors, is matched by the huge potential for savings in energy consumption, in addition to the potential for a greater share of non-hydro renewables in Russia's electricity generation. Given that current trends suggest that the only significant increase in carbon-free electricity generation in Russia is due to come from increased nuclear power generation, energy efficiency, rather than a large-scale switch to renewables, is the key to Russia's potential energy transition.

Russia in international climate politics

Russia's engagement with international climate politics: the Kyoto Protocol (1997–2012)

The United Nations Framework Convention on Climate Change (UNFCCC) was signed at the Rio "Earth Summit" in 1992. In essence, the UNFCCC acknowledged that "change in the Earth's climate and its adverse effects are a common concern of humankind."[38] In doing so, it was declared, "The ultimate objective of this Convention [is] ... stabilization of greenhouse gas concentrations in the atmosphere at a level that would prevent dangerous anthropogenic interference with the climate system."[39] Parties to the convention divided into three, partially overlapping, groups. Annex I comprised "the industrialized countries that were members of the OECD (Organisation for Economic Co-operation and Development) in 1992, plus countries with economies in transition (the EIT Parties), including the Russian Federation, the Baltic States, and several Central and Eastern European States". Annex II comprises "the OECD members of Annex I, but not the EIT Parties". Non-Annex I comprised the developing countries.[40]

In 1997, parties to the UNFCCC adopted the first legally-binding commitments to stabilize, or reduce, greenhouse gas (GHG) emissions, in the form of the Kyoto Protocol. Article 25 of the Kyoto Protocol states that the protocol will enter into force when 55 parties, which includes Annex I parties comprising 55% of total (global) CO_2 emissions in 1990, have signed and ratified the Kyoto Protocol.[41] Iceland was the 55th state to ratify the Kyoto Protocol, on May 23, 2002. By April 2004, 25 of the current EU-28, along with Norway, Switzerland, Canada, Japan, South Korea, and Ukraine had ratified the Kyoto Protocol. However, several major emitters – Australia, Russia and the United States – had still not ratified the Protocol. For it to come into effect, either Russia or the United States had to ratify the Protocol. In the event, it was Russia that brought the Kyoto Protocol into effect. With the Russian Government having signed the Kyoto Protocol on March 11, 1999, it was then ratified by the Russian Parliament on November 18, 2004. Thus, the Kyoto Protocol entered into force on February 16, 2005.[42]

The Kyoto Protocol lists the following greenhouse gases as being targeted for reduction under the scheme: carbon dioxide (CO_2), methane (CH_4), nitrous oxide (N_2O), hydrofluorocarbons (HFCs), perfluorocarbons (PFCs), and sulphur hexafluoride (SF_6). During the first commitment period of the Kyoto Protocol (2008–2012), the target for the Russian Government was to limit the collective emission of CO_2, CH_4, and N_2O (as measured in CO_2 equivalent) to 100% of the 1990 baseline level, and to limit the emission of HFCs, PFCs, and SF_6 to 100% of the 1995 baseline level.[43] Therefore, the ongoing target was to limit total greenhouse gas emissions to 3,368 million tonnes of CO_2 equivalent.[44]

For the Russian Government, this was not a strenuous target. Between 1990 and 1998, the economic decline that had accompanied the collapse of the Soviet Union had contributed to a dramatic fall in Russia's greenhouse gas emissions, from 3,368 million tonnes of CO_2 equivalent to just 2,005 million tonnes – a decline of 40.5%. Between 1998 and 2012, Russia's greenhouse gas emissions increased every year except 2009, to reach 2,297 million tonnes of CO_2 equivalent.[45] This was still 31.8% below the baseline target.

Under conditions of easily-attainable targets, ratification of the Kyoto Protocol was a no-cost policy that brought Russia the chance to promote its role in international cooperation. This stance was made evident in a statement issued by the Russian Government, to mark the ratification of the Protocol by the Russian Parliament:

> The Russian Federation proceeds from the assumption that the commitments of the Russian Federation under the Protocol will have serious consequences for its social and economic development. Therefore, the decision on ratification was taken following a thorough analysis of all factors, inter alia, the importance of the Protocol for the promotion of international cooperation, and taking into account that the Protocol can enter into force only if the Russian Federation ratifies it.[46]

Russia's ratification of the Kyoto Protocol brought further benefit: EU support for Russia's WTO accession, agreed at the EU-Russia Summit in May 2004, was at least partially dependent on Russia's ratification of the Kyoto Protocol.[47]

Why Russia did not participate in the Kyoto Protocol second commitment period (2013–2020)

At the 16th Conference of Parties (COP16) of the UNFCCC, held in Cancun, Mexico, in December 2010, the Russian Special Envoy for Climate, Alexander Bedritsky, announced, "Russia will not participate in the second commitment period of the Kyoto Protocol."[48] The statement also confirmed that Russia would continue working towards a non-binding goal of emissions 15–25% below 1990 levels by 2020.[49] The declared reason was that Russia would not commit to more legally-binding targets, if other major emitters (and economic competitors) such as China, India, and South Africa, let alone the United States, were not prepared to do so. To quote from the Bedritsky statement:

> Russia has repeatedly stated, including at the highest political level, that the adoption of commitments for the Second Commitment Period under the Kyoto Protocol as it stands now would be neither scientifically, economically, nor politically effective ... Russia consistently advocates the extension of the list of emissions-reduction-commitment countries and the inclusion of the fast-growing economies on it.[50]

Two years later, at COP18 in Doha, Bedritsky re-affirmed the position of the Russian Government:

> The Russian Federation, like a number of UNFCCC Annex-I countries, does not intend to make quantitative commitments for further GHG-emissions reduction in the second period ... On the subject of a new agreement, a strategic determinant for the Russian Federation is the active participation of all countries of the world, primarily of major GHG-emitters, in the global climate solution. While the substance of the climate commitments and the actions of developed and developing countries may differ post 2020, everything should be reflected in a single document.[51]

The desire for a global agreement, with the participation of the major emitters that currently do not have legally-binding commitments (particularly the United States, China, and India), may not be the only reason for Russia's non-participation in the second commitment period. It is also possible that the Russian government is concerned that Russia might not meet its target of emissions 25% below the 1990 baseline by 2020, and therefore wishes to keep its target non-binding.

In the eight years from 2004 to 2012, Russia's GHG emissions rose by 6.6% (142 million tonnes of CO_2 equivalent), from 2,155 to 2,297 million tonnes of CO_2 equivalent.[52] This equates to an average annual emissions growth of 0.8%. If Russia's emissions were to grow at the same pace between 2012 and 2020, they would reach 2,449 million tonnes. For comparison, a 2020 target of 25% below the 1990 baseline would equate to 2,526 million tonnes. To exceed that target, Russia's emissions would need to grow by just 1.2% per year between 2012 and 2020. The fact that the Russian Government may well regard failure to meet its emissions reduction target as a distinct possibility, suggests that there is a strong degree of scepticism about the country's ability to change the structure of its energy consumption and related GHG/CO_2 emissions.

Conclusions

Trends in Russia's energy consumption and CO_2 emissions have shown gradual growth over the past decade, under conditions of broader economic growth and stable population levels.

Russia has yet to join the most advanced economies of the world in breaking the link between economic growth and rising energy consumption. The reason for this may be found in Russia's abundant hydrocarbon resources. These resources support energy-intensive heavy industry and ensure Russia's position as a world-leading hydrocarbon exporter – a position that encourages further energy consumption through the extraction, refining, and transportation of those resources. This is demonstrated by Russia's low scores on units of GDP production per unit of energy consumed. Russia's abundant hydrocarbon resources also reduce the urgency of improving energy efficiency in both energy (heat and electricity) generation and in energy consumption. The task of upgrading Russia's district-heating boiler houses, reducing heat losses during transmission in Russia's district heating system, and insulating Russia's vast housing stock to reduce demand levels is a monumental challenge that will require huge investment, which is simply not expected in the medium-term future.

Regarding attempts to reduce CO_2 emissions from heat and electricity generation, debates over the environmental merits of replacing coal with natural gas in thermal power plants, despite the economic cost, do not apply to Russia. In Russia, natural gas already provides the source fuel for half of Russia's electricity generation, with non-CO_2 emitting hydro and nuclear power providing a further third. The "quick win" of replacing coal with gas is not viable in Russia, simply because that option has already been utilized. One promising development, in terms of the reduction of CO_2 emissions, is the rising share of nuclear power in Russia's TPES and electricity generation, at the expense of coal. Building on the 25% increase in Russian nuclear power production between 2004 and 2014, the Russian Government is keen to promote a further expansion of nuclear power production in Russia, while the state-owned RosAtom is keen to market Russian technology abroad. This would suggest that Russia's nuclear industry will continue developing in the coming decade, making further growth in the role of nuclear power in Russia's energy mix all the more likely.

When hydropower is discounted, the development of renewable energy in Russia is virtually non-existent. Despite Russia's vast latent renewable energy potential, it will not be developed on a substantial scale until it is commercially profitable to do so. This would require feed-in tariffs and electricity prices that are simply not politically and economically viable at present. Furthermore, the lack of energy import dependence, due to Russia's abundant gas supplies and well-developed nuclear industry, means that using renewables as a means of improving the country's energy security is simply not an issue for Russia.

The stability of Russia's primary energy mix, lack of investment in renewables other than large-scale hydroelectricity, and sheer size of the challenge of improving Russia's energy efficiency, means that an energy "transition" in Russia is not expected in the medium-term future. The only significant development that may impact Russia's CO_2 emissions is the increasing use of nuclear power, at the expense of coal, for electricity generation. However, in the context of abundant hydrocarbon and nuclear power resources, Russia is unlikely to engage in a dramatic change in its primary energy mix, and is equally unlikely to limit its CO_2 emissions. This, in turn, influences government policies that prioritise investment in the oil, natural gas, and nuclear power sectors, and distance Russia from international, legally-binding commitments to reduce its CO_2 emissions as part of its engagement in global climate change action.

To conclude, under conditions of rising domestic energy demand, economic constraints, and abundant supplies of traditional hydrocarbon and nuclear fuels, if Russia does embark on an energy transition, it will be a transition to dramatically improved levels of energy efficiency rather than the widespread development of renewable energy supplies.

Notes

1 BP, "Statistical Review of World Energy – Data Workbook," BP, 2015, www.bp.com/content/dam/bp/excel/Energy-Economics/statistical-review-2015/bp-statistical-review-of-world-energy-2015-workbook.xlsx; BP, *Statistical Review of World Energy* (London: BP, 2015), 10, 22, 32, 35, 36.
2 International Energy Agency (IEA), *CO_2 Emissions from Fuel Consumption: Highlights* (Paris: IEA, 2014), 36–38.
3 BP, *Statistical Review of World Energy.*
4 IEA, *CO_2 Emissions from Fuel Consumption*, 37; United Nations Framework Convention on Climate Change (UNFCCC), "Greenhouse Gas Inventory Data – Detailed Data by Party," UNFCCC, 2015, http://unfccc.int/di/DetailedByParty/Event.do?event=go.
5 World Bank, "CO_2 Emissions (kt)."
6 World Bank, "CO_2 Emissions (kt)"; World Bank, "CO_2 Emissions: Metric Tonnes Per Capita."
7 World Bank, "CO_2 Emissions: Metric Tonnes Per Capita."
8 World Energy Council, *World Energy Resources: 2013 Survey* (London: World Energy Council, 2013), 6.
9 World Bank, "World Data Bank: Russia," World Bank, 2015, http://databank.worldbank.org/data//reports.aspx?source=2&country=RUS&series=&period=#selectedDimension_WDI_Ctry.
10 World Bank, "World Data Bank: GNI per Capita, PPP."
11 Organization of Petroleum Exporting Countries (OPEC), "Table 3.22. World Natural Gas Exports by Country (Million Standard Cubic Metres)," OPEC, 2015, www.opec.org/library/Annual%20Statistical%20Bulletin/interactive/current/FileZ/XL/T322.HTM.
12 Ministry of Finance (Russian Federation), "Struktura I dinamika dokhodov," 2015, http://info.minfin.ru/fbdohod.php.
13 RosStat, "26.8. Commodity Structure of Exports of the Russian Federation," RosStat, 2013, www.gks.ru/bgd/regl/b13_12/IssWWW.exe/stg/d02/26-08.htm.
14 International Energy Agency (IEA), *Energy Policies beyond IEA Countries: Russia* (Paris: IEA, 2014), 19.
15 World Bank, "World Data Bank: GNI per Capita, PPP."
16 World Bank, "World Data Bank: Energy Use."
17 US Energy Information Administration (USEIA), "Europe Brent Spot Price (FOB)," USEIA, 2015, www.eia.gov/dnav/pet/hist/LeafHandler.ashx?n=PET&s=RBRTE&f=D.
18 M. Kuboniwa, "Diagnosing the 'Russian Disease': Growth and Structure of the Russian Economy," *Comparative Economic Studies* 54 (2012): 121–128.
19 International Energy Agency (IEA), "Russian Federation: Energy Indicators," IEA, 2015, www.iea.org/statistics/statisticssearch/report/?year=2012&country=RUSSIA&product=Indicators.
20 IEA, "Russian Federation: Energy Indicators."
21 IEA, "Russian Federation: Energy Indicators."
22 Ibid.
23 H. Tuomisto, "Nuclear District Heating Plans from Loviisa to Helsinki Metropolitan Area" (presented at the Joint NEA/IAEA Expert Workshop on the Technical and Economic Assessment of Non-Electric Applications of Nuclear Energy, Paris, France, April 2013), www.oecd-nea.org/ndd/workshops/nucogen/presentations/3_Tuomisto_Nuclear-District-Heating-Plans.pdf.
24 World Nuclear Association, "Nuclear Power in Russia," World Nuclear Association, 2015, www.world-nuclear.org/info/Country-Profiles/Countries-O-S/Russia–Nuclear-Power.
25 IEA, "Russian Federation: Energy Indicators."
26 Ibid.
27 Ibid.
28 Ibid.
29 IEA, *Energy Policies beyond IEA Countries: Russia*, 171.
30 Intergovernmental Panel on Climate Change (IPCC), *Climate Change 2014: Mitigation of Climate Change: Working Group III Contribution to the IPCC Fifth Assessment Report* (IPCC, 2014), http://mitigation2014.org/report/publication/, 748.
31 Statista, "Major Countries in Worldwide Cement Production from 2010 to 2014 (in Million Metric Tons)," Statista, 2015, www.statista.com/statistics/267364/world-cement-production-by-country/.
32 IEA, *Energy Policies beyond IEA Countries: Russia*, 171.
33 US Geological Survey (USGS), *Mineral Commodity Summaries: Nitrogen (Fixed) Ammonia* (USGS, 2015), http://minerals.usgs.gov/minerals/pubs/commodity/nitrogen/mcs-2015-nitro.pdf.

34 Swedish Forest Industries Federation, "Production and Export of Paper," November 24, 2014, www.forestindustries.se/documentation/statistics_ppt_files/international/production-and-exports-of-paper.
35 IEA, "Russian Federation: Energy Indicators."
36 IEA, *Energy Policies beyond IEA Countries: Russia*, 259.
37 Ibid.
38 United Nations, *United Nations Framework Convention on Climate Change* (United Nations, 1992), http://unfccc.int/files/essential_background/background_publications_htmlpdf/application/pdf/conveng.pdf.
39 Ibid., 9.
40 United Nations Framework Convention on Climate Change (UNFCCC), "Parties and Observers," UNFCCC, 2015, http://unfccc.int/parties_and_observers/items/2704.php.
41 United Nations Framework Convention on Climate Change (UNFCCC), *Kyoto Protocol to the United Nations Framework Convention on Climate Change* (UNFCCC, 1998), http://unfccc.int/resource/docs/convkp/kpeng.pdf, 18.
42 United Nations Framework Convention on Climate Change, "Status of Ratification of the Kyoto Protocol," UNFCCC, 2015, http://unfccc.int/kyoto_protocol/status_of_ratification/items/2613.php.
43 UNFCCC, *Kyoto Protocol to the United Nations*, 19–20.
44 United Nations Framework Convention on Climate Change (UNFCCC), "Times Series Annex I Data for Greenhouse Gas (GHG) Total – GHG Total Excluding LULUCF," UNFCCC, 2015, http://unfccc.int/ghg_data/ghg_data_unfccc/time_series_annex_i/items/3814.php.
45 UNFCCC, "Times Series Annex-I."
46 United Nations Framework Convention on Climate Change (UNFCCC), "Declarations by Parties – Kyoto Protocol – Russian Federation," UNFCCC, 2015, http://unfccc.int/kyoto_protocol/status_of_ratification/items/5424.php.
47 W. Douma, "The European Union, Russia, and the Kyoto Protocol," in *EU Climate Change Policy: The Challenge of New Regulatory Initiatives*, ed. M. Peeters and K. Deketelaere (Cheltenham: Edward Elgar, 2006), 61–62; C. Bretherton, and J. Vogler. *The European Union as a Global Actor* (Abingdon: Routledge, 2006), 104–105; S. Charap, and G. Safonov, "Climate Change and the Role of Energy Efficiency," in *Russia after the Global Economic Crisis*, ed. A. Aslund, S. Guriev, and A. Kuchins (Washington, DC: Peterson Institute for International Economics, 2010), 130–134.
48 United Nations Framework Convention on Climate Change (UNFCCC), "Statement by the Adviser to the Russian President, Special Envoy for Climate, Alexander Bedritsky, UNFCCC COP 16/CMP6 (Cancun, Mexico, 9 December 2010)," UNFCCC, 2010, https://unfccc.int/files/meetings/cop_16/statements/application/pdf/101209_cop16_hls_russia.pdf.
49 Ibid.
50 Ibid.
51 United Nations Framework Convention on Climate Change (UNFCCC), "Statement by the Adviser to the Russian President, Special Envoy for Climate, Alexander Bedritsky, UNFCCC COP 18/CMP8 (Doha, Qatar, 6 December 2012)," UNFCCC, 2012, http://unfccc.int/resource/docs/cop18_cmp8_hl_statements/Statement%20by%20Russia%20%28COP%20%29.pdf.
52 UNFCCC, "Times Series Annex-I."

11

Energy and climate transitions in Mexico

The emergence of a "política ambiental de estado"

Duncan Wood

Mexico is one of the world's most important emerging economies, a major oil producer and a country that has bet heavily on industrialization and manufacturing exports for its development strategy. It is a country with a growing population, a relatively young demographic profile, and a rapidly expanding middle class with the normal aspirations. Mexico is also the world's tenth largest emitter of greenhouse gas (GHG) emissions so it is surprising to many that Mexico stands out as an important example of a country that has chosen to exercise a leadership role in the global conversation over climate action, and in recent years has begun to enact meaningful changes in national legislation aimed at reducing the country's GHG emissions, at the same time as it has been successful in building a viable renewable energy industry.

This chapter provides an overview of Mexico's emergence as such a leader, and analyzes its approach to climate action. I propose that, while Mexico has defied those skeptics who argued that an oil-producing state could not be taken seriously as a climate change leader, achieving its ambitious GHG reduction targets will require an enormous commitment of time, resources and political capital over a long period of time and across several administrations, something that has traditionally proved difficult in Mexico. Nonetheless, the recent energy reform process in Mexico provides us with hope that there will be significant reductions in the short term thanks to the modernization of the energy sector, in particular electricity generation, as well as a new Energy Transition legislative package that was passed by the Mexican Congress in December 2015.

The strongly centralized nature of Mexican energy and natural resources policy-making plays a significant role in this story. The existence of two publicly-run energy monopolies in Mexico, Pemex (the national oil company or NOC) and the Comision Federal de Electricidad (CFE, the national electricity utility), and the capacity of the federal government to engage in national energy planning and infrastructure projects has greatly helped the effort to reduce emissions in recent years. This may seem ironic to some, as both Pemex and CFE have hardly been viewed as exemplary climate citizens in the past. However, the energy reform process in Mexico that began in 2013 is proving to be transformative not only in terms of the structure of the energy sector, but also in the incentives to both reduce emissions and improve economic competitiveness. Mexico therefore serves as an important example of a positive sum game between economic development and climate action.

Explaining Mexico's approach to climate change

For many years, Mexico did not occupy a central role in the global conversation on climate change. Although the Mexican government signed the UNFCCC in 1992, and was a signatory to the Kyoto Protocol in 1998 as a non-Annex 1 country, early government approaches focused on greenhouse gas inventories and energy and emissions intensity arguments. Projections made by the Mexican government in the early 2000s showed that the priority was clearly on economic growth and improving the energy efficiency of that growth, rather than in reducing overall emissions.

It is easy to understand why Mexico put only minimal emphasis on reducing emissions in these early years. First, Mexico's economy and government budget have depended heavily on hydrocarbons since the middle of the twentieth century and Mexicans view themselves as an "oil nation." In fact until 2004, when Pemex reached record high production of 3.4 million barrels per day, revenue from oil production was seen to be an increasingly important element in national economic growth, employment and wealth creation. The truth is, of course, that oil production has become a far less important element in the Mexican economy since the 1990s, with the rise of a manufacturing export sector that has become the envy of the world, and now dwarfs oil as a percentage of national GDP. Nonetheless, Mexico's attitudes to climate change were consistent with those of an oil-producing state until the mid-2000s.

Pulver has argued that, prior to 2006, Mexican approaches to climate change were shaped by three factors.[1] First, the government adopted a "stop-and-go" attitude to climate policy, responding largely to the international and bilateral agenda with the United States, but failing to maintain consistency throughout successive administrations. Second, PEMEX successfully opposed a more aggressive stance on emissions reductions, and third, civil society failed to generate any policy momentum in this field.

This is not to say that there was no interest in climate change in Mexico, and there was considerable attention dedicated to the issue among the Mexican scientific community in the 1990s. Of particular importance was the international recognition received by Mario Molina after his 1995 Nobel Prize given for his discovery of the hole in the ozone layer. Molina went on to create an eponymous center for the study of energy and the environment with a heavy focus on climate change.

However, it was only after 2006 that the Mexican government began to adopt a more coherent approach. A number of factors conspired to drive a change in opinion and policy. First, there was rising environmental awareness and activism due to the high air pollution levels in Mexico City, which experienced persistent critical air quality advisories in the 1990s and 2000s. Although civil society is not as active as in other countries, it has played an important role as diverse environmental groups and universities have worked alongside the scientific community and a more left-leaning government of Mexico City after 1997, to raise public awareness about environmental degradation in general, and about the harmful effects of emissions in particular. Mexico's political and social opening to the world has also helped this cause, with much greater interaction between Mexican and foreign environmental NGOs. Having achieved important advances during the last 20 years in terms of the reduction of automobile emissions, of the ozone at sea level and suspended particles in the air, all levels of government have searched for other ways to improve the local and national environment which has attracted not only international attention and praise, but has also generated confidence that Mexico could be a leader in environmental questions.

The second factor was the growing awareness that Mexico's geography makes it especially susceptible to the harmful effects of climate change. With significant parts of national territory located below the Tropic of Cancer, situated beside the hurricane factory that is the Gulf of Mexico, possessing enormously long coastlines of both Pacific and Atlantic sides of the country, and with major populations in rural areas depending on near-subsistence agriculture and living in poorly constructed housing, Mexico is more vulnerable than most countries to the effects of violent weather. In fact, the Mexican government estimates that violent weather events related to climate change have cost almost 340 billion pesos in the early years of the 21st century alone.[2]

Third, Mexico's commitment to the global climate change regime solidified over time. A signatory to the original UNFCCC as a non-Annex 1 country, Mexico's position as one of the few developing states that are also members of the OECD helped to drive a realization of the importance of shared responsibility between Annex 1 and non-Annex 1 states to reduce emissions. This socialization process is reflected in the growing size of Mexican delegations to COP meetings as well as in the increasing number of written submissions to international negotiations.[3]

The fourth factor, however, is more idiosyncratic. The election of Felipe Calderón to the Mexican presidency in 2006 heralded a major shift in national climate change approaches. His predecessor, Vicente Fox, had done little to consolidate a coherent national climate change policy but Calderón, despite being from the same party (the National Action Party or PAN), adopted an aggressive approach, both domestically and internationally on the issue. As President, Calderón cited his father's firm belief in the importance of sustainability as one of the major influences in his own political thinking.[4] The theatrical release of former U.S. Vice President Al Gore's movie, "An Inconvenient Truth" is also reckoned to have motivated him. This personal attachment to the issue drove him to try to make Mexico a climate change leader, not just in Latin America but among developing countries. This was not just a passing fancy, either: since leaving office Calderón has chaired the Global Commission on the Economy and Climate, an organization that argues against the notion that fixing the climate and emissions rules are bad for the economy, calling the choice between climate action and economic prosperity a "false dilemma."[5]

Climate change and Mexican public opinion

Mexican public opinion has reflected this shift in awareness and policy. Between 2007 and 2010, the Mexican public increased its recognition of climate change as a serious threat by 10%, from already relatively high levels (see Table 11.1). What is perhaps extraordinary about this is that most Mexicans admit that they have a very elementary understanding of the issue. A Parametria poll in 2014 revealed that 76% of Mexicans accept that they know little about climate change and another 16% say they know nothing about it.[6]

However, it is not clear that awareness of climate change or scientific knowledge of the issue makes a big difference in concern for climate. A recent poll by Pew has shown that the link between education and support for climate action is less than solid. A survey of U.S. public opinion demonstrated that political affiliation and age are much better determinants. In fact, education and scientific knowledge seem to have little relation to environmental attitudes in general in the U.S.[7]

A stronger link appears to exist between levels of emissions per capita and concern over climate. Another Pew poll highlights the fact that the Mexican public has a much higher level of

Table 11.1 Threat from global warming in Latin America (% who view global warming as serious threat)

Country	*2007–2008*	*2010*	*Change (pct. pts)*
Ecuador	69%	85%	16
Venezuela	62%	80%	18
Brazil	76%	78%	2
Chile	69%	75%	6
Colombia	65%	75%	10
Peru	58%	75%	17
Mexico	**63%**	**75%**	**10**
Argentina	71%	70%	1
Costa Rica	72%	69%	−3
Uruguay	68%	68%	0
Panama	61%	66%	5
Bolivia	51%	63%	12
Guatemala	51%	61%	10
El Salvador	51%	61%	10
Paraguay	54%	61%	7
Honduras	57%	51%	−6
Dominican Republic	46%	51%	5
Haiti	35%	18%	−17

Source: Gallup, "Fewer Americans, Europeans View Global Warming as a Threat," http://www.gallup.com/poll/147203/Fewer-Americans-Europeans-View-Global-Warming-Threat.aspx, April 20, 2011.

concern over climate change than its counterparts in either the U.S. or in Canada; whereas Canadians and Americans show a relatively low level of concern over the issue, Mexico is in the same peer group as other low per capita emitters such as Argentina, Chile, India and Vietnam.[8]

Mexico and renewable energy

The renewable energy story of Mexico is particularly encouraging. The country is blessed with considerable renewable energy potential, due to impressive wind, solar and geothermal resources. Since the mid-2000s, Mexico has engaged in an ongoing process of modernizing its energy sector, focusing on trying to raise energy production and increase efficiency. At the same time, and especially since 2005, national energy policy and private sector initiatives have driven the rise of renewable energy sources in Mexico. A number of factors combined to encourage investment in the renewables field. First, the problems of the oil and gas sector should be recognized. After the national oil company, Pemex, saw peak oil production in 2004 at 3.4 million barrels per day, production dropped precipitously over the next decade, reaching only 2.2 million barrels per day by 2016.[9] As a result, replacing the use of fuel oil for electricity generation with alternative sources became attractive to policy makers as hydrocarbons not used internally would become available for exporting. However the true potential of this approach would not be seen until the PRI administration of Enrique Peña Nieto after 2012.

In 2008, the Mexican Congress passed a new Law for the Better Use of Renewable Energy and the Financing of the Energy Transition (LAERFTE in Spanish). Significantly, the law required that the energy ministry, SENER, create new rules and offer new incentives for the renewable energy industry in Mexico. Therefore, in addition to establishing a clearer legal framework for renewable energy, the law established a modest fund of US$220 million a year for three years to fund projects related to renewable energy. Simultaneously, a National Energy Strategy put forward by the Secretary of Energy and the Works and Investment Program of the Electricity Sector (POISE) proposed by the national electricity utility, CFE set a goal to generate 35% of electricity from non-fossil sources by 2024.

Although the government and CFE have been key drivers of renewable energy in Mexico, foreign actors have also played an important role in promoting renewable energy. A number of foreign governments have engaged with Mexico on the issue, including the United Kingdom and Germany, and investors from those same countries along with Spain have been active in developing and operating wind power plants in the Mexican southwest. However, it is fair to say that the United States has been the most closely involved in Mexico. In the 1990s, the U.S. government financed solar power projects in Mexico through the United States Agency for International Development (USAID). Later, USAID promoted the use of wind energy in Oaxaca by sponsoring a survey of the wind power resources in the state in 2002.[10] This initiative laid the foundation for the development of the energy program in the region of La Ventosa and was exhaustively used to attract investors to the project.[11] In 2008 and 2009, USAID also played a role in developing public education programs centered on the financing for renewable energy projects and underlined the lessons that should be learned from the regulatory frameworks used in the United States of America for renewable energy.[12]

By 2014, Mexico had made impressive progress on renewable energy. 22% of all electricity generated came from non-fossil fuels, with hydro-electric power by far the biggest component. Of the total of 16,295 MW of total renewable installed capacity in 2014, Mexico had 11,632 MW of hydro capacity, accounting for 18% of total installed generation capacity. Hydroelectricity supplied about 15% of Mexico's electricity generation in 2014.[13] In addition to large hydroelectric plants, located predominantly in the south of the country, the CFE and private firms are also focusing efforts on completing a number of smaller projects (<30MW). It is expected that these projects will bring another 289 MW of hydroelectric power to the system by 2016.[14]

However, non-hydro renewable power in Mexico offers an equally, if not more, encouraging story than traditional hydro. Prior to the energy reform in 2013, Mexico was the world's fifth largest producer of geothermal power, with almost 1,000 MW of installed capacity, and the passing of the geothermal law and the Round 1 auction of geothermal fields to private bidders offers the prospect of significant additions to that total.

Of critical importance, however, in the renewable energy story in Mexico has been the role of the private sector.[15] The decision by some of Mexico's leading businesses, including cement giant Cemex, mining conglomerate Grupo Bal, Walmart de Mexico S.A., and Coca Cola bottlers FEMSA to reduce their carbon footprint by engaging in self-supply contracts with private renewable energy providers allowed for the rapid expansion of wind power in Mexico up to more than 2,000 MW of installed capacity by the end of 2012.[16] This was helped greatly over the past decade by the high price of electricity for Mexican industrial consumers. Both the states of Oaxaca and Baja California have been centers of wind energy investments, and Energy Secretary Pedro Joaquin Coldwell announced in 2015 that the government hopes to bring in $14 billion in investment in wind energy by 2018 to boost total installed capacity from 2,000 MW to 12,000 MW.[17] Although this seems ambitious, between 2010 and 2013 Mexican wind power saw an almost seven-fold increase in installed capacity. To make it possible, however, there will need to be concomitant investments in transmission capacity to get the electrons to market. Mexico's opening of its electricity will be a crucial component in facilitating such investment.

Although solar power is still largely undeveloped as a resource in Mexico, recent years have seen growing interest and a number of significant projects. Until recently, the main attraction for solar investments was in rooftop PV installations for residential applications. Due to the sliding tariff used by CFE, large consumers of electricity who would normally have to pay the high domestic consumption (or DAC) rate, found that they could achieve major savings on their electricity bills by installing solar panels, thereby gaining access to the lower tariff reserved for smaller consumers. However, since 2012 there has been renewed attention to large-scale solar generation, especially in the north of the country. Mexico has an extraordinary solar resource, with a national insolation average of around 5 kilowatt hours per square meter per day (5kWh/m^2/day). In northern states such as Sonora, Chihuahua and Durango, the solar resource is of extremely high quality, with around 6kWh/m^2/day. It is expected that, with increasing attention to solar, and with lower costs due to mass production of solar panels and improved technologies, solar power will experience a boom in Mexico in the near future.

Energy policy and climate under the Calderón administration: modernization and mainstreaming

Throughout the Calderón administration, the President repeatedly referenced the need for both national actions and international collaboration to address the issue. At home, the PAN government was able to secure legislation in the Congress to help develop renewable energy, and sought to limit the emissions from electricity generation, and improve energy efficiency. Much of this was foretold by the publication of Mexico's 2007 National Strategy on Climate Change (ENCC), which laid out the executive branch's perspective on climate action and provided insights into future legislative priorities. Although traditionally a hydrocarbons-heavy energy producer, the early years of the 21st century saw Mexico begin to focus its attention on alternative energy sources.

Pulver has provided an excellent overview of Mexico's climate change policy up until 2012 and shows that the Calderón administration enacted a number of legislative initiatives that highlighted the president's commitment to climate action. The administration made an impressive early commitment to increasing the share of renewable energy in the electricity generation mix, as well as reducing overall emissions. However, the culmination of these efforts came in the last year of the Calderón government, with the Congressional passage of a General Law on Climate Change (GLCC), which made the following ambitious commitments:

- The government would create an inter-ministerial commission on climate change.
- A new climate fund would be created to help mitigation efforts.
- An emissions market would be created.
- The government would establish emissions measurement, reporting and verification requirements.
- The government would reduce and then reverse the trend of deforestation.
- The government would increase the percentage of renewable energy in the electricity generation mix to 35%.
- The government would reduce fossil fuel subsidies.

Perhaps the single most important component of the GLCC was the creation of the inter-ministerial commission on climate change. This body allows for a mainstreaming of climate policies across government ministries and policies, driving greater coordination and cross-fertilization of ideas.

Throughout the Calderón administration, international financing for Mexico's climate action strategy remained a priority. Whether it was through the Green Climate Fund (GCF), through Clean Development Mechanism (CDM) financing for methane avoidance, renewable energy or energy efficiency projects, or through foreign investment in renewable energy projects, international money has been behind much of the progress. It is important to note that Mexico became a major beneficiary of the CDM, with the second largest number of CDM projects in Latin America after Brazil, and fourth in the world (with 127 CDM-financed projects by 2013).

Alongside renewable energy (see above for details), the Calderón administration took the lead in emphasizing energy efficiency as a strategy that was justified both in terms of its climate effects and its economic benefits. In 2009 the Mexican government launched a national energy efficiency plan, coinciding with the transformation of the National Council on Energy Savings (CONAE) to the National Council on the Efficient Use of Energy (CONUEE). This plan, named National Program for the Sustainable Use of Energy (PRONASE) developed in conjunction with the consulting firm McKinsey, evaluated different opportunities for energy efficiency gains in Mexico and identified seven strategic areas where the government could make significant progress: lighting, domestic appliances, transport, green buildings, cogeneration, industrial motors and water pumping.

The PRONASE plan evaluated each of these opportunities on a cost-benefit calculation and identified lighting as the most cost-effective way to launch a national energy efficiency program. The program focused on the transition from incandescent light bulbs to more efficient, fluorescent bulbs. Beginning with government buildings, and then encouraging homeowners to switch over by offering subsidies for the purchase of the new bulbs, the Mexican government, working closely with CFE, was able to secure rapid progress. In a two-year period, the program exchanged 45.8 million light bulbs (and gave away almost 23 million light bulbs), achieving energy savings equivalent to taking nearly 600,000 cars off Mexican roads for a year.

The second phase of the PRONASE focused on appliances and offered subsidies for consumers who replace older, less efficient domestic appliances such as fridges and washing machines with newer models. By emphasizing the economic benefits of doing so, and by offering financial incentives, the government was able to convince a large number of consumers to make the switch, with nearly 2 million appliances being replaced under the program by the end of the administration. The Mexican government also collaborated with U.S. authorities to develop a national efficiency certification standard, known as FIDE, and to harmonize the standard with the United States' Energy Star standard.

Also of major significance was the decision by the Calderón administration to end the practice of subsidizing gasoline prices in Mexico. Though widely reviled by the public, the monthly "*gasolinazos*" of raising fuel prices by reducing the fuel subsidy succeeded in focusing attention on fuel conservation and the energy efficiency of automobiles. This practice was continued under the Peña Nieto administration until the gasoline subsidy had finally been eliminated in anticipating of the opening of the downstream market in 2016.

The Peña Nieto administration and energy reform: discovering a "green lining"

The major achievement of the Peña Nieto administration to date has been the successful execution of a transformative energy reform that opens oil, gas and electricity sectors to private and foreign participation. The heavy media focus on oil production in this reform has meant that many observers have neglected the climate change policies of the government since 2012. Indeed most analysts displayed considerable skepticism about Peña Nieto's commitment to climate action in the early months of the government,[18] but the administration seized on the progress made by its predecessor and developed policy ownership on the issue. Despite a main focus in energy policy on rejuvenating the hydrocarbons industry, in the first three years of the PRI government we have seen an impressive commitment to building on existing laws and strategies, and continuing to make major international commitments on GHG emissions reduction.

The energy reform itself, a Constitutional reform approved by the Mexican Congress in December 2013, with secondary (implementing) legislation following in August 2014, represented a paradigm shift for Mexico's energy sector. From a closed, state-dominated, monopolistic paradigm, Mexico's energy sector has rapidly been transformed into an open, market-oriented system. This is allowing for greater efficiencies, for experimentation with new technologies and processes, and access to capital.

While it is true that the administration has been most concerned with attracting investment into the oil and gas sector in Mexico, a concomitant focus on both climate change and renewable energy has established strong political credentials for the government in international negotiations. What's more, the administration has succeeded in offering both environmental and business benefits from the reform process thus far.

The first suggestion that the President would not neglect the renewable energy sector came in the announcement of the energy reform package in the fall of 2013. Alongside a major shakeup of the hydrocarbons and electricity sectors, the President announced a new geothermal energy law that offers concessions to private firms to develop the country's significant geothermal resources. Although there have been suggestions that the intention here was to offer business opportunities to political allies anxious to get involved in the sector, the new law has already attracted interest from national and foreign investors.[19]

The most important contribution towards enabling emissions reductions in Mexico, however, came with the government's commitment to build new gas pipeline infrastructure to reduce natural gas shortages and to lower the cost of electricity generation. In the final years of the Calderón administration, natural gas consumers in Mexico were hit by severe shortages of gas as Pemex both cut production of the fuel due to low prices, and increased gas consumption at its refineries to save costs versus its traditional fuel source, fuel oil. The shortages hit the manufacturing sector hard in 2012, with firms forced to shut down production several times due to commodity shortages. Companies were therefore obliged to turn to more expensive options, such as liquefied natural gas (LNG) or liquefied petroleum gas (LPG) to make up for the shortfall. These problems, combined with the issue of already higher electricity prices in Mexico for industrial consumers (compared to the United States) were cited by both government and private sector as major obstacles for national economic competitiveness.

In response the government moved ahead with existing plans, drawn up under the Calderón administration, to build new gas pipelines across the U.S.-Mexico border to bring in more plentiful, cheaper gas from American producers. The most important of these pipelines, the Los Ramones project, is predicted to bring in sufficient new gas to Mexico to satisfy national demand until 2018. Other pipelines, such as the Waha-San Elizario (to connect with the planned San Isidro-Samalayuca gas conduit in the northern border state of Chihuahua), are important in satisfying demand in different geographic regions of the country.[20]

The new fuel source has made possible an accelerated transition in generating capacity in Mexico as the CFE has modernized its plants, dramatically reducing the use of much higher polluting fuel oil. Furthermore, it has enabled CFE to lower its generating costs, in turn allowing them to lower electricity costs to their customers.

The other aspect of the energy reform that will greatly aid the long-term reduction of emissions from the electricity sector is the freeing of the electricity market itself, to allow private investment in all aspects of generation. The new investments in generation capacity that are likely to come into Mexico in the next few years will almost all focus on either natural gas or renewables as their fuel source. This offers Mexico the opportunity to increase its energy supply without significantly raising its emissions levels and, in the long term, to reduce emissions below current levels.

In addition to the modernization of the energy sector, the Mexican government made a major statement of climate action when the Congress approved the creation of a carbon tax in 2014. The tax, which was substantially reduced through negotiations in the legislature, sets an approximate average price for carbon at $US3.5/tCO_2e$. Natural gas was exempted from the tax, and a sliding scale was used, with coal and oil coke the most heavily penalized. The Mexican government estimates that the tax will raise around US$1 billion per year for the Fnance Ministry.

Despite considerable progress on multiple fronts, there is a notable weakness in the Peña Nieto government's approach to climate action thus far. The administration announced that the national energy efficiency program (PRONASE) would be widened beyond the five strategic areas identified during the Calderón administration, but it has yet to establish a plan for implementation or even for general principles. The PRONASE website has instead focused on criticizing the previous administration's approach, accusing it of being too limited. However, energy efficiency is an emerging issue in trilateral talks with the United States and Canada, through the North American Energy and Natural Resources Ministers' meeting, and the North American Climate Change and Energy Working Group, where discussions have taken place about harmonizing energy efficiency standards.

Mexico and the international climate regime 2006–2015

At the international level, Mexico has gradually emerged as a leader among developing nations. Although it was a hesitant participant in the early years of the UNFCCC, focusing on the need for industrialized states to commit to GHG emissions reductions, Mexico acquired a role as a developing country leader under Calderón. In fact, President Calderón began to develop Mexico's credentials as a climate action leader as early as 2007. At the Bali COP 13, the administration unilaterally offered to reduce emissions by 50% over 2002 levels by 2050. This offer received favorable international reactions, although at least one climate NGO expressed some skepticism over the government's ability to implement and follow through on such a commitment.[21]

The Bali summit was important in another way, however, as President Calderón dedicated himself to the COP process. Bitterly disappointed by the failure of the Copenhagen COP 15, Calderón made a commitment to ensure that the process would get back on track during the COP16 meeting in Cancun, and worked closely with the British government to develop a strategy for achieving consensus in 2010.[22] Despite major tensions between the leading negotiating parties, Mexican foreign minister Patricia Espinosa employed her decades of experience in international diplomacy and pulled off a final agreement at the summit to ensure that the talks would continue on track in later years.[23] By extending the period for talks to replace the Kyoto Protocol, it is no exaggeration to state that the Mexican government's deft handling of negotiations paved the way for the eventual agreement of a post-Kyoto scenario at the Paris COP21 talks in December 2015.

Another major achievement of Calderón in the international climate regime was the creation of a GCF during COP16. The GCF was envisioned by Calderón to be a source of financing from all nations for all nations. A key element of its design was to make accessing climate financing easier and more transparent. But the most important element of the GCF initiative was the notion that both industrialized and developing countries would share the burden of financing climate transition, an idea that had never before been agreed upon in international talks. Since Cancun, this idea has been substantially modified but the GCF highlighted Calderón's leadership in the field and his government's ability to generate consensus.

The administration of Enrique Peña Nieto, to the surprise of many observers, has continued to put Mexico at the center of the global climate regime as a leading developing country. In anticipation of the 2015 Paris COP21 meeting, Mexico became the first developing nation to publish its Intended Nationally Determined Contribution (INDC) in April 2015, following on from the United States submission in March of the same year. This reflected a growing realization that the United States and Mexico need to adopt compatible approaches to climate action, given the highly integrated nature of their economies.

In the lead in to the Paris talks, Mexico under Peña Nieto again adopted a leadership role. On March 28, 2015, Mexico became the first developing country to deliver its INDC to the UNFCCC, in anticipation of the Paris COP21 meeting later that year. Mexico committed to unconditionally reduce its emissions of GHGs by 25% below baseline emissions by 2030 (Table 11.2), in line with existing national legislation, and by 50% by 2050. The Peña Nieto administration also embraced the possibility of a 40% reduction by 2030, if certain international conditions are met (mostly commitments by other states and the availability of financing).

This impressive commitment is based on a number of elements. What is most intriguing about Mexico's INDC is that it focuses heavily on black carbon, a component found most commonly in soot from diesel, coal, fuel oil and biomass. When one examines the main sources of black carbon GHG emissions in Mexico, it is clear that transportation and industrial activity are the major culprits, and it remains to be seen how the Peña Nieto government plans to bring about such a dramatic reduction (see Table 11.3). However, the shift from fuel oil to natural gas in electricity generation will greatly facilitate reductions in GHG emissions and a continued focus on energy efficiency and renewables must be part of the equation.

In addition to these mitigation commitments, a large part of Mexico's INDC focused on climate change adaptation efforts. The three main lanes of the Peña Nieto government's approach to adaptation lie in improving resilience in vulnerable communities, strengthening ecosystems and achieving a 0% deforestation rate by 2030, and putting in place specifically-designed strategic infrastructure and productive systems.

Despite the grand ambitions of Mexico's commitment to the Paris COP21 talks, the plans have been criticized by a number of analysts. While some have applauded the Mexican government for including an "economy-wide emissions reduction goal and the specification of an unconditional *and* a conditional reduction,"[24] criticisms have focused on the lack of detail in the implementation of the plans, as well as what has been perceived as a less aggressive renewable energy target (5% of total generated by 2018) than the one first put forward by the government (8.2% by 2018). By extension, concerns have emerged that Mexico will be able to reach its medium-term target of 24% of generation from renewables by 2024.

However, shortly after Mexico made its Paris commitments, the national Congress passed a new Energy Transition Law (Ley de Transición Energética or LTE). The law contains ambitious targets for renewable energy use in Mexico, setting a goal of 25% of electricity generation from clean sources by 2018, 30% by 2021, 35% by 2024, 45% by 2036, and 60% by 2050. In addition, the Mexican Congress refused to bow to pressure from the national steel producers' association, CANACERO, to include natural gas as a "clean" energy source, meaning that the steel industry will be forced to seek alternative forms of energy for its production processes. Overall, the LTE provides the potential to reduce Mexico's emissions by between 95 and 115

Table 11.2 Mexico's greenhouse gas emissions by sector ($MtCO_2e$), 2013 and 2030

Emission category	*GHG emissions 2013*	*GHG emissions goal 2030*
Transport	174	218
Electricity generation	127	139
Residential and commercial	26	23
Oil and gas	80	118
Industry	115	157
Agriculture and livestock	80	86
Waste	31	35
LULUCF*	32	0
Total emissions	665	776
LULUCF absorption	−173	−14
Total	492	762

Source: Government of Mexico, "Intended Nationally Determined Contribution: Mexico", April 2015, http://iecc.inecc.gob.mx/indc.php.

*Land Use, Land Use Change and Forestry.

Table 11.3 Mexico's black carbon emissions ('000 tons), 2013 and 2030

Emission category	*Black carbon emissions*	*Black carbon emissions goal 2030*
Transport	47	10
Electricity generation	8	2
Residential and commercial	19	6
Oil and gas	2	3
Industry	35	41
Agriculture and livestock	9	10
Waste	<1	<1
LULUCF	4	4
Total emissions	125	75
LULUCF absorption	0	0
Total	125	125

Source: Government of Mexico, "Intended Nationally Determined Contribution: Mexico," April 2015, http://iecc.inecc.gob.mx/indc.php.

$MtCO_2e$ by 2030, aproximately 50% of the proposed reductions under the INDC. The law also lays out details of Clean Energy Certificates (CELs), modeled on the California system, that will help to incentivize renewable energy, setting a target of a minimum of 5% CEL use by power producers by 2018.

Conclusion

Since 2006, Mexico has emerged as a climate action protagonist among developing countries. Initially this leadership was due to the personal beliefs of PAN President Felipe Calderón, but it quickly became institutionalized in Mexican government policy as the government developed a succession of plans and strategies. These activities culminated in the GLCC of 2012, a crucial milestone in Mexican climate action, as it became the legal framework inherited by the new PRI-ista government of President Enrique Peña Nieto. Of central importance in this institutionalization process is the creation of an inter-ministerial commission on climate change, which is promoting the mainstreaming of climate change across government agencies.

Mexico's commitment to climate change mitigation has been hailed in international forums, especially in the form of its leadership of UNFCCC talks. This climate leadership by Mexico is seen as important not only in building coalitions (seen most clearly at the Cancun talks), but in strengthening the argument that all countries, not only the Annex 1 states, should contribute to the global climate action agenda.

Of much greater significance, however, is the fact that Mexico has followed up on its international commitments with a coherent and cross-cutting domestic climate agenda, composed of aggressive promotion of renewable energy, a modernization of electricity infrastructure and a switch to natural gas as the primary fuel source, and a package of laws that promote a lower emissions economic model. The passing of a carbon tax law, alongside the creation of Clean Energy Certificates, and the continuation of a commitment to energy efficiency strategies, should facilitate the achievement of emissions stabilization by the end of the 2020s and then steep reductions into the 2030s and beyond. This is particularly impressive given Mexico's relatively youthful demographic profile and a rising middle class that has expectations of a higher standard of living.

Mexico's climate action profile can therefore be seen as an example for other major developing economies. It has embraced both industrialization and sustainable development, and sees conformity with international norms as a crucial component in making this a sustainable proposition and the international climate agenda remains central to pushing Mexico to follow through on its commitments.

Notes

1 Pulver, Simone, "Climate Politics in Mexico in a North American Perspective," paper presented at "Climate Change Politics in North America," Woodrow Wilson International Center for Scholars Washington, DC, May 18–19, 2006.
2 Government of Mexico, "Intended Nationally Determined Contribution: Mexico," http://iecc.inecc.gob.mx/documentos-descarga/2015_indc_ing.pdf.
3 Pulver, Simone, "A Climate Leader? The Politics and Practice of Climate Governance in Mexico," in David Held, Charles Roger and Eva-Maria Ng, eds, *Climate Governance in the Developing World*, Cambridge: Polity Press, 2012.
4 Cited in "Reconocen vida y obra de Luis Calderón Vega, máximo cronista de Acción Nacional," El Sol de Morelia, www.oem.com.mx/elsoldemorelia/notas/n1432037.htm.
5 "The Onearth 10," OnEarth, Natural Resources Defense Council, December 2015, www.onearth.org/tags/felipe-calder%C3%B3.
6 Animal Politico, "8 de cada 10 mexicanos sabe poco del cambio climático," www.animalpolitico.com/2014/06/pesar-de-la-alta-preocupacion-por-cambio-climatico-8-de-cada-10-mexicanos-sabe-poco-del-fenomeno/, June 9, 2014.
7 Cary Funk, Lee Rainie and Dana Page, "Americans, Politics and Science Issues," Pew Research Center, July 1, 2015, www.pewinternet.org/2015/07/01/americans-politics-and-science-issues/.
8 Stokes, Bruce, Richard Wike and Jill Clarke, "Global Concern about Climate Change, Broad Support for Limiting Emissions," Pew Research Center, November 5, 2015, www.pewglobal.org/2015/11/05/global-concern-about-climate-change-broad-support-for-limiting-emissions/.
9 Pemex statistics, 2015, www.pemex.com.
10 Feinstein, Charles and Juan Mata, 2006, *Mexico – Large Scale Renewable Energy Development Project: environmental assessment*, Government of Mexico, http://documents.worldbank.org/curated/en/2006/04/6769002/mexico-large-scale-renewable-energy-development-project-environmental-assessment-vol-1-2.

11 Wood, Duncan, "Wind Energy Potential in Mexico's Northern Border States," Woodrow Wilson International Center for Scholars, Washington, DC 2012, www.wilsoncenter.org/sites/default/files/Border_Wind_Energy_Wood.pdf.
12 Ibid.
13 Energy Information Agency, "Mexico," www.eia.gov/beta/international/analysis.cfm?iso=MEX, September 21, 2015.
14 Ibid.
15 WRI, 2006, "Mexican Industry Takes Voluntary Action Against Climate Change; Government Gives Public Recognition," Press release, World Resources Institute, Mexico City.
16 Wood, 2012, op. cit.
17 OEM en Linea, "Anuncia Joaquín Coldwell inversión en energía eólica de 14 mil mdd entre 2015 y 2018", www.oem.com.mx/oem/notas/n3669108.htm#sthash.6KBG7kza.dpuf.
18 Teixeira, M., "Mexico's climate law to face challenge under new president." Reuters.com, July 24, 2012, www.reuters.com/article/us-mexico-climate-policy-idUSBRE86N0A220120724.
19 David Jiménez, Javier Félix and Raquel Bierzwinsky, "A New Geothermal Framework for Mexico," www.chadbourne.com/A-New-Geothermal-Framework-For-Mexico_projectfinance, May 2015.
20 "Consortium including Slim's Carso Energy wins contract for Texas pipeline," Fox News Latino, http://latino.foxnews.com/latino/news/2015/01/29/consortium-including-slim-carso-energy-wins-contract-for-texas-pipeline/.
21 Alexis Madrigal, "Bali Meeting Ends; Mexico Emerges as a Leader on Climate Change," http://archive.wired.com/science/planetearth/news/2007/12/mexico_climate.
22 Hilen Gabriela Meirovich, 2014, "The Politics of Climate in Developing Countries: The Case of Mexico," PhD dissertation, Georgetown University.
23 The story of that summit has become a case study for the Mexican Foreign Service on how to negotiate a multilateral agreement.
24 Climate Action Tracker, "Mexico," http://climateactiontracker.org/countries/mexico.html, consulted January 15, 2016.

12
South Africa's pragmatic transition

Robert E. Looney

Introduction

South Africa accounts for slightly under two-thirds of Africa's total power-generating capacity.[1] The country is richly endowed with sources of energy generating resources, particularly coal. With its large coal reserves, South Africa was able to take advantage of its comparative advantage in developing an energy intensive mining sector through the provision of low cost electricity. The resulting economic development model has been characterized as a powerful "minerals-energy-complex."[2] For years, this model served the country well, generating relatively high rates of economic growth with price stability and more than adequate levels of foreign exchange needed to supply essential imports.

In recent years, however, this energy paradigm has come under increasing criticism as being no longer economically or environmentally sustainable.[3] After years of underinvestment in the nation's power supply, demand for electricity began to noticeably outstrip supply in 2008.[4] Subsequently, the country has experienced sharply rising power tariffs and frequent periods of load-shedding, brownouts and even blackouts.

Energy shortfalls and rising electricity costs are particularly detrimental to the mining industry, which uses 15% of all electricity for drilling and pumping water from mines. In December 2015 one of the country's largest mining companies, Anglo American, announced[5] it would reduce its staff by over 50% and sell off many of its assets in an attempt to remain profitable. Moody's rating agency has downgraded its debt to junk status.[6]

The decline in mining, with mineral exports historically making up around 60% of total exports, has put additional pressure on the already large current account deficit. The manufacturing sector also suffers[7] from the adverse effects of power-shortfalls, with foreign investment falling off due to investors' perception of deterioration in the country's business climate. As things stand, energy in South Africa is increasingly seen as a major constraint on growth, new firm development and broad based employment creation. Recent research suggests energy shortages are one of the major factors impeding South Africa's micro, small and medium firms' growth.[8] Studies[9] for Sub-Saharan Africa suggest that these economies suffer a fall in their potential growth of nearly 2% because of inadequate power supplies.

Normally the South African government would have responded to the energy crisis in a conventional manner – massive investments in new coal generating capacity.[10] And to a certain extent this has been the case, although the new plants are falling behind their completion schedules. In energy trilemma terms the country would, as in the past, opt to increase its energy affordability and energy security at the expense of environmental sustainability. However, increased concern in recent years over greenhouse gas emissions and the consequences for the country of increased global warming have forced attention not only on energy affordability but also on water supplies and environmental sustainability.[11]

Currently South Africa ranks[12] 130 out of 130 countries on the World Energy Council's index of environmental sustainability. This stems from the country's extremely high greenhouse gas emissions due to the country's almost complete reliance on coal for electricity generation. The South African government is well aware of the environmental situation, and has made a number of international pledges[13] to reduce carbon emissions. In addition to an improved environment, South Africa is also trying to provide affordable energy to as many low income households as possible. Some 15% of the population still lacks access to modern energy services.[14]

With the decline in the country's mining sector due to falling commodity prices, rising costs, labor strife, and falling investment, South Africa is finding a new development model will be necessary if growth and employment creation are to be restored. Simultaneously making major changes in its development strategy and energy priorities will present major challenges. The sections that follow attempt to identify the complex factors that are likely to come to bear as South Africa attempts its energy transition away from coal and a new development model away from mining.

Energy patterns and trends

As might be expected South Africa's energy trilemma pattern shares several similarities to the four other (Saudi Arabia, Canada, Russia and Mexico) carbon-producing countries examined in the current volume. But because the country is a coal producer and the others largely reliant on oil, a number of notable differences also occur.

Relative to other carbon-producing countries, South Africa scores considerably below the others in overall energy performance defined in terms of balancing the three main dimensions of the energy trilemma: energy security, energy equity and environmental sustainability. Nor has there been an improvement on this key energy dimension over the available data period, 2011–2015 (Table 12.1). As with the other carbon-producing countries, South Africa, as noted, demonstrates a priority of energy security and energy equity at the expense of environmental sustainability.

The deterioration in the country's energy equity score in recent years is reflective of the rapid rise in energy prices brought on by the growing gap between electricity supplies and demand. As discussed below, the recent improvement in energy security can be attributed to the encouraged expansion of renewable fuels to help fill the country's electricity shortfalls.

These general patterns are supported by another trilemma-type assessment of the energy sector, the World Economic Forum's Energy Architecture Performance Index (EAPI).[15] The EAPI provides a measure of the relative performance of the energy sector in national energy systems. The EAPI consists of eighteen indicators reflective of progress in three trilemma-like sub-indexes: (a) economic growth and development, (b) environmental sustainability, and (c) energy access and security.

In the EAPI framework, economic growth and development captures the impact of the country's energy sector on the country's economic expansion. Environmental sustainability

Table 12.1 The energy trilemma: carbon-producing countries

	2011	*2012*	*2013*	*2014*	*2015*	*2011/2015*		*Score*
						Difference	*Average*	
South Africa								
Energy performance	93	97	93	96	93	0	94.4	
Energy security	52	55	43	42	30	22	44.4	B
Energy equity	73	75	78	85	87	−14	79.6	C
Environmental sustainability	129	129	128	129	130	−1	129	C
Saudi Arabia								
Energy performance	57	57	57	67	54	3	58.4	
Energy security	32	38	45	68	49	−17	46.4	B
Energy equity	18	14	12	7	7	11	11.6	A
Environmental sustainability	124	124	124	125	120	4	123.4	D
Canada								
Energy performance	8	9	8	4	9	−1	7.6	
Energy security	2	2	1	1	1	1	1.4	A
Energy equity	2	2	2	2	2	0	2.0	A
Environmental sustainability	61	66	60	56	71	−10	62.8	C
Russia								
Energy performance	53	48	46	40	43	10	46.0	
Energy security	1	1	2	2	15	−14	4.2	A
Energy equity	65	57	61	44	37	28	52.8	B
Environmental sustainability	102	102	99	104	108	−6	103.0	D

	2011	*2012*	*2013*	*2014*	*2015*	*2011/2015*		*Score*
						Difference	*Average*	
Mexico								
Energy performance	51	47	38	37	55	−4	46.0	
Energy security	28	35	29	30	37	−9	31.8	B
Energy equity	62	52	47	43	61	1	53.0	B
Environmental sustainability	71	73	75	74	80	−9	74.6	C

Source: World Energy Council, Energy Trilemma Index, 2015, https://www.worldenergy.org/data/trilemma-index/country/.

depicts the impact of energy production and consumption on the environment. Finally, energy access and security measures the degree to which the country's energy supply is secure, accessible and diversified. Scores for each are assigned on a scale from 0 to 1 with the EAPI a composite of the three sub-indexes.

The EIPA and related sub-indexes for South Africa show several interesting patterns relative to other carbon-producing countries (Table 12.2). On the overall EAPI summary index, South Africa scores the second lowest, above Saudi Arabia. On the other hand, South Africa ranks above the group average in terms of economic growth and development and environmental sustainability. Energy's relatively high economic impact in South Africa probably stems from the fact that the country's energy sector has more linkages with the domestic economy than is usually the case for oil based economies. However, because of its oil import dependency South Africa scores considerably below the group average on the energy access and security dimension.

In the last several years (2013–2015) South Africa has made the greatest improvements in its weakest area, energy access and security. Progress occurred to a lesser extent in environmental sustainability, while the country suffered a decline in the energy sector's contribution to economic growth and development. No doubt rising energy tariffs and shortfalls account for this decline and highlight the importance of expanded energy production if the country is to break out of its current economic slump.

Both the World Energy Council's Energy Trilemma Index and the World Economic Forum's EAPI index represent the culmination of past patterns of both energy usage and production. In this regard the composition of South Africa's energy mix shows some similarities and differences with those found in the other energy-producing countries. As expected all four countries have a predominance of fossil fuels in energy use, with South Africa's averaging in the high 80% level, with a slight upward trend since 2000 (Table 12.3).

Combustible renewals and waste account for slightly over 10% of the country's total energy use with a downward trend of 7.3% between 2000 and 2012. This was considerably less of a decline than found in Saudi Arabia, Russia and Mexico. Because it has only one nuclear plant, the 30-year-old 1,800 MW Koeberg facility near Cape Town,[16] alternative and nuclear energy only accounts for around 2.5% of South Africa's energy mix and this has been declining simply because nuclear generating capacity has remained constant.

At least through 2012 renewable energy has also been declining as a share in South Africa's final energy consumption. This source accounted for 18.5% in 2000, but by 2012 this had fallen to 16.9% for a decline of 8.5%. However as discussed below, this trend is likely to be reversed in the near future as the country is turning to renewables as a way to close the current electricity shortfall.

Other aspects of South Africa's energy environment relevant for future policy discussion include the fact that the country's energy exports as a percentage of energy use (Table 12.4) have been falling[17] rather sharply since 2000. In 2000 the country exported about a third of its energy, largely coal, but by 2012 this had fallen to 18.6% for a decline of 44%.

Of the energy-producing countries surveyed here, South Africa had by far the greatest CO_2 intensity (Table 12.4) measured in terms of kg per kg of oil equivalent energy use. In other words, South African energy use to date is one of the worst in terms of contributing to global warming. While there has been a slight downward trend in CO_2 intensity, South Africa lags considerably behind the other energy-producing countries in reducing greenhouse gas emissions.

South Africa's energy use per capita is considerably below the average for the energy-producing countries examined here. No doubt this stems largely from the fact that per capita incomes are

Table 12.2 Global Energy Architecture Performance Index 2013–2015

Country	*Country rank*		*EAPI index*		*Economic growth and development*		*Environmental sustainability*		*Energy access and security*	
	2015	*2013*	*2015*	*2013*	*2015*	*2013*	*2015*	*2013*	*2015*	*2013*
South Africa	*66*	*59*	*0.58*	*0.54*	*0.59*	*0.60*	*0.53*	*0.49*	*0.65*	*0.54*
Saudi Arabia	112	82	0.47	0.46	0.39	0.30	0.19	0.28	0.82	0.78
Canada	25	23	0.69	0.63	0.59	0.61	0.61	0.47	0.89	0.82
Russia	39	27	0.66	0.61	0.60	0.58	0.50	0.54	0.80	0.71
Mexico	55	39	0.62	0.59	0.57	0.61	0.54	0.50	0.76	0.67
Average	59.4	46.0	0.60	0.57	0.55	0.54	0.47	0.46	0.78	0.70
% Change: 2013–2015										
South Africa			*7.41*		*−1.67*		*8.16*		*20.37*	
Saudi Arabia			2.17		30.00		−32.14		5.13	
Canada			9.52		−3.28		29.79		8.54	
Russia			8.20		3.45		-7.41		12.68	
Mexico			5.08		-6.56		8.00		13.43	
Average			6.71		1.48		3.95		11.36	

Source: Global Energy Architecture Performance Index, *Report 2015*, December 2014.

Note: Scores are on a scale of 0 to 1.

Table 12.3 Energy producing countries: composition of energy mix

	200	*2005*	*2010*	*2011*	*2012*	*201_*	*200–2012 % Change*
Fossil fuel energy consumption (% of total)							
South Africa	84.3	87.1	87.5	87.1	87.0		*3.2*
Saudi Arabia	100.0	100.0	100.0	100.0	100.0		*0.0*
Canada	76.8	75.4	74.4	73.5	73.4	72.[illegible]	*−4.4*
Russia	91.1	90.7	90.7	91.1	91.1		*−0.1*
Mexico	86.8	87.9	89.3	89.2	90.2	89.[illegible]	*3.9*
Combustible renewables and waste (% of total energy)							
South Africa	11.6	10.7	10.2	10.4	10.7		*−7.3*
Saudi Arabia	0.0	0.0	0.0	0.0	0.0		*−26.8*
Canada	4.6	5.0	4.8	4.9	4.9	5.0	*6.3*
Russia	1.1	1.1	1.0	1.0	1.0		*−11.8*
Mexico	6.2	5.3	4.8	4.5	4.5	4.6	*−27.6*
Alternative and nuclear energy (% of total energy)							
South Africa	3.2	2.4	2.4	2.7	2.6		*−17.8*
Saudi Arabia	0.0	0.0	0.0	0.0	0.0		*0.0*
Canada	19.8	20.4	21.7	22.7	23.3	24.[illegible]	*17.5*
Russia	7.8	8.4	8.5	8.1	8.1		*3.3*
Mexico	7.0	6.9	6.0	6.3	5.6	6.3	*−20.2*
Renewable energy consumption (% of total final energy consumption)							
South Africa	18.5	16.9	16.9	17.1	16.9		*−8.5*
Saudi Arabia	0.0	0.0	0.0	0.0	0.0		*−23.6*
Canada	20.5	21.5	20.7	20.7	20.6		*0.4*
Russia	3.5	0.36	3.3	3.2	3.2		*−7.2*
Mexico	12.4	10.7	10.0	9.6	9.4		*−24.9*

Data from database: World Development Indicators. Last updated: December 22, 2015.

Table 12.4 Energy-producing countries: aspects of energy use

	2000	*2005*	*2010*	*2011*	*2012*	*2000–2012 % Change*
Energy imports, net (% of energy use)						
South Africa	−33.3	−23.1	−14.8	−15.0	−18.6	*-44.0*
Saudi Arabia	−386.2	−365.9	−186.7	−232.9	−212.1	*-45.1*
Canada	−48.2	−48.6	−57.4	−61.4	−67.1	*39.3*
Russia	−57.9	−84.6	−83.8	−78.0	−76.0	*31.2*
Mexico	−52.1	−44.7	−21.4	−16.2	−19.1	*-63.3*
CO_2 intensity (kg per kg of oil equivalent energy use)						
South Africa	3.4	3.1	3.2	3.2		*0.3*
Saudi Arabia	3.0	3.2	2.9	2.9		*-3.7*
Canada	2.1	2.1	2.0	1.9		*-9.7*
Russia	2.5	2.5	2.5	2.4		*-2.7*
Mexico	2.6	2.6	2.5	2.5		*-3.8*
Energy use (kg of oil equivalent per capita)						
South Africa	2,483.3	2,710.5	2,808.8	2,752.2	2,674.8	*7.7*
Saudi Arabia	4,574.5	4,952.6	6,600.0	6,184.1	6,789.2	*48.4*
Canada	8173.7	8,378.8	7,390.4	7,366.6	7,225.7	*-11.6*
Russia	4,224.3	4,541.0	4,925.1	5,165.8	5,283.4	*25.1*
Mexico	1408.6	1537.1	1486.0	1525.4	1543.3	*9.6*
Energy intensity level of primary energy (MJ/$2011 PPP GDP)						
South Africa	250.2	243.5	232.4	223.9	216.2	*-13.6*
Saudi Arabia	5.5	5.4	6.2	5.5	5.8	*5.7*
Canada	9.2	8.7	7.6	7.5	7.3	*-20.7*
Russia	320.7	250.6	227.3	228.9	226.8	*-29.3*
Mexico	320.7	250.6	227.0	228.9	226.8	*-29.3*

Data from database: World Development Indicators. Last updated: December 22, 2015.

relatively low with some individuals not yet receiving access to electricity. South Africa's precipitate energy use closest resembles that of Mexico with both showing a steady increase in per capita energy use.

Finally, South Africa's energy intensity level is quite high, but still in the same range as that found in Russia and Mexico. Still, South Africa remains by far the most energy intensive economy in Africa, contributing to the country's now chronic power shortfalls. For the longer term, this ratio is likely to fall as the country continues to move away from mining and towards a more diversified manufacturing and service oriented economy.

In sharp contrast to the other energy-producing countries surveyed here, South Africa generates its electricity almost exclusively with coal. Coal is responsible for nearly 94% of South Africa's electricity generation, and this has increased slightly (0.8%) since 2000 (Table 12.5). The next largest coal generating country is Russia at 15.6% and Mexico at 15.0% in 2012, but both countries have been experiencing a relatively rapid decline (2000–2012) in coal usage for electricity generation – 16.9% for Russia and 21.6% for Mexico.

Next to Saudi Arabia, renewables in South Africa account for the lowest share in electricity generation for the countries surveyed here. Percentage wise there has been a 41.1% increase (2000–2012) in electricity generated by renewables, but in 2012 this source accounted for only 1.0% of electricity generated in South Africa.

One of the South African government's main goals has been the spread of electricity access to most of the population. The program has been quite successful with access to electricity increasing by 29.2% from 66.1% in 2000 to 85.4% by 2012. While most urban residents now have access, the rural areas are lagging with access there increasing by 80.2% from 37.1% with access in 2000 to 66.9% by 2012.[18] These additions took place without additional power stations being built. The rapid increase in household access to electricity has significantly contributed to the surge in electricity demand that is currently outrunning available supplies.

Energy plans and resource shifts

The energy situation in South Africa has created several paradoxes. In November 2015 a headline[19] appeared arguing that the recent hiatus in load-shedding was actually bad news because it was a reflection of the stagnation that has been plaguing the South African economy. Between 2014 and 2015 South Africa's peak demand for electricity has shrunk between 1,500MW and 3,000MW because of the economic stagnation that has set in. Even more worrying the peak power demand in 2011 was 3,6000MW whereas in 2015 peak demand was only 3,0000MW. Clearly electricity shortages have taken a tremendous toll on the economy.

Empirical research[20] has identified a statistically significant causal relationship linking South Africa's economic growth to the country's energy supply, with energy consumption a determinant of economic growth. Clearly the country will have to find ways to rapidly expand the supply of electricity if it has any hope of restoring strong economic growth and reducing the high level of unemployment which now stands at around 22% of the labor force.

In order to expand the country's supply of energy, the South African government has developed several comprehensive energy plans, some of which focus on the shorter term, while others are oriented towards reaching longer-term targets. In doing so the South African government no longer leaves the country's power plans exclusively with Eskom, the state-owned utility company. The reasons behind this move lie largely with the size of the task involved in revamping the country's energy system. It also is an attempt to overcome some of the limitations associated with Eskom's coal-orientation and the company's energy stewardship since the days of apartheid.

Table 12.5 Energy-producing countries: aspects of electricity

	2000	*2005*	*2010*	*2011*	*2012*	*2000–2012 % Change*
Electricity production from coal sources (% of total)						
South Africa	93.1	94.6	94.2	93.8	93.8	0.8
Saudi Arabia	0.0	0.0	0.0	0.0	0.0	0.0
Canada	19.4	16.2	13.8	12.2	10.0	−48.3
Russia	19.9	17.3	16.0	15.5	15.7	−20.8
Mexico	9.5	13.6	12.0	11.5	11.7	22.7
Renewable electricity output (% of total electricity output)						
South Africa	0.7	0.7	1.0	0.9	1.0	41.1
Saudi Arabia	0.0	0.0	0.0	0.0	0.0	0.0
Canada	60.6	59.6	61.3	62.0	63.2	4.4
Russia	18.7	18.2	16.1	15.8	15.6	−16.9
Mexico	19.9	15.6	17.6	15.9	15.0	−24.6
Access to electricity (% of population)						
South Africa	66.1		82.7		85.4	29.2
Saudi Arabia	90.9		94.1		97.7	7.5
Canada	100.0		100.0		100.0	0.0
Russia	100.0		100.0		100.0	0.0
Mexico	98.0		99.0		99.1	1.1
Access to electricity, rural (% of rural population)						
South Africa	37.1		64.1		66.8546	80.2
Saudi Arabia	86.9258		90.1		92.8546	6.8
Canada	100.0		100.0		100.0	0.0
Russia	100.0		100.0		100.0	0.0
Mexico	92.5		97.5		97.2	5.1

Data from database: World Development Indicators. Last Updated: December 22, 2015.

Eskom

Eskom and the coal industry began to prosper in 1948 when the apartheid regime came to power. At the time the apartheid government viewed the country's vast coal reserves and the provision of cheap energy as the key to rapid growth. The coal reserves also provided energy security against the anticipated imposition of UN sanctions over the country's racial policies. With the country's post-war economic boom, Eskom grew into the world's fourth-largest power utility as it implemented a strategy that gave preference in electricity deliveries to factories, mines, and white households.

Eskom's expansion stopped in 1985 as anti-apartheid sanctions began to cripple the economy and reduce the demand for electricity. By the time the African National Congress took power (1994) the economy had contracted to the point that Eskom's reserve margin had increased to nearly 30%, or twice the international average.

Understandably Eskom's management viewed the company's backup capacity as adequate for years to come. But post-apartheid governments adopted policies that represented a major shift from their predecessors. Most importantly, these governments placed a high priority on connecting neglected black households, schools and clinics to the grid. In a relatively short period, hookups jumped from 5 million to 12 million.

At the same time, the end of sanctions opened up the economy to foreign investment, and growth, particularly in manufacturing, surged. From 1985–1993 GDP growth had averaged 0.73% per annum with private consumption increasing at an annual average rate of 1.27% and investment declining at an average annual rate of 3.02% per annum (Table 12.6). Manufacturing also contracted at an average annual rate of 0.35%. In sharp contrast, with the end of apartheid, from 1994–2007 GDP accelerated to 6.66% per annum, private consumption rose to 4.46% per annum and investment rose to 7.22% per annum. Manufacturing also expanded sharply at an annual average rate of 3.50% per annum.

Eskom, as often is the case with monopolies, was unprepared for the shift in priorities towards bringing electricity access to communities for the first time during a period of sharply accelerating economic growth. It wasn't until May 2007 that the utility finally reacted to the country's dwindling reserves by authorizing the construction of the Medupi and Kusile mega coal-fired plants to add 10,000MW to the grid.

Originally scheduled to be up and running and providing ample capacity by 2011, both plants are way over budget and years behind schedule. Inadequate planning, skills shortages, labor disruptions and technical issues have hindered the arrival of both coal plants, causing significant delays. The Medupi power station was not supplying the grid until one unit came online in August 2015. The plant will not be at full capacity until at least 2018. The completion of

Table 12.6 South Africa: growth sub-periods (period average, constant price percentage)

	1985–1993	*1994–2007*	*2008–2015*
GDP	0.73	3.66	19.1
Private consumption	1.27	4.46	2.03
Investment	-3.02	7.22	2.43
Agriculture	4.56	1.68	2.44
Manufacturing	-0.35	3.50	0.34
Services	1.16	4.21	2.57

Source: Compiled from EIU South Africa data base.

Kusile's first 800MW unit has been pushed back from 2011 to the first half of 2017, with the plant complete by 2021.

Eskom's other problems stem from the manner in which it responded to the blackout and load-shedding crisis of 2008. At that time, Eskom's management opted for a strategy of "keeping the lights on at all cost." But postponing maintenance for seven years led to an increase in plant breakdowns. After running its 25- and 30-year-old plants at full capacity for seven years, trouble began in January 2015 with multiple equipment breakdowns. Afterwards, Eskom juggled 21 days of rolling blackouts in the first three months of the year as it struggled with plant shutdowns. This stifled production, and manufacturing fell 2.4% in the first quarter of that year.

The country's electricity shortages took a heavy toll on the economy. Specifically, ESKOM's power-shedding program in early 2015 cost the economy between R 20,000m. (US $1,600m.) and R 80,100m.[21] ($6,500m.) per month. Analysts estimate shortfalls in power supplies have cost R 300,000m. ($24,700m.) over the seven years prior to 2015. Other estimates suggest power shortages reduced the potential growth of the economy during 2007–14 by 10%, while also costing more than 1 million jobs.[22]

By the end of August 2015, the country had experienced 99 days of load-shedding leading to significant drops in manufacturing and mining in particular. South African GDP contracted 1.3% in the second quarter of 2015. It was estimated that power cuts had cost the economy between $1.7bn and $6.8bn per month, depending on the load-shedding stage implemented.[23]

Eskom's problems have been amplified due to the accelerating maintenance requirements of its ageing plants, most of which have been run continuously since the 2008 crisis. As things stand, the company is seriously burdened with debt, and based on the Medupi and Kusile experiences, has shown little capability in managing major infrastructure investments.

Integrated Resource Plan (IRP)

After the 2008 load-shedding crisis, the South African government realized the country would experience a recurrent pattern of energy shortages and crisis if it did not take a more active role in planning, particularly for the period following completion of the Medupi and Kusile plants. At the time this was expected to have occurred by 2011 or 2012.

To reduce dependence on coal and improve energy security, the government pledged to increase exploration to find domestic gas feedstock, identified as critical, in the National Development Plan (NDP). The authorities also developed and released a 20-year Integrated Resource Plan (IRP)[24] in 2010 to diversify the energy mix by increasing nuclear, natural gas and renewable power generation capacities.[25] In this regard, the IRP marks a clear departure from the country's traditional reliance on coal-fired power. In 2011, the government also released its Renewable Independent Power Producer Procurement Program in 2011.[26]

The IRP is revised every two years to take into account changes in the energy market and new technological developments. The implementation of the IRP has been a critical component of the country's push to improve supply, opening up space for private sector-led renewable energy. The plan calls for coal-fired electricity to account for less than 15% of all new generation capacity added through to 2030, and for the commodity to provide under 50% of the total grid capacity. Renewables are set to take up 42% of all new generation capacity, which would see the construction of 9.6 GW of nuclear energy and 11.4 GW of renewables, including solar and wind by 2030.

The IRP anticipates that natural gas will also play a larger role, through both imports by pipeline and proposed liquefied natural gas terminals, possibly as well as from domestically produced sources such as the country's shale gas and coal-bed methane reserves. It is also anticipated that the lives of Eskom's ageing coal-based generators will not be prolonged by retro-fitting pollution control mechanisms. This will require additional new generation capacity to come on-stream, as these stations are to be mothballed from 2018 until 2050.

Shifts of these magnitudes away from coal towards other sources of energy are difficult even under the most favorable of circumstances. Clearly much will depend on the amounts of investment available from both the public and private sectors along with the costs of expansion. On the surface, South Africa would appear to have a number of viable options. A closer look, however suggests a limited range of choices that are economically and/or politically feasible.

Coal

As noted, the IRP marks a clear departure from the country's traditional reliance on coal-fired power. While at first glance this shift may seem to be based largely on concerns over global warming, future profitability in coal projects may be playing a much larger role.

Currently South Africa possesses the world's ninth-largest recoverable coal reserves, with 3.4% of the global total as of the end of 2015.[27] Current coal production comes largely from the mature Witbank, Highveld and Ermelo fields in the Central Basin near the eastern border with Swaziland. Coal production was 145.3m tons of oil equivalent (MTOE) in 2013, increasing to 147.7 MTOE in 2014 after having experienced a brief 0.8% contraction from 2012 to 2013.

While coal has historically proved to be an easily accessible input for South Africa's power sector, this may not be the case for much longer. In 2012 Eskom was anticipating that it would consume 4bn tons of coal from then until 2040.[28] However, investment in new coal projects has been slow to materialize. Eskom is concerned over the possibility of a 17m-tonne coal shortfall at its Matla, Tutuka and Hendrina power stations in 2015 and at the Kriel and Arnot power stations in 2016. In June 2015, Eskom officials began to predict that the company could start running out of coal in 2015. Indeed, by 2020 it could see a deficit of between 60m and 100m tons per year.[29]

With their reserves expected to run out in less than a decade, the industry has turned its attention to inland coalfields in the Limpopo Province and around the resource-rich Waterburg Basin. Commercial viability for exploration depends on overcoming water, transportation and infrastructure constraints. The poor financial health of mining companies – due principally to rising power costs, shortfalls in productivity and steep declines in global coal prices down to $50 per metric tonne – could inhibit large-scale capital investments in exploration programs and greenfield projects. Another factor that may curtail investment is new rules on Broad-Based Black Economic Empowerment (BBBEE). A new supplier of coal now has to have a 50% plus one share empowered.[30] This is up from the 26% figure incorporated in the Minerals and Petroleum Resources Development Act.[31]

Shale

Following the shale boom in the US, unconventional exploration and production has increased around the world and has met with some success in South Africa – although it has not been without its challenges. In 2013 the EIA estimated South Africa's technically recoverable shale gas reserves to be 390 tcf, the eighth-largest reserves in the world.

The Whitehill formation, located in the Karoo Basin and thought to be among the most prolific reserves in the country, holds an estimated 36 tcf of recoverable shale gas, or roughly 30 times that of PetroSA's Mossgas project. In 2009 and 2010 PASA awarded technical cooperation permits to four international energy companies to conduct geological surveys of potential shale reserves in different areas of the Karoo Basin. Following this, the government placed a 19-month ban on fracking as it conducted environmental impact assessments amid concerns over its potential impact on aquifers and ecosystems. The ban was later lifted and the government announced its intention to proceed with the issuing of licenses once the shale regulations are published. The Department of Energy will process existing applications first and will issue new licenses in three years. However, water will remain an issue if only in the form of a premium price that producers will have to pay, given the limited supplies throughout the prime shale bearing districts.[32]

The drop in oil prices has slowed shale activity worldwide, including in South Africa. Shell, which holds three of the five shale blocks, announced in March 2015 the redeployment of its senior staff to outside of South Africa.

Nuclear

South Africa has had 30 years of experience in producing nuclear power, and given the country's pressing need for additional power supplies, nuclear power has a number of strong advocates.[33] Officially the government, based on its 2010 IRP, would like to have 9,600MW of new nuclear power by 2030. President Zuma confirmed the country's commitment to nuclear power in his February 2016 State of the Nation Address[34] when he announced that the country hopes to install 9,600MW of nuclear power over the next 15 years.

However, nuclear advocates face a number of questions concerning the economics of a major expansion in South Africa's generating capacity. In addition to the usual safety concerns, these include the construction costs, length of time between approval and full generating capability, the impact such a large increase in capacity will have on electricity prices, and the terms that the country is negotiating with potential builders. More fundamentally, with Eskom's new projects coming on line in a few years, and the recent surge in investment in renewables, it is not clear that additional nuclear plants will even be needed.[35] Proponents of the program believe this nuclear expansion will unlock South Africa's industrialization potential and secure its energy security. At an estimated cost of between $60bn and $100bn, critics believe it will bankrupt the state and permanently endanger the country's credit rating.

To date South Africa has had exploratory talks with five countries, China, Russia, France, the U.S. and South Korea on collaboration. However, these talks have been very opaque, leading many in the country to suspect corruption and kickbacks to the country's political elite.[36]

Given the uncertainty surrounding nuclear energy in South Africa, major decisions should, no doubt, be postponed until it is determined whether renewable energies have the potential to provide sufficient base-load power in addition to that coming on line during the next several years.

Renewable energy

Until quite recently, and given the country's abundant deposits of coal, renewable energy had played a very minor role in South Africa's energy mix. Yet in a very short period of time, the country has become one of the fastest growing incubators of renewable energy.

This somewhat surprising development came about as a result of two quite unrelated developments. First, the 2008 energy crisis and Eskom's need to rapidly expand the energy supply, and second the fact that South Africa has the largest carbon footprint in Africa: it's responsible for 40% of the total CO_2 emissions in the continent. At the COP 15 climate conference, the country pledged to reduce its carbon emissions 34% by 2020, and 42% by 2025.[37] By the 2015 Paris COP 21 South Africa was pledging to "peak, plateau, and decline" by 2030.[38] How best to manage both tasks? The South African government concluded the only feasible solution was to expand the supply of renewable energy.

As early as 2009, the government had begun exploring the best way to proceed with renewable energy development. Initially the use of feed-in tariffs (FITs)[39] seemed the best for renewable energy, but ultimately the country settled for a system of competitive tenders. Because of Eskom's financial constraints the government had little choice but to turn to the country's independent power producers (IPPs) for additional amounts of clean power.

To the surprise of many, the resulting program, the Renewable Energy Independent Power Producer Procurement Program (REIPPPP),[40] has been a major success[41] proving to be an efficient and innovative approach to a country-specific renewables policy; it relies on private sector actors – as opposed to the South African government – to realize renewable energy projects. To this end, the policy uses a very clear international bidding process known as renewable energy auctions through its Independent Power Producer Program, which has been praised by the International Renewable Energy Agency (IRENA). Furthermore, the policy not only reduces borrowing and funding burdens that would have otherwise fallen on the South African government, but also promotes off-grid and small-scale renewable projects that the public utility company ESKOM has little incentive to pursue. The policy also aims to reduce corruption risks known to be prevalent in such projects.

The workings of the program are straightforward and transparent. The process is set up to procure renewable energy through bid windows for the IPPs. Each bid window specifies the maximum amount of power that can be purchased for a set price.

And what are the results? The first phase, Bid Window 1, was initiated in 2011 and saw a total of 1,190.34MW auctioned. Of those projects, 99.4% (or 1,182.62MW) have been brought to a commercial stage. This outstanding success – in just a year – led to second, third and fourth bidding windows being announced in 2012, 2013 and 2014 respectively. In total, the government has awarded 3,634.42MW worth of projects. According to IRENA, the number of bidders at the renewable auctions has increased by 50%, while solar and wind costs have shrunk by 39% and 23% respectively. Furthermore, since 2011 – when the first bid was announced – the private sectors, comprising both domestic and international players, have invested a total of $13 billion in South Africa's renewable energy sector.[42]

The South African government has contracted for 3,725MW of renewable energy by 2016. The bidding process will continue until the required amount of energy is available. So far, if recent progress can be sustained the program is on track to meet the renewable target for 2030.[43]

While boosting renewables is the program's main goal, it also pays attention to the country's rampant unemployment and economic inequality. (13% of the South African population does not have access to electricity.) REIPPPP has thus encouraged partial community ownership of the renewable projects, and a share of generated revenue is being diverted towards enterprise development, including hiring locals and transferring skills.

Conclusions

The South African experience is one of pragmatism under stress. Facing a major short-fall in electricity, a government that in the past has shown ambivalence toward the private sector, designed an innovative program to draw on that sector's expertise and financing. The success of the government's REIPPPP has surprised most observers and is drawing wide-spread attention as an option to improve both energy security and environmental sustainability without major sacrifices in energy affordability.

The transferability of REIPPPP to other settings will depend largely on the willingness of local and international private sectors and financiers to actively participate. The South African case suggests a necessary requirement will be one where a well-designed and transparent procurement process exists, there is a reasonable level of profitability, and where the government is willing to mitigate key risks.

The South African experience also sheds additional light on the energy trilemma. Specifically, by improving incentives or introducing innovative institutional changes previously conflicting objectives can actually be made complementary.

Notes

1 Government of South Africa, Department of Energy, "Basic Electricity" www.energy.gov.za/files/electricity_frame.html.
2 Lucy Baker, Peter Newell and Jon Phillips, "The Political Economy of Energy Transitions: The Case of South Africa," *New Political Economy* 19:6, 791–818
3 For example, Glenn Ashton, "Towards a more resilient South African energy system," *BDlive*, April 14, 2014, www.bdlive.co.za/indepth/greeningyourbusiness/2014/04/14/towards-a-more-resilient-south-african-energy-system.
4 "How South Africa can transition to a less energy-intensive economy," The Conversation, July 7, 2015, https://theconversation.com/how-south-africa-can-transition-to-a-less-energy-intensive-economy-44240.
5 Alexander Wexler, "Anglo American Job Cuts Strike at Core of South Africa's Mining Industry," *Wall Street Journal*, December 15, 2015.
6 John Gapper, "Anglo American failed to adapt to a booming China economy," *Financial Times*, February 18, 2016.
7 Mark Allix, "Load shedding is squeezing the life out of SA's factories," *BDlive*, July 15, 2015, www.bdlive.co.za/business/industrials/2015/07/15/load-shedding-is-squeezing-the-life-out-of-sas-factories.
8 A.M. Mthimkhulu and M.J. Aziakpono, "What impedes micro, small and medium firms' growth the most in South Africa? Evidence from World Bank Enterprise Surveys," *South African Journal of Business Management* 46, no. 2 (2015):, 15–27.
9 T.B. Anderson and C.J. Dalgaard, "Power outages and economic growth in Africa," *Energy Economies* 38 (2013): 10–23.
10 Alistair Anderson, "South Africa 'needs coherent energy policy,'" *BDlive*, February 19, 2013, www.bdlive.co.za/business/energy/2013/02/19/south-africa-needs-coherent-energy-policy.
11 Tien Shiao and Andrew Maddocks, "Finding solutions for South Africa's coal-fired water and energy problems," *The Guardian*, September 18, 2014.
12 World Energy Council, Energy Trilemma Index: South Africa, 2015, www.worldenergy.org/data/trilemma-index/country/south-africa/2015/.
13 South Africa's Intended Nationally Determined Contribution (INDC), http://www4.unfccc.int/submissions/INDC/Published%20Documents/South%20Africa/1/South%20Africa.pdf.
14 World Bank, Development Indicators, http://databank.worldbank.org/data/reports.aspx?source=world-development-indicators.
15 The most recent Index can be found at: www.weforum.org/reports/global-energy-architecture-performance-index-report-2015.
16 A description of the facility and its operation can be found at: www.capetown.gov.za/en/electricity/Documents/HowKoebergWorks.html.
17 Negative sign, top group, Table 12.4 signifies net energy exports.
18 World Bank, World Development Indicators, 2015.
19 James-Brent Styan, "Why no load-shedding is bad news," *Finweek*, November 12, 2015, 4.
20 Nicholas M. Odhiambo, "Electricity consumption and economic growth in South Africa: A trivariate casualty test," *Energy Economics* 2009, 31, 635–640.

21 "Eskom blackouts cost SA as much as R80bn p/m," www.enca.com/money/eskom-blackouts-cost-sa-much-r80bn.
22 "South Africa weighing up nuclear power," Oxford Business Group, April 29, 2015. www.oxfordbusinessgroup.com/news/south-africa-weighing-nuclear-power.
23 "99 days of load shedding in 2015 hits SA economy hard," *businesstech.co.za*, August 26, 2015, http://businesstech.co.za/news/business/96823/99-days-of-load-shedding-in-2015-hits-sa-economy-hard/.
24 A copy of which can be found at: www.energy.gov.za/IRP/irp%20files/IRP2010_2030_Final_Report_20110325.pdf.
25 Government of South Africa, Department of Energy, Renewable Energy IPP Procurement Programme (REIPPP) for South Africa, August 2014, http://sastela.org/wp-content/uploads/2015/10/DoE-REIPPPP-for-South-Africa.pdf.
26 The Independent Power Producer Procurement Programme, www.ipp-projects.co.za/.
27 According to the BP Statistical Energy Review 2015. www.bp.com/content/dam/bp/pdf/energy-economics/statistical-review-2015/bp-statistical-review-of-world-energy-2015-full-report.pdf.
28 David McKay, "Politics hurting SDA's energy economy," *Finweek*, December 11, 2014.
29 Barry Baxter, "South Africa: Eskom and domestic coal supply," *World Coal*, June 11, 2015, www.worldcoal.com/special-reports/11062015/South-Africa-Eskoms-woes-and-domestic-coal-supply-coal2404/.
30 Laura Cornish, "Eskom: Insisting on 50% BEE will send coal suppliers abroad," *BizNews.com*, March 13, 2015, www.fin24.com/BizNews/Eskom-Insisting-on-50-BEE-will-send-coal-suppliers-abroad-20150316.
31 Republic of South Africa, Department of Mineral Resources, "Mineral Resources," www.dmr.gov.za/publications/summary/109-mineral-and-petroleum-resources-development-act-2002/225-mineraland-petroleum-resources-development-actmprda.html.
32 Tjasa Bole-Rentel and Salem Fakir, "The water and shale gas nexus: Who needs Who?" *Daily Maverick*, February 14, 2016.
33 Terence Creamer, "Nuclear advocate urges South Africa to consider emerging reactor designs," *Creamer Media's Mining Weekly*, November 5, 2015.
34 "South Africa to procure nuclear energy at affordable scale: Zuma," *Reuters*, February 11, 2016.
35 Hartmuit Winkler, "Why South Africa should not build eight new nuclear power stations," *Mail & Guardian*, November 5, 2015.
36 Joe Brock, "South African nuclear power plan stirs fears of secrecy and graft," *Reuters*, August 14, 2015.
37 "South Africa to cut carbon emissions by 34%," *BBC News*, December 7, 2009
38 "Paris tracker: Who pledged what for 2015 UN Climate pact?" www.climatechangenews.com, February 5, 2016.
39 Anton Eberhard, Joel Kolker and James Leigland, *South Africa's Renewable Energy IPP Program: Success Failures and Lessons*, Washington, DC: World Bank, 2014.
40 Renewable Energy Independent Power Producer Program (REIPPP), International Energy Agency, www.iea.org/policiesandmeasures/pams/southafrica/name-38785-en.php.
41 Dirk De Vos, "Renewable energy – can we continue to REIPPPP what we sow?" *Daily Maverick*, May 7, 2015.
42 "IRENA guide: renewable energy auctions drive down costs," *IRENA*, June 23, 2015, www.esi-africa.com/news/irena-guide-renewable-energy-auctions-drive-down-costs/.
43 Joe Gurowsky, "South Africa's Innovative Energy Policy," Foreign Policy Association, July 25, 2014, http://foreignpolicyblogs.com/2014/07/25/south-africas-innovative-energy-policy/.

Part III

Energy transitions in the intermediate carbon-producing/consuming countries

13

The politics behind the three Es in China

Economic growth, energy security and environmental protection

Xu Yi-Chong

Energy is a bundle of issues not only because it consists of a variety of sources each of which has its distinct issues, but also because there is an inseparable relationship between energy and development, energy and environment, energy and social and economic justice and energy and security – these are the entangled three E issues: economic growth, energy security and environmental protection. Energy has fuelled industrialization, modernization and urbanization. It is a necessary ingredient for modern society – no country has developed without adopting modern (commercial) energy and the more developed a country is, the higher its energy consumption. Not all countries are rich in natural energy endowment and access to it is a concern. Energy thus is a national as well as global issue: wars have been waged on its behalf; people are deprived of development opportunities due to lack of it; companies are enriched from discovering, producing and trading it; and the environment is threatened by burning it. Energy is a political issue as choices have to be made and making choices pits one group of interests against another.[1] Energy companies seek to benefit from high energy prices which affect end-users adversely. We like to maintain and improve the comfort of our living (driving SUVs and keeping electric appliances on stand-by) but do not want to see power plants, hydro dams, transmission grids, or uranium disposal sites in our backyard. While development of renewable energy is urgent and needs more resources, traditional fossil fuels refuse to retreat from their dominant positions. Energy exporting countries want to have the market power over production and marketing while energy importing countries hope to ensure adequate supplies at an affordable price. Developing countries need better access to modern energy in order to reduce poverty and want few restrictions on their development opportunities while few who take modern energy for granted are willing to sacrifice their economic well-being or standard of living by limiting or reducing their energy consumption. These contradictory demands exist in all countries. Yet, how they are balanced varies significantly across countries, depending to a large extent on the availability of natural resources, the stage and speed of economic and social development, popular demands, and the capacity of government to manage competing demands.

This chapter discusses the changes in energy policies in China in the past three decades, especially since 2002, and explains why specific policies were adopted, how climate change has shaped recent policies and how different interests are balanced. The challenges facing China regarding energy security and climate change can be categorized into three: structural, technical, and political. While to address energy-related security and climate change issues requires a fundamental change in our approach to producing and consuming energy and therefore structural change in economy and the way of life of its citizens, it is also at the core of the difficulties in dealing with the intertwined challenges of energy security, economic and social development and climate change. There is no one 'right' or 'wrong' way to meet these multifaceted challenges. This chapter focuses on three issues: (a) structural challenges regarding energy security and environmental pollution; (b) policies and measures undertaken by the government to deal with these challenges; and (c) the politics involved in developing and implementing the policies. It argues that (a) to address these challenges requires not only short-term and immediate adjustment in its economic development but, more important long-term structural changes in its economy which are much more difficult to accomplish; (b) policies that offer new opportunities can be implemented more easily than those requiring structural changes; and (c) contrary to the common assumption, the Chinese government has limited capacity to push through the necessary measures for a sustainable development.

First we need to clarify what we are talking about in terms of energy. The global discussion of 'energy security' is dominated by the concerns of access to energy resources, especially oil, in part 'because oil is a global commodity that can be shipped at a cost that is low relative to its value [and] the price of oil is essentially determined by the world market regardless of where it is produced.'[2] Yet, for most countries and most of the time, energy security is about providing citizens with safe, reliable and cost-efficient modern energy (electricity and heat predominantly). For China, access to secure supply of oil became an issue only in the 21st century. Its change from an oil exporter to an importer in the early 1990s did not draw much attention from policy makers in Beijing or the international community until the early 2000s when rapid urbanization took place.[3] Since then demand for oil has climbed up steadily while the country's oil reserves did not change (1.1% of the world total). In contrast, the main domestic energy issue has been how to meet rising electricity demand. There is a close correlation between electricity consumption and economic and social development: New York state with a population of 19.5 million in 2010, for example, consumed nearly the same amount of electricity as the whole Sub-Sahara (except South Africa) with a population of 791 million.[4] Electricity consumption per capita in China in 1981 (258 kwh) was lower than that of Ghana in 1968 (300 kwh). The Chinese story about energy development and its associated climate change issues in the past three decades has been about how to produce and consume electricity.

Electricity can be generated from a variety of sources, such as water, sun, coal, oil, natural gas, wind, and geothermal. As electricity cannot be economically stored and must be supplied in real time to meet fluctuating demand, different sources of generation require different management of the whole system. Renewable sources of electricity generation (solar and wind) are as available as intermittent and non-dispatchable. The more diverse sources a country has, the easier it is for the country to increase the share of renewables. China has four main sources of natural energy – coal, water (hydro), wind and sun. The discussion of the energy and energy-related climate change issues has to keep this in mind.

Structural challenges

Before 1980, energy development in China was severely restricted by central allocation of resources and the consequences were that all energy sectors were among the world's least efficient ones and most electricity was consumed by industries. In the first two decades of the reform (1980–2000), as GDP quadrupled, energy consumption only doubled, thanks to a rapid decline of energy intensity. If energy intensity had remained the same as it was in 1977, China would have consumed double the amount of energy in 1980–1996 as it actually did and its CO_2 emissions would have surpassed the US in 1990.[5] In the late 1990s, suddenly the perennial power shortages that had dogged the Chinese economy disappeared after 1997 as a result of a combination of slowdown in heavy industry in response to slowing economic growth and state-owned enterprises (SOEs) reforms that led to shut-down of many inefficient small and medium-sized SOEs. The government banned investment in power generation for three years while resources were channelled to upgrading transmission and distribution infrastructure in urban and rural areas at the end of the 1990s. So far, the story about energy and energy security, energy and development and energy and environment was not too bad in China, with rapid movement toward a market-based system (dual pricing systems, enterprise reform, etc.). Unintended consequences of market reforms were the disappearance of some incentives for improving efficiency, such as elimination of energy quotas, elimination of tax breaks for efficiency with the new tax code (1994), and elimination of energy loan subsidies.

These unintended consequences had a long-term impact once economic growth picked up its speed in the early 2000s in combination with the structural change in society. First, the urbanization process placed tremendous pressure on energy and environment in part because energy consumption per capita in urban versus rural areas is about 4:1. For example, electric appliances started entering urban households only after the mid-1980s: by the mid-1990s, about 80% of the households in urban areas had washing machines and colour TV sets and by the end of the 1990s, 80% of the urban households also had refrigerators. Room air conditioners and computers started entering urban households in the 21st century, and at a much faster speed. Electrical appliance ownership in rural areas lagged urban households by 10–20 years. Second, from the mid-1990s onwards, about 15 million people every year moved to urban areas and by the early 2000s, the speed of urbanization escalated (Table 13.1). Population in Beijing, for instance, was 8.7 million in 1978; it doubled by 2005 to 15.4 million and reached over 23 million in 2015. Other major cities followed a similar pattern of population expansion: '85% of China's direct carbon emissions are from cities.'[6] Yet, as Professor Kenneth Lieberthal pointed out, when a 'cumulative population of over 400 million "middle class" people that are scattered around in a sea of over 800 million people who live very much in developing-country conditions,'[7] it is impossible and simply wrong to tell the 800 million people not to move or not have the living standard the other 400 million have. Pressure on energy and all its associated issues was structural and tremendous.

Third, several power shortages returned to China in 2002 when economic growth picked up its speed with China's entry into the WTO. Energy-intensive industries particularly expanded fast. There were three implications for energy and energy related issues: (i) energy consumption by industries as a share of total energy consumption in China was and still is much higher than most countries in the world. Among OECD countries, for instance, the share of electricity consumed by industries fell from 47.2% in 1972 to 30.8% in 2013. In China, industries

Table 13.1 Percentage of population residing in urban areas in China, selected years

	1980	*1995*	*2000*	*2005*	*2013*
Urban population (%)	19	31	36	43	53

Source: World Bank Data, and United Nations, *World Urbanisation Prospects 2014 Revision* (New York: United Nations, 2014), 21.

continued to consume a lion's share of the total electricity (74% in 2014). (ii) China was the world's large producer of most energy-intensive highly-polluting industries – in 2014, it produced 49.4% of the world's steel, 56.6% of cement, 50% of flat glass, and 38.5% of paper/pulp. Even though more than two-thirds of each of these products was consumed domestically, they still made China a virtual energy exporter. Consequently, about 25% of China's carbon emissions are caused by manufacturing products that are consumed abroad.[8] (iii) These Chinese industries fell behind other countries in energy efficiency. According to IEA, energy use per unit of output across eight sectors was about 48% higher than the best practice in the world.[9]

China's oil demand doubled from 1.7 to 3.4 million bpd between 1985 and 1995. It doubled again, reaching 6.8 million bpd by 2005, with the result that in 2005 China imported 2.46 million bpd – or about 40% of its oil needs. In 2005, 94% of overall energy supply came from domestic sources, but it was predicted China's dependence on energy imports would reach 20%. This prediction alarmed Chinese political leaders as it was believed that growing dependence on imports meant increased insecurity and it would also impose greater burdens on the economy. It also alarmed the world and global energy prices responded.

Given that these three factors were behind rapid growth of energy demand – urbanization, industry-centred and export-oriented economy, and low energy efficiency, the problems of rising energy demands and their associated environmental problems were 'not the usual complaints: an overabundance of cheap coal or a reckless disregard for the environment.'[10] Rather they reflect the fundamental structure of the economy. Adding to these structural issues is the low natural energy endowment. In 2014, China had about 1.1% of the world's oil reserves, 1.8% of world natural gas, and 12.8% of world coal reserves.[11] Heavy reliance on coal for electricity generation was primarily because coal was widely distributed in China, easy to mine and coal was cheap. The consequences of burning coal for electricity were apparent: (a) rapid depletion of coal reserves that raised concerns of energy security, and (b) significant impact on the natural environment – water, soil and air. See Table 13.2 for China's CO_2 emissions.

Managing the twin challenges

Energy shortages and their associated environmental problems have been discussed in policy making ever since the reform started in the late 1970s. In the first two decades, the focus was on increasing modern energy production and consumption – primarily electricity – because direct coal burning at the time not only had very low energy efficiency rates but also produced a large quantity of ashes and particulates. For example, in the early 1980s, the concentration of particulates of sulphur dioxide (SO_2) and nitrogen oxides (NO_x) in the air in the centre of Beijing exceeded the national standard by two to four times, and in Beijing the average dust (total suspended particulate) levels were about seven times greater than the US air quality standard.[13] In rural areas, biomass fuel use caused serious hillside soil erosion, excessive water runoff, deforestation, and declines in soil fertility.[14] To prevent further increases in already unacceptably high levels of urban air pollution, as well as to economize on fuel, China invested heavily to replace decentralized and uncontrolled burning of coal in households and enterprises with centralized, large-scale, environmentally controlled combustion in order to produce cleaner forms of energy (gas, electricity, steam, hot water) for distribution to final users.[15] Its total primary energy supplies more than doubled between 1979 and 2000 (236%) and then almost doubled again between 2000 and 2007 (193%). Electricity consumption grew four and half times between 1979 and 2000 (532%) and then one and half times between 2000 and 2007 (245%).

Table 13.2 CO_2 emissions per capita in China, world and OECD countries (t CO_2/capita)[12]

	2002	*2003*	*2004*	*2005*	*2006*	*2007*	*2008*	*2009*	*2010*	*2011*	*2012*
China	2.55	2.89	3.65	3.88	4.27	4.57	4.91	5.13	5.43	5.92	6.08
World average	3.89	3.99	4.18	4.22	4.28	4.38	4.39	4.29	4.44	4.50	4.51
As % of world average	65	72	87	92	100	104	112	120	122	131	135
OECD average	10.96	11.08	11.09	11.02	10.93	10.97	10.61	9.83	10.10	9.95	9.68
As % of OECD average	23	26	33	35	39	42	46	52	54	59	63

Source: International Energy Agency, *Key World Energy Statistics, 2015* (Paris: Organization for Economic Co-operation and Development, 2015).

Table 13.3 Wind power generation capacity in China (MW) and increase from the previous year, 2002–2014

	2001	*2002*	*2003*	*2004*	*2005*	*2006*	*2007*	*2008*	*2009*	*2010*	*2011*	*2012*
MW	404	470	568	765	1,272	2,559	5,871	12,024	25,828	44,733	62,364	75,324
%		16.0	20.9	34.7	66.3	101.0	129.0	105.0	115.0	73.2	39.4	20.8

Source: Global Wind Energy Council (GWEC), *Global Wind Report 2010*, April 2014, 39.

Before 2000, few talked about greenhouse gas (GHG) emissions in China. Even the World Bank took the view that economic and social development, and poverty alleviation, could not have been accomplished without people having access to modern energy. It had assisted China greatly in its expansion of electricity generation capacity by providing financial assistance, making connections with multinational companies to invest in China, and pushing and advising electricity reforms. During this period (1980–2000), energy intensity (energy consumption per unit of GDP) improved significantly 'by default' – that is, China had started the reform from a very low base in terms of living standards and economic performance. Technology change was the most important contributor to the reduction in energy intensity – in two decades, its economy quadrupled while its energy consumption doubled. By 2000, energy intensity had declined by 65% compared to 1980.

In 2002–2012, the gross national income in China nearly tripled in constant dollar terms and more than quintupled in current dollar terms, while the total energy consumption a little more than doubled. In the first five years of this period, several power shortages led to a rapid build-up of power generation capacities and many small-sized power plants went into operation. They were the main contributors to worsening the environment. In 2007, the State Council, the Chinese equivalent of a cabinet, adopted 'China's Energy Conditions and Policies', which acknowledged the worsening conditions of the environment, yet drew the conclusion that development was the only way for its survival; development could not take place without energy; China did not have adequate energy endowment; priority must be given to conservation, efficiency, technological break-through and international cooperation. 'It strives to build a stable, economical, clean and safe energy supply system, so as to support the sustained economic and social development with sustained energy development.'[16] In other words, measures to deal with climate change threats would be part and parcel of the broad efforts to secure energy supplies.

Three economy-wide policies were adopted as the government's 'green initiatives':

1. A *clean energy standard* mandating that 15% of China's primary energy come from non-fossil sources by 2020. China currently gets around 9% of its energy from these sources.
2. An *efficiency target* mandating a reduction in energy intensity of 20% below 2005 levels by the end of 2010. China had reduced energy intensity by around 13% by July 2009.
3. A *carbon target* mandating a reduction in carbon intensity of 40–50% below 2005 levels by 2020.

These economy-wide policies were reinforced by an array of sector- and technology-specific policies, targets and incentives. Most policies and measures designed to achieve the target of 15% of non-fossil energy sources were implemented more successfully than those to achieve energy efficiency and carbon reduction. Those concerning the first set of policies included specific measures to encourage development of non-fossil energy sources – wind, solar and nuclear. The Renewable Energy Law adopted in 2005 and a target set by the State Council for the five major power generation companies to have at least 3% of their generation capacity from renewable resources by 2020 provided further incentives for renewable development in China.[17] The central government also issued other taxation and fiscal policies to facilitate renewable and low-carbon energy development.[18] Value-added tax reductions and rebates for wind generators and imported materials used in wind turbine manufacturing were the major incentives for many firms to invest in wind generation capacity. Indeed, rapid expansion of wind and solar PV generation capacity in China brought down the cost of renewable energy worldwide (Table 13.3).[19]

By 2010, China had also become the world leader in installed solar hot water systems and indeed about 70% of the world's solar heating systems were in China. About 10% of Chinese households used solar water heaters. China also expanded its nuclear energy programme. It started its nuclear energy programme quite late and its first nuclear power plant went into operation only in 1991. In 2007, the Chinese government revised its initial target of 40GW installed nuclear capacity by 2020 to 60–70GW. By 2015, there were 31 nuclear reactors in operation and another 21 under construction, producing about 2.4% of the country's electricity. All in all, despite these rapid developments in renewable sources of electricity generation, they made little difference to the heavy dependence on coal.

Another two important measures to deal with the challenges of energy security and climate change were construction of large power generation plants that are much cleaner and more efficient and of advanced ultra-high voltage transmission grids. In the past decade China moved away from the construction of subcritical power plants, which accounted for around 95% of the additional power plants in the early 2000s, towards more efficient ultra-supercritical power plants; half of the additions in 2014 were supercritical and 85% of the new ultra-supercritical plants in the world were built in China (Table 13.4).[20]

The second measure was to construct ultra-high-voltage AC (1000kv) and ultra-high-voltage DC (±800kv) systems to connect power generation in remote areas to load centres with relatively low line losses and more efficiency. These infrastructures particularly helped integrate remotely located large-scale wind and solar generation bases to load centres with high-population density and high demand for electricity. Both measures of larger and more technologically advanced power plants and high voltage transmission grids involved serious investment but they were encouraged by international organizations to improve energy efficiency.

In addition, since the early 2000s, subsidies to energy had been dramatically reduced and energy prices in China increasingly reflected actual costs.[21] 'Three tax-related measures – corporate income tax, deductions, vehicles and fuel taxes, and export taxes – have been used to promote energy efficiency in China in recent years.'[22] Other taxation and fiscal policies were also adopted to encourage investment in low-carbon energy resources and to reduce energy consumption of those energy-intensive sectors. Together these constitute what the Chinese have called a 'recycling' industrialized economy. 'The aim of these measures is to increase energy efficiency and the share of renewable energies in order to cut energy costs, increase energy security … [and] overcome the negative effects of energy scarcity on economic growth.'[23]

Table 13.4 Coal capacity by technology, 2014

	OECD		*China*		*India*		*Rest of world*	
	GW	*%*	*GW*	*%*	*GW*	*%*	*GW*	*%*
Total	647		864		174		238	
Subcritical	415	64	529	61	149	86	158	67
Super-critical	147	23	205	24	25	14	28	12
Advanced	85	13	130	15			51	21

Source: International Energy Agency, *World Energy Outlook 2015* (Paris: Organization for Economic Co-operation and Development, 2015), 331.

Politics in balancing energy and environment

'To its credit', stated the World Bank, 'China's government fully recognises that [its development] trends cannot continue indefinitely and therefore is committed to building a resource-saving and environmentally friendly society as a stated national policy'.[24] The government in Beijing expressed its desire to move away from being the world's factory to an economy with larger high-tech and services sectors. It invested heavily in developing low-carbon energy sources and improving energy efficiency as a way to ensure sustainable energy and environmental security. As Kenneth Lieberthal stated in front of the US Senate Foreign Relations Committee:[25]

> China's rate of growth of carbon emissions, especially since 2002, has been extremely steep and pollution problems in China, I think, are rightly viewed as severe. Most Americans seem to believe that China is, therefore, ignoring its carbon emissions while pursuing all-out economic growth. But the reality is that the leaders in Beijing have adopted serious measures to bring growth in carbon emissions under control, even as they have tried to maintain rapid overall expansion of GDP.

Acknowledging the importance of climate change and energy security is the first necessary step towards the adoption of proper policies and measures to deal with the twin challenges. However to translate these policies and measures into actions and produce immediate and desired results requires more than an acknowledgement of the problems or even the right diagnoses of the problems. It requires the capacity of the government to develop one set of consistent policies and to implement them accordingly. Such capacity first of all needs an institutional structure that can prescribe 'the rules of the game', provide both incentives and restrictions for economic players, and can facilitate cooperation among various energy sub-sectors, between the central and provincial government, and among enterprises and the public. More importantly, they need institutional capacities to make policies a political reality.

In China, policy and institutional failures were a major cause of these environmental and resource-use problems. China is regarded as a centralized state, with a hierarchical structure of the central, provincial, municipal and other levels of governments. The Chinese Community Party (CCP) dictates this 'party-controlled government' in whichever way it sees fit.[26] The Party-state is 'the absolute power centre' in Chinese politics. In political reality, the government has not been able to develop either a coherent set of policies regarding energy, security, development and environment; nor does it seem to have the institutional capacity to implement policies accordingly. These are the two sides of the same coin of governing. Two key functions of all governments are coordination and resolution of disputes. If we agree with David Lampton that China 'has gone from being ruled by strongmen with personal credibility to leaders who are constrained by collective decision-making, term limits and other norms, public opinion, and their own technocratic characters,'[27] it is important to understand the collective decision making – who is involved in the process, who coordinates, how coordination takes place, and how interests are balanced. The CCP may be in power, yet, at each stage of the 'seemingly endless process of making policy in China, there are struggles – over resources, over power, over ego.'[28]

More than a dozen central government ministries or commissions share responsibilities over energy policy making in 2003–12. The National Development and Reform Commission (NDRC) emerged in 2003 from the previous macroeconomic planning agencies. It was given the primary responsibility for energy coordination. That all ministries and commissions were at the same bureaucratic level of NDRC, accountable to the State Council, meant that leadership would be needed and NDRC's ranks would not be sufficient. In 2005, the State Council created the Energy Leading Group – a supra-ministerial coordinating body headed by the Premier of the State Council. The Leading Group consisted of around 20 ministers from Finance, Foreign Affairs, Industry, Commerce, Land and Resources, the National Environmental Protection Administration (now the Ministry of Environmental Protection), the Ministry of Water Resources, the State Administration of Coal Safety, etc. In addition, as all major energy companies were state-owned under the supervision of the State-owned Assets Supervision and Administration Commission (SASAC), SASAC was part of the Leading Group too.

Each of these players had its distinct organizational interests; they differed in their mandates, in their ways of looking at things, in their political networks, and in their organizational culture and operational tradition. SASAC, for instance, wanted to consolidate energy firms and encouraged them to become the dominant players in their given sectors by acquiring weaker and small players, while NDRC demanded breaking up the monopolistic or dominant positions of energy companies in the name of encouraging competition. Yet, this was pursued as a result of the tilted power balance in favour of large state-owned energy corporations. As it was commented then 'the energy corporations initially served as a vehicle to resolve increasingly blurred rights and claims between central and local government control over energy assets, and also to attract foreign technology and financing to develop domestic resources under tight credit market conditions.'[29] With the SOE reforms in the 1990s and especially after commercialization and corporation, these energy companies transformed from government agencies to corporations whose first priority was to maximize their interest, including profit maximization. Indeed, they were expected to do so and assessed and awarded for doing so. When these SOE energy corporations were given the autonomy to operate as commercial entities, the NDRC's energy bureaucracy lost much of its expertise and access to information to the large SOEs. Insisting that economic stability would need some macroeconomic 'control', NDRC kept some of its key functions, among which were those of project-approval and price-setting.

Interests of provinces meanwhile added fuel to the chaotic situation of energy investment and energy production. For example, in the first half of the 1990s, the central government decentralized the loss-making coal industry – large ones to provinces and small-sized ones were in practice 'privatized' – as it did not have the resources to support the industry.[30] Small-sized and unregulated coal mines boomed, and their production jumped, so did their fatality rate, which was more than 10 times that of the large state-owned coal mines in 1998–2003 and remained 5–7 times for the rest of the first decade of the 21st century.[31] Together with waste coal mining and its associated pollution in soil, water and air, the coal industry drew not only domestic but international attention. The central government issued many 'orders' to close down small and dangerous coal mines, but had little impacts as they were supported by provincial and local governments. There were similar developments in electricity generation as subcritical coal-fired power plants were built regardless of environmental consequences, again supported primarily by provincial local governments to meet rising electricity demand and also to achieve higher GDP growth (discussed below).

The chaotic situation of energy production, rising dependency on oil imports, worsening environmental problems and conflicts among several energy sub-sectors 'forced' the central government to restructure its decision making apparatus to achieve greater coordination and more centralized policy making. In March 2008, the 11th National People's Congress established the National Energy Commission to replace the National Energy Leading Group. The State Council created the National Energy Administration (NEA) to replace the Energy Bureau of the NDRC. The NEA was given the mandate to coordinate energy policies and their implementations. The NEA was larger than the old energy department of the NDRC, with

240 positions allocated rather than 30, and had a higher status – vice ministry. Its first chairman was a senior official at the State Planning Commission, a veteran of both SOE and energy reforms, and at the minister level. The NEA might have been given a broad range of responsibilities, but little authority: it was responsible for planning, coordinating and governing activities of the oil, gas, coal and power industries. Yet, few people expected much from this newly created NEA not least because it was not clear how turf battles among sub-energy-sectors could be managed, but more importantly because NEA was neither 'independent' (placed under the NDRC, but bureaucratically half a level higher than the NDRC's departments) nor given much authority. NDRC maintains all key authorities – project approval and price setting. The NEA could propose changes to energy prices and be consulted on overseas energy investment project approval. The creation of the NEA was a compromise between the NDRC that did not want to give up its control over the energy sectors and those who called for a mega-ministry of all energy sectors. Since it is not a full ministry, it was unable to coordinate actions of relevant government agencies. It also lacks human capital, authority and instruments to bring in line the actions of all state-owned energy corporations.

While the chairman of the NEA might be the deputy minister of the NDRC, the NEA itself did not have independent authority over any energy policy issues. Its first chairman, Zhang Guobao, from the very beginning openly admitted that he was a powerless minister, without authority to coordinate, approve projects and regulate price regulation. He liked to tell people:

> One needs 5 licenses to operate a coal mine. None of the authority of issuing them is in the hand of the NEA. I lead an energy bureau which does not have any authority over pricing. When they increased petro price, I was not even told. I went to the Premier to complain. The Premier joked: it only showed our good work of confidentiality. When wind power expansion went too fast and much exceeded the transmission capacity to absorb it, the NDRC and 10 other government agencies issued a document on slowing down wind power development, the NEA was not even informed.[32]

There was a loud voice in 2007–08, calling on the central government to establish a mega-ministry equivalent to the US Department of Energy, giving it sufficient resources and authority to coordinate and enforce polices. It never happened. Meanwhile, various energy departments of NDRC turned their authority of project approval and price setting into rent-seeking opportunities. When the current government took over in 2013, nearly all senior officials of these departments were found guilty for being corrupt, taking bribery and some even more serious crimes.

In the midst of chaotic energy development and production, the environment suffered at least as much as energy resources. In March 2008, the State Council upgraded the State Environmental Protection Administration into a full-ministerial agency: the Ministry of Environmental Protection (MEP). This was a clear indication that environmental concerns attracted serious interest within the central government. Nonetheless, the MEP was neither given sufficient resources nor authority in making and enforcing environmentally related policies. For example, the MEP was allotted 2,600 positions with only 300 working out of the headquarters in Beijing. In contrast, the US Environmental Protection Agency has 17,000 employees with nearly 9,000 working in Washington, DC. MEP was dwarfed by both energy companies (large and small) and local governments; for them making profits and maximizing local interests were the first priorities, and both had much more resources at hand in resisting the MEP. More importantly, the MEP was not a heavy-weight player among central government agencies. Throughout the first decade of the 21st century, officials at the MEP (and its predecessor), including its minister, warned about the neglect of the environment when economic decisions were made. The central government set a GDP growth at 8%, not as a ceiling, but as a floor target, reflecting a longstanding policy of 'bao ba' (保八) or, 'protecting the eight'. Annual GDP growth of 8% was believed to be sufficient to absorb the new workforce entering the market each year and thereby to maintain political stability. Governors of provinces were assessed and rewarded for their GDP growth.

When the economy slowed down affected by the global financial crisis, for instance, different parts of the government talked about different things: the Ministry of Finance emphasized the importance of boosting government revenue and controlled spending to balance the budget. Its minister cautioned against overspending while there was a clear sign that revenue would go down for 2008–09. The newly merged Ministry of Industry and Information Technology talked about the necessary restructuring of the economy as some industries were burdened with over production capacity. The MEP again raised its warning that pollution had deteriorated so quickly that it threatened the key components of Hu Jintao's 'harmonious society' and suggested balancing economic growth and environmental protection. For the State Council, keeping a minimum 8% GDP growth was the first priority. Even though it talked about 'green development', 'green' was subject to 'growth', and despite both experts and policy makers talking about 'rebalancing the economy' toward domestic consumption and reducing overcapacity, few knew how to do it without hurting economic growth. The government did what it knew how – by following its policies of the Asian financial crisis (1998–99) – to 'hit the downturn' quickly, with the greatest force it could muscle, targeting those areas which would have direct impact on the economy, as the Australian government at the time decided to 'go hard, go early and go household' to have the maximum impact.

The big stimulus package of 4 trillion yuan (the equivalent of US$586 billion) contained 'ten priorities' including: to speed up the development and construction of (1) social housing, (2) rural infrastructure, and (3) rail, road and airports, (4) health and education, (5) to improve environment and ecosystems, (6) to speed up economic structural readjustment and engage in innovation, (7) to speed up earthquake reconstruction, (8) to raise rural and urban resident incomes, (9) to reduce taxes for enterprises and encourage their technology innovation, and (10) to strengthen its financial support for economic growth. While $30 billion of the package was dedicated to 'green' projects, emphasis in allocating the fund was clearly in favour of 'development' rather than 'green': transport and electricity infrastructure took the lion's share of the package (38%); one-quarter of the package went to earthquake reconstruction; and the rest was split among affordable housing (10%), technology innovation (9%), rural infrastructure (9%), environment (5%), and health and education (4%). With the financial incentives, 'investment in China's clean energy sector was 60% higher than in the United States ($54.4 billion versus $34 billion) … and in 2009, China surpassed the United States in total installed clean energy production capacity.'[33] Yet, the environmental impact of the rest of the package was much greater than that of the 'green' energy as energy-intensity and polluting industries continued to expand despite many of them having already shown signs of over-capacity. Indeed these developments and the spreading environmental pollution led MEP officials, including its minister, to openly criticize the government for 'the lack of respect for the environment as the country carries out its economic stimulus plan.'[34]

The MEP was not the only central agency that held different views on how to balance economy, development, energy security and environment. Nor were these differences unique in China. Thus the key functions of all government are 'coordination and resolution of disputes'. The Chinese political system displays two seemingly contradictory phenomena: 'overcentralization' and decentralization: it is overcentralized as 'its leaders are overloaded' with demands to make decisions and it is over-decentralized as decisions are made by so many players. In the US a system of checks and balances was created by design with a deep suspicion of government while the checks and balances existed in the Chinese system by default with a deep entrenched expectation 'that senior leaders are indispensable agents of dispute resolution across an infinite range of issues'.[35] Governing in China is like 'Whac-A-Mole' – top leaders are constantly pulled to all directions without institutionalized capacities as those government agencies that are supposed to 'assist' can seldom agree on major issues and on cooperation.

The situation continued: the Ministry of Commerce preferred to see continuing export growth while the MEP and NEA called for changing the export-oriented economic structure. The Ministry of Science and Technology would like to see the development of domestic low-carbon energy technologies, the MEP and NEA push for adoption of the most advanced technologies, domestic and international. While the NEA would like to see China's energy companies invest overseas to ensure direct access to energy resources, the Ministry of Foreign Affairs issues cautions about the diplomatic impacts of Chinese investments in certain countries and regions. The policy gap between the central and provincial governments is even larger. In sum, as the World Bank pointed out, it would be very difficult to adopt a set of coherent energy strategies and implement a national approach towards energy security and climate change, 'if different arms of government send conflicting signals'.[36] The weak governing capacity was also 'reflected in lack of adequate policy research to assist the government in making sophisticated policies'.[37]

In contrast with fragmented government agencies, large state-owned enterprises enjoyed oligopolistic positions in their respected fields, obtained substantial financial wealth and had significant political influence in shaping the policies in their favour. They are the modern corporations seeking to expand their profits by taking advantage of the market while the market was neither well monitored nor properly regulated. The abandonment of the old annual, five-year and long-term national plans that specified detailed material targets, based on which resources were allocated[38] would require better capacity for government to coordinate in order to achieve some degree of coherence in its policies. That was lacking. For instance, instead of providing long-term projections and well-planned national strategies based on the projections, senior officials at the energy departments of the NDRC and NEA were busy using their power and authority for rent-seeking. Energy companies did their own things: in less than 2–3 years after the government adopted policies to encourage renewable energy, both wind and solar sectors suffered over-capacity in some places where infrastructure was not in place. The solar manufacturing industry produced mainly for exports and sank into financial holes when global financial crisis dried up export markets. Rapid expansion of installed wind capacity was not matched by the similar expansion in transmission and distribution capacity, and electricity generated by wind farms could not be sent to load centres (known as curtailment). By 2010, the total installed wind capacity (44.7GW) was 6–10 times the initial projected target (3–7GW), creating insurmountable challenges for the T&D infrastructure. In addition, there was a concentration of these wind capacities, thanks to, at least, policies of another powerful government ministry – the Ministry of Finance. By 2011, five provinces and regions hosted over 70% of the country's total wind capacities – Inner Mongolia, Liaoning, Hebei, Jilin and Heilongjiang. Except Hebei province, they were all electricity surplus producers to begin with and had already developed more coal-fired thermal generation capacities than they could consume. Unwilling to close down thermal capacities which would affect local economies, they expanded their wind capacities due to the subsidy policies put in place by the Ministry of Finance. Without adequate T&D infrastructure, curtailment of wind power became serious financial and political issues.

In sum, in the past decade, the central government had repeatedly said that its policy priority was to 'rebalance' the economy by shifting its export-dependent to domestic consumption-driven economic development, to change from quantitatively rapid growth to a qualitative superior growth, and to have a better redistribution system to reduce inequality. These broad policy outlines were believed to provide a 'shared framework for understanding problems, even as vigorous debate raged about the amplitude and rigor of specific policy measures.'[39] In the aftermath of the global financial crisis, few knew how to have a high GDP growth rate while rebalancing the economy. Most economists including those at the World Bank then believed that China should not allow the economy to slow down and spending would keep China and the world out of trouble. The guiding principle in 2008–10 was summarized into 16 Chinese characters – control inflation, adjust structure, promote reform, and increase revenue while reducing spending (控制赤字、调整结构、推进改革、增收节支). The stimulus package, however, supported none of these objectives. Without 'rebalancing' the economy, energy-related climate change issues worsened.

Recent developments

Official reports found that 20% of China's arable land, more than 60% of its underground water, and 33% of its surface water are polluted. The World Bank and the State Council's Development Research Centre estimated the costs of this environmental degradation reached approximately 10% of GDP in 2008, representing a significant drag on the economy.[40] Furthermore, air pollution contributed to 17% of all deaths (or 1.6 million people) in China between April 2014 and March 2015 according to estimates by the US-based research non-profit Berkeley Earth.[41] The current government coming to power in early 2013 made its pledge to fight the 'war of air pollution'. It announced an Action Plan on Prevention and Control of Air Pollution, with bans on new coal-fired capacity in key regions and absolute reductions in coal use in eastern provinces in 2012 to 2017. To prepare for the UN conference on climate change in December 2015, the NDRC prepared a set of policies and actions on climate change in 2012, which was upgraded in November 2014, following the pledge the Chinese government made at an APEC summit in Beijing in September that year. It promised to peak CO_2 emission by 2030 and to raise the share of non-fossil energy to 20% by 2030. It also promised that by 2030, non-fossil fuel would reach 20% of the total primary energy consumption, and CO_2 emission per unit of GDP would be down by 60–65% from the 2005 level.[42]

The State Council formally adopted the Energy Development Strategic Action Plan (2014–2020) (能源发展战略行动计划 2014–2020) in 2014. The plan contains four objectives: (a) to improve energy efficiency and conservation, (b) to pursue low-carbon green energy (to cap coal consumption at about 420 million tonnes, accounting for no more than 62% of the total energy consumption by 2020, and to increase renewable energy capacity: between 2016 and 2030, China would have to add 10GW of nuclear capacity, 15GW of hydro, 30GW of solar PV and 40GW of wind power capacity), (c) to ensure self-sufficiency (85% of energy will come from domestic sources by 2020), and (d) to adopt innovation-led energy development. To achieve these objectives, the Plan sets two basic conditions: further reform in energy sectors, and improvement of energy policies and regulations.

China has been leading in investment in renewable power generation, in hydro, wind, solar PV and solar water heating capacity. By the end of 2014, it hosted a quarter of the world's renewable electricity generation capacity, including 280GW hydropower.[43] In 2014, the total renewable power generation capacity of wind, solar PV and biomass in China was 50% more than that in the US and double that in Germany. Nuclear, wind and solar PV together generated only 5.5% of total electricity.[44] These targets and actual data tell several important features of the energy industry in China: (a) size matters;[45] (b) speed matters – institutional development did not catch up with the rapid economic growth and development; and (c) momentum matters – industries have been on a treadmill for high growth; it became very difficult to slow down without having significant impact on firms as well as the society as a whole. The recent economic structural changes and their related economic slowdown might be an opportunity for those inefficient and polluting power generators to withdraw from the market. Instead, slow economic growth in the past two years did not lead to equivalent slowdown in investment in electricity generation – signs of overproduction worried many in the industry.

The trend has been downwards (Table 13.5), but investment has not gone down, as one of the five national SOEs in generation complained:

> There is apparent over-capacity in generation. We are running fewer and fewer hours. But we are under significant pressures from provinces to build more coal-fired power plants. If we refuse, the province will take the licence back and give it to local power companies, which are less efficient and care less about environment.[46]

Many scholars who study Chinese politics and economics have all recognized the power of provinces and local governments.[47] The State Council in late 2015 reemphasized that 'no one should build coal-fired thermal plants without proper approval'. It seems feeble as the State Council has little capacity to enforce it or to monitor it. This was already shown in the previous policies: in late 2005, the Chinese leaders set an ambitious target to reduce the country's energy per unit of GDP by 20% and cut total CO_2 emissions by 10% in 2006–10. The following 12th Five-Year Plan (2011–15) called for a 16% reduction in energy intensity and a 17% reduction in carbon intensity. To achieve the goals, mandatory sub-targets were subdivided and assigned to provinces and lower administration levels and to administrators of key national programmes, with accountabilities for delivery. While it was reported that most provinces met their targets,[48] quantitative results in reducing the overall air, water and soil pollution were disappointing.[49]

Self-reporting systems always raised the issue about the accuracy of the reports from provinces. More importantly, it was the capacity of the central government to ensure adequate and proper implementation of its policies. Often central government agencies promulgate new national regulations and policies, set national targets, establish national standards, design and oversee implementation of a series of policies. Without a well-established operational legal system, it is difficult to regulate economic activities.

Conclusion

In China, environmental and climate issues have been incorporated in energy policies and development. Indeed, China has expanded its low-carbon energy sources significantly since 2003 and they are all developed in the name of ensuring adequate energy supplies and minimizing environmental and climate change threats. Meanwhile, fundamental structural changes in its economy are difficult to come by because of (a) fragmented government agencies competing for agenda, (b) the government lacking the capacity to adopt coherent and consistent policies and to implement them accordingly, and (c) slow creation of an operational legal and regulatory system. Policy makers in China are struggling simply to figure out what is unfolding on the ground while looking for points of leverage to achieve their desired changes. Given that the two issues cover a far wider range of implications on the economy and society, actions taken on one issue may or may not have the positive results on the other. Addressing energy-related

Table 13.5 Power plant utilisation rate (hours), 2001–2014

	2001	*2002*	*2003*	*2004*	*2005*	*2006*	*2007*	*2008*	*2009*	*2010*	*2011*	*2012*	*2013*	*2014*
Thermal	4,900	5,752	5,767	5,991	5,865	5,612	5,344	4,885	4,865	5,031	5,305	4,982	5,021	4,706
Hydro	3,129	3,289	3,239	3,462	3,664	3,393	3,520	3,589	3,328	3,404	3,019	3,591	3,359	3,653
Average	4,588	4,860	5,245	5,455	5,425	5,198	5,020	4,648	4,546	4,650	4,730	4,579	4,521	4,286

Source: China Electricity Council, ‘Summary of the Electricity Industry in 2014,’ China Electricity Council, 2015.

development, security, and climate change threats requires the political commitment of governments, wide participation of enterprises and the public and technological innovation. It has been a great challenge for leaders in Beijing to balance long- and short-term development and balance the diverse interests of urban and rural populations, coastal and interior regions, and the elite and the masses.

To manage these issues, the current government came to power, promising to tackle the problems of 'top design' of government institutions – 顶层设计, as Xi-Li team emphasized when they took over the government in 2013. Yet it remains to be seen what 'top design' might be for the energy industry. A mega ministry had been suggested since the beginning of the 21st century, but is still an illusion rather than the reality. After taking office, the current government quickly abolished the SERC, folded it into the NDRC and assigned it a very junior partnership role to the NDRC – 'participate in planning national energy development, suggest power tariff adjustments, suggest further electricity reform', etc.

Notes

1 J. E. Chubb, *Interest Groups and the Bureaucracy* (Stanford, CA: Stanford University Press, 1983); S. Haggard, and M. D. McCubbins, *Presidents, Parliaments, and Policy* (New York: Cambridge University Press, 2001).
2 J. Bordoff, M. Deshpande, and P. Noel, 'Understanding the Interaction between Energy Security and Climate Change Policy', in *Energy Security*, ed. C. Pascual and J. Elkind (Washington, DC: Brookings Institution Press, 2010), 212.
3 Bo Kong, *China's International Petroleum Policy* (Santa Barbara, CA: ABC Clio, 2010).
4 IEA, *World Energy Outlook, 2010* (Paris: OECD, 2010).
5 J. E. Sinton, M. D. Lavine, and Q. Y. Wang, 'Energy Efficiency in China: Accomplishments and Challenges', *Energy Policy*, 26, no. 11 (1998): 813–829; J. E. Sinton, and D. G. Fridley, 'What Goes Up: Recent Trends in China's Energy Consumption', *Energy Policy*, 28, no. 10 (2000): 671–687.
6 Zhu Liu, *China's Carbon Emissions Report 2015* (Cambridge, MA: Harvard Kennedy School, Belfer Centre for Science and International Affairs, 2015), 10.
7 Kenneth Lieberthal, *US-China Clean Energy Cooperation: The Road Ahead* (Washington, DC: Brookings Institution, 2009), 7.
8 Liu, *China's Carbon Emissions Report 2015*.
9 International Energy Agency (IEA), *World Energy Outlook 2007: China and India Insights* (Paris: Organization for Economic Co-operation and Development, 2007).
10 William Chandler, and Holly Gwin, *Financing Energy Efficiency in China* (Washington, DC: Carnegie Endowment for International Peace, 2008), 6.
11 BP, *Statistical Review of World Energy* (London: BP, 2015), www.bp.com/en/global/corporate/energy-economics/statistical-review-of-world-energy.html.
12 International Energy Agency, *Key World Energy Statistics, 2004* (Paris: Organization for Economic Co-operation and Development, 2004).
13 Xu Yi-chong. *The Politics of Nuclear Energy in China* (Basingstoke: Palgrave Macmillan, 2010).
14 World Bank, *Energy Sector Management Assistance Program* (no.101/89) (Washington, DC: World Bank, 1989).
15 World Bank, *China: Long-Term Development, Issues and Options* (Washington, DC: World Bank, 1985).
16 State Council of the People's Republic of China, *China's Energy Conditions and Policies* (Beijing: Information Office of the State Council of the People's Republic of China, 2007), 11.
17 World Bank, *Developing a Circular Economy in China: Highlights and Recommendations* (Washington, DC: World Bank, 2009); World Bank, *Winds of Change: East Asia's Sustainable Energy Future* (Washington, DC: World Bank, 2010); World Bank, *World Development Report 2010: Development and Climate Change* (Washington, DC: World Bank, 2010).
18 Eighth Senior Policy Advisory Council, *Tax and Fiscal Policies to Promote Clean Energy Technology Development* (Beijing: Eighth Senior Policy Advisory Council, 2005).
19 Joanna I. Lewis, *Green Innovation in China* (New York: Columbia University Press, 2013); Kelly Sims Gallagher, *The Globalisation of Clean Energy Technology* (Cambridge, MA: MIT Press, 2014).

20 Conventional coal-fired power plants, which use water boilers to generate steam that activates a turbine, have efficiency of about 32%. Supercritical and ultra-supercritical power plants that operate at temperatures and pressures above the critical point of water have energy efficiency of above 45%. Supercritical and ultra-supercritical power plants also require less coal per-kilowatt-hour electricity generated, leading to low emissions, including carbon dioxide and mercury, higher efficiency, lower fuel costs per kilowatt of electricity.
21 IEA, *World Energy Outlook 2007.*
22 Nan Zhou, Mark D. Levine, and Lynn Price, 'Overview of Current Energy Efficiency Policies in China', *Energy Policy* Z 38, no. 11 (2010): 15.
23 C. Richerzhagen, and I. Scholts, 'China's Capacities for Mitigating Climate Change', *World Development* 36, no. 2 (2008): 312.
24 World Bank, *Developing a Circular Economy in China: Highlights and Recommendations* (Washington, DC: World Bank, 2009).
25 Lieberthal, *US-China Clean Energy Cooperation*, 6.
26 Kerry Dumbaugh, and Michael F. Martin, *Understanding China's Political System* (R41007) (Washington, DC: Congressional Research Service, 2009).
27 David M. Lampton, *Following the Leader* (Berkeley, CA: University of California Press, 2014), 76.
28 Lampton, *Following the Leader*, 79.
29 Edward S. Steinfeld, Richard K. Lester, and E. A. Cunningham. 'Greener Plants, Greyer Skies: A Report from the Frontlines of China's Energy Sector,' *Energy Policy*,, 37, no. 5 (2008): 1809–24.
30 E. Thomson, *The Chinese Coal Industry* (New York: Routledge Curson, 2003); Huaichuan Rui, *Globalisation, Transition and Development in China: The Case of the Coal Industry* (New York: Routledge Curzon, 2004); T. Wright, *The Political Economy of the Chinese Coal Industry* (New York: Routledge, 2012).
31 Jianping Zhao, and Xu Yi-chong, 'The Shenhua Group', in *The Political Economy of State-owned Enterprises in China and India*, ed. Xu Yi-chong (Basingstoke: Palgrave Macmillan, 2012), 29.
32 Quoted in Wu Ge, 'Zhang Guobao in Their Views,' *China Economic Weekly*, 1 March 2011 (in Chinese).
33 Michael D. Swaine, *America's Challenge* (Washington, DC: Carnegie Endowment for International Peace, 2011), 263.
34 It is interesting to note that according to Michael Swaine, China was more effective in managing climate change issues than the United States because 'China's top official (currently Xie Zhenhua) is both chief climate negotiator and plays a major role in moving China toward a less carbon-intensive path, where as the top U.S. official (currently Todd Stern) is strictly a negotiator' (Swaine, *America's Challenge*, 271). The frustration of Xie about his inability to push through environmental issues among other government agencies was apparent.
35 Lampton, *Following the Leader*, 50.
36 World Bank, *Developing a Circular Economy in China*, 28.
37 Ibid., 23.
38 T. Kambara, and C. Howe, *China and the Global Energy Crisis* (Cheltenham: Edward Elgar, 2007).
39 Barry Naughton, 'A New Team Faces Unprecedented Economic Challenges', *China Leadership Monitor*, 26, 2 September 2008: 1.
40 World Bank, *China 2030: Building a Modern, Harmonious, and Creative Society* (Washington, DC: World Bank, 2013), 39, 233.
41 R. A. Rohde, and R. A. Muller, 'Air Pollution in China: Mapping Concentrations and Sources', *PloS one*, 10, no. 8 (2015): e0135749.
42 National Development and Reform Commission, *China's Policies and Actions on Climate Change* (Beijing: National Development and Reform Commission, 2014).
43 REN21, *Renewables 2015: Global Status Report* (REN21, 2015), www.ren21.net/wp-content/uploads/2015/07/REN12-GSR2015_Onlinebook_low1.pdf, 30.
44 China Electricity Council, 'Summary of the Electricity Industry in 2014' (China Electricity Council, 2015), www.cec.org.cn/guihuayutongji/gongxufenxi/dianliyunxingjiankuang/2015-02-02/133565.html.
45 Lieberthal, *US-China Clean Energy Cooperation.*
46 Interview in Beijing, 15 December 2015.
47 A. L. Wang, 'The Search for Sustainable Legitimacy: Environmental Law and Bureaucracy in China', *Harvard Environmental Law Review* 37, no. 2 (2013): 365–440; A. L. Wang, 'Chinese State Capitalism

and the Environment', in *Regulating the Visible Hand*, ed. B. L. Liebman, and C. J. Milhaupt (New York: Oxford University Press, 2015).

48 Liu, *China's Carbon Emissions Report 2015.*

49 Robert P. Taylor, G. J. Draugelis, Y. Zhang, et al., *Accelerating Energy Conservation in China's Provinces* (Washington, DC: World Bank, 2010).

14

The USA's energy and climate transition

Partial success without a plan

Robert E. Looney

Introduction

The US energy picture is unique in a number of ways. Until recently replaced by China, the US was the world's largest consumer of energy. It is also a major producer of energy. In sharp contrast to most other developed industrial countries, US output of natural gas and oil has boomed in recent years, largely as a result of technological advances in shale technology.[1]

For all of the country's energy positives, there is one glaring negative – the country has never formulated what might be considered a comprehensive energy policy. Henry Kissinger is said to have observed: "The Energy Policy of the United States is not to have an Energy Policy,"[2] Other policy critics contend that the US actually has too many individually focused plans and programs. These are more often than not limited in scope, with little consideration given to their impacts outside their intended beneficiaries. The result is a morass of competing and conflicting outcomes.

To one critic of US energy policies, John Deutch, former Director of Energy Research at the Department of Energy the US government has been unable to formulate and sustain an effective energy policy because:[3] (1) the authorities have tended to opt for popular but unrealistic goals, (2) public attitudes often push politicians to short-term stop-gap policies rather than longer-term permanent solutions, and (3) there is an absence of objective quantitative analysis in planning, policymaking, and administration of government programs.

Many free market advocates assert that the country's energy sector is dynamic precisely because of the lack of comprehensive governmental planning. They argue that rather than constraining, bureaucratic rule making, markets have unleashed the private sector's creative talents to innovate and allocate energy related capital effectively.

Their contention is borne out by the US rankings in the World Energy Council's Trilemma Index[4] (Figure 14.1) that gauges country progress in energy equity (affordable energy), environmental sustainability, and energy security. From 2011 the starting date of the Index into 2015, the US has ranked first in the world in affordable energy. Largely as a result of the market driven surge in shale-based oil and gas the US also improved its standing in energy security,

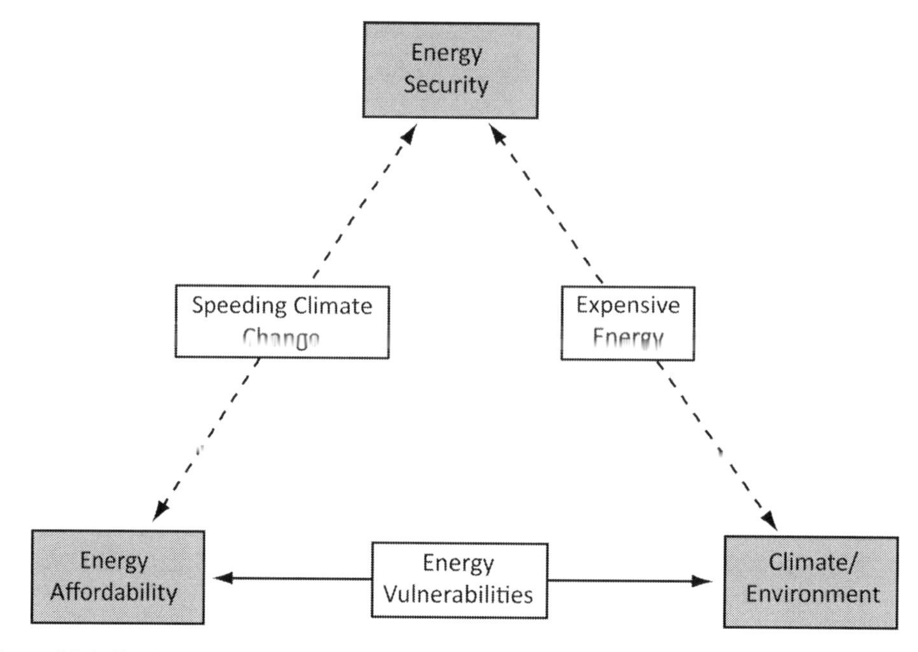

Figure 14.1 Classic energy trilemma

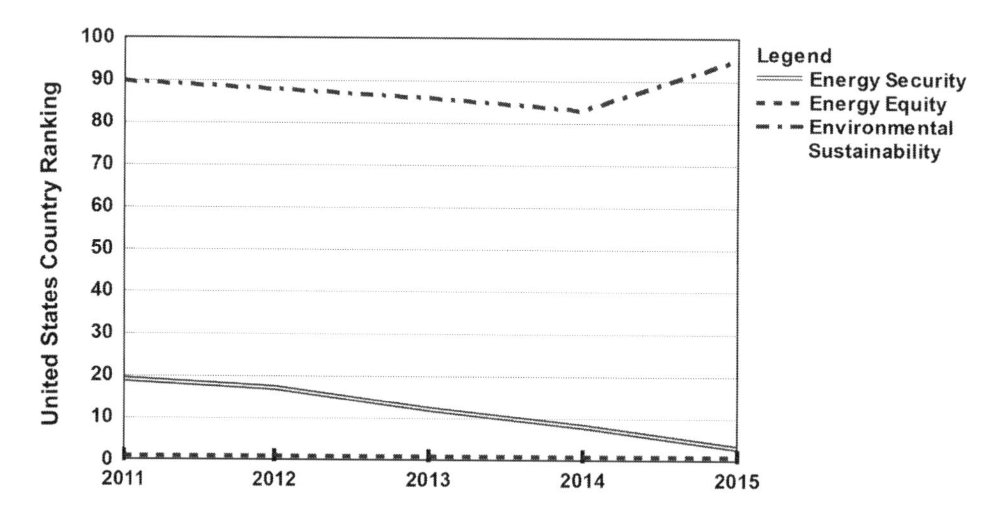

Figure 14.2 United States ranking on the Energy Trilemma Index

moving from 19th in the world in 2011 to 3rd by 2015. On the other hand, in the area of environmental sustainability where markets greatly underprice carbon, the United States ranked 95th in 2015 (Figure 14.2).

As US priorities along with those in the rest of the world began shifting toward environmental concerns, especially those associated with climate change, policy makers confronted challenges never faced before. Typically in the US energy policy making is driven by the need to respond to a dramatic event such as the oil price shocks in the 1970s, or the oil price increases in the years leading up to the global financial crisis in 2008–09.

In the years before major concerns over climate change, US policymakers were able to design programs and policies that were fairly straightforward. In energy trilemma terms they were able to move ahead with actions that did not involve large trade-offs with other national objectives. Coal was always there to guarantee cheap available energy. In policy circles this environment and associated problems are usually dubbed "Tame Problems."[5] Typically, tame problems involve a clear objective and the causal factors underlying the problem are simple and widely understood. Energy security a problem? – fill up the Strategic Oil Reserve. Gas prices too high? – put on price controls until new sources of supply can be developed.

Thanks to coal, piecemeal energy policies that did not distort markets were relatively effective in assuring energy security without significantly increasing the price of energy. The two nodes of the trilemma energy security and affordability were easily achieved if at the cost of the environment and climate change.

Unfortunately, climate change or rather its deterioration does not produce market signals that firms can act on to increase their profits. Nor is it a dramatic shock that governments feel compelled or are under immediate pressure to respond to. One ramification is that a high percentage of Americans do not see climate change as a serious problem and a danger to the US.[6] In part this simply reflects an ideological barrier as many conservatives see climate change as simply a left-wing urbanites' cause and tend to dig in against it.[7]

In this environment, along with the evidence that coal is a major contributor to greenhouse gases, US authorities are now confronted with what is usually referred to as a Wicked Problem.[8] Wicked problems have no definitive formulation due to their extreme complexity. Any solution causes feedback effects and tradeoffs, often making the policy environment extremely contentious. Wicked problems are in a constant process of evolution and thus are never completely solved.[9] Any solution causes feedback effects which may negate other efforts at resolving or lessening the problem. As such, solutions are usually only fleeting. Wicked problems as in the case of climate change involve many diverse groups with sharply competing interests. Often a potentially infinite solution set exists, making policy consensus on a course of action almost impossible.

Compounding the wicked energy problem in the US is the fact that most of the optimal policies in aiding the country's energy transition are unavailable. This has forced policymakers to choose from a number of "second best" (and "third best") policies. Specifically, while a carbon tax would be the most efficient way to shift usage away from greenhouse gas emissions, the government has been forced to price carbon indirectly, or subsidize wind, and solar power, but this does nothing to discourage the continued usage of fossil fuels.

After examining the US's relative position in the energy trilemma, the following sections trace US efforts at limiting climate change. They show that despite the fact that the US has never been able to develop a comprehensive energy/climate plan, and often selects conflicting policies, the country appears, thanks to fortuitous developments, to be on a successful path of transition towards a sustainable increase in climate and energy security.

Trilemma patterns

The current US situation with regard to climate change, economic security and economic affordability can be best appreciated through comparisons with other parts of the world. For purposes of classification, the US falls in a group of countries that both produce and consume large amounts of energy, with energy imports accounting for between 5% and 30% of total energy supplies (see Chapter 1, Table 1.1).

The World Energy Council Trilemma Index represents an attempt to roughly quantify the three components of the energy trilemma for the purpose of not only identifying country priorities between the three, but also identifying which countries have made the most progress in each area.

As noted previously, historically US priorities and the outcome of market forces have positioned the country with high marks in energy security and energy equity (affordability). America's average score on these two dimensions would rank the country at around 6th in the world in terms of attainment. This score is 82 country rankings ahead of the country's progress in environmental sustainability which is the most extreme difference in the countries included in the trilemma index (Table 1.2).

China has a similar pattern, although that country ranks considerably below the US in all three dimensions. The difference between China's two highest priorities, energy security and energy equity and its progress in environmental sustainability is 74 country rankings. In the case of Poland the difference is 55 country rankings, while for India it is 30 country rankings.

The UK presents a much less extreme pattern. While energy security and affordability are clearly that country's priority areas, the difference between their average and that of energy sustainability is only 9 country rankings. In Brazil's case energy security and environmental sustainability are the top priorities with their average only 4.6 country rankings ahead of energy equity.

These patterns suggest that energy producers with a large component of coal production such as the US, China, Poland and India are much more likely to encounter a "hard trilemma" where improvements in the environment, at least in the short run are likely to bring down their rankings in energy security and/or energy affordability. Countries like the UK with environmental concerns and not reliant on coal, and the financial ability to develop other energy sources such as nuclear face a "soft trilemma" where balance between the three energy dimensions is attainable without much sacrifice in any particular area. The same should be the case for Brazil, but that country, largely dependent on hydro power, has not been able to expand supplies of electricity in line with the growing demand. This represents more of a policy failure rather than inherent hard trilemma trade-off. When the country's large off-shore oil discoveries begin production, the country should score much better in the affordability area.

US efforts towards energy and climate security – the Bush Administrations

With increasing international pressure stemming from its failure to sign the Kyoto Protocol[10] (2001), the US, for the first time, began to confront a hard energy trilemma. Throughout his administration (January 2001 to January 2009) President Bush followed a strategy based on long-term CO_2 goal-setting without mandatory enforcement.

Instead, US emphasis was on encouraging the adoption of new technologies via additional research and development (R&D) funding and partnerships. Specifically, in 2002, President Bush outlined a policy of reducing the ratio of emissions intensity to GDP over ten years.[11] In 2006 and 2007, he applied this approach to global negotiations, arguing that a series of international meetings were needed to set "aspirational goals" for emissions reductions.[12] In October 2007, he convened a "climate summit"[13] in Washington to begin this task, but no agreement was reached on long-term objectives.

Finally in 2008 the Bush Administration shifted from focusing on reducing emissions intensity by 2012, to targeting a halt in the growth of US emissions by 2025. Moreover, the administration's projections for meeting this goal anticipated 17 years of increasing emissions (currently growing by a few percentage points per year), so the 2025 level would be significantly above current levels. It also did not specify an actual target; even large and repeated year-on-year increases would be consistent with this goal as long as the pace of increase dropped to zero in 2025.[14]

In addition to setting this distant target, the Bush administration outlined a preferred method of reaching the goal. Ideally it would involve all major emitters, including China, India and Russia; it would focus on encouraging innovation in low-carbon technologies while making nuclear power and coal part of any "solution." The idea was to provide a smooth transition for the economy while undertaking an "honest assessment of the costs, benefits and feasibility"[15] of any emissions reduction effort.

The administration's strategy was to achieve progress in CO_2 emissions, but not if it involved imposing the costs of improving efficiency and reducing emissions on industry or the public, or significantly tightening US regulation. In this regard, the administration strongly opposed mandatory approaches which they feared might sharply increase costs, reduce US competitiveness, or stifle the diffusion of new technologies.

While the Bush Administration was steadfast in its emissions position, momentum was building in the US for greater efforts towards averting climate change. First, Congress began debating legislation for the creation of a US domestic cap-and-trade program along the lines of the EU Emissions Trading Scheme (ETS).[16] In addition, several of the states, California and those in the north-east, began considering trading initiatives.[17]

Second, in the area of regulation, a slowly moving battle in the courts was developing between the Environmental Protection Agency (EPA) – which as an executive agency falls under the Bush White House – and various states and interest groups that were attempting to use a pre-existing statute, the Clean Air Act, to force the EPA to regulate greenhouse gases.

In an April 2007 ruling, the US Supreme Court found that the Clean Air Act does, in fact, allow the EPA to regulate greenhouse gases, and said that the Agency could only avoid taking remedial action on emissions "if it determines that greenhouse gases do not contribute to climate change" or otherwise provides some "reasonable explanation." However, the Court did not impose a timetable for such a determination, and the EPA has chosen to proceed slowly.

Little progress towards environmental sustainability and the reduction of CO_2 emissions was made during the Bush years. At the time, the United States was the only major developed economy to have remained significantly uncommitted to international climate policy. In contrast to previous administrations that made significant improvements to energy security and in the provision of affordable energy. Ideologically, the Bush Administration was unable to design a credible climate change policy. It was also unable to cope with the increased complexity associated with climate change. Instead the Administration reverted to its free-market orientation, and the hope that new technologies would provide a low-cost solution to the CO_2 problem. However free markets only provide solutions and new technologies if they are profitable, and the administration did little to assure this outcome.

US efforts towards energy and climate security – the first Obama Administration

By the end of the Bush administration climate change had moved from being a niche environmental concern to one with broad national and international interest. The shift in US leadership was expected to give climate change a high priority and thus shift the US approach to international climate negotiations.

President-elect Barack Obama had been a consistent supporter of a US cap-and-trade system. During the campaign for the presidency he also proposed setting long-term goals of a 20% reduction in emissions (below 1990 levels) by 2020 and 80% reduction by 2050.[18] He also expressed a desire to engage in international negotiations through the UN Framework Convention on Climate Change (UNFCCC) and G8+5 processes.[19] At the same time there was concern in environmental circles that the nation's economic problems stemming from the lingering financial crisis of 2008 would divert his attention to more pressing issues.

These concerns were quickly addressed by the Administration's actions concerning the environment. In this regard, much of the President's climate agenda was integrated into the administration's response to the on-going economic and financial crisis. As part of the administration's fiscal stimulus, provision was made for a significant expansion of expenditures directed toward the creation of "green jobs" and the development of "green technologies." Included in the package were: 20 billion dollars in continued tax credits for renewable energy; 15 billion dollars for building mass transit projects; more than 5 billion dollars for energy R&D projects; 5 billion dollars for the weatherization of houses; and 500 million dollars to train workers in conducting efficiency improvements.[20]

Another sign of the administration's seriousness at combatting climate change was the appointment of Steven Chu as secretary of energy and John Holdren as White House science adviser. Dr. Chu, a Nobel laureate, was the director of the Lawrence Berkeley National Laboratory in California. John Holdren was a long-time energy technology and climate science expert, former president of the American Association for the Advancement of Science (AAAS), and professor at the Kennedy School of Government at Harvard.[21]

Other tangible signs of the importance given climate and environmental issues by the Obama administration were signaled by the signing of executive orders reversing Bush decisions on automobile fuel efficiency. In this regard the administration instructed the Department of Transportation to issue guidelines to bring the US auto fleet fuel efficiency standard to 35 miles per gallon by 2020 or earlier.

Instructions were also given to the EPA to consider granting a waiver to California and 13 other states that would allow them to regulate greenhouse gas emissions from automobiles. Such waivers, which would cover nearly half the US auto market, would likely drive even greater changes in the overall market – and were long opposed by the industry. For example, the California rules would seek to cut vehicle emissions by 30% by 2016.[22]

Looked at from a broader perspective, the administration's energy and climate policies can be seen as integral parts of an overriding strategy of linking the goals of: (1) remaking the US energy and transportation systems; (2) reducing carbon dioxide emissions that lead to climate change; and (3) attempting to ensure that this process provides domestic economic benefits through job creation.

In pursuit of these political and policy objectives, the Obama administration initiated a major tactical shift away from the Bush administration. Specifically, the administration showed its willingness to (1) re-engage with international environmental negotiations, both on climate and other topics; (2) direct federal agencies to identify and pursue regulatory changes that would reduce emissions or encourage low-carbon energy technologies; (3) allow latitude to states that wish to pursue more aggressive policies on emissions – in a manner similar to his approach towards auto fuel-economy standards.

The administration's move towards greater state involvement reflected the political reality that it is now rare in the United States for major policy changes to emerge initially on the federal level. Since the 1970s, most successful national reforms (such as welfare reform) were based on policies originally developed at the state level and then extended by Congress to the rest of the country. In contrast, in cases where major changes were first introduced at the federal level, these were largely unsuccessful (e.g. the Kyoto climate policy).

In short, the Obama administration's initial policy initiatives were intended to establish a critical link between economic recovery and climate change. As opposed to previous failed top-down approaches, the administration opted for a bottom-up approach to regulatory change.

The first major piece of legislation addressing climate under the Obama administration was the American Clean Energy and Security (ACES) Act of June 2009.[23] Although it had little support from Republicans, the ACES's significance lay in the fact that it represented the legislation passed by the House of Representatives that set limits on greenhouse gas emissions. Specifically, its goals were: (a) to cut emissions by 83% below 2005 levels by 2050; (b) to establish a renewable energy standard (percentages) for electric utilities; (c) subsidies for carbon capture and storage demonstration plants; and (d) additional financial assistance for advanced vehicle technology and battery research.[24]

A key concession that helped force ACES through the House was an agreement to largely grant free carbon credits to energy intensive businesses for the first 15 years of the scheme, which helped to mute opposition.[25] One important missing element in the bill was a significant commitment to nuclear power.

While President Barack Obama pushed for a similar bill in the Senate, little progress was made. Failure in the Senate illustrates the extreme difficulty in passing broad sweeping legislation at the national level. Ironically, business pressure and lobbying may not have been the deciding factor. Historically, US businesses have been fairly unified in opposition to a GHG emissions cap. Such limits were seen as a burdensome cost, and constraint on economic growth. Yet this monolithic view is changing. Some companies simply prefer policy certainty on the issue, so they can plan multi-billion dollar capital investments and build generating capacity.[26]

One factor that may have prevented the bill's passage in the Senate was the attachment many senators or representatives had to specific technologies. Those from coal producing states tended to favor carbon capture and storage (CCS), while liberal Democrat legislators tend to favor renewables such as wind and solar power. Conservatives are often inclined to favor nuclear power while representatives and senators representing coal, gas and oil rich states advocate expanding domestic production as the proper solution. Reconciling these diverse views has proved extremely difficult if not impossible.

Another stumbling block and part of the wicked problem aspect of climate change is uncertainty over the costs involved in reducing green-house gas emissions. However costs are critical in garnering public support for legislation in the area. In 2009, the year the ACES was passed by the House of Representatives, the country was divided on a cap-and-trade policy with 52% of the public supporting the proposed scheme. However 58% of respondents backed cap-and-trade if it raised their monthly electricity bills by 10 dollars or less per month, but support dropped to just 39% if electric bills increased by 25 dollars per month.

Unfortunately independent estimates of the impact of bills like the ACES often produce dramatically different estimates of future costs and energy price increases – it all depends on the assumptions built into the analysis. Most studies at the time showed a wide range of price estimates to achieve the ACES target emissions reductions. In its base case, the Department of Energy's Energy Information Administration (EIA) found costs at 32 dollars per metric ton by 2020 and 65 dollars per ton by 2030. However, with rapid deployment of nuclear power and CCS technology, and with the widespread availability of international offsets, the study estimated that 2030 costs could be as low as 41 dollars per ton. Under a less favorable scenario, costs could be nearly five times higher – 191 dollars per ton by 2030.[27]

In sum, economic modeling at the time suggested the costs to the economy of imposing a cap-and-trade scheme could be anywhere from manageable to painful, depending upon key assumptions themselves subject to a wide variation in probabilities. Even a greater variation in results was produced by studies estimating the likely economic benefits associated with the planned ACES reduction in CO_2 emissions. With no clear and unambiguous statement of future costs and benefits, most legislation in the US is dead even before it is put to a vote.

Summing up, there are three potential ways to begin reducing greenhouse gas emissions: (1) setting norms and standards (i.e., a "command-and-control" policy); (2) a carbon market (i.e, "cap-and-trade," which is embodied in the EU's current Emissions Trading Scheme, and was the centerpiece of the failed ACES program); and/or (3) through a carbon tax.

During President Obama's first term, the administration attempted both to establish a carbon market through the ACES, and cut emissions through a command-and-control approach – by mandating a doubling of fuel economy standards in cars and trucks. The former failed spectacularly, while the latter succeeded – and, at the same time helped increase US energy security by reducing imports.

The possibility of a consensus on a carbon tax received a boost in July 2012, when a conference organized by the American Enterprise Institute brought together conservative and liberal analysts and scholars and achieving a surprising degree of consensus on the benefits of such a tax. There were also hopes that tax negotiations during on-going fiscal reform talks could be "greened," with some of the necessary new revenues being achieved through a carbon tax. However, as with many discussions in Washington, good ideas alone don't carry much weight.

Nevertheless, despite any significant Congressional legislation in the climate area, between 2009 and 2012, the end of President Obama's first term in office, the US made progress in reducing greenhouse gas emissions. CO_2 intensity declined at an average annual rate of 3.0% while CO_2 emissions per capita declined at an average annual rate of 1.6%. In part the CO_2 emissions stem from the administration's "stealth" climate policy, in the form of encouraging utilities to switch from coal to gas-fired power plants (which are half as carbon intensive). In 2011 for example, energy produced via gas increased by 3% while energy produced from coal fell 6%.[28] Other factors contributing to the reduction in greenhouse gases were the administration's measures to increase the percentage of renewables in the final energy supply.

US efforts towards energy and climate security – the second Obama administration

At the start of his second term in office, President Obama renewed his administration's efforts to reduce the country's greenhouse gas emissions. This apparent revival of a priority buried after the failure of cap-and-trade legislation in 2009 comes after the previous year's historic drought, the devastation caused by Hurricane Sandy, and record high national average temperatures in 2012. The key question was how much meaningful regulatory action is possible in the absence of new legislation, given that the Republican controlled House of Representatives would reject any new climate change bill.

As a result, President Obama has been forced to resort to a series of "second best"[29] actions to combat climate change. Early on in his second term the president initiated his Climate Action Plan,[30] a framework incorporating cuts in carbon usage, formulating responses to climate change impacts, and participating at international forums to develop cooperative arrangements leading to the reduction in greenhouse gases.

Another objective of the plan is to indirectly reduce CO_2 emissions by encouraging and promoting renewable energy and efficiency. Given the stand-off in Congress, the Climate Action Plan largely resorts to the use of executive orders to reduce greenhouse gases directly: (1) The Clean Power Plan, (2) Standards for Heavy-Duty Engines and Vehicles, (3) energy efficiency standards, and (4) economy-wide measures. Although not explicit in the Plan, the administration's strategy includes actions in specific cases like the Keystone Pipeline where the administration feels an adverse effect on greenhouse gases was likely.

The philosophy of the Climate Action Plan's regulatory orientation is to seek to establish partnerships between the federal government and states, permitting flexibility towards reaching EPA-mandated emission rate-based carbon dioxide goals – through improvements to existing power plants, lower-carbon alternative generating capacity or energy efficiency schemes.

Of these measures, the EPA's actions toward the coal sector will probably have the greatest effect on greenhouse gases. In June 2014 the EPA under the Climate Action Plan introduced a plan to reduce carbon emissions from the power generation sector by 30% from 2005 levels by 2030. The new regulations do not represent a comprehensive carbon reduction plan. Instead they address only the power generation sector. The EPA reports that power generation is the primary source in the United States for GHG emissions. Specifically this sector accounts for 32% of the total, and carbon dioxide is the primary GHG, at 82% of US emissions.[31]

The plan will not include a federally-mandated program of carbon reduction. It will instead set carbon emission rate targets for each state for existing fossil fuel plants. States must submit implementation plans for EPA approval between 2016 and 2018. The regulations on new power plant construction are so stringent that coal-fired plants cannot meet them. Specifically coal plants are now mandated to capture and store 20–40% of carbon they produce.[32]

Utility companies and states whose political climate is opposed to the EPA (e.g., Texas) or that are dependent on the coal industry (e.g., West Virginia) will challenge the plan in court. Such lawsuits are unlikely to succeed. In Massachusetts v. EPA in 2007, the Supreme Court ruled that carbon dioxide falls under the EPA's jurisdiction under existing Clean Air Act (CAA) statutory authority. Therefore, the EPA has promoted the health benefits of GHG reduction (one of the main reasons for the CAA), arguing that there will be 140,000 fewer asthma attacks in children by 2030.

The Supreme Court ruled in EPA v. EME Homer City Generation[33] in April that states do not need a second opportunity to file a state implementation plan if the first is rejected by the EPA for not meeting federal guidelines. This gives the EPA ultimate authority on the issue; by inviting state partnerships, it has avoided the legal issue of federal mandates. This should prevent most challenges against the core of the EPA's new rules from being upheld.

How effective are the new regulations likely to be in reducing greenhouse gases? Coal plants are the oldest within the US fossil fuel power sector – with an average age of 42 years – compared to an average of 14 years for gas-fired plants. Since coal plants are mostly nearing the end of their lifespan and natural gas plants emit about half of the carbon while becoming increasingly cheap, it is likely that most coal plants will be phased out by the rule's 2030 target. What will happen to coal? Unless similar actions are taken by other coal consuming countries, US coal will simply be diverted to export markets. Given that global emissions are what really matter, the regulations are not nearly as effective as they might appear at first glance.

On the other hand, the plan is likely to lead to greater investment in renewable energy; states targeting reduced carbon emissions per MWh will have an incentive to subsidize or prefer new installations in wind or solar. Investment is likely to lead to technological innovation, and public support for home installation of solar panels (especially if households are allowed to sell power to the grid) could spread the technology rapidly.

The Administration's climate strategy over its last several years will likely be one of increased EPA regulation. The president has directed several departments and agencies, most notably the EPA, to issue directives and regulations that would impose progressively lower ceilings on GHG emissions. For example, on June 10, 2015 the EPA announced that it will begin to craft rules to limit greenhouse gas (GHG) emissions from airplanes.[34]

As with the cuts to carbon emissions, none of these regulations followed passage of new legislation – all relied on expansive interpretations of existing statutory authority for executive agencies. The strategic political problem that induced Obama to direct agencies to promulgate all these wide-ranging regulations is his lack of a working majority in Congress.

Further complicating the issue, Democrats from consumer states (such as Missouri, which relies on coal for 80% of its energy) worry about the impact of higher energy prices that may result from new emissions controls. In states that produce oil and coal, Democrats have already paid a price, as in West Virginia, Ohio and Pennsylvania, especially in the state legislatures.

In the international arena Obama has been willing to pay the price of electoral vulnerability as part of a larger strategy to implement international agreements that can make a serious dent in emissions by demonstrating US credibility. Even so, international agreements are not likely to be approved by Congress. Consequently, Obama's White House will probably rely on a variant of the arms control agreements known as parallel unilateral policy declarations (PUPDs)[35] in which the parties to an accord do not sign a binding international treaty, or conclude an executive agreement, but rather simultaneously declare that they will each unilaterally implement a limitation.

While it is very unlikely appeals will succeed, states can slow enforcement whether or not Congress and courts act against the EPA's new regulations. On the other hand states may back national regulation because national standards prevent one state from exporting its pollution downstream to another. States may also participate in national intergovernmental programs to avoid losing out on available funds.

When a regulatory regime promises significant costs and where costs are highly concentrated, it is a blueprint for state resistance. In many states (especially those dominated by Republicans), governors, state legislatures and attorney generals will oppose the new rules. In 32 states, officials are readying to resist that part of the regulations.[36] Coal-burning states are starting to pass laws that provide for cost-benefit analyses for each new power plant. While it is too early for many of the assumed benefits of the administration's regulatory regime to be felt, the country's declining energy sustainability ranking in the World Energy Council's Energy Trilemma Index from 90th in 2011 to 95th in 2015[37] is not encouraging.

The EPA strategy is to reduce state opposition by getting more of the corporate sector on board. This strategy has worked with coal sludge, methane reduction in pipelines and natural gas leaks, sea drilling rules and miles per gallon increases mandated for automobiles.

The price of obtaining such cooperation involves some dilution of rule-making authority and targets, which is likely to occur in each individual state plan over the next two years. This also involves a great deal of industry self-reporting. Good faith efforts at reporting in many industries are often lacking while information is received late and incomplete; penalties for non-compliance or uneven compliance are often token. They will not be as effective as mandatory regulating.

Market factors will continue to be more significant than government action in determining the mix of energy in the electrical grid. Government policy is more likely to be successful when it involves distributive politics, such as loans, loan guarantees, subsidies for research and development and guaranteed procurement of product.

To the extent that regulation is promoted, it is more likely to be successful when it co-opts the major energy and emissions producers, or water and air polluters. They often have a stake in promoting national standards. This is a more costly way to proceed if one simply counts up federal expenditures and guarantees, but may be far cheaper in the long run, when counting up the societal and economic costs of failing to lower emissions through development of new technologies and energy sources.

In sum, neither Congress nor the judiciary will overturn the principle that the EPA may regulate greenhouse emissions. However, weak enforcement capability at the national level, the reliance on state officials when a majority of state governments are controlled by the Republican Party and the pattern of self-reporting by industry will slow implementation. Markets and price effects will have a greater impact in the short-term than regulations. Finally, the risk of executive orders is that they can always be rescinded or dismantled by a new administration.

Alternative strategies

As we reach the end of the Obama presidency the limitations of the administration's top-down regulatory approach suggest that other avenues may be more effective in combatting climate change. Several possibilities include (1) a bottom-up strategy where cities and/or states take more of the initiative in designing and implementing their own climate plans, and (2) a reoriented national strategy that builds on the US shale oil and gas boom in a manner that eliminates many of the harsh trade-offs inherent in the hard trilemma. Inherent in the first strategy is an economic security dimension associated with climate change, while in the latter climate change is seen more from a national security perspective.

Bottom-up approach

Jeffrey Sachs contends that "[i]n the US, because Congress has been stalemated and lot of our politicians bought by the oil, coal, and gas sector, things haven't happened at the federal level except through the regulatory apparatus. The real action has been happening at the state and city level."[38]

In combatting climate change US local jurisdictions often have the advantage of being more homogenous politically. They are also likely to see the effects of climate change first hand – droughts in California, coastal erosion in Florida, intense storms along the Atlantic coast – and thus have a much easier time forging consensus as to the best way to combat climate change. In addition to begin to get Republican political support for combatting climate change these jurisdictions are able to shift climate discussions away from problematic long-run global warming concerns and towards more immediate issues – economic security.[39] While economic security is most often thought of on an individual basis – the ability to achieve a stable income or standard of living into the future, on a regional or local basis economic security simply means the ability of that jurisdiction to follow its choice of policies to develop in the manner desired.

As noted, much of the local effort to avert climate change is taking place at the city level. While there are obstacles to cities cutting emissions, there is a growing political impetus at the city level for mitigation initiatives, in part driven by a growing realization that cities can be the best social unit for driving innovation on tackling climate change.

Cities have advantages tackling climate change compared to state governments: (1) they are large enough to put into practice pilot programs, yet sufficiently close to communities to be faster and more effective in meeting public aspirations; (2) they are better placed for central planning for water, energy and waste management, and establishing citywide building codes; (3) they can drive immediate changes in public transportation and longer-term changes in land-use planning; and (4) they can offer local level subsidies for renewable sources of energy or loans to make buildings more efficient.

In addition, cities often find cooperating with each other easier than with federal or even state jurisdictions who are often restricted by the demands of geopolitics. For example, over 1,000 US cities have made commitments to meet or exceed the Kyoto Protocol targets of GHG emissions (6% below 1990 levels by 2012), even though the Protocol has never been ratified at the federal level.

The Pew Foundation estimated that voluntary actions from cities and local governments in the United States cut over 23 million tons of GHGs in 2009. The Pew study calculates that such actions from around 150 cities and counties could meet 10% of the goal President Barack Obama offered (but never ratified in the Senate) of reducing emissions by 17% below 2005 levels by 2020.[40]

US cities have also been quite innovative in reducing greenhouse emissions:[41] (1) Boston has introduced the country's first green building code; (2) Ashville, North Carolina, city employees started a "compressed workweek" in which they worked ten hours per day, four days a week, to save energy on travel and heating; (3) Santa Monica aims for net zero emissions energy by 2020 mainly by using energy efficiency measures and solar power; (4) Chicago has raised local taxes to allow it to invest more than 330 million dollars in a flood control program over the next ten years.

Sometimes sub-national entities enter into agreements with national governments. This was the case with a 2013 agreement between California Governor Jerry Brown and Chinese President Xi Jinping. It commits them to joint efforts on combatting climate change, including Californian authorities sharing policy approaches with Shenzhen to help it develop an emissions trading system.[42] In contrast to the Obama administration's early failure[43] to initiate a national cap and trade system, California[44] had little difficulty in introducing a sophisticated system to reduce carbon.

Reoriented efforts at the national level

While the US has not been able to develop a comprehensive plan or even a national consensus towards an approach capable of guiding the country's transition to a lower carbon environment, consistent with international standards for reducing greenhouse gases, the EPA efforts along with several other initiatives may still be capable of putting the US on a sustainable path to a low carbon economy.

For one thing, the country is no longer facing a hard trilemma where improvements in the environment must come at the expense of either (or both) energy security and affordable energy. Break-throughs in shale technology have fundamentally changed the US energy picture from one of long run decline to one of being a major competitive exporter of oil and gas.[45]

It is likely that US shale gas production will continue increasing through 2014. However, the path will not be smooth. Currently shale oil and gas are encountering the effects of the dramatic fall in oil prices beginning in the fall of 2014. In 2014, US oil output grew by 16.2%, in large part due to fracked shale plays in North Dakota and Texas. Production growth will not be as strong in 2015 and 2016 – 8.1% in 2015 and 1.5% in 2016, due to lower prices and thus falling profitability.[46]

Cuts to capital expenditure have started to hit certain shale regions. The US rig count has dropped from 1,473 to 825[47] over the past year, although many of those retired were older and less efficient rigs. Producers have increased efficiency by adopting new techniques such as "pad drilling" (multiple wells drilled in tight clusters with fewer rigs). There is considerable confidence within the industry that further efficiency gains are possible and that even if prices are depressed for several years the shale industry, while slowing down, will remain viable and competitive.

With the continued expansion of US shale oil and gas assured, the hard energy trilemma facing US policy makers will soften and may even go away. Oil expansion increases energy security, while the gas will be increasingly used in power generation reducing CO_2 levels associated with coal. In this situation US policymakers will be faced with what might be called a virtuous trilemma (Figure 14.3). With abundant energy, economic security replaces energy affordability.

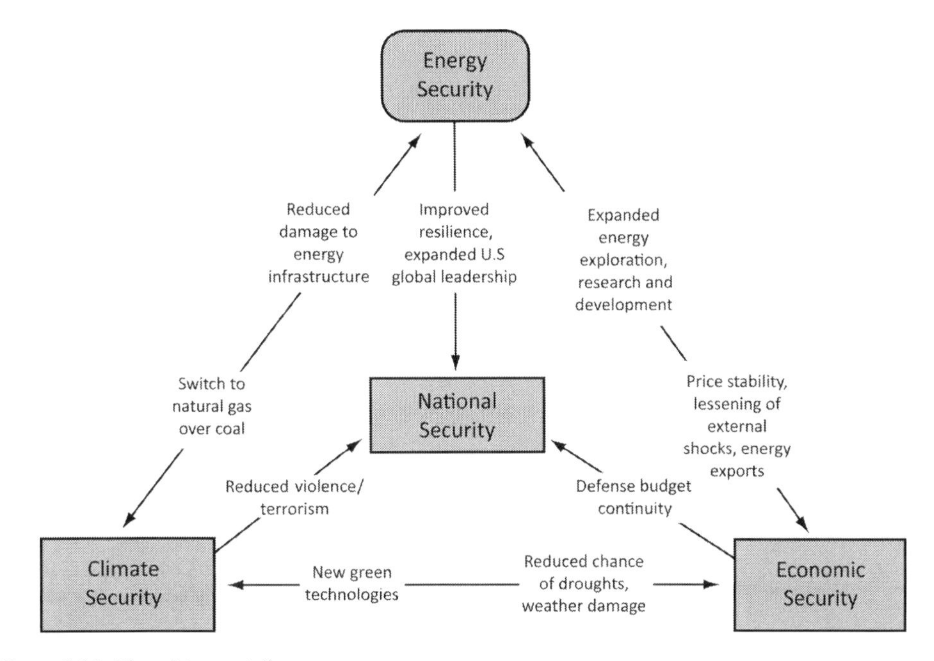

Figure 14.3 The virtuous trilemma

Other changes from the conventional trilemma include national security as a goal that is enhanced by improvements in energy security, climate security, and economic security. Regarding national security, the traditional US view has been to look at national security largely in military and intelligence terms – the growth in potential advisory defense budgets, nuclear proliferation, cyber security threats, and the rise of rogue nuclear states. However broader definitions of national security are coming into the picture. For example, the 2010 National Security Strategy[48] includes economic instability along with climate change as security threats. Similarly, the 2010 Quadrennial Defense Review draws a number of connections between climate change and conflicts, both now and increasingly in the future.[49]

In the US context, rather than a series of conventional trilemma tradeoffs, the virtuous trilemma will be increasingly centered on a series of complementarities: increased energy security enhances economic security, in turn improved economic security contributes to climate security which in turn improves energy security, with improvements in each contributing positively to national security. These shifts from the conventional trilemma are reflective not only of the effects of the shale oil and gas boom but also of a number of recently observed phenomena and empirical relationships.

The lesson from cities is that when the effects of climate change are close at hand it is much easier to gain support. At the national level National Security is an encompassing concept that often unites parties that are often opposed on a wide spectrum of issues.

A number of linkages in the virtuous trilemma are straightforward and in need of no further elaboration. Improved economic security contributes directly towards funding those expenditures contributing to national security. Improved climate security improves energy security through reducing the threat to the energy grid or energy facilities stemming from excessive heat, or violent coastal storms.[50] Other critical linkages are less obvious and in need of further elaboration.

Climate security – national security

US Republicans would be more supportive of efforts at combatting greenhouse gases if climate change were seen as a critical component of national security. There is growing evidence that this is the case. In his definitive study of the subject,[51] Daniel Moran sees "climate change less as a direct threat than as an additional source of stress on the sinews of public life, which cause fragile governments to fall or may provide a new impetus for a range of violent outcomes, ranging from social upheaval to aggressive war."[52] In this vein Gwynne Dyer a military historian, notes that "in a number of the great powers, climate-change scenarios are already playing a large and increasing role in the military planning process."[53]

For its part, the Pentagon is integrating climate change threats into all of its "plans, operations, and training" across the entire Defense Department, signaling a comprehensive attempt to tackle the impacts of global warming.[54] In its 2014 Climate Change Adaptation Roadmap[55] the Pentagon details its strategic blueprint to address climate change, calling it a "threat multiplier" that has the power to "exacerbate" many of the challenges the US faces today, including "infectious diseases and terrorism."

Empirical studies confirm climate's impact on conflict. For example a recent study conducted at Stanford University found[56] that shifts from normal weather patterns are likely to increase the risk of conflict. Nowhere is this pattern more visible than in the Middle East and Africa. As Secretary of State John Kerry has observed, "It's not a coincidence that, immediately prior to the civil war in Syria, the country experienced its worst drought on record. As many as 1.5 million people migrated from Syria's farms to its cities, intensifying the political unrest that was just beginning to roil and boil in the region."[57] In a similar vein the rise of Boko Haram in northern Nigeria can be traced in part to climate change[58] and the drought in the Lake Chad region that disenfranchised[59] large segments of the population, impoverishing the region[60] and created an expanding pool or recruits to radical causes. It provided the right environment for ISIS to expand enough to become a global threat.[61]

The danger of attempting to link climate change to violence, terrorism or other acts of violence is that one can easily overstate the case and thus lose credibly. Instead it is best to consider climate change as one element of several elements that increase the likelihood of conflict. As Peter Gleick, President of the Pacific Institute observes,[62] "As the climate changes, as water systems fail, as energy reliability vanishes, these factors are piled on top of the misery already in play. Will they cause conflict? I think that's the wrong question. Will they increasingly influence the risks of conflict and war in some regions? Unquestionably." This alone should qualify climate change as a national security issue.

Energy security – national security

Increasingly there are situations where improved energy security and national security are complementary. The US shale boom is clearly a case in point. In the military area the classic link between energy security and national security is a lessening imperative to maintain and preserve unstable regimes in hostile areas. For their part the US military are developing a spectrum of new fuels, technologies and new usage patterns in strengthening their own operations which also enhance US national security.[63]

In a broader context, economic power enhances the ability of the US to shape world developments in line with its interests. For example, during the period of sanctions against Iran, increased oil supplies and lower energy prices stemming from the US shale boom aided US efforts to finalize a comprehensive nuclear treaty with that country.

In a similar manner the shale boom and the likelihood of increased gas exports has likely contributed to the United States' ability to form a coalition of Asian nations willing to form a Trans-Pacific Partnership (TPP), a key element in the US efforts to assure a peaceful Pacific Basin.[64]

In the case of the Caribbean Central America – expanding LNG exports to the region would not only contribute to US energy security, but it would also contribute to national security through creating a more stable region with improved economic performance.

Energy security – economic security

Increased energy security stemming from the shale oil and gas boom will also have some clear benefits through assuring improved economic security. Sudden spikes in energy prices have been associated with nearly all of the post-World War II recessions. At other times, rapid increases in energy prices have had a retarding effect on economic growth and job creation. Households are left with less to spend on goods and services in general, causing slack in the economy, resulting in declines in productive investment.

Before the shale boom energy prices were increasingly variable.[65] This variability often created a vicious circle whereby energy price variability increased uncertainty about future energy prices which, in turn, deterred investments in conventional types of energy. With lower future supplies, sharp increases in oil prices occurred then economic growth picked up. Insofar as volatile energy prices reduce investments in domestic alternatives they exacerbate the initial instabilities by concentration in less stable regions. Because oil supplies can be increased quickly and incrementally from new shale wells, there is a dampening effect on prices "not possible from high cost long development period conventional oil wells."[66] In effect, this property of shale means the United States may replace Saudi Arabia as the world's swing producer.

Conclusions

In its November 2015 survey of climate change, the *Economist* magazine noted that "It is often said that climate change is an urgent problem. If that were true, it might be easier to tackle. In fact, it is a colossal but slow-moving problem, spanning generations."[67] As a body of clear scientific evidence as to the link between house gases and climate change accumulates, climate skeptics in the US are likely to decline in number and influence. However broad political support for climate change may still be hard to garner because it is still viewed as a situation where the costs are up front and the benefits far into the future.

In classic trilemma terms the costs of reducing greenhouse gases are either increased energy vulnerability or expensive energy. A broader national security view of efforts to reduce climate change, a Virtuous Trilemma, suggests broad complementarities flowing from efforts at reducing climate change. With the advent of the US shale oil and gas boom, rather than costly tradeoffs between climate change, energy security and energy affordability, these complementarities produce immediate benefits in terms of improved energy security, economic security and national security. Most importantly, shale gas greatly reduces the costs of transitioning to a lower-carbon economy where new green technologies can assume a dominant role in power generation.

Because of the shale boom, US energy policy has shifted from being a wicked problem back to one that is relatively straightforward. Hopefully, this development along with the growing city and state bottom-up efforts at combatting climate change will provide the momentum that enables the US to take advantage of its fortuitous energy boom. Rather than all or nothing approaches like the Keystone Pipeline, government policy at the federal level can still be highly effective in this regard as it moves away from its current regulatory approach toward market-based[68] solutions that balance growth with concerns for the climate.

Notes

1. As documented in Michael Levi, *The Power Surge: Energy, Opportunity and the Battle for America's Future* (Oxford: Oxford University Press, 2013).
2. Quoted in David Bailey and David bookbinder, "America's Accidental Energy Policy: Small is Doable," Niskanen Center, October 30, 2015. https://niskanencenter.org/blog/americas-accidental-energy-policy-small-is-doable/.
3. John M. Deutch, *The Crisis in Energy Policy* (Cambridge, MA: Harvard University Press, 2011), Kindle location 150–232.
4. World Energy Council, Energy Trilemma Index, www.worldenergy.org/data/trilemma-index/.
5. Joseph Bentley, "The Challenge of Taming Wicked Problems," October 30, 2015. http://tamingwickedproblems.com/.
6. "Pool: Democrats say climate change a bigger threat than ISIS," *The Week*, August 29, 2014. http://theweek.com/speedreads/447166/poll-democrats-say-climate-change-bigger-threat-than-isis.
7. 70% of Democratic voters saw evidence of man-made climate change in recent weather patterns, whereas only 19% of Republican voters did. A similar, though smaller, divide was found in Britain.
8. "Welcome to the age of 'wicked' problems. Or, why fighting climate change is so difficult," Citizen Action Monitor, June 28, 2012. https://citizenactionmonitor.wordpress.com/2012/06/28/welcome-to-the-age-of-wicked-problems-or-why-we-cant-solve-climate-change/.
9. Lynne Chester, "Does the Polysemic Nature of Energy Security Make it a 'Wicked' Problem?" *International Journal of Social, Behavioral, Educational, Economic and Management Engineering*, 3 (2009): 6.
10. "Bush: Kyoto Treaty Would Have Hurt Economy," *NBC News*, June 30, 2005. www.nbcnews.com/id/8422343/ns/politics/t/bush-kyoto-treaty-would-have-hurt-economy/#.VmD0O7grK00.
11. "Analysis of President Bush's Climate Change Plan," Center for Climate and Energy Solutions, February 2002. www.c2es.org/federal/executive/george-w-bush-climate-change-strategy.
12. Michael Fletcher and Juliet Eilperin, "Bush Proposes Talks on Warming," *The Washington Post*, May 31, 2007. www.c2es.org/federal/executive/george-w-bush-climate-change-strategy.
13. Alan Zarembo and Thomas H. Maugh, "Frustrated EU may boycott Bush's summit on climate," *Los Angeles Times*, December 14, 2007. http://articles.latimes.com/2007/dec/14/science/sci-bali14.
14. Sheryl Gay Stolberg, "Bush Sets Greenhouse Gas Emissions Goal," *New York Times*, April 17, 2008. www.nytimes.com/2008/04/17/washington/17bush.html?_r=0.
15. CQ Transcripts, "Bush Remarks on Climate," *Washington Post*, April 16, 2008. www.washingtonpost.com/wp-dyn/content/article/2008/04/16/AR2008041603084.html.
16. Richard Conniff, "The Political History of Cap and Trade," *Smithsonian Magazine*, August 2009.
17. Michael Hilzik, "Emissions cap-and-trade program is working well in California," *Los Angeles Times*, June 12, 2015. www.latimes.com/business/hiltzik/la-fi-hiltzik-20150613-column.html.
18. Margaret Kriz, "Shades of Green," *National Journal*, June 21, 2008. www.aspeninstitute.org/sites/default/files/content/docs/ee/2008EnergyForum_Kriz_(Nat_Jour_-_Shades_of_Green).pdf.
19. Bernice Lee and Michael Grubb, "The United States and Climate Change: From Process to Action," in Robin Niblett ed. *America and a Changed World: A Question of Leadership* (London: Wiley-Blackwell, 2010).
20. Dylan Matthews, "Did the stimulus work? A review of the nine best studies on the subject," *Washington Post*, August 24, 2011. www.washingtonpost.com/blogs/ezra-klein/post/did-the-stimulus-work-a-review-of-the-nine-best-studies-on-the-subject/2011/08/16/gIQAThbibJ_blog.html.
21. Jonh Broder and Andrew Revkin, "Hard Task for New Team on Energy and Climate," *New York Times*, December 15, 2008. www.nytimes.com/2008/12/16/us/politics/16energy.html?_r=0.
22. James Martinez, "EPA gives California and 13 other states waiver to regulate greenhouse gas emissions," *Motor Authority*, July 1, 2009. www.motorauthority.com/news/1033620_epa-gives-california-and-13-other-states-waiver-to-regulate-greenhouse-gas-emissions.
23. Center for Climate and Energy Solutions, "The American Clean Energy and Security Act (Waxman-Markey Bill)." www.c2es.org/federal/congress/111/acesa.

24 Tom Mounteer, "Comprehensive Federal Legislation to Regulate Greenhouse Gas Emissions," Environmental Law Institute, 2009. www.paulhastings.com/docs/default-source/PDFs/2130.pdf.

25 David Ingram, "Explanation of What the Cap & Trade Bill Means," *Houston Chronicle.* http://smallbusiness.chron.com/explanation-cap-trade-bill-means-3838.html.

26 PMR, "Preparing for Carbon Pricing," January 2015. www.thepmr.org/system/files/documents/PMR%20Technical%20Note%209_Case%20Studies.pdf.

27 Energy Market and Economic Impacts of H.R. 2454, *The American Clean Energy and Security Act of 2009* (Washington, DC: EIA, August 2009). www.eia.gov/analysis/requests/2009/hr2454/pdf/sroiaf(2009)05.pdf.

28 Jonathan L. Ramseur, *US Greenhouse Gas Emissions: Recent Trends and Factors* (Washington, DC: Congressional Research Service, November 24, 2014) www.fas.org/sgp/crs/misc/R43795.pdf.

29 Xavier Labandeira and Pedro Linares, Second-best Instruments for Energy and Climate Policy WP 06/2010, *eforenergy.com.* www.google.com/url?sa=t&rct=j&q=&esrc=s&source=web&cd=1&ved=0ahUKEwiL1ICP_MLJAhVSwGMKHYHpDq8QFggcMAA&url=http%3A%2F%2Feforenergy.org%2Fdocpublicaciones%2Fdocumentos-de-trabajo%2FWP06-2010.pdf&usg=AFQjCNHD-JaWf5PyJtK-cgTb-kq9I7EEDg.

30 Center for Climate and Energy Solutions, President Obama's Climate Action Plan, www.c2es.org/federal/obama-climate-plan-resources.

31 *EPA Proposes First Guidelines to Cut Carbon Pollution from Existing Power Plants/Clean Power Plan is flexible proposal to ensure a healthier environment, spur innovation and strengthen the economy* (Washington, DC: EPA, June 2, 2014). http://yosemite.epa.gov/opa/admpress.nsf/bd4379a92ceceeac8525735900400c27/5bb6d20668b9a18485257ceb00490c98!OpenDocument.

32 Brad Plumer. "How Obama's Clean Power Plan actually works – a step-by-step guide," *Vox*, August 5, 2015. www.vox.com/2015/8/4/9096903/clean-power-plan-explained.

33 EPA v. EME Homer City Generation, L.P. Harvard Law Review, November 10, 2014. http://harvardlawreview.org/2014/11/epa-v-eme-homer-city-generation-l-p/.

34 Coral Davenport and Jad Mouawad, "E.P.A. to set New Limits on Airplane Emissions," *New York Times*, June 2, 2015. www.nytimes.com/2015/06/03/business/energy-environment/epa-to-set-new-limits-on-airplane-emissions.html?_r=0.

35 Benito Muller, Wouter Geldhof and Tom Ruys, *Unilateral Declarations: The Missing Link in the Bali Action Plan*, European Capacity Building Initiative, May 2010. www.law.kuleuven.be/iir/nl/onderzoek/opinies/ecbiUDsfinal.pdf.

36 Peter Glaser, Carroll McGuffey and Hahnah Gaines, "EPA's Section 111(d) Carbon Rule: What if States Just Said No?" *Federalist Society*, November 2014. www.eenews.net/assets/2015/06/08/document_cpp_16.pdf.

37 www.worldenergy.org/data/trilemma-index/country/united-states-of-america/2015/.

38 Quoted in Jeremy Deaton, "American Cities Keep Fighting Climate Change While Contress Stonewalls," *grist.com*, November 13, 2015 http://grist.org/climate-energy/american-cities-keep-fighting-climate-change-while-congress-stonewalls/.

39 Cf. Keith W. Cooley, "Energy Security is National Security," in Sheila Ronis, *Economic Security: Neglected Dimension of National Security* (Washington, DC: National Defense University Press, 2011).

40 Rita Beamish, "Cities, States Brace for Global Warming Fallout," The PEW Charitable Trusts, December 23, 2014. www.pewtrusts.org/en/research-and-analysis/blogs/stateline/2014/12/23/states-cities-brace-for-global-warming-fallout.

41 Michael J. Coren, "Where the Action Is: Climate Goes Local; Cities Local Governments Confront Global Challenge," Yale Climate Connections, September 14, 2010.

42 "China's first carbon-trading program shows commitment to address climate change," Environmental Defense Fund, June 18, 2013. www.edf.org/news/chinas-first-carbon-trading-program-shows-commitment-address-climate-change.

43 John M. Broder, "'Cap and Trade' Loses its Standing as Energy Policy of Choice," *New York Times*, March 25, 2010. www.nytimes.com/2010/03/26/science/earth/26climate.html?_r=0.

44 Cap and Trade Program, California Environmental Protection Agency, www.arb.ca.gov/cc/capandtrade/capandtrade.htm.

45 *Annual Energy Outlook 2014* (Washington, DC: EIA, April 2014). www.eia.gov/forecasts/aeo/pdf/0383(2014).pdf.

46 Jerin Mathew, "US oil production growth at 100-year high in 2014 due to shale boom," *International Business Times*, March 31, 2015. www.ibtimes.co.uk/us-oil-production-growth-100-year-high-2014-due-shale-boom-1494342.

47 Paul Ausick "Oil Rig Count Falls; Hedge Funds Pile in on Short Positions," 24/7 Wall St. http://247wallst.com/energy-business/2015/03/21/oil-rig-count-falls-hedge-funds-pile-in-on-short-positions/#ixzz3tOxqxTEI.

48 *United States National Security Strategy* (Washington, DC: White House, May 2010), www.whitehouse.gov/sites/default/files/rss_viewer/national_security_strategy.pdf.

49 Richard A. Matthew, "The Energy-Climate Complex: Is Climate Change a National Security Issue?" *Issues in Science and Technology* Spring 2011. http://issues.org/27-3/matthew-2/.

50 Dave Levitan, "The Extreme Weather-Warming Connection," *FactCheck.org*, June 5, 2015. www.factcheck.org/2015/04/the-extreme-weather-warming-connection/.

51 Daniel Moran ed., *Climate Change and National Security: A Country-Level Analysis* (Washington, DC: Georgetown University Press, 2011).

52 Moran, op. cit., Kindle location 109.

53 Gwynne Dyer, *Climate Wars: the Fight for Survival as the World Overheats* (Oxford: Oneworld Publications, 2011), Kindle location 91.

54 "Read DoD Report: 2014 Climate Change Adaption Roadmap," The Hill, October 13, 2014. http://thehill.com/policy/energy-environment/220577-read-dod-report-2014-climate-change-adaptation-roadmap.

55 "DoD Releases 2014 Climate Change Adaptation Roadmap," (Washington, DC: Department of Defense, October 13, 2014). www.defense.gov/News/News-Releases/News-Release-View/Article/605221.

56 Solomon M. Hsiang, Marshall Burke and Edward Miguel, "Quantifying the Influence of Climate on Human Conflict," *Science* 341 (2013), www.sciencemag.org/content/341/6151/1235367.full.pdf?sid=5c364f4b-f8a3-480b-a0e2-c66c25625ed7.

57 John Kerry, "Remarks on Climate Change and National Security," (Washington, DC: US Department of State, November 10, 2015). www.state.gov/secretary/remarks/2015/11/249393.htm.

58 Nafeez Ahmed, "Behind the rise of Boko Haram – ecological disaster, oil crisis, spy games," *The Guardian*, March 9, 2014, www.theguardian.com/environment/earth-insight/2014/may/09/behind-rise-nigeria-boko-haram-climate-disaster-peak-oil-depletion.

59 John Campbell, "Climate Change and Ethnic and Religious Conflict in Nigeria," Council on Foreign Relations, December 5, 2015. http://blogs.cfr.org/levi/2015/12/05/guest-post-climate-change-and-ethnic-and-religious-conflict-in-nigeria/.

60 Robert Looney, "The Boko Haram Economy," *Foreign Policy*, July 15, 2014, http://foreignpolicy.com/2014/07/15/the-boko-haram-economy/.

61 Mohamed Raouf, "Paris Conference must reach agreement on climate action," *The National*, November 28, 2015, www.thenational.ae/opinion/comment/paris-conference-must-reach-agreement-on-climate-action.

62 Quoted in John Knefel, "How Climate Change is Threatening Iraq's Fragile Security," *The Nation*, October 27, 2015, www.thenation.com/article/how-climate-change-is-threatening-iraqs-fragile-security/.

63 *Strategic Sustainability Performance Plan FY2014* (Washington, DC: Department of Defense), http://denix.osd.mil/sustainability/upload/DoD-SSPP-FY14-FINAL-w_CCAR.pdf.

64 Sarah O. Ladislaw, Maren Leed and Molly A. Walton, *New Energy, New Geopolitics: Balancing Stability and Leverage* (Washington, DC: CSIS, April 2014), http://csis.org/files/publication/140409_Ladislaw_NewEnergyNewGeopolitics_WEB.pdf.

65 Robert McNally and Michael Levi, "A Crude Predicament," *Foreign Affairs*, July/August 2011, www.foreignaffairs.com/articles/2011-06-12/crude-predicament.

66 Martin Wolf, "Cheap Oil Puts Humanity on a Slippery Slope," *Financial Times*, December 2, 2015, p. 9.

67 "Hot and Bothered," *The Economist*, November 28, 2015, www.economist.com/news/special-report/21678951-not-much-has-come-efforts-prevent-climate-change-so-far-mankind-will-have-get.

68 www.wsj.com/video/a-missed-opportunity-with-keystone-xl/91C135BA-914D-413B-A8A0-5AD39FB463E1.html.

15

The Great British energy transition?

Caroline Kuzemko

Introduction

The UK is often held up as an example of 'best practice' in terms of its transition towards a low carbon and secure energy system. This is not least because of the, at the time groundbreaking, Climate Change Act of 2008 which committed the UK to legally binding, long-term carbon emissions reduction targets and a series of carbon budgets. This chapter will argue, however, that beneath these ambitious government targets there has been insufficient progress in policy-making to effect profound system change. The UK's Committee on Climate Change (CCC) has recently announced that if the UK is to meet its fourth and fifth carbon budgets it urgently needs to implement new policies and send clear long-term signals to investors.[1] Indeed climate targets are only really helpful in practice if they provide enough impetus and direction for political and market actors to respond with sustainable innovations that can facilitate profound system change.

What this chapter also makes more overt is that there are different types of transition processes, and a glance across the other chapters in this book confirms this observation. It is therefore worth briefly characterizing the UK's energy transition thus far. It has tended to place as heavy an emphasis on energy security as it has on climate change mitigation and it has, furthermore, tended not to challenge traditional energy market structures such as centralized generation and transmission systems and the traditional utility model favoured by the 'Big 6' gas and electricity companies. Policymaking decisions have been influenced over time by market liberal ideas that place a heavier emphasis on the role of markets in delivering goods and services, thereby arguably reducing the range of possibilities for state action. However, where state interventions have been considered necessary, such as the new Capacity Market, these have often been in support of existing energy infrastructures. What has been important, therefore, to how the UK's energy transition has unfolded is not just the influence of political norms but also the particular configurations of pre-existing energy structures.

To argue that market liberal ideas have constrained Britain's energy policy choices is not new,[2] but to emphasize the influence of embedded energy structures is more novel. However, to do so is also a more overt articulation of the idea underpinning the structure of this book in that sections have been designed according to whether countries are carbon producing, carbon consuming or intermediate. For example, one important aspect of how the UK has so far transitioned its energy sector is that fossil fuels have historically played such an important role within its economy not just in terms of taxes paid and export revenues, but also in terms of employment. This has underpinned the possibility for traditional energy companies to play a particular role in society as well as for them to have established, formal and informal, links with policy-making actor groups.

Because of the importance of the domestic political and energy context in characterizing the current transition this chapter presents an historically informed analysis of the UK energy system. This shows how energy infrastructures developed over time, the importance of energy within the political economy, and how different political institutions influenced policy choices at points in time. The analysis is split into three phases: the post-World War II era of establishing universal access; privatization and liberalization under Conservative administrations (from 1979 to 1997); and the slow emergence of what we would recognize as today's 'low carbon' energy transition (1997 to 2015). Each phase finds its roots, and starting points, in the previous phase.

British energy in historical context

In some senses the UK's energy system has always been in transition and throughout the modern age, i.e. from the Industrial Revolution onwards, energy in its different forms has played an increasingly important role. At each stage of economic, social and technological development and in each energy transition, from coal to oil and from oil to gas, more and more energy was used. There are those that have argued that access to significant quantities of indigenous coal was a significant factor underpinning Britain's role as an economic powerhouse in the 19th century.[3] Others have made similar arguments with regard to access to indigenous oil and American imperial power in the 20th century.[4] For these reasons energy industries, first coal and later oil and gas, have historically enjoyed a central position within the British political economy.

Post-World War II: centralization, nationalization and public access

Throughout the late 19th and early part of the 20th century municipal utilities played a central role in the provision of various goods and services, and in this way the British electricity system had initially developed in a distributed manner. In the post-World War II (WWII) phase, however, this was to change and one aspect of the modern British energy system was established – its heavy centralization. In 1942 a new Ministry of Fuel and Power was established, initially with the intention of ensuring adequate energy provision for military as well as commercial and domestic purposes.[5] After the war the Ministry of Fuel and Power was kept on as a government department, but energy policy objectives and design changed significantly. The principal objective of energy policy became to produce the energy required to provide social goods and, importantly, to grow the economy which had shrunk considerably over the course of World War II. Energy was seen as a pre-requisite for economic growth, the 'rule of thumb' was that GDP growth of 3% would be built upon growth in electricity demand of around 7%, as well as essential to some of the wider aims of the newly emerging Welfare State.[6]

The state took a central role in building the network infrastructure and electricity generation plant necessary to ensure universal access, as well as later in building gas pipelines to distribute North Sea gas nationally. Many private undertakings were nationalized and municipal utilities lost their responsibility for providing energy services to local populations.[7] During this era coal remained central to the provision of heating and electricity, and the coal sector still employed many hundreds of thousands of people in mining communities across Wales and Northern England in particular. By 1969, however, with the physical infrastructure to support universal access and growing national demand in place, and with the National Coal Board and the Central Electricity Generating Board managing energy service provision, it was decided that the Ministry of Power was no longer required.

Disbanding the Ministry of Power turned out to have been badly timed given the 'oil shocks' of 1973 and 1979. Over the course of the 1970s complacency gave way to acute concern and awareness that total global energy consumption had, over the previous decades, been doubling every 15 years.[8] Although Britain had remained quite dependent on coal for basic services, it was also becoming increasingly addicted to oil – largely due to the rapid expansion of car ownership. The depth and breadth of public and political concern were unsurprising given how deeply embedded energy had become within society, the sudden quadrupling of oil prices in 1973 and the various economic and social knock-on effects across Britain. The response was a wide review of energy policy and the re-establishment, in 1974, of a Department of Energy (DoE).[9] The price shocks had caused a refocus on questions of energy supply security and were interpreted as an overt example of the dangers associated with dependence on foreign imports. As such, the policy response was to provide greater support for domestically based, nuclear and coal, energy industries as well as greater consideration of how to improve energy efficiency.

Oil and gas had, however, been discovered in the late 1960s in the UK Continental Shelf (UKCS) region of the North Sea and production from these sites started in the 1970s. With the renewed focus on domestic energy production post-1973 the government announced its intention to boost output from the UKCS with the intention of becoming 'self sufficient' by the end of the decade.[10] So although diversity in terms of source and geographic location of energy was being overtly encouraged, and at this stage also by the International Energy Agency, there ran alongside a tendency to concentrate on energy independence and on *domestic* production as an antidote to insecurities in international markets.

New Conservatives 1979 to 1997: privatization and liberalization

Ultimately, as it turned out, Britain was to become an exporter of oil and gas and this lasted for almost two decades (the 1980s and 1990s), and as a consequence it started to pay less attention to questions of energy supply security once more. This provided, yet again, a different context for domestic energy policy. But other, yet more significant, changes were also taking place in Britain during the 1980s. Margaret Thatcher was elected Prime Minister in 1979. She, and her Conservative Party, were firmly convinced of the merits of a smaller state, of markets as economic actors and of privatization and liberalization. From 1982 onwards, under the guidance of the Secretary of State for Energy Nigel Lawson, the energy sector was profoundly transformed with the emphasis on liberalization and on the privatization of ownership.[11] Energy was also radically reframed: it went from being a 'public good' of vital socio-economic importance to a service that can, and should, be provided by market actors who would, in turn, be incentivized to focus on economic efficiency through competition.[12] This was partly based on the assumption that growing competition in the sector would allow for prices to fall, thus facilitating energy affordability and protecting consumers.[13] Furthermore, it was held that security of supply would be enabled through freely trading international markets, underpinned by co-operative, market-based agreements.[14]

Although most of the focus during the Conservatives' long term in office was on enacting processes of privatization and liberalization, there was some progress on low carbon energy policy. In 1990 a new delivery programme called the 'Non-Fossil Fuel Obligation' (NFFO) was implemented. The Electricity Act 1990 had enabled the raising of a fossil fuel levy to pay for the NFFO, once permission had been received from the European Commission for this subsidy. The NFFO was open to bids from renewable generation, but it had really been set up as a means to subsidize nuclear generation given that nuclear had proven too difficult to privatize, mainly for financial reasons.[15] The legacy of commitment to nuclear, initiated post-World War II and reinvigorated in the aftermath of the 1970s oil shocks, continued but this time in an attempt to fully marketize energy. In this way, sustainable energy policy became about supporting non-fossil fuel generation, with the emphasis on nuclear, but less about specific and targeted policy to support renewable electricity generation which required different, low risk support structures like the feed-in-tariff (FiT) that had been implemented in Germany in 1990.[16]

Although the NFFO was supposed to be a 'market-mimicking', competitive programme, payments per kilowatt-hour (kWh) for the first round of the NFFO were agreed between civil servants and generators before they entered their contract bids so that little competition occurred in practice.[17] Later rounds were more competitive, but the low cost-cap put in place to reduce the average price per kWh of each round meant that many renewable plants could not qualify for support. Those renewable generators, mainly onshore wind, that managed to qualify for NFFO support then faced an uphill struggle to gain planning permission. By 1997, after seven years of operation, electricity generated from renewable sources, at just under 3% of electricity, was not much more than when the NFFO was first implemented.[18]

Slightly more progress was made in terms of buildings energy efficiency in particular. Towards the end of their time in office the Conservative government initiated an energy efficiency programme – targeted at buildings improvements with some emphasis on disadvantaged customers.[19] The programme, then called the Energy Efficiency Standards of Performance (EESoP), was started in 1994 and was jointly developed and managed by the electricity regulator, Offer, and the Energy Savings Trust.[20] This programme, and its later iteration the Energy Efficiency Commitment (EEC), did result in a quite considerable wave of new insulation in easy to treat homes.[21]

In 1992, once the government was confident that energy had been sufficiently marketized, they disbanded the Department of Energy whose very name smacked of 'economic planning' according to Margaret Thatcher.[22] Through passing responsibility for energy services to market actors, and reducing formal political organization around energy to a small Energy Directorate within the DTI, knowledge about the increasingly complex energy extractive and utility sectors ended up in the private sector.[23] Two 'independent' regulators were established, one each for gas and electricity, and these were to be paid for by energy companies. In this way the specific design principle of the new regulators was to remove regulation from the influence of 'politics' but not necessarily from private energy industry actors, especially as so much knowledge needed to regulate resided with them.[24]

It is important to note that the energy system remained heavily centralized and, after an initial proliferation of companies involved, through various mergers and acquisitions the gas and electricity sector became dominated by six large companies. Partly because of the ready availability of domestic gas from the UKCS the UK was becoming increasingly reliant on gas for cooking, heating and electricity. Coal usage had dropped from 74% of overall energy in 1960 to 18% in 1998 and gas had risen from zero to 34% in the same time period.[25] UKCS oil and gas also became important in revenue terms for the UK, with some arguing that tax revenues from oil and gas helped to underpin the stronger economy in the 1990s. Indeed, oil revenues had risen six-fold between 1979 and 1985 to one-tenth of the Chancellor's budget.[26]

In environmental terms, the switch from coal to gas for electricity generation in the 1990s resulted in a significant downturn in CO_2 emissions.[27] It is this switch along with the de-industrialization of the British economy and the economic downturn from 2008 to 2014 that have helped Britain to so far stay on track in terms of meeting emissions targets. Indeed, between 1990 and 2015 emissions have fallen by 36%.[28] With significant indigenous production of gas the UK became less focused on energy supply security, and energy became a less discussed policy issue – partly also because of a degree of confidence in the private sector to deliver energy goods and services.

1997 to 2010: a 'low carbon' energy transition emerges

In the late 1990s climate change mitigation started to appear more heavily on the political agenda, indeed the New Labour administration that came to power in 1997 had made Manifesto commitments to addressing environmental issues and to expanding the production of renewable energy. New Labour's narrative, in opposition, had also focused on a critique of the Conservatives' close dealings with energy companies, sometimes referred to as 'fat cats', whilst arguing for greater efforts to address social issues such as energy poverty. Their 1997 Election Manifesto pledged carbon dioxide emission cuts of 20% over 1990 levels by 2010 and that 10% of electricity should be supplied by renewable sources, also by 2010.[29] They claimed that they would:

> put concern for the environment at the heart of policymaking, so that it is not an add-on, but informs the whole of government …[30]

The early Labour Years: 1997–2005

In effect, however, for the first terms in office Labour's approach to energy policy was more about continuity than change, especially in its overt commitment to energy liberalization, markets, and competition. This commitment was made clear in the 1998 Competition White Paper and in an early departmental review of energy policy.[31] Indeed, in terms of policies pursued at the time, the main emphasis was on completing the liberalization of the gas sector and on merging Offer and Ofgas to form Ofgem through the Utilities Act 2000. It is significant that responsibility for energy policy remained with the 'Energy Directorate', a sub-division of the Department of Trade and Industry (DTI) which was above all mandated to achieve competitive and freely trading markets in Great Britain.[32] As such the assumption remained that climate change policy, amongst other political goals, was to be achieved through liberalized, competitive markets. This quote from a 2001 review of energy policy encapsulates the thinking:

> [m]arkets can be a more effective instrument for delivery of government policy than more traditional mechanisms.[33]

In 2002 a new support mechanism for low carbon energy, the Renewables Obligation (RO), was implemented and this was more of a market (or economic) mechanism than the NFFO. An obligation was now placed on suppliers to purchase and supply a certain amount (from 3% in 2002 to 10.4% by 2010) of generated electricity from renewables in order to gain tradeable renewable obligation certificates. There were no contracts for generation from specific projects, no price or contract length stipulated in the RO and developers of renewable energy had to negotiate with a supplier for all agreements.[34] As a result of these arrangements renewable generators had to carry yet more risk, especially given that suppliers were not keen to sign contracts for too long and renewables generators had little visibility of price beyond the short-term contract. This system tended to reward large companies that could take and manage such risks. Indeed, renewable electricity generation did expand with the support of the RO, but projects tended to be large-scale and ownership of renewables remained in the hands of large generators.

Importantly, however, Labour faced growing challenges of their climate and energy policies. In 2000 the Royal Commission on Environmental Pollution (RCEP) had been overtly critical of the over-reliance on markets to fix complicated environmental problems, such as climate change.[35] In 2002 a new review of energy policy, this time conducted by a team put in place by the Prime Minister's Office, was undertaken.[36] Although the review was initiated mainly due to a growing awareness that UKCS oil and gas production was in decline after many years of high depletion rates, it presented a yet greater challenge to existing climate policy. It suggested the adoption of more formal energy efficiency and renewable energy targets as well as a new, single government department that should be responsible for climate change, energy and transport policy.[37]

As part of its response, in the 2003 White Paper on energy, the government made some more effort to silence these critiques: most significantly in the form of newly articulated energy policy objectives. These new objectives included a commitment to reduce carbon emissions by 60% by about 2050 and to ensure that every home in Britain is adequately and affordably heated.[38] This meant that energy policy was specifically set to achieve climate-related objectives whilst previous objectives had been focused on energy supply security and the promotion of competitive markets. It is worth noting, however, that the preferred means of achieving these remained, at this stage, largely market based.[39]

A transition emerges: the energy security-climate nexus

By 2006, the politics of energy and climate change was starting to shift again, but political and energy structures remained important in influencing energy policy decisions. Pressure on the government to become more involved in energy policy had continued to mount. The argument continued that Britain's over-reliance on markets and market-based mechanisms, such as the emissions trading scheme (ETS) and the RO, left it at risk of not meeting new emissions reduction and energy poverty targets.[40] Indeed, one report claimed that, corrected for the outsourcing of energy-intensive industries and for coal to gas substitution, and adding back shipping and aviation, carbon consumption had risen almost 20% between 1990 and 2005.[41]

At the same time, however, Britain was also becoming increasingly concerned about energy security once more – not least as North Sea oil and gas flows had begun to taper off and Britain had become an importer of oil and gas again. The timing could not have been more difficult in that Britain's new reliance on imports coincided with escalating oil and gas prices, growing Chinese energy demand, the Russia-Ukraine gas transit dispute and a new emphasis on energy supply security across Europe. In response to these changing circumstances by 2006 security was being articulated as one of the 'immense' challenges facing the UK as a nation.[42]

Growing political, and to an extent popular, concern about energy security also resulted in new questions about energy policy. Another review of energy policy was undertaken,[43] and this was followed by another White Paper in 2007, just four years after the previous White Paper. Here the tone changed quite significantly: concerns about unstable foreign supplies of oil and gas led to a re-emphasis on the need for 'home-grown' energy and concerns about resource scarcity became more widespread. This new tone did not go unnoticed by climate change groups which started to argue that increasing support for renewables and growing domestic energy production were complementary strategies.[44] At the same time, emissions targets had become formalized and reinforced by Tony Blair's commitment to the EU 20–20–20 agreement. As a result of this commitment Britain now had a fixed renewable energy target, 15% of total energy, as well as an energy efficiency target that it had to meet by 2020.[45] The renewable target was going to be a big ask for the UK, given that in 2007 renewables were still less than 3% of energy.

These new targets, the combination of challenges around climate change and energy security thinking and the related politicizations of energy policy did lead to some quite significant governance changes.[46] A new government department was formed in 2008, the Department for Energy and Climate Change (DECC), with specific mandates to ensure energy security and affordability, and climate change mitigation.[47] In a break from the previous emphasis on markets and competition the newly appointed Secretary of State for Energy, Ed Miliband, started to speak openly in terms of the need for a 'strategic role for government' in the delivery of energy goods and services as well as in terms of an 'energy transition'. The gas and electricity regulator, Ofgem, received a new 'sustainability' mandate in order to allow it to take environmental considerations into account when making regulatory decisions.[48] 2008 also saw the creation of the Climate Change Act and another new body, the Committee on Climate Change (CCC). This Climate Change Act was widely held up as being the first of its kind in that it not only set legally binding CO_2 emissions reduction targets up until 2050, of at least 80%, but it also set out a series of 5 year carbon budgets to 2022.[49] The CCC was instituted in order to measure progress towards meeting carbon budgets and emissions targets, to hold the government to account and to provide advice on climate change.

However, as argued elsewhere, targets are not sufficient in driving profound, sustainable practice change – especially given the embedded and path dependent nature of energy regimes.[50] What are needed, therefore, are strategies, policy instruments and regulations that can support new ways of producing and using energy. In the aftermath of the 2008 changes to policy objectives, and in recognition that existing policy was not doing enough to support a sustainable transition, a series of new energy strategies and policies were announced. These included the first 'Low Carbon Transition Plan'; new and reasonably ambitious support programmes for energy efficiency (CERT and CESP); funding for four carbon capture and storage (CCS) demonstration plants; and a feed-in-tariff (FiT) for medium and small-scale renewable generation. The level of announced support for CCS was indicative of a desire to extend the use of coal and gas electricity generation and was more suitable to the interests of existing energy generation actors than a stronger push for renewables.

New priorities: fiscal austerity and energy security

In 2010, after 13 years in office, Labour lost the general election to a coalition of Conservatives and Liberal Democrats. Again, as had been the case when New Labour came to power in 1997, energy policy was initially marked more by consistency than by anything else. Both the Conservatives and Liberal Democrats had made Manifesto claims about their commitments to climate change – the Conservatives having dubbed the coalition 'the Greenest Government ever'. Over time, however, it became apparent that the emphasis of the coalition was as much on energy supply security as on climate change, as well as on fiscal austerity and the proper functioning of markets.

Mainly in answer to fears about insufficient generation of electricity in the future a new Capacity Market was created in 2014 as part of the wider Electricity Market Reform policy.

This Capacity Market, although theoretically open to demand side response, was mainly designed in order to enable sufficient back-up electricity generation capacity and to ensure security of supply as intermittent renewables grew as a percentage of electricity generation. Furthermore, the payments were grandfathered which meant that existing coal and gas plants could be paid for agreeing to deliver future supplies of electricity and this meant, in practice, that coal plants that might have been shut by 2022 would be able to continue operating.[51] Those that had hoped that the Capacity Market would enable greater demand side response in electricity markets were disappointed, not least newly formed aggregators, like Tempus Energy, looking to establish new business models predicated on aggregating domestic loads.[52]

This tendency to reward existing generators was reflected in decisions taken to support a new wave of nuclear electricity reactors. Nuclear, like large-scale coal and gas generation, fits well with the existing centralized electricity system and nuclear industry representatives, especially Electricité de France (EDF), had made much of nuclear's low carbon and domestic credentials. The new low carbon generation support system, Contracts for Difference (CfDs), was used to offer a fixed (strike) price of between £89.50 and £92.50/MWh for a period of 35 years for the new nuclear facility Hinkley Point C.[53] This payment was especially generous in terms of fixing the payment for such a long period – partly because EDF had made it clear that they would not invest without a clear indication of profitability. Although at this stage, despite Chinese funding, it remains somewhat unclear whether this complex and expensive deal will go ahead, if it does there will be less available funding for large-scale renewable projects. This is because the total amount of support offered per annum through CfDs is capped by the Treasury through the Levy Control Framework.[54] In addition the CfD support system, like the RO before it, requires a degree of professionalization from generators to succeed in bidding, and this tends to exclude small and medium-scale enterprises from this line of funding.[55] Partly for this reason although renewable electricity generation has grown quite significantly over the past few years, much of this growth has come from large-scale wind plants.

Another notable aspect of energy policy under the coalition, and then under the Conservatives since the 2015 election, is the degree to which economic policy has influenced decisions. The coalition's response to the 2008 crisis and the subsequent growth in the UK's public indebtedness, partly caused by bailing out troubled banks, was to pursue an overall economic policy of fiscal austerity. This initially involved significant cuts in public spending and a cross-governmental re-focus on economic efficiency. For energy, aside from nuclear investments, this became about controlling the near-term costs of the newly established 'low carbon' transition. At the same time, given the historically important role of energy in terms of tax income and employment, the government has become increasingly focused on changing the regulatory regime in order to establish conditions supportive to the maximum extraction of Britain's fossil fuels.[56]

Examples of this refocus on maintaining aspects of the existing energy system, for economic reasons, are the new tax and regulatory regime for oil and gas (including shale) and the newly established National College for Onshore Oil and Gas. As was argued at the time, the decision to better enable British fossil fuel extraction was taken partly for energy security reasons, i.e. to support domestic production, but also because of the expectation of economic returns, which in the case of shale gas were expected to be 'huge'.[57] Shale gas extraction, however, faced high levels of local opposition, which resulted in local authorities refusing to give planning permission. In response the government ruled that decision-making on planning should be taken away from local authorities and made by the Secretary of State for Energy as well as setting up long-term investment funds to be paid to areas 'hosting' shale gas developments. The commitment to shale is further supported by the argument that gas can act as a useful 'bridging fuel' within a sustainable energy transition. However opponents in the House of Commons' Environmental Audit Committee have argued that large-scale extraction of shale gas is unlikely until at least the mid-2020s, by which stage it would most likely be competing with renewables rather than coal.[58]

In sum although a nascent governance plan for a 'low carbon' transition had started to emerge by 2008, since the 2010 general election there has been arguably relatively less support for renewable energy and for energy efficiency. The previous energy efficiency schemes, CERT and CESP, had been reasonably ambitious in terms of the amount of energy they were designed to save.[59] The new buildings efficiency plan announced in 2013, which included the Energy Company Obligation (ECO), was designed with an implicit annual reduction target of only 30 TWh (versus CERT's 104 TWh savings target).[60] After the 2015 election even these embattled energy efficiency measures have been discontinued, as was the zero carbon homes policy. In addition, partly in response to local opposition to wind farms and to the continuing policy of fiscal austerity, it was decided that the relatively successful small-scale FiT and support for onshore wind will both be phased out.[61] Again, these changes have been made in order to cut the near-term costs of the transition. Although carbon budgets and the Climate Change Act commitments remain in place, British energy policy has, since 2010, become increasingly subject to a broader economic policy of fiscal austerity, increasing Treasury control over the costs and ensuring supply security.

The British energy transition: low carbon, domestic and centralized

The principal arguments underpinning this chapter are that there are different types of transitions and that political and energy institutions have tended to influence energy governance decisions as well as the types of changes that have been taking place in British energy markets. The UK stands out in comparison to other, arguably more progressive, countries in that it considers nuclear energy to be sustainable (hence the 'low carbon' transition), in that new renewable generation tends to be large-scale and large company owned and in that demand reduction and flexibility have so far taken a back seat. The previous two sections have shown how political institutions, such as the market-liberal mindset but also changes in government, and energy structures, such as the pre-existing energy industry, have tended to narrow choices down and maintain aspects of the status quo. This section will explore in more detail how certain governance choices have influenced the *nature* of the energy transition that is taking place in Britain.

Policies put in place to meet emissions reduction and renewable energy targets have supported growth in renewable electricity generation in particular. Indeed renewables had grown to 17.8% of electricity consumption by 2014.[62] It is also notable that overall gas and coal consumption have fallen over the past few years, although coal's and gas' share of electricity generation remains at 30%, whilst nuclear's share of electricity generation is also falling but remains at 19%.[63] Much of the improvement in take-up of renewable energy can be accredited to the FiT and RO support policies, although there is more argument over the degree to which energy efficiency policies have contributed to lower energy demand figures. UK final energy consumption has fallen by 8.1% since 1990.[64] Government documents assign some of the fall to milder winters,[65] albeit the sharp falls also coincide with the 2008 financial crisis and subsequent declines in economic growth. It is also notable that final consumption of electricity has been rising reasonably steadily since 1985, partly due to the sharp increase in use of electrical appliances.

These statistics point towards some changes in the UK energy system, most notably that 'low carbon' electricity's share of generation had reached 39% by 2014, however it is considered important here that many of the principal characteristics of the British system remain intact.[66] It is still a centralized system, with a bias towards large-scale generation, and residential supply markets remain dominated by the Big 6, who have maintained an ability to be influential within policy and regulatory decision-making processes.[67] It is also considered important that 'low carbon' electricity, including nuclear, is considered an important measurement, whilst other countries tend to focus on renewables when referring to sustainable energy supply. The sections below analyse these tendencies in more detail, as well as explain what implications they have for longer-term trends.

Supply not demand focused policy

Here we turn to arguments outlined elsewhere about the importance of effective demand management within a sustainable energy system transition. Governance arrangements for gas and electricity, in Britain as elsewhere, have long been designed with an emphasis on providing secure supply for consumer demand, what some refer to as a 'predict and provide' mentality.[68] This mentality was traditionally predicated on assumptions about the relationship between energy demand and economic growth. Consequently, as energy service demand has grown, an infrastructure geared toward meeting, as opposed to influencing, that demand has also grown. As we move to transition the energy sector sustainably, however, it is becoming increasingly clear that this will be easier and less costly the smaller energy demand is, whilst flexible demand is also a better route to balancing electricity markets than building more (costly) back-up supply.[69]

By contrast to these arguments, however, British energy policy has still remained for the most part supply focused in its attempts to meet carbon emissions targets. This is perhaps unsurprising for a country with a long history of indigenous fossil fuel production and with large, economically strong energy companies. It is also partly the case because the EU 20–20–20 renewable energy target was so far off the UK's position at the time, and this left policymakers with little choice but to focus on renewable generation support policies. EU 20–20–20 efficiency targets, on the other hand, were more loosely defined and are more about making sure that demand is less than it might otherwise have been.[70] By classifying nuclear electricity as 'sustainable' and committing to a new generation of nuclear plants the UK has also had to focus much of its policy-making attention on securing funding and on creating a sufficiently rewarding support scheme. Aside from these low carbon supply policies the government has, as argued above, also recently concentrated much political capacity towards securing new tax and regulatory regimes to boost domestic oil and gas industries.

Taken together all these supply-side policies, many of which support traditional energy business models, have required a high degree of political and financial capacity, potentially to the detriment of demand management innovations. The commitment of political capacity to supply measures is important given limited energy governance capacities,[71] and overall political conditions of fiscal austerity since 2010. There have been a number of energy efficiency policies over the years, some targeted at industry and some at households, with varying degrees of ambition in terms of reduction in energy use. As we saw above the most recent buildings efficiency policies, the ECO and Green Deal, were limited in their ambition and have since been scrapped. The lack of current household energy efficiency policy is significant in that so much still needs to be done. Although gas demand has fallen since 2004, reductions in the domestic and services sectors have been far lower than those in industrial and electricity generating sectors. For example, demand for space heating in the domestic sector has grown substantially since 1970.[72] It is estimated that in order to meet the UK's goal of reducing emissions by 80% by

2050 it will be necessary to reduce emissions from buildings to near zero by 2050.[73] At the same time DECC estimates that, on the basis of current policies, demand will be similar in 2030 to the position in 2013.[74]

Far less has been done to encourage demand-side response in electricity markets. Demand-side response, to reduce peak demand and improve flexibility, is considered a central tenet of an efficient and cost effective transition. This is not just because of the need to integrate intermittent renewables but also because demand response can contribute to a more cost efficient system and a more secure and affordable transition.[75] As with energy efficiency, not enough is being done on flexibility, and there are concerns that peak electricity demand will continue to increase over the coming decades thereby encouraging the construction of yet more expensive generation.[76] For example, the Capacity Market was supposed, in theory, to allow for demand-side response to be able to bid in but, thus far, companies seeking to get paid for demand-side response have found it very difficult to take part. This is partly because fossil fuel generation is being offered contracts of up to 15 years, whilst demand response companies are being offered contracts of 12 months. One company, Tempus Energy, argues that this favours polluting power stations over cheaper demand reduction and flexibility options and, on that basis, they are facing the British government in a European Court.[77] The inability so far to design policies that better enable demand-side response places the energy transition at a disadvantage in comparison to some US states, such as New York, where energy markets are being re-designed specifically with demand-side innovations in mind.[78]

Scale and centralization

Taken as a whole, British energy governance over the past few decades has also done little to destabilize the system of centralized power generation. This is not just evident in the support of large-scale nuclear generation but also in the kinds of renewable projects that have been funded. Most funds that have been made available to support new renewables, through the RO and subsequently the CfD, have been allocated to large-scale projects – hence the dominant position within renewables of on- and offshore wind. This is directly to do with the design of the RO and CfDs in that certain levels of professionalization are demanded in order to qualify for contracts either with supply companies, as was the case with the RO, or within the CfD tender process. Another aspect of renewable energy development in Britain is that new renewable power plants tend to be owned by large utility companies. This type of renewable expansion stands in strict contrast to the kind of 'distributed energy revolution' seen in other countries like Germany and Denmark. The German FiT was designed specifically so that it would be applicable to small and medium-sized renewable operators, and the distributed nature of its energy system has been further underpinned by a high degree of citizen and community ownership.

Because ownership of new renewable generation remains in the hands of large generators the economic benefits that can be gained through the somewhat generous RO system have accrued to a small number of companies.[79] This has implications for which kinds of energy companies are part of the transition, as well as for how the economic benefits of the transition are distributed. Relying on incumbent market actors to fund and drive innovations tends to mean less room for innovative, new business models to enter markets and drive change.[80] Innovative new businesses have faced barriers to entry in the UK partly due to the dominance of traditional utilities, but also due to the persistence of regulations, corporate codes and licences designed decades ago and in the context of privatization and liberalization.[81]

Concentrating the benefits of the transition in incumbent rather than new hands, whilst distributing the costs quite broadly via consumer bills, also incurs political risks. The UK energy transition is less well embedded socially especially given popular opposition to the construction of onshore wind, from which local communities receive little tangible benefit, and given rising energy bills as a result of policy costs. By contrast, in Germany renewable policy has been designed partly around the notion that local objections to unsightly onshore wind farms can be lessened if local residents know that they, and/or the local community, can benefit through part ownership of that renewable generation.[82]

Households, energy poverty and acceptance of the transition

The question of how economic costs have been distributed has also been an important component of fuel poverty in Britain, partly because it has been important within the wider politics of energy transition acceptability. Various governments have paid quite a bit of rhetorical attention to keeping the costs of the transition down, albeit analysis has shown that some policies chosen, for example the RO, have ended up costing more in practice than alternatives such as the German FiT.[83] The marked tendency to focus on near-term over long-term cost control can partly be explained with reference to growing popular opposition to rising energy bills as well as by the government's commitment to fiscal austerity.

Indeed, consumers have faced significant gas and electricity price increases over the past decade, and not just because of the growing cost of policies. Limited competition amongst the Big 6 suppliers, rising wholesale prices and a tendency by suppliers to very quickly pass on increases, but to be slow about passing on price reductions, have also impacted upon the prices paid by some customers. In addition, the 'Big 6' incumbent gas and electricity suppliers that 'inherited' customers at the time of privatization have managed to retain quite a few customers, that haven't switched supplier, on standard variable tariffs (SVTs) that are higher than those offered to customers that actively switch.[84] As a result customers on SVTs, which are often also vulnerable households, pay a higher proportion of policy costs as well as making an important contribution to suppliers' otherwise lacklustre margins.

Partly as a result of these practices and because of the ongoing economic downturn energy poverty has been on an upward trend in Britain since 2007. Specifically on the most recent available figures, and despite the 2007 commitment made to eradicate fuel poverty in the UK where practicable, in 2013 there were 4.5 million households living in fuel poverty representing 17% of all UK households.[85] National Energy Action (NEA) provide more details by suggesting that there were 41,000 needless deaths across the UK, between 2010 and 2015, directly attributable to vulnerable citizens inhabiting cold homes.[86] Predictions are that, based on current policies, the (medical) costs and suffering associated with fuel poverty are likely to increase and the National Health Service (NHS) will need to spend £22 billion treating cold-related morbidity over the next fifteen years.[87]

Part of the problem is, as already mentioned, that many vulnerable customers are on higher tariffs for electricity and for gas, but another part of the problem is that many vulnerable citizens live in badly insulated homes that are not energy efficient. Energy efficiency policies, such as CESP and the ECO, were targeted specifically at hard-to-treat homes with vulnerable and/or fuel poor inhabitants.[88] This is partly because improvements for the 'easy-to-treat' homes had to a great extent already taken place under the ESSoP, EEC and CERT programmes, but also due to increased buy-in to the argument that efficiency could improve living conditions. Ironically, however, in 2014 when rising prices were blamed by energy companies on the impact of policy costs that are passed on through bills, the government reduced the Energy Company Obligation (ECO), one of the few policies left that sought to directly address efficiency and fuel poverty.[89] As such what has proven more important is concern over short-term costs and the interests of the main energy suppliers as opposed to securing the medium-term benefit of improved housing stock and fewer winter deaths.

Conclusions

This book is about transitions to a secure and low carbon energy system. The British example has shown how contextualized energy transitions are and how contingent they are on how political and energy institutions inter-relate to privilege certain forms of energy and types of transition. The question that emerges from these observations is, however, whether the UK is predestined to pursue a centralized, supply oriented, 'low carbon' transition simply because of the types of institutions and energy regimes upon which its political systems have been built. This would present a somewhat bleak proposition given arguments about how important demand management, including distributed energy, demand reduction and flexibility, are to long-term, cost effective transitions.[90] However, it can also be argued that it is in understanding the precise character of individual country governance for sustainable energy transitions that we can identify precise impediments to change. It is only by being specific about which aspects of governance tend to constrain rather than enable sustainable changes that we can better communicate what needs to change, and what the solutions should be, in ways that are tangible to elite and wider audiences.

Notes

1 Climate Change Committee, 'Next Step towards Low-carbon Economy Requires 57% Emissions Reduction by 2030', Climate Change Committee, November 26, 2015, www.theccc.org.uk/2015/11/26/next-step-towards-low-carbon-economy-requires-57-emissions-reduction-by-2030/.
2 See C. Mitchell, *The Political Economy of Sustainable Energy* (New York: Palgrave Macmillan, 2008); C. Kuzemko, *The Energy Security-Climate Nexus: Institutional Change in the UK and Beyond* (New York: Palgrave Macmillan, 2013); F. Kern, C. Kuzemko, and C. Mitchell, 'Measuring and Explaining Policy Paradigm Change: The Case of UK Energy Policy', *Policy & Politics* 42, no. 4 (2014): 513–530.
3 S. Strange, *States and Markets* (London: Pinter, 1988).
4 D. Painter, "Oil," *Encyclopedia of American Foreign Policy*, 2, no. 3 (1997): 1–20.
5 The Ministry of Fuel and Power was renamed the Ministry of Power in 1957.
6 D. Helm, *Energy, the State, and the Market* (Oxford: Oxford University Press, 2003).
7 P. Ekins, J. Skea, and M. Winksel, 'UK Energy Policy and Institutions', in *Energy 2050*, ed. J. Skea, P. Ekins, and M. Winksel (London: Earthscan, 2011).
8 J. Chesshire, 'An Energy-Efficient Future: A Strategy for the UK', *Energy Policy* 14 (1986): 395–412.
9 Chesshire, 'An Energy-Efficient Future'.
10 P. Katzenstein, 'Conclusion: Domestic Structures and Strategies of Foreign Economic Policy', in *Between Power and Plenty: Foreign Economic Policies of Advanced Industrial States*, ed. Peter Katzenstein (Madison, WI: University of Wisconsin Press, 1978), 296.
11 Helm, *Energy, the State, and the Market.*
12 N. Lawson, *Energy Policy: The Text of a Speech Given in 1982* (Oxford: Clarendon Press, 1989).
13 S. Littlechild, 'Ten Steps to Denationalisation', *Journal of Economic Affairs* 2, no. 1 (1981): 13.
14 R. Youngs, *Energy Security: Europe's New Foreign Policy Challenge* (New York: Routledge, 2009).
15 C. Mitchell, and P. Connor, 'Renewable Energy Policy in the UK 1990–2003', *Energy Policy* 32, no. 17 (2004): 1935–1947.
16 Ibid.
17 Ibid.
18 House of Commons, Renewable Energy Statistics (SN/SG/3217), (2008).
19 Ofgem, 'Previous Energy Efficient Schemes', Ofgem, 2016, www.ofgem.gov.uk/environmental-programmes/energy-company-obligation-eco/previous-energy-efficiency-schemes.
20 Ibid.

21 Centre for Sustainable Energy, *Environmental and Social Levies: Past, Present and Future* (Bristol: Centre for Sustainable Energy, 2012).
22 Thatcher in Bob Blackhurst, 'Can we Wait for Renewables?', 2004, available at the *Foreign Policy Centre*: http://fpc.org.uk/articles/264 (last accessed 6 June 2016).
23 C. Kuzemko, 'Energy Depoliticisation in the UK: Destroying Political Capacity', *British Journal of Politics and International Relations* 18, no. 1 (2016): 107–124. doi: 10.1111/1467-856X.12068.
24 Ibid.
25 Royal Commission on Environmental Pollution (RCEP), *Energy – The Changing Climate* (London: Her Majesty's Stationery Office, 2000), 67.
26 W. Keegan, *Britain Without Oil* (Harmondsworth and New York: Penguin Books Ltd and Viking Penguin Inc, 1985).
27 Helm, *Energy, the State, and the Market*; Department of Trade and Industry (DTI), *Social, Environmental and Security of Supply Policies in a Competitive Energy Market* (London: Department of Trade and Industry, 2001).
28 Climate Change Committee, 'Next Step towards Low-carbon Economy'.
29 Labour Party Manifesto, 1997, 'New Labour: Because Britain Deserves Better', available at Richard Kimber's Political Science Resources, www.politicsresources.net/area/uk/man/lab97.htm.
30 Ibid.
31 DTI, *Social, Environmental and Security of Supply Policies*; Department of Trade and Industry, *Our Competitive Future: Building the Knowledge Driven Economy* (London: the National Archives, 1998).
32 Kuzemko, *The Energy Security-Climate Nexus*.
33 DTI, *Social, Environmental and Security of Supply Policies*, 2.
34 Mitchell, and Connor, 'Renewable Energy Policy in the UK 1990–2003'.
35 RCEP, *Energy – The Changing Climate*.
36 Performance and Innovation Office, *The Energy Review. A Performance and Innovation Unit Report* (London: Cabinet Office, 2002).
37 Ibid., 144.
38 DTI, *Energy White Paper: Our Energy Future – Creating a Low Carbon Economy* (London: HMSO, 2003), 11.
39 Kuzemko, *The Energy Security-Climate Nexus*; I. Rutledge, 'New Labour, Energy Policy and "Competitive Markets"', *Cambridge Journal of Economics* 31, no. 6 (2007): 901.
40 Mitchell, *The Political Economy of Sustainable Energy*; Sustainable Development Commission, *Climate Change Programme Review: The Submission of the Sustainable Development Commission to HM Government* (London: Sustainable Development Commission, 2005); Ivan Scrase, T. Wang, Gordon MacKerron, et al., 'Introduction: Climate Policy is Energy Policy', in *Energy for the Future: A New Agenda*, ed. Ivan Scrase, and Gordon MacKerron (Basingstoke: Palgrave Macmillan, 2009).
41 D. Helm, 'Government Failure, Rent-seeking, and Capture: The Design of Climate Change Policy', *Oxford Review of Economic Policy* 26, no. 2 (2010): 183.
42 DTI, *The Energy Challenge Energy Review Report 2006* (London: TSO, 2006).
43 DTI, *The Energy Challenge Energy Review Report 2006*.
44 Industry Taskforce on Peak Oil and Energy Security, *The Oil Crunch: Securing the UK's Energy Future* (London: Industry Taskforce on Peak Oil and Energy Security, 2008).
45 Kuzemko, *The Energy Security-Climate Nexus*.
46 Kern, Kuzemko, and Mitchell, 'Measuring and Explaining Policy Paradigm Change'.
47 Cabinet Office, *Machinery of Government. Business, Climate Change, Energy and Environment* (London: Cabinet Office, 2008).
48 Mitchell, *The Political Economy of Sustainable Energy*.
49 House of Commons Environmental Audit Committee, *Beyond Stern: From the Climate Change Programme Review to the Draft Climate Change Bill* (London: The Stationery Office, 2007), 2–3.
50 C. Kuzemko, 'Climate Change Benchmarking: Constructing a Sustainable Future?', *Review of International Studies* 41, no. 5: 969–992.
51 'Carousel: Is the Capacity Market a False Economy?', *Energy Spectrum*, no. 507 (2016, February), 1–6.
52 "Interview: Sarah Bell, Chief Executive and Founder, Tempus Energy," *Utility Week*, November 6, 2015, http://utilityweek.co.uk/news/interview-sara-bell-chief-executive-and-founder-tempus-energy/1186823#.Vlh9678cSQY.
53 Department of Energy and Climate Change (DECC), 'State Aid Approval for Hinkley Point C Nuclear Power Plant', DECC, October 8, 2014, www.gov.uk/government/news/state-aid-approval-for-hinkley-point-c-nuclear-power-plant.

54 Department of Energy and Climate Change (DECC), 'Ministerial Statement to the Lords on the Levy Control Framework', DECC, July 22, 2015, www.gov.uk/government/speeches/levy-control-framework-cost-controls.
55 'Carousel: Is the Capacity Market a False Economy?', *Energy Spectrum*.
56 Department of Energy and Climate Change, *Government Response to Sir Ian Wood's UKCS: Maximising Economic Recovery Review* (London: The Stationery Office, 2014).
57 Hancock quoted in Department of Energy and Climate Change (DECC), *D3: Opportunities for Integrating Demand Side Energy Policies*, DECC, 2014, www.gov.uk/government/uploads/system/uploads/attachment_data/file/341298/D3_Opportunities_for_integrating_demand_side_energy_policies.pdf.
58 House of Commons, *Environmental Risks of Fracking: Eighth Report of Session* (2014–15: HC/856) (London: The Stationery Office, 2015).
59 J. Rosenow, 'Energy Savings Obligations in the UK – A History of Change', *Energy Policy* 49 (2012): 373–382.
60 Ibid.
61 B. Woodman, 'Rudd's "Magic Money Tree" Risks Undermining Investments in a Low Carbon Economy', *Exeter Energy Policy Group Blog*, October 13, 2015, https://blogs.exeter.ac.uk/energy/2015/10/13/rudds-magic-money-tree-risks-undermining-investments-in-a-low-carbon-economy/.
62 Department of Energy and Climate Change (DECC), 'Chapter 6: Renewable Sources of Energy', *Digest of UK Energy Statistics* (DUKES) (London: The Stationery Office, 2015), www.gov.uk/government/uploads/system/uploads/attachment_data/file/450298/DUKES_2015_Chapter_6.pdf.
63 DUKES, 'Chapter 5: Electricity' (London: The Stationery Office, 2015), www.gov.uk/government/uploads/system/uploads/attachment_data/file/447632/DUKES_2015_Chapter_5.pdf.
64 Department of Energy and Climate Change (DECC), *Energy Consumption in the UK* (London: DECC, 2015).
65 Ibid., 8.
66 Kern, Kuzemko, and Mitchell, 'Measuring and Explaining Policy Paradigm Change'.
67 Kuzemko, 'Energy Depoliticisation in the UK'.
68 P. Warren, 'A Review of Demand-side Management Policy in the UK', *Renewable and Sustainable Energy Reviews* 29 (2014): 942.
69 *House of Commons Energy and Climate Change Committee, The Future of Britain's Electricity Networks* (2009–10. HC: 194–1) (2010): 14–16.
70 Kuzemko, 'Climate Change Benchmarking?'
71 Kuzemko, 'Energy Depoliticisation in the UK'.
72 J. Skea, M. Chaudry and X. Wang, 'The role of gas infrastructure in promoting UK energy security', *Energy Policy* 43 (2012): 202–213.
73 Centre for Sustainable Energy, *Beyond the ECO: An Exploration of Options for the Future of a Domestic Energy Supplier Obligation* (Bristol: Centre for Sustainable Energy, 2014), 3.
74 Department of Energy and Climate Change, *Digest of United Kingdom Energy Statistics* (London: The Stationery Office, 2013).
75 DECC, *D3: Opportunities for Integrating Demand Side Energy Policies*.
76 'Carousel: Is the Capacity Market a False Economy?', *Energy Spectrum*, 6.
77 'Interview: Sarah Bell', *Utility Week*.
78 C. Mitchell, 'New Thinking: NY Reforming the Energy Vision', *EPG New Thinking* (blog), August 19, 2015, http://projects.exeter.ac.uk/igov/new-thinking-reforming-the-energy-vision-an-update/.
79 C. Mitchell, D. Bauknecht, and P. M. Connor, 'Effectiveness through Risk Reduction: A Comparison of the Renewable Obligation in England and Wales and the Feed-in System in Germany', *Energy Policy* 34, no. 3 (2006): 297–305.
80 F. Geels, F. Kern, G. Fuchs, et al., 'Unleashing New Entrants Versus Working with Incumbents: A Comparative Multi-level Analysis of the Ongoing German and UK Low-carbon Electricity Transitions' (paper presented at the 6th International Sustainability Transitions Conference, University of Sussex, August 2015).
81 Competition and Markets Authority, *Energy Market Investigation: Statement of Issues* (London: Competition and Markets Authority, 2014), www.gov.uk/cma-cases/energy-market-investigation.
82 M. Lockwood, 'The Political Dynamics of Green Transformations: Feedback Effects and Institutional Context', in *The Politics of Green Transformations*, ed. M. Leach, P. Newell, I. Scoones (London and New York: Routledge, 2014).
83 Mitchell, Bauknecht, and Connor, 'Effectiveness through Risk Reduction'.

84 Competition and Markets Authority, *Energy Market Investigation.*
85 Department of Energy and Climate Change (DECC), *Annual Fuel Poverty Statistics Report* (London: DECC, 2015), www.gov.uk/government/uploads/system/uploads/attachment_data/file/468011/Fuel_Poverty_Report_2015.pdf.
86 NEA (National Energy Action), *UK Fuel Poverty Monitor: 2014–2015* (Newcastle upon Tyne: NEA, 2015).
87 Ibid., 4.
88 Centre for Sustainable Energy, *Beyond the ECO.*
89 Ibid.
90 DECC, *D3: Opportunities for Integrating Demand Side Energy Policies.*

16

Energy transitions and climate security in Brazil

Fabio Farinosi

Introduction

As of 2014, Brazil was the world's fifth largest country and the seventh largest economy.[1] The majority of its population of about 206 million inhabits the main cities – 85% of the total in urban areas, mainly in the coast.[2] Sao Paulo, its largest metropolitan region, hosts the biggest financial market in South America and one of the most important ones in the world. In 2014, the service sector accounted for 71% of the total GDP (~2.2 trillion US$), followed by the industrial sector (23.5%) and agriculture (5.5%). The country's exports largely consist of the exploitation of natural resources (iron ore, manganese, bauxite, nickel and limestone) and agriculture (soybean, sugar cane, corn, and wheat). Brazil's endowment of natural resources is extensive and sufficient to ensure the country's energy self-sufficiency.[3] Brazil is amongst the countries that draw the highest share of their energy supply from renewable resources. While the world average is a mere 13% of total energy supply, Brazil stands out as a clear outlier with over 43% of the total energy mix.[4] The contribution of renewable energy is made possible by the extensive exploitation of the complex and extremely abundant hydrological system for electricity generation. Brazil has, in fact, the world's second largest hydropower installed capacity after China. In 2014, hydropower represented about 70% of the total installed capacity and 80% of the total electricity produced.[5] Biomass, ethanol, and biodiesel are also extremely important for the Brazilian energy sector. Moreover, the recent discoveries of offshore oil and natural gas reserves are expected to change the role of fossil fuels in the future energy balance.

Given the large share of renewables in the country's energy balance, the Brazilian energy sector's environmental impact in terms of CO_2 emissions is fairly limited, at least when compared to other developing countries. The main sources of emissions are non-energy sectors, such as land use change for the conversion to agriculture and livestock production of forested areas, and agriculture. This situation could change in the future if the Brazilian system fails in boosting the large economic development while at the same time keeping its energy sector eco-friendly.

Driven by robust economic growth, Brazil has more than doubled energy consumption in the past two decades (Figure 16.1) and future projections expect this trend to continue over the next years.[6] In the last decade, economic growth has been combined with initiatives aimed at including the weakest parts of the population in the development process. Social programs were

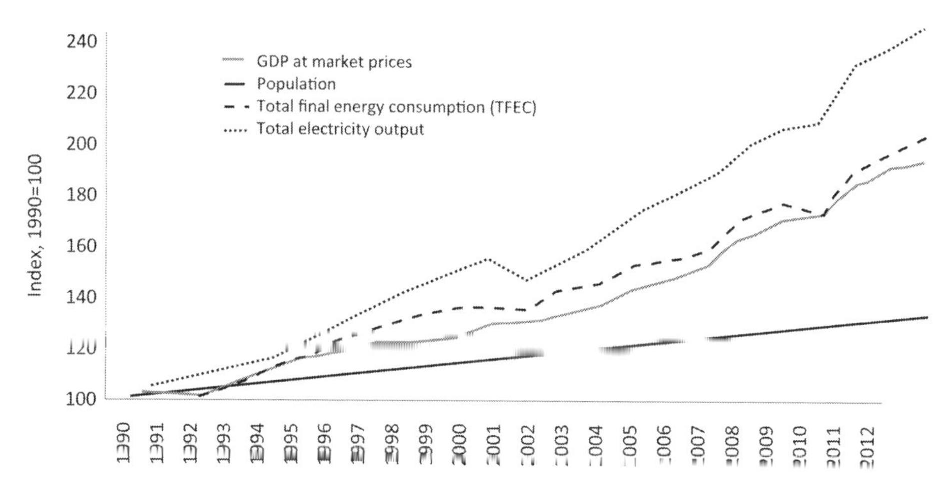

Figure 16.1 Population, income and energy consumption in Brazil (World Bank 2015) 1990–2012, 1990=100.

shaped to achieve a larger income distribution across the population. The *Bolsa familia* (family allowance) program started in 2003 and brought more than 25 million people out of poverty in a decade. This was combined with other programs like *Luz para todos* (light for all), a strategy that provided access to electricity to almost 15 million people, thereby bringing the electrification rate in the country to 99% of households, and substantially boosting the internal energy demand. Large investments in the whole energy sector, considered strategic to support future development, are currently planned by the Brazilian Government.[7] For instance, only for the electricity sector, national studies on electricity demand trends, in line with the elaborations carried out by the International Energy Agency (IEA), evaluated that Brazil needs an additional 6 GW per year in generation capacity in the next two decades.[8] Moreover, investments in fossil fueled and nuclear power generation are planned to increase, so as to diversify the system. This is crucial to minimize the vulnerabilities of the electricity sector caused by the impacts of climate variability and change on hydropower. However, the recent economic crisis, exacerbated by the corruption scandal involving Petrobras (the main Brazilian oil company), are expected to cause a delay in the ambitious investment plans illustrated in the two main documents elaborated by the Ministry of Energy (MME) and its technical office (EPE): the 30 year and 10 years energy plans.[9]

The Brazilian energy sector is extremely dynamic; nevertheless, hard challenges are expected in the near future. In this context, the chapter seeks to answer the following questions: how has Brazil historically dealt with satisfying the main objectives of energy security and economic competitiveness? Historically, the sustainability of Brazil's energy sector has mainly originated from economic opportunity more than environmental consciousness. In a rapidly expanding economy, will climate security still be a defining trait of the Brazilian energy sector? Is Brazil going to succeed in managing the interplay between energy security, competitiveness, and climate-related dynamics?

The Brazilian energy mix

The Brazilian energy mix consists of a large chunk of renewables, which is a very peculiar characteristic both comparing to the developed and other developing countries (Figure 16.2). Bioenergy sources – sugar cane derived products, solid biomass, biodiesel, and hydropower – represent

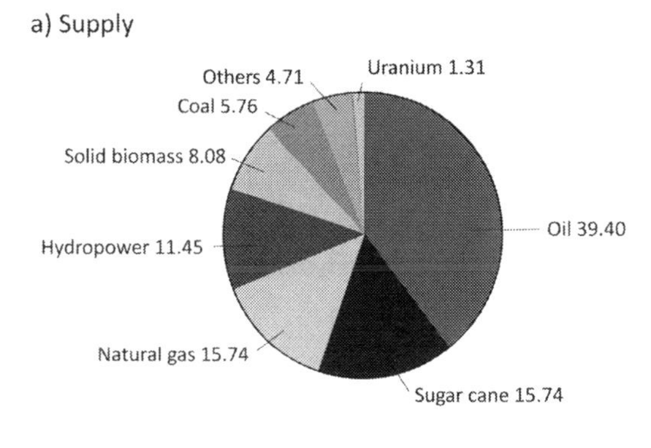

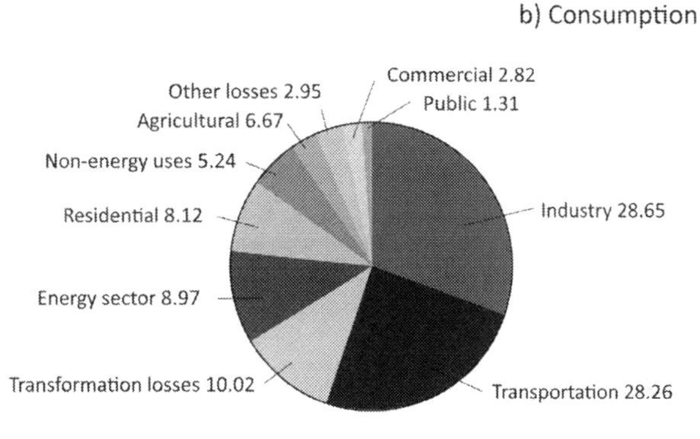

Figure 16.2 Brazil energy balance in 2014 – data in Mtoe
Source: MME, and EPE, *Balanco Energetico Nacional.*

almost half of the primary energy supply. Fossil fuel demand is mainly covered by oil, which is still the main component of the energy balance (~40%); natural gas demand is constantly increasing, but, as of 2014, its role still remains relatively small in comparison to the other, more traditional, sources in Brazil. Compared to other BRICS countries, coal has a very modest share but it remains an important source for heavy industry.

The industrial and transport sectors are mainly responsible for growing energy demand. Industrial demand has significantly increased in the period 1990–2011, by about 3.5% a year. The iron and steel industries represent an important share of industrial demand. The transport sector's demand increased at an even higher pace (about 4%) following the general economic development and in particular the growth of agriculture in the central area of the country. The main means of transportation in Brazil are via road and air. The country's railways system is still underdeveloped and transporting goods on water is significant only on the coastal areas, and much less on the inland waterways.

Residential demand grew at a significantly lower pace (about 2%), partially due to the increasing use of modern and efficient electrical appliances instead of the use of traditional biomass, and due to high energy prices.[10]

The extensive use of biofuels in the transport sector, mainly sugar cane derived ethanol and bio-diesel, finds its roots in the early 1970s when the world suffered the first oil crisis. The initial approach was based on vehicles fully powered with ethanol. In the 1990s, due to a general shortfall in ethanol supply, the demand for more flexible technology rose, until the introduction in the early 2000s of flex-fuel vehicles. This technology allows the use of both fossil and bio-fuels, making the consumer more resilient to the relative price volatility of different energy sources. Ethanol is blended with gasoline with a mandated level ranging between 18% and 25%, while bio-diesel, mainly derived from soybean oil is blended with fossil diesel with a mandated level of about 5%.[11] Liquid biofuels are not the only example of bioenergy in the Brazilian balance. Firewood is still widely used in Brazilian households, while the extensive agricultural production in the country is a source of cheap biomass for electricity and heat.[12]

The country's electricity mix (Figure 16.3) is historically dominated by hydropower. The economic growth-boosted increasing demand for electricity and the hydropower sector's high vulnerability to climate variability raised the debate about the need for a larger diversification of sources. The energy planning operations in Brazil significantly changed after the electricity crisis of the summer 2001–2002, when an unusually prolonged drought, combined with a demand growing faster than the installed capacity, led to the rationing of electricity for a significant period of time.[13] The investments in fossil fuel and biomass thermal installed capacity, in the last decade, made the system more resilient, but not enough to avoid the tangible risk of new power rationing in January 2015.[14] A deeper interconnection with neighboring countries, such as Argentina and Uruguay, has been planned to reduce vulnerability.[15] Large investments are also planned for the exploitation of other natural sources abundant in the country, like solar and wind power.

Energy security

Vulnerability of the electric system: the 2001/2002 electricity crisis and the response of the Brazilian institutions

The Brazilian electricity system is extremely interconnected. The country is divided into different regions and the demand for electricity in a specific portion of the grid is constantly coordinated with the supply capacity of the whole system. Responsible for the operation of the system is a specific technical institution called National Power System Operator (ONS – Operador Nacional do Sistema Eletrico). The real time operation of a complex system, even if not very diversified, allowed a big country such as Brazil to manage the potential shortages due to

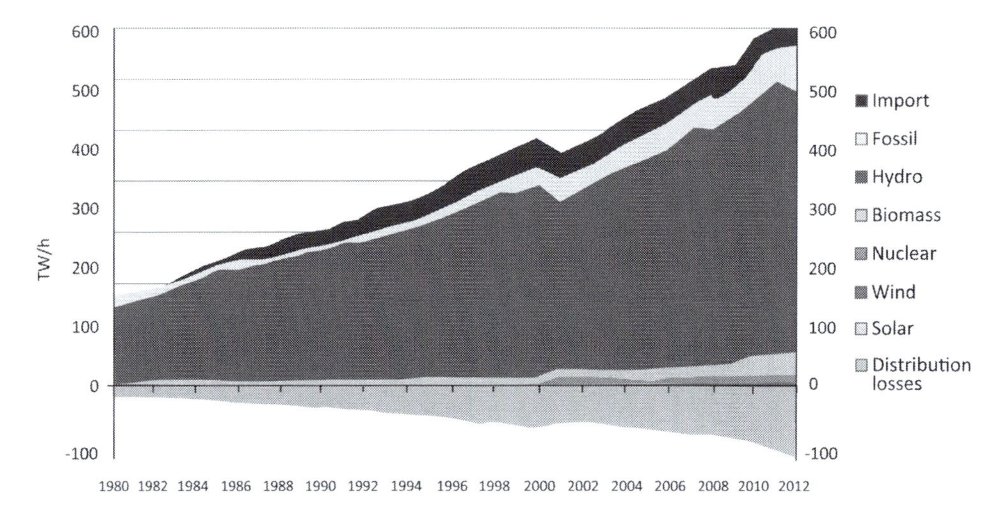

Figure 16.3 Brazil electricity mix 1980–2012
Source: Authors' elaboration based on EIA, *Brazil: International Energy Data and Analysis*, 2015.

climate variability in a portion of the country – with the correct management of stored resources in other regions. For instance, if the north-east of the country was affected by a prolonged drought, the ONS could satisfy the demand from this region using some of the operating reserve stored in the large reservoirs in the south. This system worked fairly well until the end of 1990s. In this period, the generation capacity consisted of more than 85% hydropower (Figure 16.4), but the size of the system was relatively big with respect to demand, and therefore resilient to spatial and temporal weather related shocks. The situation changed with the economic growth that sensibly boosted the demand for electricity (Figure 16.1). Increasing demand in the 1990s was not supported by adequate investments in new generation capacity. In the early 2000s, a prolonged drought noticeably reduced the operating reserves of the large hydropower reservoirs in the south and south-east of the country. In the summer 2001–2002, high temperatures boosted electricity demand. The system was operating at maximum capacity, but this was not enough to satisfy demand. Between July 2001 and February 2002, the electricity available was rationed to 80% of the historical consumption.[16] The privatization process started in 1996 with the establishment of the National Electric Energy Agency (ANEEL), but was drastically interrupted and the government re-established its total control of the electricity sector. In 2004, in order to ensure security of energy supply, control energy prices and promote access to electricity for the entire population, the Federal Government reformed the electricity sector. Three new organizations were created: the Energy Research Bureau (EPE) in charge of long term planning; Electric Sector Monitoring Committee (CMSE) aimed at monitoring the security of the electricity supply; and the Chamber of Commerce of Electric Energy (CCEE), aimed at managing the internal electricity market.[17] A new pattern of development for the electricity sector was created. New generation capacity was installed and strategic diversification of the sources promoted. The contribution of fossil fuels, especially natural gas, and solid biomass became more significant (Figure 16.3). New investments in nuclear generation were also planned. As of 2014, only 4% of Brazil's electricity generation came from nuclear. Two reactors are active in the state of Rio de Janeiro, a third one is currently under construction and

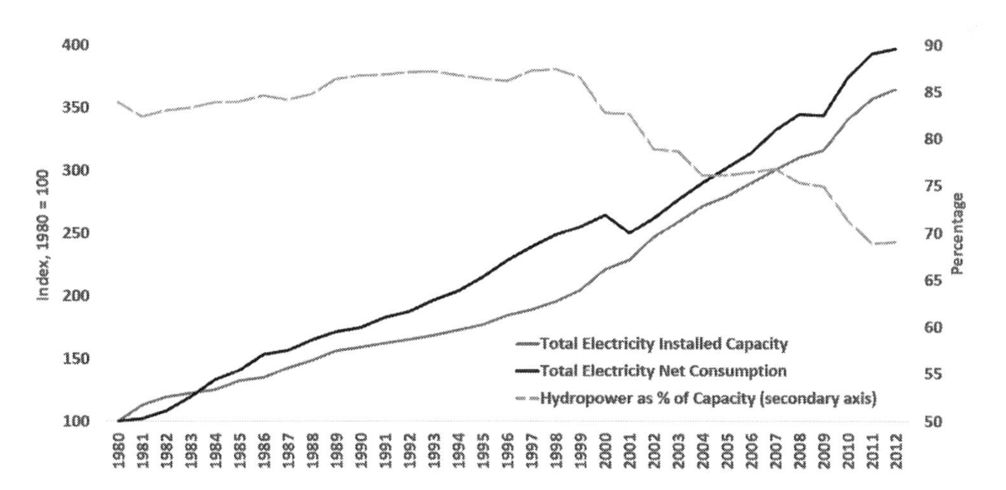

Figure 16.4 Trends in electricity consumption, total installed capacity, and share of capacity represented by hydropower. Period 1980–2012

Source: Authors' elaboration based on US Energy Information Administration (EIA), "International Energy Statistics," 2015, EIA Database, http://www.eia.gov/cfapps/ipdbproject/IEDIndex3.cfm#.

expected to be operative in 2018; seven more plants are planned for 2030. In 2008, the country also signed an international nuclear cooperation agreement with Argentina.[18]

The actions adopted after 2003 significantly boosted the generation capacity and made the matrix less dependent on hydropower, which saw its share decrease from 85% to 69% (Figure 16.4). These actions reduced the vulnerability of the Brazilian electricity system, but its resilience to shocks still needs to be increased: between 2011 and 2014 more than 180 blackouts were recorded;[19] moreover, in the period 2014–2015, a new significant drought hit the country with the risk of renewed emergency and new electricity rationing.[20]

Hydropower still protagonist in the country's electricity generation

Even though the recent offshore discoveries of fossil fuels may turn Brazil to a leading force in the oil sector, hydropower remains the main pillar of the country's energy strategy.[21] As of 2015, Brazil has developed only one-third of its estimated hydropower potential. The total potential estimated for the country's rich and complex hydrological system is approximately 245 GW.[22] According to the 2012–2022 Development Plan, issued by the MME in collaboration with the EPE, the hydropower installed capacity is expected to increase from 85 to 119 GW (about 40%).[23] The IEA estimates that 42 GW of additional installed hydropower capacity is to be developed in the period 2021–2035.[24] More recent economic projections, depressed by the economic crisis in 2015 will probably push the Brazilian authorities to postpone, at least in the short term, part of the ambitious investments in new generation capacity.

The reasons driving the large-scale exploitation of hydropower potential are various: hydropower represents a clean, extremely flexible and economically convenient source of electricity production. Moreover, the construction of dams and reservoirs presents several positive impacts both in the short run in terms of employment, and in the longer run in terms of water supply for human activities and flow regulation.[25] Notwithstanding the large potential hydropower has, this option is also strongly dependent on meteorological variability and climate trends. Increasing concerns are animating the debate about the vulnerability of hydropower technologies to climate change and the possibility for this important renewable source to sustain its future development.[26] Moreover, in order to minimize environmental and social impacts, prospective new reservoirs are relatively small in size (mainly run-of-the-river technology), a trend which is expected to significantly increase the vulnerability of the hydropower sector to climate change.[27] Most of the hydropower potential of Brazil is in the Amazon basin, a critical area both from the environmental and social points of view. The development of these new installations is therefore subject to increasing constraints aimed at minimizing the social and environmental impacts that such new infrastructures may cause. This is the case, for example, of Belo Monte. This hydropower plant, the second largest in Brazil (the third in the world), is currently under construction on the Xingu river, one of the tributaries of the Amazon. Tensions with the local population and international concerns about the impact of such project on the valuable ecosystem's diversity significantly limited the possibility to construct a reservoir able to buffer the intra- and inter-annual variability of the streamflow. The plant's peak capacity of 11,233 MW is expected to be fully exploited only between February and May, when the Xingu reaches its maximum flow. The resulting average capacity factor for this plant is expected to be about 40%, only a bit more than half of the performance of the Itaipu installation in the last decade (77%).[28] On the other hand, climate and land use changes are expected to seriously impact the precipitation and runoff patterns in Brazil. All these factors seriously threaten the country's energy security, with potentially serious challenges for economic and social development. A recent report issued by the Secretariat for Strategic Affairs (SAE), an institution aimed at long-term planning of economic and social development, brought attention to the possible impacts of climate change on Brazilian hydropower generation[29] quantifying the losses in the range between 7% and 30% against the historical generation.[30] Another characteristic of the planned expansion of the hydropower installation is that the majority of the new plants are designed with run-of-the-river technology. This means that the productivity of the hydropower sector would be increasingly affected by the seasonality of the river flows: a big challenge for the system's operation. International and local technical institutions suggest that this problem could be offset by investing in the country's wind energy potential, estimated at about 140 GW and characterized by an inverse seasonality as compared to the hydrological resources.[31]

Electricity transmission grid

Brazil is the world's fifth largest country and, due to its vast landmass, the generation of electricity is not always located close to the consumption site. This is particularly true, for instance, in the case of the new hydropower installations Belo Monte and San Antonio, and for the planned large development in the Tapajos river basin. These infrastructures are located in the Amazon and Cerrado areas, but the electricity produced is destined for the big industrial and urban centers in the south-east of the country (Figure 16.5). Therefore, huge investments on the transmission grid are needed to reduce the relatively high distribution losses (Figure 16.3).

Traditional sources: oil and natural gas

The Brazilian energy mix also largely consists of fossil fuels. The transportation sector drives Brazilian demand for oil and oil derived products, while the diversification of the electricity generation sources, with minor contribution from the transport sector, drives the demand for natural gas.

The state-controlled Petrobras is the major actor in the fossil fuel sector, controlling the exploration, production, and refinement activities. The company had a monopoly of the market until 1997, but even after market liberalization the entrance of new competitors has been slow and made possible mainly in partnership with Petrobras.[32]

In 2014, the country produced about 2.95 million barrels of oil per day, mainly from the offshore fields in the states of Rio de Janeiro, Sao Paulo, and Espirito Santo.[33] In the past decades, demand for oil was constantly higher than the production, the liberalization of the market in 1997 and the permits for exploration and production (E&P) issued by the government to international oil companies boosted production to reach level of consumption for the first time in 2006 (panel (a) in Figure 16.6). The US Energy Information Agency estimated that the production could be stably higher than consumption in 2016, but this estimate could be affected by the recent corruption scandal that hit Petrobras and the whole Brazilian financial market.[34]

The country's refining capacity (2.4 million b/d of crude oil in 17 sites in 2014) is constantly increasing, but demand grew faster than the refinement capacity. Moreover, the Brazilian refineries do not have the technology to process heavy crude, so part of it is exported unprocessed and light crude is imported. Large investments had been made in the past few years to expand the refinement capacity (panel (b) in Figure 16.6). In the state of Pernambuco, at the end of 2014, the first units of a US$20 billion refinery, Abreu e Lima, started production. The expansion of this refinery was planned to start thereafter, but the owner of the industrial site, Petrobras, announced the postponement of the project's second phase due to the financial difficulties caused by the 2015 scandal. Another big project was under development in the state of Rio the Janeiro, Complexo Petroquímico do Rio de Janeiro (Comperj). It was supposed to be

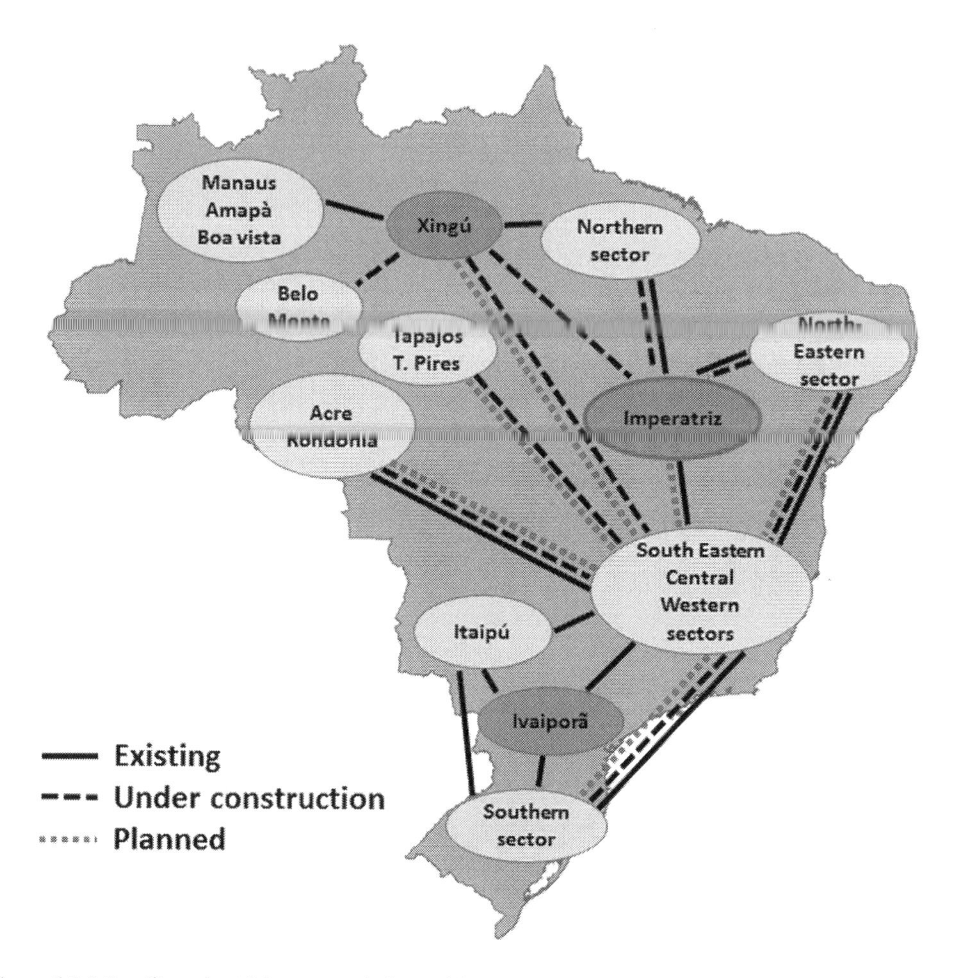

Figure 16.5 Brazilian electricity transmission grid
Source: Drawn by the author, based on MME, and EPE, *Plano Decenal de Expansao da Energia (2024)*.

operational in 2016, but the investment was temporarily suspended in 2015.[35] Two additional projects at the early stage of construction in the states of Ceará and Maranhao (Premium I and Premium II) were canceled.

Demand for natural gas in Brazil has historically been very low, almost negligible (Figure 16.7). It increased significantly due to the increasing demand for electricity and for the diversification of the electricity generation sources. Petrobras operates the national distribution grid covering mainly the south and south-eastern, and the north-eastern portions of the country. The two pipelines were connected only in 2010. Petrobras also owns the three existing regasification terminals. Natural gas is produced mainly in the coast of the Rio de Janeiro and Maranhao states. Bolivia, Argentina (connected to the land pipeline), Qatar, Spain and Trinidad (LNG) are the main suppliers of the imported natural gas.[36]

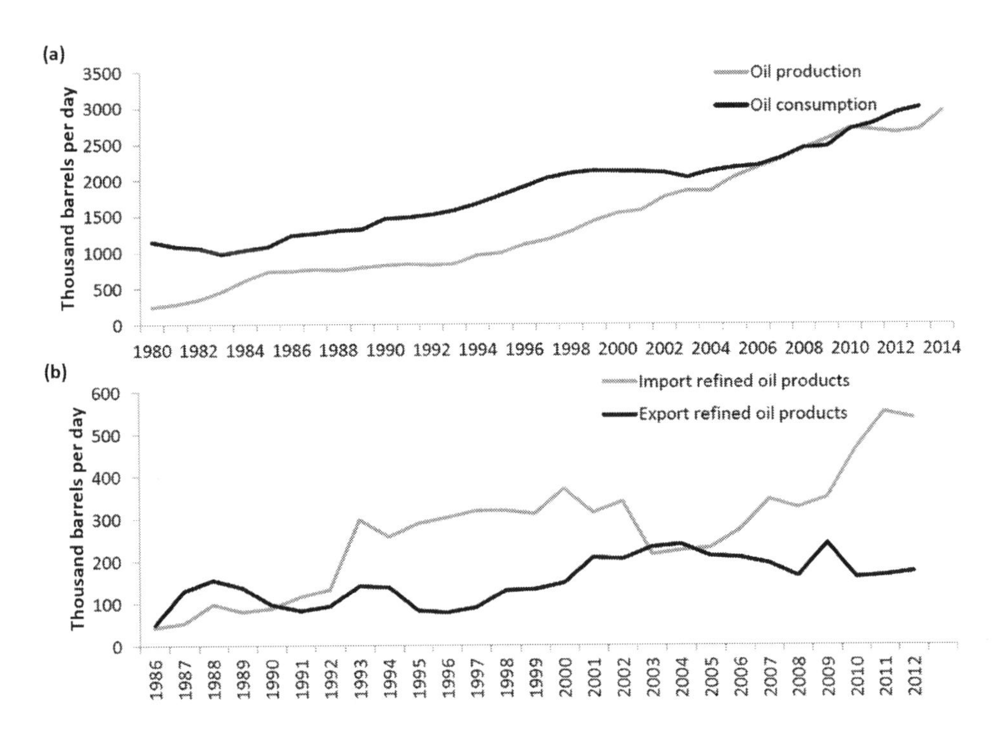

Figure 16.6 (a) Oil demand and internal production, (b) refined oil products import and export 1980–2014

Source: Authors' elaboration based on EIA, "International Energy Statistics," 2015.

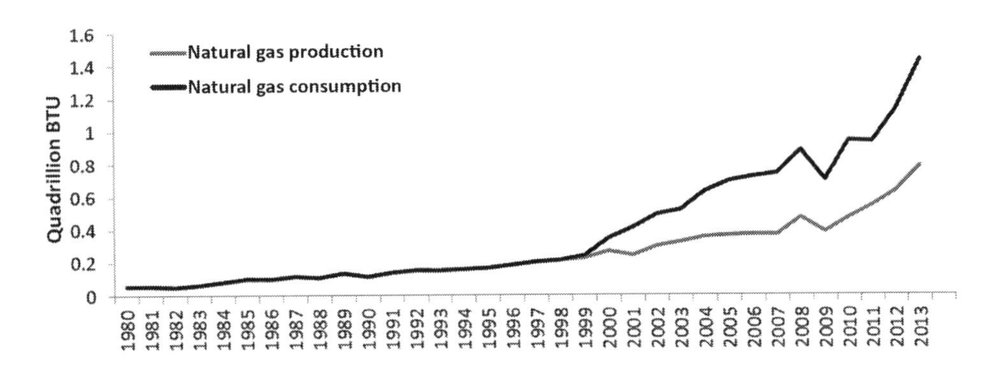

Figure 16.7 Natural gas demand and internal production 1980–2014.

Source: Authors' elaboration based on US EIA, 2015, *Brazil: International Energy Data and Analysis*, www.eia.gov/beta/international/analysis.cfm?iso=BRA.

New discoveries and the future of fossil fuels in Brazil

Brazil has been subject to major operations of exploration for inland and offshore hydrocarbons over the past thirty years. Large fossil reserves were recently discovered in the oilfields off the coasts of the states of Rio de Janeiro, Sao Paul, and Espirito Santo. The Brazilian Agency for Oil and Gas (ANP) reported proven reserves for about 16.2 billion barrels[37] of oil and about 470 billion m^3 natural gas (Figure 16.8). The geology of the area of interest is made of a 2,000

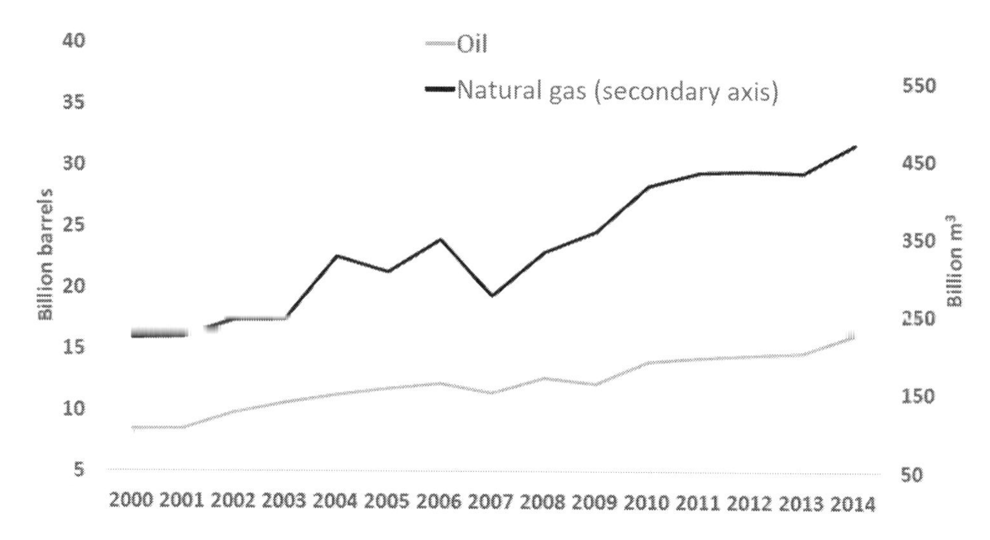

Figure 16.8 Natural gas and oil proved reserves 2000–2014.
Source: Authors' elaboration based on EIA, *Brazil: International Energy Data and Analysis*; ANP, "Dados Estatisticos."

meter thick compressed salt and rock layer. Oil reserves are located partially above the salt layer (postsalt oil reserves) and mainly below (presalt oil reserves). The presalt oil reserves, the exact quantity of which is still unknown, could be substantial in volume and make Brazil one of the world leaders in fossil fuel production. The crude is of good quality and the gas to oil ratio is about 250–300 m^3 of gas per m^3 of oil.[38] The extraction of presalt oil is extremely difficult due to the technical challenges involved in such a deep perforation, and it requires huge investments. The extraction of presalt oil reserves started in 2009 and represented in 2014 about 25% of the total production. Petrobras became the world leader in the production of very deep oil reserves through the use of floating production, storage, and offloading (FPSO) facilities.[39] The corruption scandal now hitting the company could affect the future presalt reserves exploitation, at least in the short term. Moreover, the second half of 2015 was characterized by very low quotations of crude oil: not a very favorable market condition for the profitable commercialization of the presalt oil, since its cost of extraction was estimated between US$41 and US$57 per barrel.[40]

Economic competitiveness

While analyzing the Brazilian energy sector, it is possible to paint the profile of a country that in a relatively short time period has managed its huge and variegated natural resources to become a major player in the global economic scene. Part of the story could be defined as successful, another less so, but it is above-all important to underline how the country and its institutions are facing the challenges of fast development with a peaceful and democratic approach, maximizing social inclusion and minimizing the socio-environmental impacts. Many issues have been identified under the energy security perspective in the previous section. The majority of these issues are carefully analyzed and approached by the competent Brazilian technical institutions with the short and medium term development plans mentioned above: Decadal Energy Development Plan (Plano Decenal de Expansao de Energia), National Energy Plan 2030 (Plano

Nacional de Energia 2030).[41] The main messages could be summarized as need for huge investments in all the specific components of the energy system aiming at achieving: increasing production of fossil fuels; increasing refining capacity; increasing electricity generation capacity; diversification of sources; enhancing the distribution infrastructures; deeper integration of the energy system with neighboring countries. Huge investments that could not be possible without the involvement of the private sector: the country and its institutions have been facing the challenges of balancing the internal economic interests with the attractiveness for international capital.

In 2007, the Brazilian government launched the Growth Acceleration Program (Programa de Aceleração do Crescimento). In its second phase (2011–2014), the program allocated about US$250 billion to the energy sector. The rationale of this investment program is clearly to boost the sector and attract private capital. Foreign investors in the energy sector face a complex mix of structural, bureaucratic and economic difficulties, often referred to as *Brazil Cost* (Custo Brasil), which make investments in the country less attractive.[42] This problem has many consequences: the slow implementation of development plans; the vulnerability of the country's economic development to shocks in the energy and energy-related sectors; high final costs of energy hindering investments in other economic sectors. As for the last point, an example comes from electricity consumer prices in the country. Per capita electricity consumption in Brazil remains very low relative to other similar countries, like South Africa or China.[43] As of 2012, the price paid for electricity by the Brazilian industrial sector was about $178 per megawatt-hour (MWh), while the domestic consumption was priced about $237/MWh. Industrial energy prices were higher than average prices in Europe and almost four times the prices in the United States. This significantly constrained the development of most of the energy intensive industries (chemicals, iron and steel, glass, ceramics, aluminum, and pulp and paper).[44] An example of the energy sector's and, in turn, the national economy's vulnerability to shocks, comes from the close connections that the national economy has with the two main actors of Brazilian energy management: the state-controlled Eletrobras and Petrobras. In particular, the latter has been involved in a corruption scandal in 2015 that destabilized the entire country's financial sector.

Petróleo Brasileiro S.A. (Petrobras) and the 2014–2015 corruption scandal

Petrobras is the largest company in the southern hemisphere; it was founded in 1953 and held the monopoly of the fossil fuels sector in Brazil till the 1997 privatization reform. The company is the major investor in the presalt exploration and production activities. The majority of the foreign companies that entered the market after the 1997 liberalization still operate in partnership with this state-controlled company: this, on the one hand, is justified by technical reasons given that Petrobras is the leader in presalt exploration and production, but also because the partnership with the state-controlled company to a large extent facilitates the interaction with Brazilian institutions. In 2014, Petrobras was investigated in Brazil and in the United States for bribery and money laundering. The investigation hit the management of the company and top members of the Brazilian institutions. The scandal had direct and indirect impacts on the company's operations: first, it cost more than US$8 billion; second, the inability of the company to get its financial statements certified kept Petrobras from accessing the international capital market. All this happened in a moment when the economic exposure of the company was particularly high: with about US$110 billion debt, it was rated as the most indebted company in the world in 2013; while its valuation passed from US$200 billion in 2011 to US$27 billion in December 2015.[45] The main consequence of this crisis was the immediate reduction of the company's short term investment plans: investments decreased by about 17% between 2013 and 2014, by an additional 25% in the period 2015/16.[46] In 2015, the 5-year investment plan set the oil and gas production targets (national and international) to 3.7 million barrels of oil equivalent per day in 2020, markedly lower than the targets of 5.0 and 4.0 million b/d set respectively in 2013 and 2014.[47] This situation had a great impact on Brazil's economy and its credibility in the international financial market. The country's sovereign debt reached 62% of GDP and its rating was labeled as "junk" by international rating agencies.[48] The Brazilian Real touched its 10 year low value compared to the US dollar, the inflation rate skyrocketed, and the unemployment rate rose significantly.

Fossil fuel exploration and production licensing and the "local content" requirement

Before the presalt discoveries and after the 1997 market liberalization, all the companies were allowed by Brazilian law to win concessions for exploration and production. In 2010, with the intent to maximize the national gains deriving from large presalt discoveries, the regulation was changed. While for the postsalt reserves the concession system did not change, a new state controlled company, Pré-Sal Petróleo SA, de facto a subsidiary of Petrobras, was created to manage presalt oil and natural gas exploration and production. The second step of the reform was to provide capitalization to Petrobras through 5 billion barrels of unlicensed presalt reserves in exchange for a larger share of the company. In practice, for the postsalt reserves the concession-holder owns the oil produced after the payment of royalties and taxes, without any partnership (at least officially) with the state-controlled company. For the presalt reserves Petrobras is formally involved in each concession with a minimum share of 30%.[49]

Another limitation to the economic profitability and development of presalt reserves concerns the "local content" requirement that the Brazilian government imposes in every concession issued. Practically, the concession winner is requested to purchase a share (up to 65%) of the components used for the exploration and production from national industries. This could seriously limit the development of ultra-deep-water exploration and production, where the technological component is crucial.[50]

Similar considerations hold for the natural gas market. From the legal point of view, after 1997, private companies could participate in the operations of each stage of natural gas production and distribution. However, Petrobras remains the main operator of the sector, controls the pipeline operator importing Bolivian gas and owns the existing regasification facilities. In 2013, of the 27 companies distributing natural gas to the final consumer, 21 were partially owned by Petrobras.[51]

Licensing for electricity generation capacity

Many studies underline the need for simplification of the licensing process for the installation of new generation in the Brazilian electricity sector, especially regarding new hydropower installation.[52] In order to achieve more effective, faster and cheaper licensing, five aspects were highlighted: the need for clear distinction between federal and state government competencies; the introduction of a dedicated dispute resolution mechanism in the environmental licensing process; the possibility to activate the process of licensing for different projects in the same river basin; clearer specifications of the Environmental Impact Assessment contents; a more careful consideration of the uncertainties linked to each of the projects, in its financial, environmental, and socio-economic aspects.[53]

Climate security

As highlighted in the previous sections, the Brazilian energy matrix is extremely friendly to the environment. The main source of Brazil's GHG emissions is represented by the change in land use. The deforestation rate was a serious concern between the 1990s and the early 2000s. After the reforms in the early 2000s, namely, the Action Plan to Prevent and Control Deforestation in the Amazon, the Action Plan to Prevent and Control Deforestation and Fire in the Cerrado, and the Low-Carbon Agriculture Plan, emissions linked to land use change dropped significantly. The sector responsible for the largest share of emissions in the recent past is that of agriculture.[54]

Regarding climate security, it is important to distinguish between the potential impacts of climate change on the Brazilian energy sector, and the impact of the energy sector on climate.

Climate change impacts on the Brazilian energy sector

The main impacts that climate change could have on the Brazilian energy sector are:[55]

- alterations of the hydrological cycles with consequent possible reduction of the hydropower output;[56]
- increasing intra- and inter-annual variability of the run-of-the-river hydropower output;[57]
- uncertain impact on the potential windpower capacity, decline[58] or increase;[59]
- reduction of the output of thermal power plants estimated at about 2%; and[60]
- uncertain impact on bioenergy: no change/slight increase of sugar cane production in the south, possible decrease of soybean in the Cerrado.[61]

Energy related carbon dioxide emissions

It is important to underline that Brazil's economy is one of the most efficient in the world, especially if compared to the other rising economies of the BRICS. As of 2011, the International Energy Agency estimated that Brazil emitted about 0.18 tonnes CO_2 per thousand dollars of GDP produced. The value is considerably lower than other developing countries like China (~1.0 ton/$1000 GDP), India (~0.98 ton/$1000 GDP), Russia (~0.82 ton/$1000 GDP); but also compared to developed countries like the United States (~0.35 ton/$1000 GDP), those in the EU (~0.21 ton/$1000 GDP), and Japan (~0.2 ton/$1000 GDP).[62] However, it has to be highlighted that with the increasing demand for fossil fuels in the Brazilian energy matrix, the energy related emissions have been rapidly increasing in the past 20 years reaching the threshold of 500 million metric tons (Figure 16.9).

In the recent 2015 Paris Conference of Parties (COP), the Brazilians pledged a 37% reduction of carbon dioxide emissions by 2025 relative to the levels of 2005, with a further indicative target of 43% by 2030 (Intended Nationally Determined Contribution – INDC[63]). This ambitious goal could, indeed, be achieved if the future energy matrix respects the planned share of renewable sources.[64] The MME and the EPE in the last 10-year development plan made a calculation of the possible emissions of the future Brazilian energy balance as calculated using their generation and consumption projections (Table 16.1). Attending to these projections, the energy sector should be able to maintain its level of emissions under the critical threshold fixed for respecting the Brazilian government pledges.[65] However, a critical point that needs to be raised concerns the fact that the technically rigorous and well documented energy planning

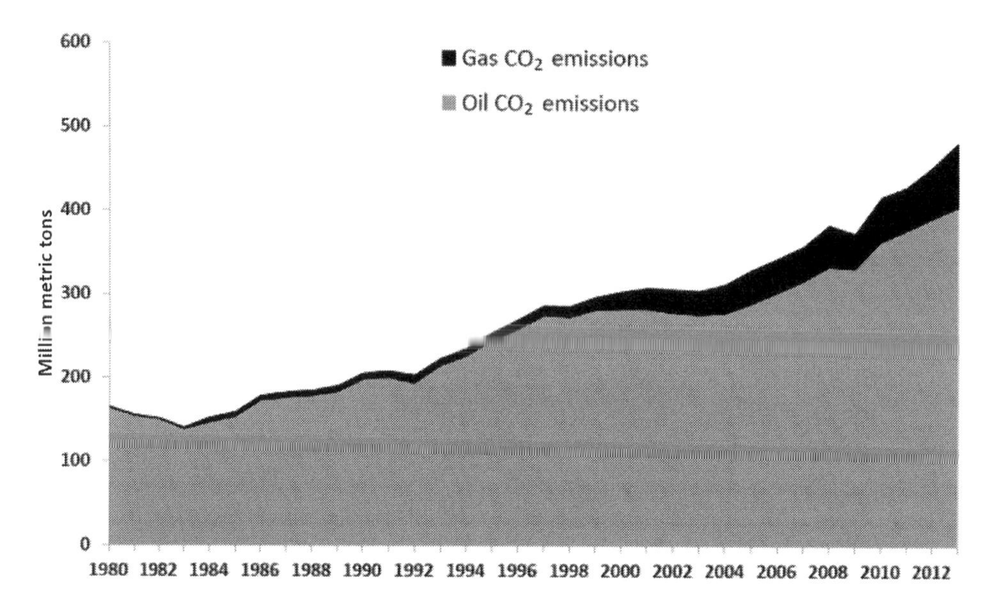

Figure 16.9 Natural gas and oil carbon dioxide related emissions 1980–2014
Source: Authors' elaboration based on EIA, *Brazil: International Energy Data and Analysis* (Washington, DC: EIA, 2015).

Table 16.1 Carbon dioxide projection for Brazilian consumer sectors

Sector	*$MtCO_2eq$*		
	2014	*2020*	*2024*
Electrical	81	46	62
Energy	32	40	45
Residential	18	21	22
Commercial	2	2	3
Public	1	1	1
Agricultural	19	20	21
Transport	222	237	269
Industry	96	113	127
Fugitive emissions	18	22	27
Total	**489**	**502**	**577**

Source: MME, and EPE, *Plano Decenal de Expansao da Energia* (2024).

developed by the Brazilian technical institutions is based on the stationarity assumption regarding future climate change. In case of high climate change impact on hydropower and thermal electricity production, for instance, the installation of new fossil fueled thermal generation capacity might be required to compensate for the losses. This aspect could partially modify the energy mix and, consequently, the emission scenarios projected in the 10-year energy plan.

Conclusions

This chapter describes the complex dynamics characterizing the Brazilian energy sector. Many critical points have been underlined under the perspectives of energy security, economic competitiveness and climate security. Brazilian institutions are managing the challenges presented by fast economic growth making huge changes in the country without compromising its social and environmental capital. The process of development has been socially inclusive and the participation of individuals in public decisions has been higher than in many other developing and developed countries in the world. Not all the strategies and actions implemented in the past two or three decades have been a full success, but definitely the whole story could be depicted as a successful one. In a few decades, Brazil became one of the top economies in the world. In less than 10 years, over 26 million people were lifted out of poverty. The income of the poorest part grew at a rate double that of the rest of the population. Indicators about education, health, infant mortality and nutrition show clear improvements even in the poorest part of the country.[66] Access to electricity, clean water and sanitation is ensured for almost the whole population, even in some of the remotest parts of the vast territory.

The challenges for Brazil are not over. In order to persist on this positive trend and maximize the socio-economic benefits deriving from it, Brazil is called on to rapidly adapt its technological, institutional, and infrastructural systems to the dynamic requirements of the globalized world. The energy sector, object of this study, is crucial in this process. The country's strengths in this regard are multiple: first, Brazil is blessed in terms of natural resources endowment; second, the country developed high competence on technical and strategic know how – the very detailed and technically exemplary planning of the energy sector is a good example of this; third, the mindset of Brazilian people is positively predisposed to the changes and to meet the challenges posed by rapid development.

In order to support the economic development and the energy security of the country, international and Brazilian technical institutions highlighted several needs, such as: increasing production of fossil fuels; increasing refining capacity; increasing electricity generation capacity; diversification of sources; enhancing the distribution infrastructures; and deeper integration of the energy system with neighboring countries. Brazilian institutions proved to have the technical and practical competence to strategically offer a solution to each of these needs. The main problems seem to be represented by the economic and financial aspects concerning the implementation of the energy strategies. The attractiveness of private capital is crucial for the timely and correct implementation of the actions shaped in the strategy. In this regard, the financial, economic, industrial and institutional sectors of the country have shown a certain level of reluctance in fully opening the national market to international competition. A certain level of protectionism is still present in the Brazilian economic structure, in particular regarding the strategic and economically crucial energy sector. A more competitive market could increase the efficiency of the energy sector in the country. Increasing competition could maximize the benefits for other economic sectors as, for instance, more affordable energy prices for the growing industrial sector of the country. Moreover, it could maximize the capital flows needed for the adaptation of the energy sector to the needs of the country. In this way, implementing the planned actions, energy security could be rapidly achieved.

Regarding the climate security of the future Brazilian energy system, the main uncertainties are linked to the impacts of climate change on the future energy production. The system heavily relies on natural and renewable sources. A larger than expected climate change impact could potentially impose changes in the planned future energy matrix and, therefore, increase the demand for traditional energy. A more accurate planning of the transport sector, the most energy intensive and largest source of GHG emissions, would be desirable for achieving climate security.

Notes

1 International Monetary Fund (IMF), *World Economic Outlook* (Washington, DC: IMF, 2013), www.imf.org/external/pubs/ft/weo/2015/01/pdf/text.pdf.
2 World Bank, “WDI – World Development Indicators Databank,” World Bank, http://databank.worldbank.org/data/home.aspx.
3 International Energy Agency (IEA), *World Energy Outlook 2013* (Paris: Organization for Economic Co-operation and Development, 2013).
4 IEA, *World Energy Outlook 2013*.
5 U.S. Energy Information Administration (EIA), *Brazil: International Energy Data and Analysis* (Washington, DC: EIA, 2015), www.eia.gov/beta/international/analysis.cfm?iso=BRA.
6 IEA, *World Energy Outlook 2015*.
7 Ministry of Energy (MME), and Energy Research Bureau (EPE), *Plano Decenal de Expansao da Energia (2024)* (Rio de Janeiro, Brazil: MME, and EPE, 2015), www.epe.gov.br/Estudos/Documents/PDE2024.pdf.
8 IEA, *World Energy Outlook 2013*; MME, and EPE, *Plano Decenal de Expansao da Energia (2024)*; Energy Research Bureau (EPE), and Ministry of Energy (MME), *Plano Decenal de Expansao da Energia (2022)* (EPE, and MME, 2013), www.epe.gov.br/PDEE/20140124_1.pdf.
9 MME, and EPE, *Plano Decenal de Expansao da Energia (2024)*; Ministry of Energy (MME), and Energy Research Bureau (EPE), *Plano Nacional de Energia 2030* (Brasilia, Brazil: MME, and EPE, 2007).
10 IEA, *World Energy Outlook 2013*; Ministry of Energy (MME), and Energy Research Bureau (EPE), *Balanco Energetico Nacional* (MME, and EPE, 2015), https://ben.epe.gov.br/downloads/Relatorio_Final_BEN_2015.pdf.
11 IEA, *World Energy Outlook 2013*.
12 IEA, *World Energy Outlook 2013*; MME, and EPE, *Balanco Energetico Nacional*.
13 IEA, *World Energy Outlook 2013*.
14 www.osul.com.br/brasil-vai-importar-energia-eletrica-do-uruguai-e-da-argentina-para-evitar-racionamento/.
15 MME, and EPE, *Plano Decenal de Expansao da Energia (2024)*.
16 IEA, *World Energy Outlook 2013*.
17 A. Livino, “Water, the Energy Sector and Climate Change in Brazil,” *ReVista Harvard Review of Latin America* (2013): 32–34.
18 http://npsglobal.org/eng/component/content/article/106-latin-america-and-caribbean/374-brazil-and-argentinas-nuclear-cooperation.html.
19 http://g1.globo.com/economia/noticia/2014/02/brasil-registra-181-apagoes-desde-2011-diz-levantamento.html.
20 http://g1.globo.com/economia/noticia/2015/01/pais-precisa-reduzir-gasto-de-energia-com-urgencia-dizem-especialistas.html.
21 IEA, *World Energy Outlook 2013*; MME, and EPE, *Plano Decenal de Expansao da Energia (2024)*; REN21, *Renewables 2013 Global Status Report* (Paris: REN21, 2013), www.ren21.net/Portals/0/documents/Resources/GSR/2013/GSR2013_lowres.pdf.
22 IEA, *World Energy Outlook 2013*.
23 EPE, and MME, *Plano Decenal de Expansao da Energia (2022)*.
24 IEA, *World Energy Outlook 2013*.
25 E. von Sperling, “Hydropower in Brazil: Overview of Positive and Negative Environmental Aspects,” *Energy Procedia* 18 (2012): 110–118, http://linkinghub.elsevier.com/retrieve/pii/S187661021200793X.
26 P. Mukheibir, “Potential Consequences of Projected Climate Change Impacts on Hydroelectricity Generation,” *Climatic Change* 121, no. 1 (2013): 67–78.
27 IEA, *World Energy Outlook 2013*.
28 Ibid.
29 Secretariat for Strategic Affairs, *Brasil 2040* (Brasilia, Brazil: Secretariat for Strategic Affairs, 2015), www.sae.gov.br/wp-content/uploads/BRASIL-2040-Resumo-Executivo.pdf.
30 Ibid., 47.
31 IEA, *World Energy Outlook 2013*; MME, and EPE, *Plano Decenal de Expansao da Energia (2024)*; MME, and EPE, *Plano Nacional de Energia 2030*.
32 EIA, *Brazil: International Energy Data and Analysis* (US Energy Information Administration, 2015), www.eia.gov/beta/international/analysis.cfm?iso=BRA.

33 Agency for Oil and Gas (ANP), "Dados Estatisticos" (Agency for Oil and Gas, 2015), www.anp.gov.br/?pg=64555%26m=%26t1=%26t2=%26t3=%26t4=%E5%3C8%26cachebust=1408326992231.
34 EIA, *Brazil: International Energy Data and Analysis.*
35 Ibid.
36 EIA, *Brazil: International Energy Data and Analysis*; ANP, "Dados Estatisticos."
37 Agency for Oil and Gas, www.anp.gov.br/?pg=42906.
38 I. Gomes, *Brazil: Country of the Future or Has its Time Come for Natural Gas?* (OIES paper NG 88) (Oxford Institute for Energy Studies, 2014), www.oxfordenergy.org/wpcms/wp-content/uploads/2014/07/NG-88.pdf.
39 EIA, *Brazil: International Energy Data and Analysis.*
40 http://insights.som.yale.edu/insights/how-does-changing-price-oil-affect-economies-around-world.
41 MME, and EPE, *Plano Decenal de Expansao da Energia (2024)*; MME, and EPE, *Plano Nacional de Energia 2030.*
42 IEA, *World Energy Outlook 2013.*
43 Ibid.
44 Ibid.
45 https://ycharts.com/companies/PBR/market_cap.
46 EIA, *Brazil: International Energy Data and Analysis.*
47 Ibid.
48 http://neweconomicperspectives.org/2015/09/credit-rating-agencies-and-brazil-why-the-sps-rating-about-brazil-sovereign-debt-is-nonsense.html.
49 IEA, *World Energy Outlook 2013*; EIA, *Brazil: International Energy Data and Analysis*; ANP, "Dados Estatisticos."
50 IEA, *World Energy Outlook 2013*; EIA, *Brazil: International Energy Data and Analysis.*
51 IEA, *World Energy Outlook 2013*; Gomes, *Brazil: Country of the Future.*
52 World Bank, *Environmental Licensing for Hydroelectric Projects in Brazil: A Contribution to the Debate* (Washington, DC: World Bank, 2008); M. M. Neves, *Difficulties in Expanding Hydropower Generation in Brazil* (Washington, DC: The George Washington University, 2009), www.aneel.gov.br/biblioteca/trabalhos/trabalhos/Artigo_Mateus_Neves.pdf; A. C. Brown, "Concessions, Markets and Public Policy in the Brazilian Power Sector," *The Electricity Journal* 25, no. 9 (2012): 67–81, www.ksg.harvard.edu/hepg/Papers/2012/Concessions and Public PolicyRevised %282%29 %282nd revision%29.pdf.
53 World Bank, *Environmental Licensing for Hydroelectric Projects in Brazil.*
54 J. Tollefson, "Brazil Reports Sharp Drop in Greenhouse Emissions," *Nature*, 2013, www.nature.com/doifinder/10.1038/nature.2013.13121.
55 IEA, *World Energy Outlook 2013.*
56 B. Hamududu, and A. Killingtveit, "Assessing Climate Change Impacts on Global Hydropower," *Energies* 5, no. 12 (2012): 305–322,. www.mdpi.com/1996-1073/5/2/305/.
57 IEA, *World Energy Outlook 2013.*
58 S. C. Pryor, and R. J. Barthelmie, "Climate Change Impacts on Wind Energy: A Review," *Renewable and Sustainable Energy Reviews* 14, no. 1 (2010): 430–437, http://linkinghub.elsevier.com/retrieve/pii/S1364032109001713.
59 A. F. Pereira de Lucena, Szklo, S.A., Schaeffer, R. et al., "The Vulnerability of Wind Power to Climate Change in Brazil," *Renewable Energy* 35, no. 5 (2010): 904–912, http://linkinghub.elsevier.com/retrieve/pii/S0960148109004480.
60 de Lucena et al., "The Vulnerability of Wind Power."
61 E. D. Assad and H. S. Pinto, *Global Warming and Future Scenarios for Brazilian Agriculture* (Aquecimento Global e Cenários Futuros da Agricultura Brasileira (São Paulo: Embrapa Agropecuária and Cepagri/Unicamp, August 2008).
62 IEA, *World Energy Outlook 2013.*
63 United Nations Framework Convention on Climate Change, http://www4.unfccc.int/submissions/INDC/Published%20Documents/Brazil/1/BRAZIL%20iNDC%20english%20FINAL.pdf.
64 MME, and EPE, *Plano Decenal de Expansao da Energia (2024).*
65 Ibid.
66 World Bank, "WDI – World Development Indicators Databank."

17

Indonesia's energy trilemma

Peter Maslanka

Introduction

As one of the largest developing and emerging countries, with a population of roughly 250 million and Southeast Asia's biggest economy, Indonesia faces three dilemmas. Coined by the World Energy Council in 2020, the energy trilemma characterizes the balancing act Indonesia is subject to regarding its energy equity, energy security, and environmental sustainability.[1] Due to the country's energy demand growing 7% per year, it is becoming increasingly difficult to achieve and ensure energy access for Indonesia's people and economy, and reduce emissions.[2] Similar to other developing countries, rising energy consumption is following the increase in gross domestic product (GDP), and consequently Indonesia is struggling to meet its domestic energy demand.

The chapter examines Indonesia's energy trilemma. First, Indonesia's energy equity policy goals are discussed, from its energy subsidies to its plan for increasing its power grid capacity. Second, its energy security is broken down to its national energy plan, as well as fossil fuel and renewable energy opportunities. Third, the country's environmental sustainability is discussed, challenged by its increasing reliance on coal for power generation. Each section provides government policy, and concludes with constraints being met by each component in the energy trilemma. Fifth, the challenges facing Indonesia's energy sector are discussed. Sixth, the risks that exist to keeping Indonesia's policies of managing its energy trilemma from being successful are examined. As Indonesia faces its energy trilemma, it appears that the government is giving more weight to both energy equity and energy security, over environmental sustainability.

Energy equity

Overview

No doubt, Indonesia recognizes strengthening its energy equity as high priority. There are two parts to energy affordability for Indonesia: expanding the electricity grid for providing more Indonesians with access to electricity, and making energy accessible to Indonesians through the use of energy subsidies.

Government policy

Presently, Indonesia subsidizes diesel and electricity. In distorting market prices, these subsidies drain the government's budget and discourage energy conservation. When energy prices are high, the problem is exacerbated, and the government pays more money for these subsidies. For years subsidies have diverted funds away from the Indonesian government. In a significant policy decision on January 1, 2015, Indonesia eliminated petrol subsidies and put a cap on the diesel subsidy. However, an additional $1.4 bn in 2015 was still earmarked for the diesel subsidy,[3] taking money away from the government that could contribute to Indonesia's long neglected infrastructure. As the diesel subsidy still exists, it uses up funds that could be spent on energy capacity growth. Further, an electricity subsidy has kept Indonesian electricity prices below the market price, and diverted government funds to pay for this subsidy. This is reported to cost the government $6.2 bn in 2015, or 5% of the total government budget.[4] Energy subsidies distorting market prices for diesel and electricity weaken Indonesia's energy security and contribute to rising carbon emissions.

In fairness, Indonesia made great strides in reducing the burden of subsidies on the government's budget. Indonesia, similar to other countries, took advantage of the steep drop in oil prices in 2014 and removed its decade-old gasoline subsidy that was previously a constant drain to the government's budget. The subsidy, originating during the Suharto era after the world's first oil shock in the 1970s, nearly took up 13.5% of the budget, but saved the government $18 bn in 2015 after it was scrapped.[5] However, in 2015 the government was accused by some analysts of flip-flopping on its subsidy removal policy, as some months it did not set gasoline to market prices.[6]

Indonesia is also slowly decreasing the electricity subsidy, and making it available only to the most impoverished citizens. The government is planning to reduce its electricity subsidy from the 2015 budget to the 2016 budget by nearly two-thirds.[7]

As Indonesia's economy is rapidly growing, it is working to increase access to electricity for all of its people as well as industry. Indonesia targets complete electrification by 2020, and the power sector is the major driver to increased energy demand. From 2015 to 2019, electricity demand is expected to increase by an average of 8.7% per year. However, Indonesia is having difficulty creating new capacity as demand rises. As of late-2014, Indonesia's total grid capacity was at 51.62 gigawatts (GW).[8] State-owned Perusahaan Listrik Negara (PLN), Indonesia's sole energy provider, is tasked with meeting the country's plan to add 35 GW worth of new power stations by 2019, however private sector financing is necessary in order to meet this target. The makeup of the 35 GW will consist of 20 GW of coal, 13 GW of natural gas, and 3.7 GW of renewables (primarily hydropower and geothermal).[9] Ministry of Energy and Mineral Resources Director General of Electricity Jarmin says that the bulk of the new power capacity will be added by private firms, 30 GW, while PLN will add 5 GW. In total, the program is expected to cost $72.5 bn.[10]

Indonesian President Joko (Jokowi) Widodo stated that his administration plans to reach 7.0% GDP growth during his five-year term. Due to a prolonged commodities slump, slowing growth in China, and the U.S. interest rate increase, in 2015 the economy averaged GDP growth of 4.7%, its lowest level in six years.[11] Increasing economic growth is at the forefront of Indonesia's policy. Despite the country's economic slump in 2015, the Economist Intelligence Unit predicts that Indonesia's economy will grow by an average of 5.5% a year from 2016 to 2020.[12]

Indonesia seeks to continue raising the wealth of its people and move from a lower income to a middle-income country, which requires more energy. By 2020 Indonesia's GDP is expected to rise by 60%, to $8,200, which will result in more Indonesians buying expensive items that use energy such as automobiles and refrigerators. Presently, the government is following its Masterplan for Acceleration and Expansion of Indonesia's Economic Development (MP3EI) to become a developed country by 2025. In line with MP3EI, President Jokowi has reiterated the need for economic policy to focus on developing Indonesia into a production-oriented economy.[13] Indonesia targets the industrial sector to contribute 40% of GDP, up from a low of 23%. As such, increasing its GDP requires more energy capacity feeding into its grid.

Constraints

Indonesia's energy equity is constrained by its energy subsidies as well as the reliance on coal for its energy grid expansion. Despite Indonesia's goals of increasing its energy equity, costly diesel and electricity subsidies still remain in place. Even though the removal of the petrol subsidy was a huge boost to the government's budget, it is unclear if the government will always keep the price of petrol at market price, as oil prices constantly fluctuate. Indonesia's massive expansion of its grid capacity, reliant on coal, weakens its environmental sustainability. In all, Indonesia appears to be placing energy affordability high on its policy to-do list as it focuses on meeting the energy demand of its growing economy and population.

Energy security

Overview

Indonesia's energy security is in transition. Historically, Indonesia has been a net energy exporter. However, in the last decade, its energy supply and demand has changed. Discussed in this section are Indonesia's energy supply and demand, energy policy, and energy opportunities.

Demand

Indonesia's energy demand is rising rapidly simultaneously with its high GDP growth. As a result of increased energy demand, energy consumption is expected to rise by almost 30% by 2020.[14] Oil is integral to Indonesia's energy needs. Wood Mackenzie, an energy consulting firm, forecasts that Indonesia's oil demand will grow from 1.6 million barrels per day (bpd) in 2014 to 2.3 million bpd in 2030, largely driven by transportation use from its growing population. Indonesia's energy demand consumption is split as follows: residential sector (37%), industry (30.5%), transport (27.6%), and commercial services (4.9%).[15]

Supply

Growth in Indonesia's domestic energy demand causes its government to focus on strengthening its energy security. As a major fossil fuel producer and exporter, Indonesia is the world's largest coal exporter (the fourth-largest coal producer), the seventh largest liquefied natural gas (LNG) exporter (the tenth largest gas producer), and the largest exporter and producer of palm oil (used for biofuel) in the world. However, formerly a founding Member of the Organization of the Petroleum Exporting Countries (OPEC), Indonesia left OPEC in 2009 as a result of its shift to becoming a net oil importer in 2004. This shift is due simultaneously to Indonesia's production declining and its demand rising. Figure 17.1 shows the drastic decrease in oil production and rise in oil consumption. Its oil production has been in steady decline since the mid-1990s, and its oil production fields are dated; over 85% of Indonesia's oil production is from

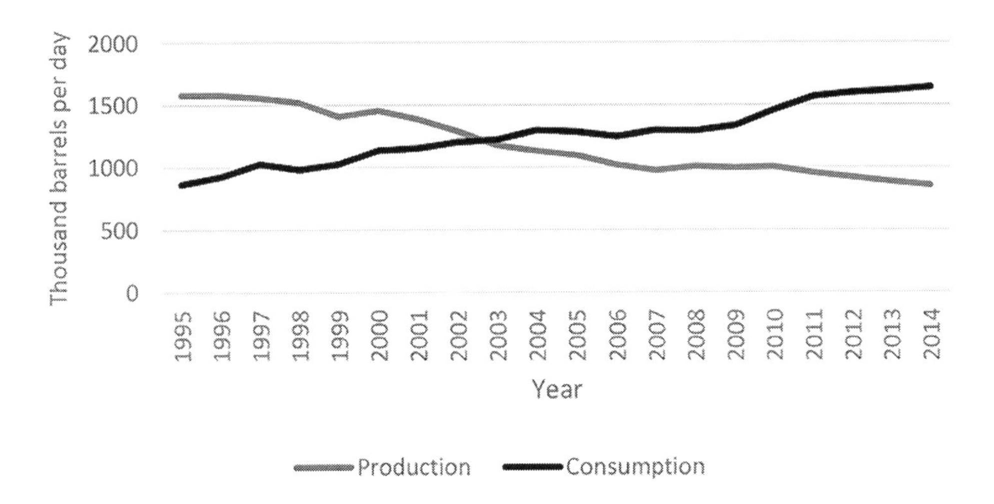

Figure 17.1 Indonesia's oil production vs. consumption
Source: BP, *Statistical Review of World Energy* (London: BP, 2015), http://www.bp.com/en/global/corporate/energy-economics/statistical-review-of-world-energy.html.

fields discovered before 1975. However, in December 2015 Indonesia rejoined OPEC, in order to strengthen its energy security from suppliers and to attract investment in its energy industry from OPEC member-countries.

Government policy

Following Parliament's approval of the National Energy Council's[16] new National Energy Plan (NEP14), Government Regulation No. 79/2014, the National Energy Council set Indonesia's most recent national energy policy in 2014. It is evident in the 2014 policy that the government strongly views energy as a scarce resource. This version replaced the 2006 National Energy Plan. The policy established four substantial policy changes reflecting a different energy security situation that the country was experiencing in 2006.

The first shift in policy is redirecting energy resources to the domestic market. As Indonesia historically has been an energy exporter to foreign markets across the globe, instead, it will redirect its hydrocarbons for domestic use in order to meet its energy needs. Second, the policy seeks to transition away from its reliance on oil and increase the use of gas, coal, and renewables. Figure 17.2 displays Indonesia's targeted energy diversification from 2012 to 2025, with oil dropping considerably and renewables significantly increasing in the share of its energy mix. In order for it to meet its target energy mix for 2025, it must increase its use of natural gas and coal by more than double, while renewable energy use must increase nine-fold.[17] NEP14 also established plans in place for energy emergency policy through the placement of energy emergency management structures and the buildup of oil buffer stocks. Finally, as Indonesia's electrification ratio currently is at 87.5%, it plans to achieve 100% electrification for its entire population by 2020.[18] By comparison, several of its regional neighbors – Singapore, Brunei Darussalam, Malaysia, Thailand, and Vietnam – have electrification ratios of 95% or higher.[19] However, achieving full electrification is a major challenge considering the geographic makeup of the country.

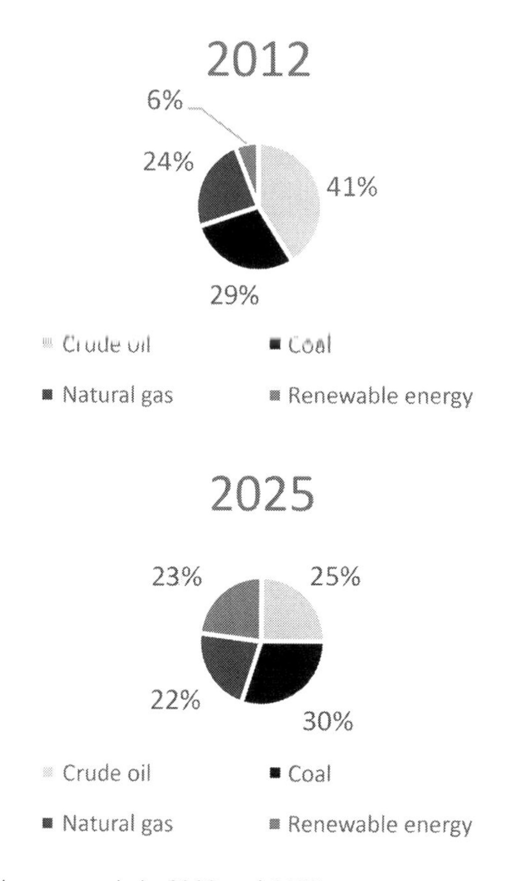

Figure 17.2 Indonesia's energy mix in 2012 and 2025
Source: International Energy Agency, 'National Energy Policy (Government Regulation No. 79/ 2014),' http://www.iea.org/policiesandmeasures/pams/indonesia/name-140164-en.php, 2014.

Fossil fuel opportunities

Indonesia's energy security is enhanced by its abundance of hydrocarbons. The country's fossil fuel opportunities include oil, gas, coal, and shale gas, as well as nuclear power. However, not all hydrocarbons are likely to have the same amount of use as others to meet Indonesia's energy needs. At current production rates, Indonesia's oil reserves are estimated at 23 years, natural gas at 59 years, and coal at 146 years.[20]

Currently Indonesia is in the process of transitioning from being an oil exporter to a net oil importer, as its oil market has been challenged with high consumption and low production. As of 2012 Indonesia had just 3.6 bn barrels of proven oil reserves. On the production side, its reserve replacement ratio stands at roughly 50%, and its production capacity suffers from lack of investment.[21] Although new project developments are limited, the Cepu Block is the only major new development, containing 600 million barrels in Central and East Java.[22]

While Indonesia's oil reserves are diminishing, its coal reserves are vast and likely to play a long-term role in its energy security. Coal has experienced strong production growth in the last decade. As displayed in Figure 17.3, in the last decade Indonesia's coal production has tripled and its consumption has doubled. Indonesia is the world's largest coal exporter, but seeks to

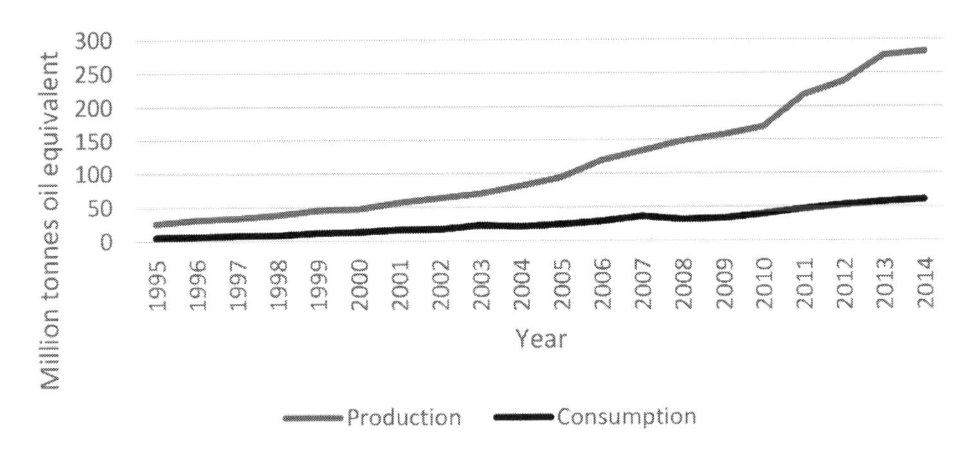

Figure 17.3 Indonesia's coal production versus coal consumption
Source: BP, *Statistical Review of World Energy* (London: BP, 2015).

redirect its reserves for domestic use. Coal is integral to achieving the government's goal for complete electrification across the archipelago.

Coal is an attractive option in accelerating Indonesia's large and rapid expansion of electricity production, as it is a domestic, readily available source. Power stations can be assembled quickly and inexpensively.[23] This is because coal is significantly cheaper for generating electricity in comparison to other energy sources, although its emissions are worse. Roughly 84.3% of extracted domestic coal is used for power generation.[24] To boost electricity generation, the government has discussed the possibility of a moratorium on the export of its coal. By comparison to oil, coal has experienced strong growth in the last decade. Coal is found throughout most of Indonesia's 34 provinces, with the most significant resources lying on the archipelago's two biggest islands, Sumatra and Kalimantan.

Although not as vast as Indonesia's coal reserves, its natural gas reserves can help meet the country's energy needs. With 103 Tcf of natural gas, Indonesia contains the world's 12th largest proven reserves.[25] However, in the past few years natural gas production has been in decline, while the country faces a shortage to meet its domestic supply. Poor infrastructure continues to make distribution for domestic consumption difficult. Its current production is found in East Kalimantan, Papua, and Sumatra, mostly offshore. Most of the reserves are found in low demand or remote areas. For example, Indonesia transports gas that has been liquefied from Papua to Lampung,[26] which is re-gasified prior to transferring the gas to customers.

In unconventional gas, Indonesia has abundant resources stemming from shale gas and coal-bed methane (CBM). The total reserves of shale gas and CMB are estimated at 570trn cubic feet and 450trn cubic feet, respectively.[27] The country hopes to develop its vast, untapped unconventional gas resources. However, regulatory issues and high cost compared to conventional gas must be overcome in order for these resources to boost Indonesia's energy supply. A report by the National Bureau of Asian Research (NBR) suggests that conventional gas will be the main source of Indonesia's new gas through 2030.[28] Still, the country has hopes to develop these resources in the long term.

Nuclear energy is tabled as a large, clean source of energy for the country to meet its future energy needs. Currently, the country operates no major nuclear reactors. On the other hand, there are three small-scale reactors: a 100-kilowatt reactor in Yogyakarta, a 250-kilowatt gdp reactor in Bandung and a 30-megawatt (MW) reactor in Serpong, in Banten. However, NEP14 states that Indonesia will only develop nuclear energy as a last resort.[29] This is largely due to nuclear energy having a negative stigma attached to it, as Indonesia is a seismically active country.

Indonesia's large area reduces its seismic density, but it is risk prone to a major off-shore earthquake damaging its inland, for example the 2004 Indian Ocean earthquake and tsunami off the west coast of Sumatra.[30] The Fukushima Daiichi nuclear disaster in 2011, primarily caused by a tsunami, is a regional example that raises widespread questions about the safety of nuclear energy. Previously, Indonesia planned to construct and operate four nuclear power plants, installed by 2025 that would supply 6 GW to its energy grid. However, this plan was cancelled.[31]

Renewable energy opportunities

The potential for renewable energy is vast in Indonesia and could provide a lot of energy to its grid. Yet, presently it does not utilize many of its renewable energy resources. Indonesia uses just 5% of its renewable capacity. NEP14 prioritizes renewable energy, as the government intends to have renewables account for 23% of its energy mix by 2025. Renewable energy sources include the following: geothermal, hydro, biomass, offshore wind, and solar (Table 17.1).

Geothermal stands out as Indonesia's best opportunity for utilizing its renewable energy resources. As an archipelagic state, it is situated on the Pacific Ring of Fire, and it contains an estimated 29,215 MW of electricity potentially generated from geothermal. Indonesia contains 40% of the entire world's geothermal energy potential, the most of any one country, but is the third largest producer.[32] However, currently Indonesia has installed just 1,341 MW of its total geothermal capacity, or roughly 5% of its geothermal potential. While most renewables supply an intermittent source of energy, geothermal is unique in that its flow of energy is constant. Indonesia's geothermal reserves are also strategically located near the highest demand areas, which includes Sumatra (13,800 MW), Java and Bali (9,250 MW), and Sulawesi (2,000 MW).[33]

Traditionally, geothermal exploration has been difficult due to legal barriers as 42% of the country's geothermal potential is located in forest conservation areas. Previous law considered geothermal exploration as a mining activity, while mining activities have been halted under a moratorium, impeding project development. However, the new geothermal law in 2014 removed geothermal exploration as a mining activity, easing future geothermal development.[34]

Table 17.1 Renewable energy: installed capacity vs. potential capacity

Source	*Installed capacity*	*Potential capacity*
Biomass	500 MW	49,810 MW
Geothermal	1,341 MW	29,215 MW
Hydro	6,850 MW	75,000 MW
Wave	0.001 MW	49,000 MW
Solar	22.4 MW	4.8 kWh/m2/day
Wind	1.87 MW	3–6 m/s

Source: Ministry of Energy and Mineral Resources (Indonesia), *2014 Handbook of Energy and Economic Statistics of Indonesia*, 2014.

Indonesia also has a large amount of potential for hydropower. With 75,000 MW of potential, this renewable source offers the largest potential for renewable energy. Like geothermal, Indonesia is far from using hydropower's full potential, as its total installed capacity is just 6,850 MW. A major contributing factor is because most of its hydropower potential is found on the less inhabited, outer-lying islands, in provinces such as Papua. However, as the International Energy Agency notes, the installed capacity could increase for medium and large-scale resources as Indonesia plans to develop industrial zones in Papua and other rural areas under MP3EI.[35]

Even more than geothermal and hydropower, Indonesia is using a minimal amount of its solar power potential. Its solar power potential is estimated at 4.8 $kWh/m^2/day$, while installed capacity is at just 22.4 MW. Solar power could be an alternative source of power on remote islands, which typically rely on costly, polluting diesel-fired generators, prone to power outages. As the archipelago is located near the equator with strong intensity from the sun, Indonesia can benefit from solar power to meet its growing demand.

Indonesia is well endowed with great biomass energy potential, which is estimated at 49,810 MW of possible capacity. Thus far, only 500 MW is being used. Indonesia's biomass potential comes from many sources, including palm, cassava, molasses, jatropha, curcas, nyamplung, and corn. Palm makes up the bulk of biomass, as Indonesia is the world's largest producer.

Of the renewable energy sources mentioned, Indonesia's potential for wind energy and wave energy is less researched, and its potential is suspected to be small. The wind in Indonesia is slow, at just three to six meters per second. Its installed capacity is just 1.87 MW, but more research is being done to see if there is any additional potential from wind that the government may not be aware of through possible installation of offshore turbines.

Instead, perhaps Indonesia could benefit from wave energy, as it has been estimated that 49,000 MW could be generated.[36] Further research is also being done to get a better grasp on Indonesia's wave energy potential. Since Indonesia has a large coastline with over 17,000 islands, wave energy appears as a great opportunity.

Constraints

Indonesia's energy security is being constrained by its depleting oil reserves and the expansion of its energy grid. Despite the removal of the petrol subsidy and low oil prices, Indonesia's oil demand is expected to continue to substantially increase due to a rising per capita income and robust economic growth. Oil demand is estimated to increase by nearly 50% from 2014 to 2030.[37] As Indonesia develops its energy grid, more coal will be directed towards domestic consumption. However, Indonesia's coal reserves are vast.

Environmental sustainability

Overview

From rising emissions due to increased energy, to skyrocketing greenhouse gas emissions due to Indonesia's annual haze, environmental sustainability is increasingly becoming an issue of concern to the Republic of Indonesia. As a country with the second largest biodiversity in the world after Brazil, and a coastline of 80,000 km made up of 17,000 islands, Indonesia has much at stake if global temperatures and sea levels continue to rise. Contributing to climate change, Indonesia is increasing its reliance on coal, as 60% of its new power generation will come from that energy source. Further, it continues to emit large amounts of CO_2 due to the burning of its forests and peatlands, creating a haze that is a regular irritant to its regional neighbors.

Exacerbated by Indonesia's worst-ever haze on record, in 2015 it overtook Japan as the world's fifth largest CO_2 emitter.[38] Not only did its emissions rise, the 2015 haze cost the country roughly $16 bn in economic losses, or 1.9% of its total GDP according to the World Bank.[39]

In the UN's report, *Climate Change and its Possible Security Implications Indonesia*, it cites five devastating results for the world's largest archipelago. As 60% of all Indonesians live in the low-lying coastal areas, its population is at risk from the submergence of its cities. This could lead to massive internal and external migration of Indonesians. Home to over half of Indonesia's population, Java in particular is already one of the world's most densely populated islands. The report found that temperature increases over 2.5°C would drop agricultural productivity and decrease incomes. Rice yields would drop leading to higher imports, and decrease agricultural revenue between 9% and 25%, directly hindering Indonesia's fight against poverty. A loss of biodiversity is a risk to Indonesia's agriculture, fishery, and forestry. Due to the possibility of shifting seasonality, climate change could lead to uncertain weather patterns, largely affecting water availability. Last, human health could be negatively impacted due to an increase in infectious diseases.[40]

Indonesia's emissions are rising fast due to fossil fuel use, and are expected to double within the next 25 years (Figure 17.4). The bulk of the emissions are projected to come from the power sector and transportation. In September 2015, the country set targets to reduce greenhouse gas emissions by 29% by 2030, and reduce emissions by a further 41%, if it receives $6 bn in international assistance. However, the World Resources Institute (WRI), a leading environmental think-tank doubts Indonesia's ambitions or reality to meet this goal, citing vagueness in the country's emissions reduction plan.[41]

Government policy

As the government considers climate change and other important environmental topics, Indonesia has taken steps to reduce climate change. The 2009 Environmental Law strengthened local governments, civil society, and non-governmental organizations that work to mitigate

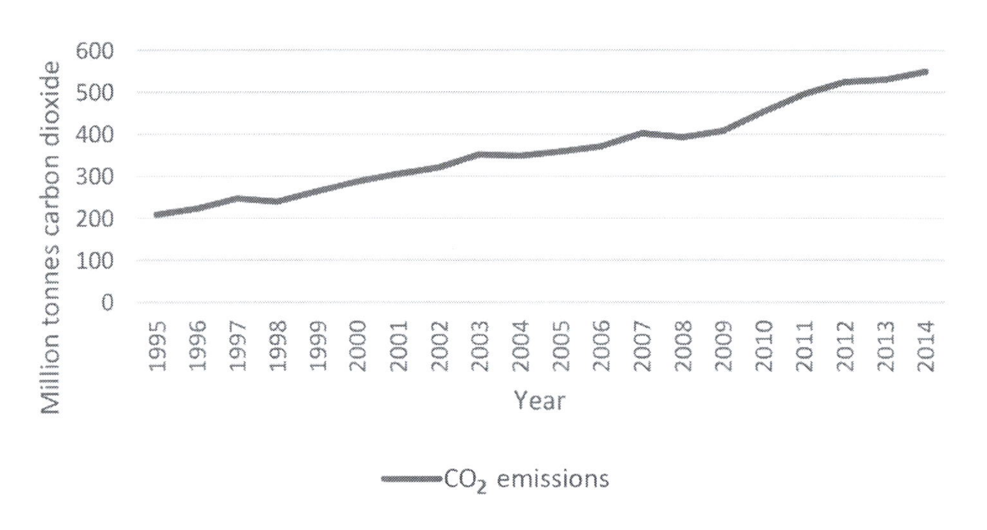

Figure 17.4 Indonesia's CO_2 emissions
Source: BP, *Statistical Review of World Energy*.

environmental pollution and damage. The law includes stricter penalties than previous laws, and imposes strict liability to companies that cause environmental damage from the use of hazardous material. Further the law requires that companies perform environmental audits as well as an Environmental Management statement, an Environmental Management Efforts-Environmental Monitoring Efforts Report or an Environmental Impact Assessment (AMDAL). Importantly, all mining companies must issue an AMDAL, carry out an environmental risk analysis, and obtain an environmental permit for mining operations.[42]

A few years after the passing of the 2009 Environmental Law, Indonesia introduced the presidential decree for NAPRGG. Presented by Indonesia's president at the time, Susilo Bambang Yudohoyono, NAPRGG set Indonesia's first GHG emission target at 26% unilaterally, and 41% if sufficient international assistance was provided, although the goal barring international assistance was increased to 29%, and extended until 2030. NAPRGG set the framework for all provinces to develop their own emissions-reduction plans. To keep development across the archipelago in line with reducing emissions, BAPPENAS was appointed as co-coordinator and issued guidelines at the national and sub-national level.[43]

To tackle emissions stemming from deforestation and forest degradation, Indonesia became a member country of the REDD and REDD+ programs. As most of Indonesia's GHG emissions are land- and forest-based, largely from the annual haze, these programs play a significant role in achieving its emissions reduction target. In order to help curb environmental damage in the country, in May 2011 the government enacted a two-year moratorium on new mining licenses for land that contains peatland and natural forest. As coal mining and palm oil contribute to Indonesia's emissions, its participation in these programs is an important step to reducing emissions from these two sources.[44]

Constraints

Due to rising energy demand and the country's high economic growth rate, Indonesia faces a dilemma in having the capacity to curb its emissions and protect its biodiversity. In strengthening its energy security, and simultaneously reducing emissions, the country will be able to achieve its policy and nearly quadruple its renewable energy supply by 2025. However, at 30% of Indonesia's energy mix targeted for coal 2025, the country appears likely to remain dependent on that source for energy.[45] Of the country's ambitious electricity capacity initiative to install 35 GW through 2019, 20 GW will be coal-powered.[46] In addition, carbon capture and storage (CCS) does not seem to be a viable solution. A study conducted by the World Bank found that CCS would reduce power output by almost one-third, and nearly double the price of electricity.[47] Therefore, while its targets for GHG reduction are ambitious, its reliance on coal for its energy mix seems counterintuitive.

Challenges

Despite Indonesia's many opportunities for a robust energy security, the government faces many challenges to implementing its desired policy. These include corruption, decentralization, lack of coordination between government agencies, a difficult investment climate, and infrastructure.

Corruption

Indonesia's endemic corruption is a constant hindrance to its energy security. The country's Energy and Mineral Resource Ministry and Pertamina, the state-owned company active in the upstream and downstream oil, gas, and geothermal energy sectors, have a history of misusing state funds. High-profile government officials from the previous administration, former Energy and Mineral Resource Minister Jero Wacik, and former head of Indonesia House of Representatives Commission VII on energy, Sutan Bhatoegana, are on currently on trial after being named graft suspects by the Corruption Eradication Commission.[48] However, reducing corruption is at the forefront of President Jokowi's policy agenda.

In an effort to clean up Indonesia's "oil and gas mafia," Jokowi replaced all of the directors in the Energy and Mineral Resource Ministry as well as Pertamina in late-2014. Further, an oil and gas reform team was established – consisting of academics, anti-corruption proponents, and government officials – providing a number of recommendations for cleaning up Pertamina and the oil and gas industry at large. Indeed, Transparency International's 2015 Corruption Perceptions Index (CPI) ranked Indonesia 88 from a total of 168 countries – an improvement of 19 places up from 107 in the 2014 CPI.[49] Still, corruption remains a problem entrenched in the country.

Decentralization

The decentralization of Indonesia's government further challenges its energy security trilemma in making policy synchronization between the national and local governments difficult. A product of the country's transition to democracy, in 2001 it transferred decision-making power and budgetary resources away from the central government, to provinces, regencies, cities, and villages.[50] As a result, decentralization has provided local governments special rights and responsibilities related to its economic, energy, and climate policies. Local government grants land rights, as well as issuing concessions permits, and licenses for coal mining and renewable energy projects, among others. For example, regional governments receive 15% of net revenues from oil and 30% net revenues from gas.[51] Decentralization has increased the number of stakeholders involved in policy-making, decreasing government efficiency.

Lack of coordination between government agencies

Complicating Indonesia's ability to execute uniform policy for its economic, energy, and environmental needs is the lack of coordination amongst government agencies. At the cabinet level, there are also concerns arising from a lack of trust. For example, in August 2015 Vice President Jusuf Kalla and Coordinating Maritime Affairs Minister Rizal Ramli were disagreeing over Indonesia's energy policy to add 35 GW of new electricity capacity by 2020. Following a cabinet reshuffle in August 2015, on Rizal's second day on the job he commented that Indonesia's goal was unrealistic.

Difficult investment climate

Indonesia lags behind its regional neighbors in the ability to conduct business inside the country, negatively impacting its ability to attract investment for energy production and power generation. The World Bank's 2016 Ease of Doing Business Index ranked Indonesia 109. In comparison, other countries in Southeast Asia ranked as follows: Singapore (1), Malaysia (18), Thailand (49), Vietnam (90), and the Philippines (103).[52] This index considers factors such as starting a business, dealing with construction permits, and getting electricity, among others.

Following economic headwinds and the difficulty conducting business inside the country, the government enacted policies in order to boost its economy by attracting investment and create a more business-friendly climate. The Indonesia Investment Coordinating Board established a one-stop service in order to speed up the amount of time for investors to get business licenses. Further, in late-2015 to early-2016 Indonesia released a series of targeted stimulus packages with the aim of boosting its economy and liberalizing it to foreign investment. The economic stimulus package targeted increasing the country's industrial competitiveness, accelerating electrification, cutting red tape, and lowering fuel prices, amongst others.[53]

Infrastructure

Indonesia's energy mix goals are weakened by its lack of infrastructure investment. The large expansion of renewable energy is reliant on investment in the transmission grid, as well as creating new power capacity from geothermal, solar, and wind.

Due to a lack of gas infrastructure the Energy and Mineral Resources Ministry's oil and gas directorate general estimates that it needs $32.42 bn[54] to finance the necessary investment. As of late-2014, the country has a total of 12,034 km of gas pipeline dedicated to open access pipeline, upstream pipeline, downstream pipeline, and private-use pipeline. By 2025 Indonesia plans to increase its total pipeline amount to 27,273 km.[55]

As a result, improving Indonesia's infrastructure is one of the Jokowi administration's key policy goals. Driven by a cut in the fuel subsidies, more capital is being allocated towards infrastructure for power plants, roads, ports, and dams. Reflecting the surge in infrastructure spending, Indonesia's infrastructure budget grew from 9.5% in the 2014 budget to 15% in the 2016 budget.[56]

Risks[57]

Compounding Indonesia's energy trilemma are risks, which include supply shocks that can increase the price of energy such as natural disasters geopolitics, and piracy. Additional risks include adverse global market/pricing conditions, and electricity outages. Natural disasters present perhaps the biggest risk to Indonesia's energy security.

Natural disasters

Due to its location, Indonesia is prone to natural disasters. As Indonesia is situated on the ring of fire, this presents the potential for disruptions to its energy security from volcanic eruptions, floods, earthquakes, and tsunamis. The most recent natural disaster was the 2004 tsunami that hit Indonesia's northern-most Aceh province. It killed 200,000 Indonesians and caused $4.4 bn in damage.[58] Floods regularly occur during the rainy season, from December to March. In the capital, Jakarta regularly floods every year, which cost roughly $380m in losses in 2014.[59] This is particularly disruptive to business and energy development because Jakarta is the heart of Indonesia's business and economic activity.

Geopolitics

Indonesia is also prone to geopolitics affecting its energy sector from the South China Sea disputes as well as turmoil in petro-states. China claims virtually all of the South China Sea as its own, becoming a Chinese lake. Indonesia's Natuna gas field, located in the Natuna waters from its Natuna island chain is at risk from Chinese aggression. It is one of the largest recoverable undeveloped gas fields in the world, at 46 Tcf, or 40% of Indonesia's total natural gas reserves.

The Natuna gas field is located within Indonesia's exclusive economic zone (EEZ). However, it overlaps with China's nine-dashed line map.[60] Tensions are rising in this area as China is increasingly assertive in the SCS, from placing an oil rig in waters claimed by Vietnam, to island building in waters contested with the Philippines. These tensions in the South China Sea could disrupt Indonesia's energy production and transportation.

Geopolitics far from Indonesia's shores could disrupt its oil imports, at a critical time when its oil consumption continues to grow. In 2014 Indonesia imported roughly one-third of its oil from the Middle East, and 18% from Nigeria.[61] The country receives the majority of its oil imports from petro-states located in volatile areas of the world.

Piracy

Indonesia is also prone to piracy along the Malacca Strait. Providing passage to one-third of the world's shipping, it is one of the major chokepoints for oil. Japan, Taiwan, and South Korea receive 75% of their oil imports from Africa and the Middle East from this strait, while China receives 37% of its total demand.[62] Pirates tend to target oil and palm oil tankers, which could take away from government revenues. In 2005 piracy was so high that Lloyd's Market Association of London designated the Malacca Strait a "war risk zone," although the situation improved a few years later. However, since 2014 there has been a dramatic rise in attacks, with Southeast Asia overtaking Somalia as the world's piracy hub. In 2014 there were 107 attacks and attempted piracy attacks – a 700% increase in five years.[63] Indonesia has limited funds and a large coastline; guarding against piracy in this strait is challenging. Piracy could disrupt its energy imports and exports.

Adverse global market/pricing conditions

Adverse global market/pricing conditions could also affect Indonesia. Oil and gas prices, currently at historical lows, do not necessarily help Indonesia's economy because it reduces the government's potential profits from oil and gas exports. Due to historically low global oil prices from a global oil supply glut, Pertamina's oil revenue decreased by 20% in 2014 compared to 2013, dropping from $3.06 bn to $2.4 bn, which lowers Indonesia's government expenditures as well. Due to a fall in other commodities, China's slowing economic growth has decreased government revenues on coal and palm oil. As Indonesia still heavily relies on commodities for its government's finances, and is one of the world's top exporters of these commodities, this weakens the government's development of its energy sector.

Conclusion

Indonesia appears to emphasize its energy equity and energy security over environmental sustainability. The difficulty in managing its energy equity, energy security, and environmental sustainability is not unusual for one of the world's largest developing countries. The country is in the process of substantially expanding its energy grid capacity, in order to meet its growing energy demand due to its fast expanding economy and growing population. However, this expansion is being done largely by exploiting coal-fired power plants, adding to its carbon emissions. Yet, Indonesia must meet the immediate needs of its population.

Notes

1 "Energy trilemma" refers to a country's difficulty in balancing energy security, i.e. managing energy supply and the ability to meet current and future demand; energy equity, i.e. the accessibility and affordability of energy across a populace; and environmental sustainability (supply- and demand-side energy efficiencies and development from renewable and low-carbon sources).

2 Pradeep Tharakan, *Summary of Indonesia's Energy Sector Assessment*, Asian Development Bank, 2015, www.adb.org/sites/default/files/publication/178039/ino-paper-09-2015.pdf, 5.

3 Reuters, "Indonesia to scrap petrol subsidy and float prices," *Financial Times*, December 31, 2014, www.ft.com/intl/cms/s/0/4c00ccd8-90ba-11e4-8134-00144feabdc0.html#axzz4AjiphdXN.

4 International Institute for Sustainable Development, *Indonesia Energy Subsidy Briefing* (September 2014), www.iisd.org/gsi/sites/default/files/ffs_newsbriefing_indonesia_sep2014_eng.pdf.

5 Sharon Chen, and Herdaru Purnomo, "Indonesia Doubles Transport Budget on $18 Billion Fuel Gain," *Bloomberg Business*, January 5, 2015, www.bloomberg.com/news/articles/2015-01-05/indonesia-doubling-transport-budget-with-10-billion-fuel-saving.

6 Ben Otto, "Indonesia's Widodo Stalls Plan to Raise Fuel Prices," *The Wall Street Journal*, May 15, 2015, www.wsj.com/articles/indonesias-widodo-stalls-plan-to-raise-fuel-prices-1431676582.

7 Cahyafitri, Raras, "Govt to give electricity subsidies only to poor," *Jakarta Post*, September 21, 2015, www.thejakartapost.com/news/2015/09/21/govt-give-electricity-subsidies-only-poor.html (accessed on February 14, 2016).

8 Perusahaan Listrik Negara, *Annual Report 2014*, Perusahaan Listrik Negara, 2014, www.pln.co.id/eng/?p=55, 16.

9 Jong, Hans Nicholas, "Govt told to shif to renewable energy as coal prices fall," *Jakarta Post*, 10 May 2016, www.thejakartapost.com/news/2016/05/10/govt-told-shift-renewable-energy-coal-prices-fall.html (accessed on June 5, 2016).

10 Jarman, "Interview: Indonesia's Electricity Objectives," *The Oil & Gas Year*, October 21, 2015, www.theoilandgasyear.com/interviews/indonesias-electricity-objectives/.

11 Saifhulbahri Ismail, "Indonesia's Economy Expected to Grow 4.7% in 2015: World Bank," December 15, 2015, www.channelnewsasia.com/news/business/indonesia-s-economy/2348248.html.

12 Economist Intelligence Unit, "Indonesia," 2016, Economist Intelligence Unit, http://country.eiu.com/indonesia.

13 Albert Nonto, "A New Road Map for Industrialization in Indonesia," *Jakarta Globe*, August 10, 2015, http://jakartaglobe.beritasatu.com/archive/new-road-map-industrialization-indonesia/.

14 *Economist* Intelligence Unit, "Indonesia," 5.

15 International Energy Agency, 'Indonesia 2015,' www.iea.org/bookshop/704-Indonesia_2015 (accessed on January 18, 2016), 2015.

16 Established in 2007, the National Energy Council sets the government's national energy plan. It requires Parliament's approval.

17 Natalie Bravo, Clara Gillispie, Mikkal E. Herberg, Hanan Nugroho, Alexandra Stuart and Nikos Tsafos. National Bureau of Asian Research, "Indonesia: A Regional Energy Leader in Transition," www.nbr.org/publications/issue.aspx?id=326 (accessed on January 18, 2016), December 2015.

18 Kementerian Koordinator Bidang Perekonomian (Coordinating Ministry for Economic Affairs), "Paket Kebijakan Ekonomi IX," January 27, 2016, www.ekon.go.id/ekliping/view/siaran-pers-paket-kebijakan.1960.html.

19 Ayomi Amindoni, "AGO Forms Team to Monitor Power Project," *Jakarta Post*, January 7, 2016, www.thejakartapost.com/news/2016/01/07/ago-forms-team-monitor-power-project.html.

20 Tharakan, *Summary of Indonesia's Energy Sector Assessment*, 7.

21 *Economist* Intelligence Unit, "Indonesia," 4.

22 International Energy Agency, "Indonesia 2015," www.iea.org/bookshop/704-Indonesia_2015 (accessed on January 18, 2016), 2015.

23 Neil Gunningham, *Managing the Energy Trilemma: The Case of Indonesia*, Australian National University, 2010, http://papers.ssrn.com/sol3/papers.cfm?abstract_id=2342925, 13.

24 International Energy Agency, "Indonesia 2015," www.iea.org/bookshop/704-Indonesia_2015 (accessed on January 18, 2016), 2015.

25 Energy Information Agency, "Indonesia Analysis," www.eia.gov/beta/international/analysis.cfm?iso=IDN (accessed on January 19, 2016), October 7, 2015.

26 Papua is Indonesia's Eastern-most province. Lampung is a province located on the southern-most tip of Sumatra, near West Java.
27 *Economist* Intelligence Unit, "Indonesia," 11.
28 Natalie Bravo, Clara Gillispie, Mikkal E. Herberg, et al., *Indonesia: A Regional Energy Leader in Transition*, National Bureau of Asian Research, 2015, www.nbr.org/publications/issue.aspx?id=326m, 43.
29 International Energy Agency, "Indonesia 2015," www.iea.org/bookshop/704-Indonesia_2015 (accessed on January 18, 2016), 2015, p. 25.
30 Peter D. Hendrickson, and Sharon Squassoni, *Earthquakes and Nuclear Power*, Center for Strategic Studies, 2011, http://csis.org/blog/earthquakes-and-nuclear-power.
31 Rangga Prakoso, "Indonesia Vows No Nuclear Power Until 2050," *Jakarta Globe*, December 12, 2015, http://jakartaglobe.beritasatu.com/business/indonesia-vows-no-nuclear-power-2050/.
32 International Energy Agency, "Indonesia 2015," www.iea.org/bookshop/704-Indonesia_2015 (accessed on January 18, 2016), 2015.
33 International Energy Agency, "Indonesia 2015," www.iea.org/bookshop/704-Indonesia_2015 (accessed on January 18, 2016), 2015.
34 Raras Cahyafitri, "Legal Barrier to Geothermal Development Removed," *Jakarta Post*, August 24, 2014, www.thejakartapost.com/news/2014/08/27/legal-barrier-geothermaldevelopment-removed.html.
35 International Energy Agency, "Indonesia 2015," www.iea.org/bookshop/704-Indonesia_2015 (accessed on January 18, 2016), 2015.
36 *Economist* Intelligence Unit, "Indonesia," 28.
37 "Indonesia Energy Market Outlook 2015–Oil," *Wood Mackenzie* (January 2015): 1–18, www.woodmac.com/reports/energy-markets-indonesia-energy-markets-outlook-2015-oil-26682777.
38 "Indonesia's Forest-fire Haze," *The Economist*, November 6, 2015, www.economist.com/blogs/graphicdetail/2015/11/daily-chart-3.
39 Herdaru Purnomo, "Haze Crisis Cost Indonesia Almost 2% of GDP, World Bank," *Bloomberg Business*, January 20, 2016, www.bloomberg.com/news/articles/2016-01-20/haze-crisis-cost-indonesia-almost-2-of-gdp-world-bank-says.
40 United Nations, "Climate Change and its Possible Security Implications Indonesia," https://sustainabledevelopment.un.org/content/dsd/resources/res_pdfs/ga-64/cc-inputs/Indonesia_CCIS.pdf (accessed on January 24, 2016), year unknown.
41 Suzanne Goldenberg, "Indonesia to Cut Carbon Emissions by 29% by 2030," *The Guardian*, September 24, 2016, www.theguardian.com/environment/2015/sep/21/indonesia-promises-to-cut-carbon-emissions-by-29-by-2030.
42 International Energy Agency, "Indonesia 2015," www.iea.org/bookshop/704-Indonesia_2015 (accessed on January 18, 2016), 2015.
43 Ministry of Energy and Mineral Resources (Indonesia), "2014 Handbook of Energy and Economic Statistics of Indonesia," 2014. Ministry of National Development Planning/National Development Agency, "Guideline for Implementing Green House Gas Emission Reduction Action Plan," www.google.com/url?sa=t&rct=j&q=&esrc=s&source=web&cd=2&ved=0ahUKEwia3aWZnLfKAhXFQSYKHbZSCZwQFggrMAE&url=http%3A%2F%2Franradgrk.bappenas.go.id%2Frangrk%2Fimages%2Fdocuments%2FBuku_Pedoman_Pelaksanaan_Rencana_Aksi_Penurunan_Emisi_GRK_English.pdf&us=AFQjCNHbIpimGFY4K-jGNkZX9U5-HcKY0w&sig2=u1E1f9pLfLqRdIou4MLeqA&cad=rja (accessed on January 19, 2016), 2011.
44 International Energy Agency, "Indonesia 2015," www.iea.org/bookshop/704-Indonesia_2015 (accessed on January 18, 2016), 2015.
45 International Energy Agency, "National Energy Policy (Government Regulation No. 79/2014)," www.iea.org/policiesandmeasures/pams/indonesia/name-140164-en.php (accessed on June 13, 2016), 2014.
46 Energy Information Agency, "Indonesia Analysis," www.eia.gov/beta/international/analysis.cfm?iso=IDN (accessed on January 19, 2016), October 7, 2015.
47 Masaki Takahashi, *The Indonesia Carbon Capture Storage (CCS) Capacity Building Program: CCS for Coal-fired Power Plants in Indonesia*, World Bank, 2015, http://documents.worldbank.org/curated/en/2015/06/24810099/indonesia-indonesia-carbon-capture-storage-ccs-capacity-building-program-ccs-coal-fired-power-plants-indonesia, 2–3.
48 Hans Nicholas Jong, "Sutan Gets 10 Years, But No Link to Jero," *Jakarta Post*, August 20, 2016, www.thejakartapost.com/news/2015/08/20/sutan-gets-10-years-no-link-jero.html.

49 Transparency International, *Corruption Perceptions Index*, Transparency International, 2015, www.transparency.org/cpi2015.
50 Decentralization divided Indonesia into 34 provinces; 99 cities; 410 regencies; 6,543 districts; and 75,244 villages.
51 International Energy Agency, "Indonesia 2015," www.iea.org/bookshop/704-Indonesia_2015 (accessed on January 18, 2016), 2015.
52 World Bank, "Ease of Doing Business Rankings," 2016, World Bank, www.doingbusiness.org/rankings.
53 Indonesia Investments, "Indonesia Unveils Seventh Stimulus Package," Indonesia Investments, December 4, 2015, www.indonesia-investments.com/news/todays-headlines/indonesia-unveils-seventh-economic-stimulus-package/item6257.
54 The total breakdown is as follows: $13 bn gas station, $8.5 bn pipeline, $8 bn liquefaction and regasification, $2.5 bn city gas, and LPG $0.42 bn.
55 Cahyafitri, Raras, "Indonesia needs $32.42 billion for gas infrastructure," *Jakarta Post*, www.thejakartapost.com/news/2015/09/16/indonesia-needs-3242-billion-gas-infrastructure.html (accessed on January 19, 2016), September 16, 2015.
56 Siwage Dharma Negara, "Indonesia's 2016 Budget," Institute of Southeast Asian Studies, 2016, 4, www.iseas.edu.sg/images/pdf/ISEAS_Perspective_2016_3.pdf (accessed February 14, 2016) January 19, 2016.
57 This section draws upon Peter Maslanka, "Securing Indonesia's Energy Future," *Journal of Energy Security* (June 23, 2015), www.ensec.org/index.php?option=com_content&view=article&id=577:securing-indonesias-energy-future-&catid=126:kr&Itemid=395.
58 Paul Brown, "Tsunami Cost Aceh a Generation and $4.4bn," *The Guardian*, January 21, 2015, www.theguardian.com/world/2005/jan/22/tsunami2004.internationalaidanddevelopment.
59 Cory Elyda, "BNPB Wait for Governor to Act on Flooding," *Jakarta Post*, November 19, 2016, www.thejakartapost.com/news/2014/11/29/bnpb-wait-governor-act-flooding.html.
60 Clive Schofield, and Ian Storey, *The South China Sea Dispute: Increasing Stakes and Rising Tensions*, Jamestown Foundation, November 2009, www.academia.edu/2062121/The_South_China_Sea_Dispute_Rising_Tensions_Increasing_Stakes, 38.
61 Energy Information Agency, "Indonesia Analysis," www.eia.gov/beta/international/analysis.cfm?iso=IDN (accessed on January 19, 2016), October 7, 2015.
62 Jacob Pedersen, "China Leads Peers in Resolving Malacca Energy Shipping Dilemma," *The Wall Street Journal Blog*, May 13, 2013, http://blogs.wsj.com/indonesiarealtime/2013/05/13/china-leads-peers-in-resolving-malacca-energy-shipping-dilemma/?mg=blogs-.
63 Patrick Winn, "Strait of Malacca is World's New Piracy Hotspot," *NBC News*, March 27, 2014, www.nbcnews.com/news/world/strait-malacca-worlds-new-piracy-hotspot-n63576.

18

Egypt

The challenge of squaring the energy–environment–growth triangle

Robert Springborg

Egypt's struggle to square the energy–environment–growth triangle is important in its own right and for its broader implications. The "echo effect," whereby an increase in fertility rates combined with an unprecedented number of women being of childbearing age, is driving a renewed, rapid population increase. Whereas 2.1 million children were born in 2008, four years later 2.6 million were born.[1] At projected growth rates, Egypt by 2050, having as many as 140 million citizens, will be more populous than either Russia or Japan.[2] Egypt accounts for more than one-fifth of all oil and two-fifths of all gas consumed in Africa, making it the continent's largest fossil fuel consumer in 2013.[3] While Egypt's per capita carbon dioxide emissions are well below the average of OECD countries, they are high by the standards of lower middle income countries and rising at one of the world's fastest rates, doubling between 1990 and 2010.[4] Egypt's energy efficiency of production has not improved over the past two decades, making it one of the world's least energy efficient producers.[5] Having been in 2005 the world's ninth largest exporter of natural gas, earning more than half its export income from fossil fuels, its exports and earnings plummeted after 2010, rendering the country a net fuel importer from 2013.[6] An increasing natural gas shortage has been the major contributor to ever more frequent electric power outages throughout the country, further slowing production in the energy intensive fertilizer, cement, steel, and ceramics industries, while disrupting the daily lives of virtually the entire population. By early 2015 steel factories were operating at about 40% of normal capacity and fertilizer production had dropped by 70% over the preceding year.[7] Heavily dependent upon energy intensive products for its exports, Egypt's trade deficit increased by almost one-fourth in the year ending March, 2015, as goods exports fell by some 14% in that period.[8] After 2010 the rate of growth of GDP slumped to an annual average of less than the rate of population increase, resulting in growing impoverishment and unemployment.

Egypt, in sum, has failed to effectively utilize its endowment of fossil fuels or to develop alternative energy supplies in pursuing more rapid, sustained economic growth. Its rising population and carbon emissions, to say nothing of degradation of the nation's water, soil, and air, combined with its ever greater reliance on imported foodstuffs, pose threats both to the country's and the world's well-being. Egypt is, therefore, a country of global importance in that its failure thus far to square the energy–environment–growth triangle poses threats to itself and to others. Moreover, it serves as an exemplar of the particular challenges posed to achieving beneficial trade-offs between energy, environmental protection and economic growth in other lower middle income countries with at least some endowment of hydrocarbons coupled with growing populations, many of which, like Egypt, suffer from poor governance. Indeed, the number of countries falling into this category is growing particularly rapidly as a result of the technologically driven diversification of energy supplies, which has made numerous African, Asian and Latin American lower middle income countries fossil fuel exporters in recent years. Egypt is historically ahead of this curve, having one of the oldest oil industries in the Middle East, so its management of the energy–environment–growth triangle may be especially indicative of broader trends.

Unfortunately the general impact of Egypt's exploitation of its fossil fuel reserves, coupled with it occupying a substantial role in the partially integrated Middle East and North Africa (MENA) oil economy, has been to confirm the hypothesized "oil curse," claimed by Michael Ross and others to be particularly prevalent in countries with poor governance.[9] Egypt, like other MENA countries, has suffered for some two generations from the Dutch Disease affliction of an overvalued currency, driven not only by the impact of hydrocarbon exports, but also by the high labor market clearing wage rate resulting from a large and relatively overpaid civil service, and by prevailing wages in the region's labor importing countries, most notably in the Gulf and Libya. One predictable, devastating consequence of Egypt's Dutch Disease has been to slow the growth of manufacturing industry, the only sector of the economy capable of absorbing large numbers of entrants into the labor market.

A more fine grained, political economy analysis of Egypt suggests that additional, largely political factors have reinforced the Dutch Disease or played an altogether independent role in preventing effective utilization of windfalls from exploitation of oil and gas reserves. Those factors result from the Republican state founded by the military in the wake of its 1952 coup being "fierce but brittle," as Nazih Ayubi aptly characterized it.[10] This dualism is reflected in the state's weak extractive capacity, as manifested by low rates of direct taxation, savings and investment, all of which tended to decline from the Nasser through the Sadat and then the Mubarak eras. The Republican state has been mired in a more or less permanent fiscal crisis, with temporary relief provided by windfalls from geo-strategic or hydrocarbon rents, such as those that accrued from regaining oil fields in the Sinai and Gulf of Suez after the 1979 peace treaty with Israel, from joining the anti-Iraq U.S.-led coalition in 1991, or from the gas boom that lasted some half a dozen years from 2002–3. Fierce but brittle and weak, the state lacked the administrative capacity and legitimacy required to impose reasonable and equitable personal and corporate income taxes. It also feared the consequences of seeking to impose hardship on its potentially restive population by directing revenue into savings and investments rather than consumption. Revenue was thus deployed disproportionately to prop up the oversized state and its sprawling civilian and military bureaucracies; to provide food and energy subsidies from which ultimately over 90% of the population benefitted; to be channeled into illegal capital flight which amounted to some 2% of GDP from 2003 until after the fall of the Mubarak regime; and increasingly under Sadat and Mubarak, to cultivate crony capitalists with whom the president and his entourage shared rents.[11] Those rents were allocated to personal consumption and to bolstering the political machine through which the regime could sustain its hybrid authoritarianism, for which liberal trappings, such as elections, required ever greater patronage resources.[12] Allocated to regime maintenance rather than human, physical or other developmental purposes, hydrocarbon rents reinforced the Dutch Disease caused in significant measure by Egypt being caught up in the MENA oil economy.

In addition, extraction of those rents and their allocation to political purposes, to say nothing of elite corruption, required that the financial and even technical management of the oil and gas sectors be opaque. Control of those sectors by the presidency and the military, which competed for the rents they generated, was hidden from public view. The nationalized companies involved in Egypt's upstream and downstream oil and gas industries could not be substantially upgraded as their role was primarily that of extraction of rents rather than development of capacities. Having first commenced production of oil well before World War II, Egypt remained more heavily dependent upon foreign oil and gas companies than most other Arab states in which fossil fuel extraction began much later.[13]

Heavy pollution of air, water and soil has resulted from state policies driven by rent seeking and populist motives, coupled with weak regulatory capacities. The concentrations of heavy particulate matter and other toxins in Cairo's air are among the highest in the world, resulting primarily from the use of subsidized diesel fuel for transport and the siting of polluting industries, including cement and steel plants, in the metropolitan area. Modern sewerage systems do not exist in most rural areas, despite high population concentrations. As a result contaminated wastewater commonly intermixes with irrigation water, creating dangerously high levels of bacterial pollution in soils and the crops grown in them, frequently rendering those crops unacceptable in export markets. Contamination and cessation of delivery of drinking water in both urban and rural areas are regular occurrences resulting from inadequate treatment and distribution facilities and poor maintenance of them. The Environmental Protection Agency is essentially toothless. Although Egyptians are well aware of pollution and its dangers, over-centralized government ensures that they have no access channels through which to register their discontent or impact policies.[14] Egypt's marine environments on the Mediterranean, Gulfs of Suez and Aqaba, and Red Sea coasts have been degraded by the destruction of coral reefs, overfishing, and introduction of irrigation run off and sewerage water. The country is facing a range of profound environmental challenges that would require both effective governance and billions of dollars of expenditure to address. Neither is presently available.

These then in broad outlines are the outcomes of Egypt's essentially unsuccessful efforts to resolve its energy–environment–development trilemma. That the country faces economic, energy and environmental crises is not an overstatement, as is suggested by the rhetoric and actions of the military government that seized power in the summer of 2013. Confronting a collapsing economy with attendant fiscal and foreign exchange crises, aggravated by increasingly severe shortages of oil, gas and electricity, the new regime has gathered virtually all political and economic powers into the hands of the military as if the nation were facing a decisive battle, which indeed it may well be. But this raises the question of whether the military is the appropriate institution to lead efforts to resolve the economic and energy crises. As for the environmental one, the military has clearly signaled that environmental protection will have to be sacrificed to save the nation's economy, as for example by authorizing coal burning power generation in close proximity to both urban and touristic areas, and by excessive exploitation of fossil water reservoirs for military-led land reclamation projects. The military sees the nation facing a dilemma, not a trilemma. The remainder of this chapter will discuss how it is confronting the economic and energy challenges, provide a preliminary assessment of the prospects for success, and draw some tentative conclusions about the consequences of military rule for effectively addressing the economic, energy, environmental trilemma.

Economic challenge and responses

Following his election as President in May, 2014, President Sisi mandated a set of macroeconomic policy reforms seemingly taken from the IMF playbook and probably intended to prepare the ground for a new standby agreement with that institution in order to open up desperately needed new lines of credit. By the time he assumed the presidency almost a year after the military's coup, Egypt's benefactors in the Gulf, key of which were Saudi Arabia, the UAE and Kuwait, were already suffering from donor fatigue after having provided some $20 billion to Egypt largely for the purpose of saving it from the Muslim Brotherhood. Having accomplished that with the coup led by General Sisi against the Brotherhood government in June–July 2013, these Gulf Arabs obviously wanted Egypt to stabilize its economy to entice new donors and lenders to share the financial burden. The resulting macroeconomic reforms, including reduction of energy subsidies, better targeting of food subsidies, efforts to reduce the fiscal deficit by imposing new taxes, gradual depreciation of the currency, and inducements to Mubarak era cronies to repatriate flight capital and personally return to Egypt, had within a year reduced the magnitude of the economic crisis, but not resolved it. In February, 2015, following Article IV consultations in Cairo, the IMF delegation stated that "Egypt's structural and monetary reforms were starting to produce a turnaround in the economy."[15] It predicted that GDP growth would rise from 2.2% in 2013/14 to almost 4% in 2014/15 and 5% in "the medium term." The rather measured tone of the IMF report was further emphasized by its estimate that the fiscal deficit, which was 14% of GDP in 2013–14, would not drop below 8% of GDP prior to 2019, with inflation remaining above 7% "over the medium term." The IMF added that the tax base remained narrow, that the current account balance continued to depend too heavily on transfers from the Gulf, that foreign currency reserves of $15 billion were inadequate, and that the civil service and public sector wage bill of LE178 billion in 2014, a 24% increase over the previous year, was too high.[16] In the 2015 budget that figure increased yet further, to LE218 billion, or a full quarter of all budgetary expenditures. The large fiscal deficit, government debt amounting to more than 100% of GDP, and continuing heavy borrowing needs coupled with the declining willingness of domestic financial institutions to buy government bonds and treasury bills, prevented Sisi's reforms from lifting Egypt's low bond rating, which Moody's Investors Services retained at Caa1.[17] Egypt's poverty and unemployment rates continued the steady climb that had commenced in earnest in 2010/11.

Given the magnitude of economic challenges facing the country, the Sisi government's reforms of 2014–15 were tepid. They were aimed at gradual macroeconomic stabilization rather than at resolving the maladies responsible for longstanding economic underperformance. Indeed, the government seemed to care relatively little about addressing structural obstacles to improved economic investment and growth, which are probably best summed up in the indicators reported in the World Bank's annual *Doing Business Report*. In the 2015 edition, Egypt ranked 112th out of 189 countries on the Ease of Doing Business Index, an improvement of just one place over the previous year. On the critical measure of "enforcing contracts," its performance remained unchanged at 152nd place.[18] This measure is embedded in a broader cluster of indicators that constitute the core of the Fragile State Index, which provides a ranked comparison of the stability of overall political economies. That stability is in turn both a result of and contributing factor to economic growth. On that Index Egypt's standing deteriorated steadily from 2011 to 2014, with the worst performance being on the indicator "human rights and the rule of law." Egypt's score deteriorated dramatically over that period, reaching 9.7 out of the worst possible score of 10 on that indicator by 2014. Of the 178 countries ranked in 2014, Egypt was the 31st most fragile, with its ranking on human rights and the rule of law the most discrepant from the global average of the dozen indicators from which the Fragile State Index is constructed.[19] Egypt's business climate cannot be improved substantially unless and until progress is made on establishing the rule of law, the foundation upon which not just dispute resolution, but overall governmental effectiveness, hence economic performance, rests. The Sisi government's disinterest in substantive reform reflects its choice for political repression rather than inclusion, coupled with its rejection of a free market economy in favor of a military dominated command economy.[20]

"Sisinomics" is a reversion to Nasserist "Arab Socialism," with three variations.[21] First, Sisi's military has been put in charge not just of the state administration and much of the public sector, as Nasser's was, but through its three major holding companies and other economic enterprises it has come to be the country's single largest provider of economic goods and services. Second, Nasser's import substitution industrialization has been modified by a search for external partners, most successfully so far with government owned enterprises from the key GCC states, to team with military owned or dominated enterprises to produce goods or provide services primarily for domestic markets. These partnerships are obviously intended by the Gulf participants as means to strengthen their emerging geo-strategic ties with the Egyptian military while supporting its repression of the various forces, whether jihadi, peaceful Islamist, or liberal reformist, that are also challenging the Gulf states. The Sisi government clearly hopes that other authoritarian states, most notably China and Russia, will follow the GCC example of responding to the mix of geo-strategic and economic blandishments, most especially in the strategically vital Suez Canal Zone. Finally, Sisinomics is providing rather more space for crony capitalists than did the Nasser prototype, especially for cronies with direct dealings with the military.

Excluded from Sisinomics are concerns with and support for small and medium enterprises, or for opening markets more generally. The intent is not to liberalize the political economy, but to reconfigure and expand the limited access order the military inherited. It is to be the new gate keeper, deciding national economic priorities and who shall implement them. The military is also to be the guarantor of the social contract, directly providing subsidized goods from its own food, fuel and service companies. Egyptians are to be made aware of their personal and their country's material and security dependence on the military. Sisinomics promises a command economy similar to that of Nasser's ill-named Arab Socialism, which in effect was Arab Statism. Like its predecessor, Sisinomics is replete with extravagant showcase projects, bevies of fawning ministers at ribbon cutting ceremonies, and tawdry deals with cronies presented as mobilization of national capital, with the new addition of providing "opportunities" for Egyptians to combine their nationalist sentiments with personal financial interests by extending credit to the military.[22] As was the case under Nasser, the showcase projects are announced out of the blue, preceded by few if any economic feasibility studies and no public discussion. The model as a whole is presented in nationalist, indeed militant rhetoric, including claims that military production will double to $8.5 billion by 2019 and that Egypt's assertion of economic and political independence has thwarted "international forces" that "thought they could establish a new regional system to give them more influence."[23]

Sisinomics is thus a model that has already been tried and failed, not just in Egypt but elsewhere. Command economies are inherently incapable of dealing with the complexities of globalized markets and competition, even if the commanders really know economics, finance, and business management. Alas, Egypt's military, tasked now with running much of the economy, has little such expertise. Moreover, this mission creep into the economy comes at a time when the military is already over tasked with trying to counter a domestic quasi-insurgency; to support and even supplant the Ministry of Interior's efforts to control restive crowds of demonstrators; to defend its borders against a host of challenges arising in Gaza, Libya and the Sudan; to construct civilian political window dressing to cover but not undermine its dominant role; and to devise and manage the country's foreign policy. Already doing too much with too little capacity, but wary of sharing its powers and authority with civilian actors, especially those autonomous from the state, the military simply cannot make or implement effective economic policy. Among its most difficult challenges is fashioning a policy that provides sufficient energy to the domestic economy at costs affordable at the micro and macro levels, strikes a reasonable balance between different economic interests, and which does not jeopardize other sectors or the broader environment.

Energy policy

The most serious domestic challenge facing the military upon its seizure of power in 2013, other than imposing security, was resolving the energy crisis that it inherited from the Mubarak era and which had intensified in the interregnum. From 2010 Egypt became a net importer of fossil fuels as gas production failed to keep pace even with domestic demand, to say nothing of meeting export commitments contracted for with Israel in 2008 and with the international oil/gas companies that had developed Egypt's gas fields. Domestic oil consumption had overtaken production some years earlier, despite rising figures for proven reserves. Egypt's two LNG plants, owned and operated by BG and the Spanish-Italian Union Fenosa Gas (UFG), were intended to liquefy gas for export, but fell idle in 2012 due largely to declining production from fields that were not adequately maintained as a result of non-payment to their operators.[24] Egypt's energy intensive industries, including steel, cement, fertilizer and ceramics, were forced to restrict output and commence a search for new energy sources, key to which was coal, amidst declining profits.[25] Electricity generating plants, intended to burn natural gas, turned increasingly to heavy fuel oils. Electricity supply nevertheless fell increasingly behind demand, which rose at 7.5% annually in the last six years of the Mubarak era, resulting in intensifying outages.[26] The policy commitment of 2008 to produce 20% of electricity by renewables (8% hydro, 12% solar and wind) by 2020 had not been acted upon.[27]

The energy situation, in sum, had increasingly come to resemble that in the early and mid-1970s, when electricity blackouts were assessed by Washington to be posing a political threat to their new ally, President Sadat. As a result, USAID committed more than $1 billion to constructing generating plants and overhauling the much degraded distribution network. After that extensive upgrade, which ended in the early 1990s, generating capacity was not substantially increased during the remainder of the Mubarak era. The new Sisi regime thus faced the threat of a political backlash similar to that which had confronted Sadat some forty years earlier. In Sisi's case, however, he did not have the support of a US ready, willing and able to overhaul the country's electrical power system. Cash transfers and emergency fossil fuel shipments from his main Gulf backers could and did provide only temporary solutions.

After winning the presidential election in May, 2014, President Sisi immediately turned his attention to both the demand and supply sides of the energy crisis. On the demand side subsidies for transport fuels and for gas and electricity supplied to industry were reduced, causing the price of some grades of transport fuels to rise by 70% overnight. But introduction of the long promised smart card system for subsidized methane, the primary source of energy for the poor, was delayed yet again. Almost a year later no further subsidy reform had occurred. An informed source noted that the limited subsidy reform "would only have a marginal effect on narrowing the deficit."[28] And indeed, in the 2015 budget energy subsidies of 27% of total government expenditures were only exceeded by the cost of debt servicing (28%), so despite the steep decline in global oil and gas prices, the fiscal drag exerted by energy subsidies remains unabated.[29]

Resolution of the immediate supply crisis was sought through leasing from Norway a ship-based regasification unit to process LNG imported from Algeria, the Gulf, and elsewhere.[30] The law banning the use of coal for electricity generation was voided by executive action in April,

2014, accompanied by announcements that some seven coal-fired generating plants were being purchased from China and that the military as well as private investors would be building several more.[31] A year later Tharwa Investments, a firm with strong links to the military, announced that it had signed an MOU with the Minister of Electricity to build an $11 billion, 6 gigawatt coal-fired power plant, which it claimed would be the world's largest.[32] Discussions were commenced with Israel to reverse the flow in the Egypt–Israel pipeline so that instead of the former supplying the latter, Israel would draw upon its newly developed offshore fields to supply Egypt, with possible subsequent use of the two idle LNG plants on Egypt's Mediterranean Coast to liquefy Israeli gas for export to Europe.[33] In May, 2015, the government granted permission to private Egyptian companies to import Israeli gas.[34] In the first half of 2015 Egypt paid about one-half of the more than $6 billion owed to foreign companies involved in exploiting its gas reserves and commenced negotiations for revisions of prices for gas from existing fields and to establish those for new ones. This resulted in across the board price rises of some 40% on average and still higher for gas from offshore fields, thus almost overnight converting Egypt's gas price from being below global averages at less than $3 per million BTUs, to an average of about $5, thus higher than the prevailing $3 price in North America and not much below that for new European gas delivery contracts. All of the gas to be produced under these renegotiated and new contracts was to be purchased by Egypt for domestic consumption, so the era of cheap energy and gas exports has clearly come to an end.[35] In February, 2015, during a visit to Cairo by Russian President Putin, it was announced that the long delayed construction of a nuclear power plant west of Alexandria would be commenced under a joint venture between Russia and the Egyptian military. In order to stimulate investment in electricity generation, most notably in BOOT (build, own, operate, transfer) and other forms of public/private partnerships to construct generating facilities, including those from renewable sources, the cabinet approved a draft law in February, 2015, that would in theory permit private companies to transmit and sell electricity directly to consumers, thus bypassing the state owned Egyptian Electric Holding Company. This was coupled with a stated commitment to increase electricity prices, a further incentive to attract investors. Yet whereas negotiations over gas extraction and sales resulted in firm, contractual commitments, those over electricity remained as preliminary agreements, with the private parties concerned obviously wary of the Government's pledges, most notably that regarding the feed-in tariff to the grid. Since some government representatives were speaking of increases being phased in over five years and then being capped at 80% of nominal costs, it was clear that considerable policy ambiguities remained in the electricity sector, ambiguities resulting primarily from the government's wariness in raising electricity prices for consumers.[36]

Environmental and associated political issues were swept aside in the rush to overcome energy shortfalls. When the Minister for the Environment objected to burning coal, particularly if adjacent to residential and touristic areas, she was replaced. The new minister argued that coal was a clean fuel posing no environmental threat and, in any case, China used a lot of it.[37] Protests against construction of the nuclear power plant by local tribesmen and owners of touristic facilities were met with strong responses by security forces, coupled with offers of financial compensation. Protests against importation of gas from Israel were brushed aside. No specific mention was made of the commitment made under Mubarak to produce one-fifth of the nation's electricity by renewables in 2020. In March, 2015, preliminary agreements were signed with Siemens to produce parts for wind generators and with Saudi and Emirati firms to build solar and wind generating facilities, but no contracts had been forthcoming by the time of this writing (August, 2015). The contribution made by solar and wind power to the nation's electrical grid was registered as less than 1% in 2014–15.[38]

It remains to be seen whether these varied and dramatic steps to increase the country's supplies of energy will succeed. The precipitate global drop in oil and gas prices from June, 2014, in theory at least works in Egypt's favor as it almost overnight has become a substantial net energy importer. Negotiations over a long term commitment to Israel to utilize the two LNG plants are suggestive of the fact that neither the government, nor Israel, nor the gas companies concerned, anticipate that Egypt will again become a gas exporter for many years, if ever. Egypt thus squandered in a decade a significant endowment of that fossil fuel, leaving in place neither fixed nor liquid capital nor appropriate expertise with which it could meet the energy challenge it now faces. That challenge is particularly severe in the 2015–17 period. The availability of substantial Israeli gas does not commence prior to the last year of that period, which is also the earliest date upon which completion is expected of construction of the initial new power generating capacity already contracted.[39] The nuclear power facility will not be ready before 2020. The government has yet to make firm contractual commitments to increase the supply of renewable energy, a hope that can only be realized if it commits to substantially higher prices to consumers or an increase in targeted subsidies that would then attract private investors.[40] For the next two to three years, therefore, Egypt is facing severe power shortages and heavy demands for financing fuel imports and construction of generating facilities. Competition between industrial and public users of electricity and gas is already intense, with both feeling the pain from shortages and price increases. Finding a balance between them that sustains production in the economically vital energy intensive industries, but does not cause other hardships and possibly mass demonstrations due to outages or high prices, will be a major challenge. Since 70% of petrol and 60% of natural gas subsidies presently go to the wealthiest 10% of the population, with the three bottom income deciles receiving only one-quarter of even "their" subsidies, which are for the butane gas cylinders that provide their primary energy source, there clearly are socio-political reasons to shift the benefit of subsidies down the social ladder.[41] But the poor have little voice in a regime which rests far more heavily on the upper middle class and the wealthy, the latter of which dominate the energy intensive industries upon which the country has depended heavily since the onset of the gas boom in 2002–3. As presently constituted, the government and political system more generally do not have the capacities to fine tune energy policy to try to meet contending objectives of growth, equity, and political stability.

Conclusion

Egypt is caught between slow growth, rapidly expanding energy needs and an environment under increasing stress. The Sisi government's resolution of this trilemma has been to ignore environmental consequences, whether of release of carbon into the world's air or of more pollution of Egypt's air, land, and water, in a rush to accelerate economic growth and energy supply. The model it has chosen to meet both objectives is what one might expect from a military run government. It is a modified version of the very same model that the military adopted on seizing power more than sixty years ago. Its prospects for achieving a sustained rate of growth in per capita GDP seem dim, in part because of the inherent frailties of a militarized, quasi-command economy, and in part because of rapid population growth coupled with declining hydrocarbon extraction. Unable to grow sufficiently rapidly to substantially increase per capita wealth when it was a net energy exporter, Egypt will have far more difficulty doing so when it is a net importer with what in a decade will be more than 20 million more mouths to feed. As for meeting its energy needs, that is basically impossible in the short term, so the economy and population will suffer from energy deficits for two to three years as a result. Over the medium and long hauls the situation is likely to improve, but the cost of expanding energy production and consumption will have to be borne by an economy that is already struggling and a population of which almost half is now living on less than $2 daily.[42] So the question of who is going to pay for increased energy supplies will become ever more pressing. Governmental resources cannot sustain simultaneously even the now reduced energy subsidies and a semblance of fiscal balance. Much of the population cannot afford electricity, diesel, gas or methane at true market prices. Energy intensive processing, formerly the most profitable, export oriented sector of the economy, is losing its competitiveness as the prices it pays for energy inputs are rising to prevailing global levels. Since the military, through its command economy and the docile civilian polity it has installed, is clearly responsible for making these choices, it will reap the benefits if they work well enough for enough of the population, and pay the price if they do not. Whatever that outcome proves to be, in the meantime Egypt will contribute a growing share to global carbon emissions as it fails to take adequate advantage of its potential to produce wind and solar energy.

Notes

1 Caroline Krafft, and Ragui Assaad, "Beware of the Echo: The Impending Return of Demographic Pressures in Egypt," *Policy Perspective*, no. 12 (2014, May).
2 Enas Hamed, "Egyptian Official: Poverty Main Cause of Overpopulation," *al-Monitor*, September 11, 2014, www.al-monitor.com/pulse/fr/contents/authors/enas-hamed.html.
3 US Energy Information Administration, *Country Analysis Brief: Egypt* (Washington, DC: US Energy Information Administration, 2014).
4 Egypt's rapid population increase is coupled with an even more dramatic rise in total carbon dioxide emissions, which rose from 75,944 thousand metric tons in 1990 to 204,776 in 2010, proportionately the world's largest increase. World Bank, "World Development Indicators" (Washington, DC: World Bank, 2015), http://wdi.worldbank.org/table/3.8?tableNo=3.8. See also Hossein Razavi, *Clean Energy Development in Egypt* (Tunis: African Development Bank, 2012).
5 Between 1990 and 2011 GDP in purchasing power parity dollars produced from a kilogram of oil equivalent rose from $10.5 to $10.9, compared to an average increase for lower middle income countries from $4.8 to $8.0. Razavi, *Clean Energy Development in Egypt*; World Bank, "World Development Indicators." See also Bassam Fattouh, and Laura el-Katiri, *Energy and Arab Economic Development*, UNDP Research Paper Series, United Nations Development Programme, 2012.
6 Robert Springborg, "Gas and Oil in Egypt's Development," in *Handbook of Oil Politics*, ed. Robert Looney, 295–311 (Abingdon: Routledge, 2012).
7 Ahmed Kotb, "Grinding to a Halt," *Al Ahram Weekly*, April 16, 2015, http://weekly.ahram.org.eg/News/10986/18/Grinding-to-a-halt.aspx.
8 "Current Account Deficit Widens to US$8.4 Billion," *Mada Masr*, June 7, 2015, www.madamasr.com/news/economy/current-account-deficit-widens-us84-billion.
9 Michael Ross, *How Petroleum Wealth Shapes the Development of Nations* (Princeton, NJ: Princeton University Press, 2012). For similar assessments of the relationship between oil wealth and governance, see Robert Looney, ed., *Handbook of Oil Politics* (Abingdon: Routledge, 2012).
10 Nazih N. Ayubi, *Overstating the Arab State: Politics and Society in the Middle East* (London: I. B. Tauris, 1995).
11 In 2014, Global Financial Integrity reported that since 2003 illegal external capital flows from Egypt amounted to $3.8 billion annually. Leaked documents from the Swiss subsidiary of HSBC revealed in early 2015 that Egypt was the 20th ranked country of citizenship of holders of private dollar accounts, the purpose of which was to avoid taxes and/or conceal illegally gained assets. Mariam Rizk, "Egypt 20th Ranked Country with most Swiss HSBC Bank Accounts: Leaked Report," *Ahram Online*, February 10, 2015, http://english.ahram.org.eg/NewsContent/3/12/122645/Business/Economy/Egypt-th-country-with-most-Swiss-HSBC-bank-account.aspx.
12 On the persistent fiscal crisis in Republican Egypt, see Samer Soliman, *The Autumn of Dictatorship: Fiscal Crisis and Political Change in Egypt under Mubarak* (Palo Alto, CA: Stanford University Press, 2011).

13 On the relative underdevelopment of the Egyptian oil and gas industries, see Springborg, "Gas and Oil in Egypt's Development."
14 Nicholas S. Hopkins, and Sohair Mehanna, "Living with Pollution in Egypt," *The Environmentalist* 23 (2003): 17–28; Nicholas S. Hopkins, Sohair Mehanna, and Salah el-Haggar, *People and Pollution: Cultural Constructions and Social Action in Egypt* (Cairo: American University Press, 2001); Nicholas S. Hopkins, and Sohair Mehanna, "Social Action against Everyday Pollution in Egypt," *Human Organization* 59, no. 2 (2000): 245–254.
15 Cited in "IMF Says Egypt's Reforms Starting to Spur Turnaround in the Economy," *Ahram Online*, February 11, 2015, http://english.ahram.org.eg/NewsContent/3/12/122790/Business/Economy/-IMF-says-Egypts-reforms-starting-to-spur-turnarou.aspx.
16 "Egypt's Budget Deficit Hit 12.8% in FY2013/14: Finance Ministry," *Ahram Online*, November 8, 2014, http://english.ahram.org.eg/NewsContent/3/12/115058/Business/Economy/Egypts-budget-deficit-hit–in-FY-Finance-ministry.aspx.
17 Paul Rivlin, "Egypt's Economy: Sisi's Herculean Task," *Middle East Economy* (Tel Aviv University) 4, no. 11 (2014).
18 Deya Abaza, "Egypt Ranks 112th in World Bank's Annual Doing Business Report," *Ahram Online*, October 29, 2014, http://english.ahram.org.eg/NewsContent/3/12/114251/Business/Economy/Egypt-ranks-th-in-World-Banks-annual-Doing-Busines.aspx.
19 http://ffp.statesindex.org/2014-egypt.
20 Those in the civilian governmental façade, including the prime minister, are left the task of mouthing the slogans of economic reform à la the World Bank, but they sound profoundly optimistic, even hypocritical. So, for example, Prime Minister Ibrahim Mehleb claimed in September 2014, that Egypt would become one of the top 30 economies in the world in production, competitiveness, and citizen satisfaction. See Niveen Wahish, "The Pursuit of Growth," *Al Ahram Weekly*, September 11, 2014, http://weekly.ahram.org.eg/News/7247/17/The-pursuit-of-growth.aspx.
21 For analyses of the central role of the military in "Sisinomics," see Stephan Roll, "Al-Sisi's Development Visions: Projects and Power in Egypt," *SWP Comments* 26 (2014, May); and Cherine Chams-el-Dine, "Fragile Alliances in Egypt's Post-Revolutionary Order," *SWP Comments* 46 (2014, October).
22 The largest showcase projects include digging of a second canal parallel to the existing Suez Canal, constructing 1 million low cost housing units, reclaiming 4 million acres of desert and building a nuclear power facility west of Alexandria. The first one is underway; the second, being implemented by the UAE's Arabtec construction company, is due for completion in 2020, but has yet to produce a single unit, having faltered as a result of financial complications arising in the partnership between the military and the UAE company formed to implement the project; the third has yet to commence and probably never will due to major doubts about the availability of either fossil or Nile water; while the fourth was given new momentum by the agreement signed in February 2015 between Presidents Putin and Sisi for the Russians to provide technical assistance. It is interesting to note that Presidents Nasser, Sadat and Mubarak all announced major reclamation schemes, the first two at the very outset of their presidencies. None of these schemes was successful and all were subsequently abandoned in whole or part. That Sisi's reclamation project is likely to meet a similar fate is suggested in commentary in Mona El-Fiqi, "Reclaiming Land," *Al Ahram Weekly*, November 20, 2014, http://weekly.ahram.org.eg/News/7777/18/Reclaiming-land.aspx; and in Walaa Hussein, "Sisi's Land Cultivation Plan Ignores Worsening Water Crisis," *al Monitor*, October 27, 2014, www.al-monitor.com/pulse/originals/2014/10/egypt-starts-to-cultivate-land.html. The ceremonial dimension of these and other projects is typically reinforced by the regime's mobilization of ministers and other dignitaries at ribbon cutting and other events. The Prime Minister and nine other ministers, for example, attended a street rehabilitation project in Islamic Cairo in September 2014. Nevine El-Aref, "Al-Muizz Restored," *Al Ahram Weekly*, September 11, 2014, http://weekly.ahram.org.eg/News/7246/17/Al-Muizz-restored.aspx. Some twenty military attachés were taken by bus to attend the opening of the Suez Canal dredging project by Admiral Mameesh, head of the Suez Canal Authority. Amirah Ibrahim, "Excitement all the Way," *Al Ahram Weekly*, September 11, 2014, http://weekly.ahram.org.eg/News/7226/17/Excitement-all-the-way.aspx. Financing of the $8 billion Suez Canal expansion was provided by "certificates" purchased primarily by Egyptians who were attracted by the 12% interest rate, a minimum 2% premium over bank savings deposits and government bonds, although the government emphasized the "patriotic" motives of investors. One of the most tawdry of the deals struck between the Sisi government and cronies was that in which Nassef Sawiris, Egypt's richest man and one of the three richest in Africa, "donated" LE2.5 billion to the "Long Live Egypt Fund," established by

President Sisi and managed by his military colleagues, in return for forgiveness of an equal amount of outstanding tax liabilities and permission for Sawiris's Orascom Construction Industries to build a $2.5 billion coal fired power station on the Red Sea coast in partnership with the UAE based Petroleum Investment Company. "OCI Donates LE2.5 Billion of Recouped Tax Settlement to Tahya Misr Social Fund," *Ahram Online*, November 13, 2014, http://english.ahram.org.eg/NewsContent/3/12/115458/Business/Economy/OCI-donates-LE-bn-of-recouped-tax-settlement-to-Ta.aspx; and Waad Ahmed, "Orascom Construction to Build $2.5 Billion Coal-fired Power Station," *Ahram Online*, November 5, 2014, http://english.ahram.org.eg/NewsContent/3/12/114791/Business/Economy/Search.aspx?Text=%20Energy%20crisis.

23 Ibrahim Alsahary, "Business Monitor: Egypt's Military Production to Increase to $8.5 Billion in 2019," *Egypt Independent*, November 8, 2014. The claim that Egypt's military production in 2014 was worth $5.2 billion as reported in this article is a huge exaggeration. For President Sisi's quote on "international forces," see "World Powers Failed in Making New Middle East: Egypt's El-Sisi," *Ahram Online*, October 28, 2014, http://english.ahram.org.eg/NewsContent/1/64/114166/Egypt/Politics-/World-powers-failed-in-making-new-Middle-East-Egyp.aspx.

24 Perrihan Al-Riffai, Julian Blohmke, Clemens Breisinger, et al., "Harnessing the Sun and Wind for Economic Development? An Economy-wide Assessment for Egypt" (Working Paper 851, Economic Research Forum, Cairo, 2014), 3.

25 Deya Abaza, "Suez Cement Sees Profits Drop, as Chairman Blames Energy Shortage," *Ahram Online*, October 30, 2014, http://english.ahram.org.eg/NewsContent/3/12/114349/Business/Economy/Suez-Cement-sees-profits-drop,-as-chairman-blames-.aspx; Waad Ahmed, "Ezz Dekheila's Profits Fall on Energy Shortage," *Ahram Online*, October 28, 2014, http://english.ahram.org.eg/NewsContent/3/12/114175/Business/Economy/Ezz-Dekheilas-profits-fall-on-energy-shortage.aspx.

26 Al-Riffai et al., "Harnessing the Sun," 4.

27 Isabel Esterman, "While Renewables Stall, Coal Powers Ahead," *Mada Masr*, December 1, 2014, www.madamasr.com/sections/environment/while-renewables-stall-coal-powers-ahead.

28 Laura El Katiri, and Bassam Fattouh, *A Brief Political Economy of Energy Subsidies in the Middle East and North Africa* (Oxford Institute for Energy Studies, 2015), www.oxfordenergy.org/wpcms/wp-content/uploads/2015/02/MEP-11.pdf. The authors note that the Minister of Planning predicted that it would be another five years before subsidies would drop to 20% of the real cost and that electricity subsidies would be retained during that period.

29 For an analysis of the 2015 budget, see Ziad Bahaa-Eldin, "Egypt: Understanding the New Budget," *Ahram Online*, July 15, 2015, http://english.ahram.org.eg/NewsContentP/4/135476/Opinion/Egypt-Understanding-the-new-budget.aspx.

30 "Egypt Signs Deal with Norwegian Firm to Rent LNG Import Terminal," *Ahram Online*, November 3, 2014, http://english.ahram.org.eg/NewsContent/3/12/114659/Business/Economy/Egypt-signs-deal-with-Norwegian-firm-to-rent-LNG-i.aspx.

31 "Egypt to Sign 7 Coal-Power deals with Chinese Companies," *Mada Masr*, December 24, 2014, www.madamasr.com/news/environment/egypt-sign-7-coal-power-deals-chinese-companies; "Report: Suez Cement Begins Burning Coal in Qattamiya," *Mada Masr*, October 29, 2014, www.madamasr.com/news/environment/report-suez-cement-begins-burning-coal-qattamiya; "Newspaper: Army Signs $700 mn Deal with Multinational Consortium for Power Stations," *Egypt Independent*, December 24, 2014, www.egyptindependent.com/news/newspaper-army-signs-700mn-deal-multinational-consortium-power-stations; "Orascom Construction Granted Land for Coal Plant on Red Sea Coast," *Mada Masr*, November 13, 2014, www.madamasr.com/news/environment/orascom-construction-granted-land-coal-plant-red-sea-coast.

32 Isabel Esterman, "After a Law Liberalizing the Electricity Sector, Egypt Scores Major Investments at Summit," *Mada Masr*, March 16, 2015, www.madamasr.com/sections/economy/after-law-liberalizing-electricity-sector-egypt-scores-major-investments-summit.

33 "Noble Energy Team in Cairo for Talks on Importing Israeli Gas, Sources," Reuters, February 1, 2015, www.reuters.com/article/2015/02/01/egypt-energy-israel-idUSL6N0VB0RF20150201; Amira Howeidy, "Egypt 'Doesn't Mind' Israeli Gas," *Al Ahram Weekly*, February 12, 2015, http://weekly.ahram.org.eg/News/10403/18/Egypt-doesnt-mind-Israeli-gas.aspx.

34 Hedy Cohen, "Egypt Okays Tamar Gas Imports – Report," *Globes*, May 21, 2015, www.globes.co.il/en/article-egypt-okays-tamar-gas-imports-1001038832.

35 "Petroleum Ministry Agrees to Pay Foreign Firms More for Gas Produced in Egypt," *Mada Masr*, March 26, 2015, www.madamasr.com/news/economy/petroleum-ministry-agrees-pay-foreign-firms-more-gas-produced-egypt.
36 Maye Kabil, "Power Deals Dominate," *Al Ahram Weekly*, March 19, 2015, http://weekly.ahram.org.eg/News/10766/18/Power-deals-dominate.aspx.
37 Esterman, "While Renewables Stall, Coal Powers Ahead."
38 Ibid. According to Esterman, Egypt has one wind project, a 250 megawatt wind farm in Zaafarana on the Red Sea Coast, commenced in 2009 but not yet operational. Of the two government supported solar projects, one in Kom Ombo and the other in Kuraymat, neither is operational. Total kilowatt hours of solar produced electricity fell by some 50% between 2011 and 2013. In 2014–15, solar's contribution to energy production was 0.1%, while wind's was 1.2%. Niveen Wahish, "Energy Pricing 101," *Al Ahram Weekly*, July 9, 2015, http://weekly.ahram.org.eg/News/12747/18/Energy-pricing—.aspx.
39 Amr Kamal Hammouda, "A Questionable Deal," *Al Ahram Weekly*, October 30, 2014, http://weekly.ahram.org.eg/News/7613/18/A-questionable-deal.aspx.
40 Esterman, "While Renewables Stall, Coal Powers Ahead."
41 Wahish, "Energy Pricing 101."
42 The Government of Egypt defines absolute poverty as per person daily income of LE10.7 (about $1.50) and near poverty as LE13.9 (about $2). The number of those living in absolute poverty rose from 16.5 million in 2010 to 21.5 million in 2013, with 49.9% of the population receiving less than LE13.9 daily by the latter date. During this same period, *Forbes* reported that the combined fortunes of Egypt's eight richest billionaires rose by 80%, to a total of $23.4 billion, or some 6% of the country's total wealth. Of these eight men, seven were from two nuclear families. Edmund Bower, "Egypt's Billionaires 80% Richer Than before the Revolution," *Mada Masr*, May 5, 2015, www.madamasr.com/news/economy/egypt%E2%80%99s-billionaires-80-richer-revolution.

Part IV

Energy transitions in the carbon consuming countries

19

Japan's energy security

Challenges, prospects, and global implications

Julia Nesheiwat

Japan has been a global leader in energy and climate reform for decades. It is no accident that Japan hosted the Kyoto Protocol, which is currently the most significant framework for mitigating climate change. By consistently leading efforts towards a greener globe, Japan had been established as an exemplary energy model through its use of nuclear technology. However, its over-reliance on nuclear technology led to an alteration in this position in March 2011. The Great East Japan Earthquake, tsunami, and Fukushima nuclear disaster eventually resulted in a total shutdown of Japan's nuclear power, which constituted a large portion of its energy mix.[1] These events, known as 3/11 because they occurred on March 11, 2011, had a deleterious effect on Japan's energy situation, and highlighted areas of weakness in its energy planning. Outside of its nuclear program, Japan lacks self-sufficiency when it comes to energy. As a resource-poor island-nation, Japan depends heavily upon imports for the majority of its energy needs. Thus, despite boasting the world's third largest economy, Tokyo ranks among the frailest countries in the world when it comes to energy security. Due to Japan's global stature, after-shocks of the 3/11 disaster have extended beyond Japan to the international community, and Japan's resolve as a global leader in climate change and civil-nuclear energy has been shaken. However, recent positive steps have been taken by Tokyo to regain footing in the energy sector. Given the energy security challenges that Japan faces, it has been critical for Tokyo to reassess its energy objectives, and it is in the global landscape's best interest to support initiatives that strengthen Japan's energy future.

Challenges to Japan's energy security

Legacy of Fukushima

Registering at 9.0 MW, the Tohoku earthquake was the strongest on record to ever hit Japan. Due to its underwater placement, a tsunami occurred.[2] The after-effects of the earthquake and tsunami caused a loss of coolant accident. Subsequently, three nuclear meltdowns were triggered, causing a widespread release of radioactive materials.

Over a year after the disastrous events of 3/11, the Japanese Diet created the Fukushima Nuclear Accident Independent Investigative Committee (NAIIC). What the NAIIC found was that the Fukushima Daiichi nuclear power plant catastrophe was due to human error. A failure in both public and private sectors to plan for foreseeable disasters and implement safety regulations and precautions in Fukushima created an otherwise avoidable situation.[3] Fear of public backlash was given as a motivating factor for such negligence. TEPCO, the private sector company that owns Fukushima and powers Japan's eastern grid, felt that by addressing such issues they would be reaffirming anti-nuclear sentiment predicated on the notion that nuclear is unsafe and opening themselves to litigation.[4] However, the events of 3/11 did far more damage to the image of nuclear power than implementing safety precautions could have.

Japanese anti-nuclear sentiment exploded in the aftermath of the Fukushima disaster and now stands as a significant challenge to Tokyo's energy security. Public opinion polls conducted by the *Asahi Shimbun* in 2012 and 2013 reveal that roughly 80% of Japanese citizens distrust the government's nuclear power plant safety measures, while 70% want nuclear power to be phased out completely.[5]

While strong nuclear opposition certainly hinders Japan's nuclear and energy policy at present, forces much stronger than public opinion are driving Tokyo's policy, and nuclear power will again be a part of the Japanese energy mix. In Japan's past nuclear accidents, similar public reactions have had limited staying power. Most notably, the 1999 Tokaimura reactor accident resulted in the death of two workers due to radiation exposure. Anti-nuclear sentiment soared in the wake of the accident but subsided within the decade and even gave way to growing favorable opinions of nuclear power.[6] Moreover, when Democratic Party of Japan (DPJ) Prime Minister Noda, buoyed by public opinion in 2012, moved toward a policy of gradual nuclear phase-out, the plans were quickly abandoned over concerns voiced by the private sector.[7] Now that the traditionally pro-nuclear Liberal Democratic Party (LDP) is firmly in power, they have already begun taking measures to reintegrate nuclear into Japan's energy mix. However, the proportion of nuclear energy will be less, which has forced Tokyo to grapple with the expansion of renewables. In the short-term, it must rely more heavily upon traditional thermal energy sources. Revisions of Japan's energy mix produce both challenges and opportunities for its future energy stability, and must be approached realistically.

Japan's current energy mix

Prior to the 3/11 disaster, nuclear power provided roughly 30% of Japan's electricity with plans to expand – making the sudden shutdown of all nuclear capacity a huge shock for Japan's energy security. Tokyo's energy self-sufficiency rate was as high as 19% with nuclear power prior to 3/11, but plummeted to a mere 4% in the wake of the disaster, climbing to a meager 6% in 2013.[8] In the absence of nuclear, Japan's energy mix predominantly consists of oil, coal, liquefied natural gas (LNG), and a small but growing percentage of renewable energy sources.

Oil represented nearly half of Japan's total primary energy supply (TPES) in 2012, and is still vitally important for the country and its transportation sector.[9] However, Tokyo is uncomfortably dependent on Middle Eastern countries for over 80% of its oil imports. The past few years have seen massive amounts of upheaval in the Middle East. Events from the Arab Spring have had widespread ramifications. Oil flow has been disrupted in some countries as a result, with unease over the future supply capabilities of others becoming a constant concern. Further, oil-field seizures by Daesh have moved significant amounts of oil from world markets to the black market. Even though Japan has expressed interest in Iran's reserves to counteract instability linked to portions of its supply in the region, current US sanctions create a barrier to substantial imports from Iran. While that could change, due to the US-Iran nuclear talks that have occurred, there is still much uncertainty. Further, over-reliance on any one country is inadvisable, as it would tether Japan to that country's status quo, resulting in disruption if any shifts occur. Diversification is key to energy security. Yet, obstacles exist to diversification.

Costly infrastructure supports the well-developed Middle East trade routes and movement away from them comes with a significant price tag. Oil resources are finite, thereby creating a zero sum scenario when it comes to imports. China's economic power and energy appetite has produced competition with Japan for the world's supply. Further, as a close US ally, Japan cannot pursue oil from sanctioned countries, such as Sudan.[10] Barriers to diversification present a major challenge to Japan's energy stability because oil will continue to be a critical part of Japan's energy needs. However, oil demand is modestly declining, thanks to higher fuel efficiencies and fuel-switching policies.[11] Long-term reduction goals for oil in Japan's energy mix will allow it greater international maneuverability, as well, by reducing dependence on foreign countries for its energy security.

Another foreign resource Japan heavily imports is coal. In 2013, coal accounted for 32% of Japan's electricity generation, second only to gas.[12]Additionally, Japan's continued need for coal, and its desire to contribute to climate change reform, have naturally led Tokyo to pursue a more efficient means to generate power from coal while cutting carbon emissions. Japanese companies such as Hitachi have made Japan a world leader in clean coal technologies, a position that also brings with it the potential of export earnings through technology sales.[13] However, clean coal currently has limited capabilities and should not be viewed as an immediate remedy to Japan's energy issues. Much of the coal currently being used to cater to Japan's energy needs does not utilize clean coal technologies and greatly contributes to its carbon footprint.

LNG is another energy source imported to feed Japan's energy consumption. Japan is by far the world's leading importer of LNG, and increased LNG imports largely made up for the loss of nuclear power post-Fukushima.[14] Unlike Tokyo's oil imports, LNG imports are well diversified. Australia, Qatar, and Malaysia are the largest suppliers. America has also increased its LNG exports to Japan, strengthening the bond between both countries. No single country provides more than 20% of total imports, though, allowing for more robust Japanese autonomy.[15] For the foreseeable future, LNG will continue to play a critical role in Japan's energy mix.

Aside from nuclear energy and fossil fuels, Japan also has a growing percentage of renewables in its energy mix. In 2011, roughly 8% of Japan's electricity generation came from renewable sources – predominantly hydropower.[16] By 2013, renewable electricity generation increased to about 13%, with gains primarily from solar, wind, and biofuels.[17] With much of Japan's hydropower capacity already developed, continued gains in renewable energy will have to come from these other sources.[18] By developing renewable alternatives, Japan can insure its long-term energy stability due to the greater amount of self-sufficiency such options provide it. Further, international technological developments have emerged that may allow renewables to take on a greater heft of the energy mix.

Japan's way forward

Energy policy

In April 2014, three years after the Fukushima disaster, Japan published its latest Strategic Energy Plan. It aimed to achieve progress on the (3E + S) framework: energy security, economic efficiency, environmental suitability, and safety.[19] However, due to the unpopularity of nuclear at the time, the role of that resource remained uncertain. The Ministry of Economy, Trade and Industry (METI) followed the Basic Energy Plan with the "Long-Term Energy

Supply and Demand Outlook" for fiscal year 2030 in July 2015. METI's proposed energy mix by 2030 to support the 3 Es entails 20–22% nuclear, 22–24% renewables, 27% LNG, 26% coal, and less than 5% oil.[20] From 2013 levels, this proposal cuts fossil fuel reliance by nearly 30%, while more than doubling renewable usage and restoring nuclear power to roughly two-thirds of its pre-Fukushima capacity.[21]

Critics have already assailed this plan as not truly working towards less nuclear dependency. However, given that pre-Fukushima plans called for 50% nuclear energy by 2030, the new target of 20–22% is a responsible goal. It represents a substantial decrease in nuclear power, while maintaining enough nuclear to keep electricity prices from soaring and simultaneously to make progress in reducing greenhouse gas emissions.[22]

Reform

Japan's energy policy has long been driven by tight government business cooperation.[23] Its vertically integrated power market has resisted badly needed reforms. Yet, in spite of the vested interests of the METI, *Keidanren* (Japan Business Federation), *Denjiren* (Federation of Electric Power Companies), and the LDP, tangible steps are now being taken towards electricity system reform.[24] The Strategic Energy Plan maintains that the government's hand will still work to ensure a desirable energy mix, but extensive reforms to promote competition and fully liberalize power generation mark a significant shift in Japanese policy.[25] As the Organization for Cross-regional Coordination of Transmission Operators (OCCTO) was created to more efficiently and economically manage the electricity supply nationwide, the first reforms were enacted in April 2015.[26] Full retail competition is scheduled to go into effect in April 2016, and further reforms and liberalization by 2020.[27]

Increasing the share of renewables

To support its proposed goal of more than a 20% share of renewable energy by 2030, Tokyo instituted a feed-in-tariff (FIT) in July 2012 to help offset costs and encourage the introduction of more renewable power sources. Lifted by a surge in solar power production, Japan added roughly 10 GW of renewable energy capacity from July 2012 through March 2014.[28] Significant investments in solar and the government's traditionally amenable stance towards solar power make recovering energy from the sun an important renewable energy source for Japan, yet rapid renewable expansion is constrained by the extra price burden on consumers and the economy, and the need for improved transmission and distribution systems and cross-regional power interconnections.[29]

Climate change mitigation efforts

Once a global leader in climate change mitigation efforts and broker of the 1997 Kyoto Protocol agreements, Tokyo's commitment has faltered since the 3/11 disaster.[30] However, its recent prominent role in the twenty-first session of the Conference of the Parties (COP21), a major international climate change initiative, has signaled its willingness to regain footing in the global arena. In its Intended Nationally Determined Contribution (INDC), a pledge submitted prior to the conference, Japan lays out its plan to hold climate change increase to no more than the 2 degrees Celsius objective. Specifically, it aims to reduce its greenhouse gas (GHG) emissions by a minimum of 50% by 2050 through the development of low carbon technologies and other forward minded actions.[31]

Of Japan's current GHG emissions, 90% originate from energy based CO_2. As such, most of Japan's reductive efforts must come from the energy sector. The INDC states that they hope for a 25% reduction of energy-originated CO_2 by 2030 when compared to 2013. Japan's energy mix for 2030 as listed in the INDC is ambitious. Oil drops to an ambitious 3%, while renewables increase their share to between 22% and 24%.[32] Nuclear represents a symmetrical portion, in line with Japan's long-term nuclear goals. Japan has already begun making headway on its nuclear aspirations, and hopes to increase its renewables through its Joint Crediting Mechanism (JCM). Japan's JCM invests in the development of carbon reducing technology and allows for the diffusion of such technology to developing countries. Concurrently, it quantitatively tracks Japan's overall contribution to GHG reduction through these technologies and their diffusion and adds it to their overall reduction number.[33] The prominence of the JCM in Japan's INDC, as well as Japan's pursuance of it during the COP21, signal that it will continue to be a key part of Tokyo's climate change mitigation efforts.

Even more remarkable was Japan's willingness to exceed its initial pledge by accepting the final COP21 terms, which aim to hold world temperature change to no more than 1.5 degrees Celsius beyond postindustrial levels. The results of the COP21, which saw all 195 participating countries accept ambitious climate change goals, cement it as one of the major climate change reform efforts in history. Japan's significant role in the COP21 demonstrates its commitment to moving beyond Fukushima and reintegrating into the international energy sphere.[34]

Technological developments

A global leader in energy technologies, Japan's continued investment in energy research and development aims to bolster energy security while generating cleaner, more efficient power. For the long term, Japan is working to further diversify its energy options with potential domestic options. Technological advancements may allow hydrogen and methane hydrates to become important energy sources, the latter of which is recoverable in substantial amounts off Japan's coast.[35]

Energy conservation

Japan is a world leader in energy conservation efforts, and continually increasing energy efficiency undergirds Tokyo's plans for the future. Since the oil crisis of the 1970s, Japan has increased its energy efficiency, measured as primary energy used per GDP, by roughly 40%.[36] Looking forward, by 2030 Japan aims to reduce energy consumption by 10% from 2013 levels by legislating building efficiency standards, promoting LED lighting and energy-saving appliances, and improving automobile gas mileage.[37]

Global implications

Global aftershocks of Fukushima

The effects of the 3/11 disaster were not confined to Japan alone, but impacted global energy markets and security. Largely as a result of Japan's increased appetite for (NG, investment in global LNG projects dramatically expanded.[38] Dramatic increases in demand lead to volatility in any market, and energy markets are no exception. Favorable public perceptions of nuclear power also dropped 5–10% in most countries and strengthened the already simmering nuclear opposition in Germany to prompt Chancellor Merkel to commit to phasing out nuclear power by

2022.[39] The delay in nuclear technologies globally only caused countries to remain reliant on traditional energy resources, such as oil, leaving them beholden to the political developments of oil exporting countries. In the case of Germany, this is especially significant as it is widely speculated that their reliance on Russian oil neutered their response to Russia's actions in Ukraine. Global public safety concerns have also led to extra costs and delays in the construction of nuclear power plants as many countries have adopted more stringent safety regulations.[40]

While the disaster may not significantly alter global nuclear energy projects in the long term, delays and fears in the short term will result in countries turning to high-carbon emitting fossil fuels, as the development and wide-scale implementation of affordable renewable sources is still premature.[41] Not only does this negatively impact climate change efforts, but it leads to a global reliance on finite resources. When it comes to resources such as oil, this is especially problematic given the fragile political situations of many oil exporting regions. Black market oil, which largely supports organized crime and terrorist organizations, could be seen as an attractive option to certain nations in such a scenario, as well.

Japan's domestic nuclear issues negatively affect US interests through its widespread global implications. Tokyo has been a key US partner and global leader in civil nuclear energy research and development (R&D), promotion of nuclear safety, and climate change mitigation efforts.[42] Emerging economies across the globe are expressing interest in nuclear energy, Without Japan-US leadership in enumerating technology and safety standards in the civil-nuclear sector, those bridge building opportunities will be missed. Other countries may step in to fill the void, but likely with less concern for nonproliferation of nuclear weapons.[43] In such a situation, it could shift regional power dynamics away from Tokyo, leading to even more dramatic global consequences. Accordingly, a non-nuclear Japan is a big loss for the international community. Minimal nuclear power has forced Japan to rely heavily on fossil fuels and revise and delay its goals for greenhouse gas emissions reduction. Until Tokyo regains its nuclear footing, it will contribute far less to critical global initiatives for civil nuclear energy technology, nonproliferation, and safety and climate change issues. However, Japan's inclusion of nuclear energy in its energy mix for long-term planning signals that it is on the road to recovery.

Japan's recent progress in securing its energy future is a welcome global development. An energy insecure Japan would be a fragile state, struggling to address national security and geopolitical issues. As a major US ally and regional pole, a reduction in stature for Japan would significantly alter global power dynamics. Its long-term objectives underscoring self-sufficiency seek to correct the vulnerability Japan has experienced in the past few years.

Fossil fuels were used by Japan heavily in the wake of Fukushima to supplement energy needs. To accommodate this, Japan has turned to countries such as Russia to supply them with the resources necessary for their energy consumption.[44] Japan has reluctantly complied with US-led sanctions against Russia, but Tokyo has long been a major importer of Russian natural gas and crude oil, and Prime Minister Shinzo Abe is understandably warming to the idea of restoring Russian relations and investing further in Russian energy development.[45] METI's report even advocated strengthening relations with Russia.[46] While regional cooperation can be a positive in certain respects, using Russian oil as a crutch diminishes Japan's ability to stand on its own. In turn, it provides developing energy sectors in Asia with a more narrow set of choices for growing their energy infrastructure. Importantly, it also complicates an otherwise strong relationship between the US and Japan.

Given the desire to build stronger security ties within the Asia-Pacific region and counterbalance China's rise, the US-Japan alliance is of great importance. Although Tokyo has been beholden to America for its security provisions, Japan is likewise beholden to its energy needs, and therefore further attention to Japan's energy insecurity is important. To that end, plans are already in the works to export LNG from the continental United States to Japan starting in 2017.[47] A more energy-secure Japan is not only a stronger US ally, but an important regional leader. And a more energy-secure Japan is a stronger leader within the international community on critical global issues like civil-nuclear energy and climate change. As many global conflicts center on resource scarcity, especially in the context of energy, this is also a welcome development in the international security landscape. Japan's re-emergence on the world stage at COP21 is an important global development. To ensure that Japan maintains its position in energy, there are steps the international community should take to aid its recovery.

Global investment in Japan's energy sector should be a priority. Japan has been a long time producer of green technology, and it benefits everyone to encourage its further development and diffusion. Additionally, efforts should be undertaken to aid Japan in diversifying its current energy mix, especially in the fossil fuel sector. Incentivizing Japan's regional energy coordination, through its JCM programs, as well as helping it track such mechanisms successfully, is another way in which the international community can strengthen Japan's energy security.

Conclusion

Energy security is a global issue. An energy insecure Japan has too many negative ramifications for the international community for it to go unaddressed. Recent steps taken by Japan will allow for progress in its post-Fukushima recovery, and its participation in the historic COP21 is a massive step forward. However, it is in the world's interest to continue supporting Japan's path to regaining its role in the global energy landscape, as it still needs help on its road to rehabilitation.

Notes

1 Julia Nesheiwat, and Jeffrey S. Cross, "Japan's Post-Fukushima Reconstruction: A Case Study for Implementation of Sustainable Energy Technologies," *Energy Policy* 60 (2013): 509.
2 Ibid., 510.
3 "Fukushima Report: Key Points in Nuclear Disaster Report," *BBC World News*, July 5, 2012, www.bbc.com/news/world-asia-18718486.
4 Justin McCurry, "Fukushima Disaster Could Have Been Avoided," *The Guardian*, October 15, 2012, www.theguardian.com/environment/2012/oct/15/fukushima-disaster-avoided-nuclear-plant.
5 "Asahi Poll: 80% Distrust Government's Nuke Safety Measures," *Asahi Shimbun*, March 13, 2012, http://ajw.asahi.com/article/behind_news/social_affairs/AJ201203130031; "Why Wait for the Election? Abe Should Address Nuclear Power Issues," *The Asahi Shimbun*, February 19, 2013, http://ajw.asahi.com/article/views/editorial/AJ201302190028.
6 Vlado Vivoda, *Energy Security in Japan: Challenges after Fukushima* (Surrey: Ashgate Publishing, 2014), 134, 142.
7 Ibid., 139.
8 International Energy Agency [IEA], *Japan – Overview* (IEA, 2014), www.iea.org/media/countries/slt/JapanOnepagerAugust2014.pdf; Masakazu Toyoda, *Energy Security and Challenges for Japan* (Tokyo, Japan: Institute of Energy Economics, 2012), http://eneken.ieej.or.jp/data/4856.pdf, 3.
9 International Energy Agency [IEA], *Energy Supply Security 2014* (IEA, 2014), www.iea.org/media/freepublications/security/EnergySupplySecurity2014_Japan.pdf, 276.
10 Vivoda, *Energy Security in Japan*, 68–70; IEA, *Energy Supply Security 2014*, 276.
11 Sylvie Cornot-Gandolphe, and Carole Mathieu, *Japan's Energy and Climate Policy: Towards Dispelling the Uncertainties* (Institute Français des Relations Internationales Centre Energie, 2015), www.ifri.org/sites/default/files/atoms/files/noteenergiejapon.pdf, 16.
12 IEA, *Japan – Overview*.
13 Vivoda, *Energy Security in Japan*, 109–111; Cornot-Gandolphe, and Mathieu, *Japan's Energy and Climate Policy*, 13.

14 Vivoda, *Energy Security in Japan*, 71; International Gas Union, *World LNG Report – 2014 Edition* (International Gas Union, 2014) www.igu.org/sites/default/files/node-page-field_file/IGU%20-%20World%20LNG%20Report%20-%202014%20Edition.pdf, 9–10.
15 IEA, *Energy Supply Security 2014*, 284.
16 Vivoda, *Energy Security in Japan*, 145.
17 IEA, *Japan – Overview*.
18 Vivoda, *Energy Security in Japan*, 151.
19 Ministry of Economy, Trade and Industry (METI), *Strategic Energy Plan* (METI, 2014) www.enecho.meti.go.jp/en/category/others/basic_plan/pdf/4th_strategic_energy_plan.pdf.
20 Akira Yanagisawa, "Discussions on the Energy Mix," *IEEJ e-Newsletter* no. 61, May 15, 2015, http://eneken.ieej.or.jp/en/jeb/150515.pdf, 3; "Plan Sets Out Japan's Energy Mix for 2030," *World Nuclear News*, June 3, 2015, www.world-nuclear-news.org/NP-Plan-sets-out-Japans-energy-mix-for-2030-0306154.html.
21 *World Nuclear News*, "Plan Sets Out Japan's Energy Mix for 2030."
22 World Nuclear Association, "Nuclear Power in Japan," World Nuclear Association, May 22, 2015, www.world-nuclear.org/info/Country-Profiles/Countries-G-N/Japan.
23 Vivoda, *Energy Security in Japan*, 14.
24 Ibid., 24.
25 METI, *Strategic Energy Plan*, 60–62.
26 Ministry of Economy, Trade and Industry (METI), "Inauguration of the Organization for Cross-regional Coordination of Transmission Operators (OCCTO)," METI, April 1, 2015, www.meti.go.jp/english/press/2015/0401_02.html.
27 Ministry of Economy, Trade and Industry (METI), "Electricity System Reform," METI, April 24, 2015, www.meti.go.jp/english/policy/energy_environment/electricity_system_reform/index.html.
28 Cornot-Gandolphe, and Mathieu, *Japan's Energy and Climate Policy*, 14.
29 Vivoda, *Energy Security in Japan*, 169; Cornot-Gandolphe, and Mathieu, *Japan's Energy and Climate Policy*, 14.
30 Cornot-Gandolphe, and Mathieu, *Japan's Energy and Climate Policy*, 18.
31 United Nations Framework Convention on Climate Change (UNFCCC), *Submission of Japan's Intended Nationally Determined Contribution*, UNFCCC, July 17, 2015, http://www4.unfccc.int/submissions/INDC/Published%20Documents/Japan/1/20150717_Japan's%20INDC.pdf.
32 Ibid.
33 Ibid.
34 Helen Briggs, "Global Climate Deal: In Summary," *BBC World News*, December 12, 2015, www.bbc.com/news/science-environment-35073297.
35 Cornot-Gandolphe, and Mathieu, *Japan's Energy and Climate Policy*, 16–17.
36 Hidemasa Nishiyama, *Japan's Policy on Energy Conservation* (Ministry of Economy, Trade and Industry, 2013), http://eneken.ieej.or.jp/data/4749.pdf, 4.
37 Institute of Energy Economics, Japan, *Japan: Government Estimates the Effects of Energy Conservation, Various Energy-Saving Measures to be Implemented* (Institute of Energy Economics, Japan, 2015), http://eneken.ieej.or.jp/data/6016.pdf.
38 Masatsugu Hayashi and Larry Hughes, "The Fukushima Nuclear Accident and its Effect on Global Energy Security," *Energy Policy* 59 (2013), 104, http://web.mit.edu/mission/www/m2018/pdfs/japan/policy.pdf.
39 Judy Dempsey, "How Merkel Decided to End Nuclear Power," *The New York Times*, August 13, 2011, www.nytimes.com/2011/08/13/world/europe/13iht-germany.html; Hayashi and Hughes, "Fukushima Nuclear Accident," 106.
40 Hayashi and Hughes, "Fukushima Nuclear Accident," 104.
41 Ibid., 106.
42 Emma Chanlett-Avery, Mark E. Manyin, Ian E. Rinehart, et al., *Japan-U.S. Relations: Issues for Congress* (RL33436) (Washington, DC: Congressional Research Service, 2015), www.hsdl.org/?view&did=765132, 14.
43 Mikkal Herberg, Edward Lincoln, and Michael Wallace, "Japan's Energy Security: Outlook and Implications," National Bureau of Asian Research, 2012, www.nbr.org/downloads/pdfs/ETA/ES_Japan_roundtable.pdf, 5.
44 US Energy Information Administration, "Japan," US Energy Information Administration, July 30, 2015, www.eia.gov/beta/international/analysis.cfm?iso=JPN.

45 Yuka Hayashi, "Abe, Putin Agree to Improve Ties Despite Sanctions: Ahead of the Asia-Pacific Economic Cooperation Summit, Japan and Russia Meet to Discuss Joint Efforts," *Wall Street Journal*, November 9, 2014, www.wsj.com/articles/abe-putin-meet-ahead-of-apec-summit-1415560145.

46 METI, *Strategic Energy Plan*, 31.

47 Chanlett-Avery et al., *Japan-U.S. Relations*, 16.

20

Transitions to energy and climate security in Thailand

Adam Simpson and Mattijs Smits

Introduction

This chapter examines the transition of Thailand to energy and climate security as an energy consuming country. As an emerging economy in Southeast Asia with a democratic history, albeit one afflicted by persistent authoritarianism, environmental activists and civil society have played a significant role in the development of public energy discourses and, to a lesser extent, government policies. Governance in Thailand tends to oscillate between direct military rule and more competitive elected governments. A coup in May 2014 resulted in the current military regime, which appears unlikely to surrender power to democratic forces anytime soon. Nevertheless, energy policy over the last two decades has remained largely impervious to changes in government, although much of the good work on developing renewable energy markets is unravelling under the current government.

Thailand is highly susceptible to climate-induced weather extremes: it is one of the top ten countries in the Global Climate Risk Index 1993–2012.[1] It is highly dependent on regular monsoon rains for its food production and increased monsoonal variability, resulting in both flooding and drought, is already beginning to have severe impacts on its economy; these impacts are only likely to increase. It is in this context that we consider Thailand's energy politics and the interactions between the concepts of energy security, modernity and sustainability in the development of its energy policies.

From a critical perspective, energy and climate security should both be considered key components of a broader environmental security, where issues of justice preponderate.[2] While many activist groups promote energy security models that adhere to this approach, government policies have largely adopted a state-centric energy security model that, since the 1990s, has been augmented by neoliberal policies allowing foreign investment and the entrance of private players. In some ways this fracturing of the market has provided space for small scale producers and the renewable energy sector to flourish, but it has also privatized profits while socializing risks.[3]

At the same time it has begun to outsource the environmental pollution and risks associated with large dams and fossil fuel extraction by developing a slate of these projects in its poorer neighbours, primarily Laos and Myanmar, and importing the resultant electricity. Due to all these factors, in the policy trilemma that frames this volume Thailand has largely focused on energy security and economic competitiveness, particularly when the benefits have accrued to industries owned and controlled by elite networks, at the expense of climate security and environmental justice.

This chapter discusses these issues in the following six sections. It begins with an examination of the links between the historical and contemporary concepts of energy security, modernity and sustainability in Thailand and the broader region. It then considers the risks to Thailand of a warming climate and the effects on monsoon variability. The next section examines the dynamics of authoritarianism and environmental activism in Thailand, which is followed by an analysis of Thailand's current energy policies and trajectory. The final main section explores activism over Thailand's fossil fuel energy projects focusing on two case studies: the unsuccessful campaign against the Trans-Thai Malaysian (TTM) Gas Pipeline; and the successful campaign against the Bo Nok coal-fired power plant. The conclusion then considers Thailand's progress in the transition to climate and energy security.

Energy security, modernity and sustainability in Thailand

The history of energy, and energy security more specifically, is strongly related to ideas about modernity. Modernity itself is a big and contested issue, both in the public discourse as well as in academia.[4] The term will here be explored with reference to two ideal-type positions: (1) as a philosophical or epistemological condition; and (2) as a distinct historical or empirical instance.[5] The term is often used in light of the second position, as a specific state to strive for or which has already taken place. In the context of Southeast Asia, the point of reference for modernity as a distinct instance is often various Western countries (including Japan), but also increasingly other Asian countries, such as South Korea and China.[6] By contrast, the first position is not related to a specific time or place, but rather reflects a process of social change which can include specific human-environment interactions.

These abstract ideas about modernity can be helpful to think through and analyse ideas about energy and energy security.[7] Starting with energy security, modernity is often invoked to mean increasing amounts of energy or the use of 'modern' production and use of energy, such as electricity. Throughout history, ideas about what is sufficient energy and what are modern forms of energy have constantly changed. Initially, modern forms of energy security were mainly focused on providing enough energy for the elites in local areas. The broad industrialization of economies has resulted in energy security being primarily considered through the lens of the nation state although it has also expanded to include transboundary energy systems, including electricity grids, gas pipelines and global flows of fossil fuels, such as coal and oil.[8] Critical approaches to energy and environmental security have begun to link human and energy security although the dominant discourse remains wedded to state-centric notions of energy security.[9]

The relationship between modernity and energy security is not one-directional; rather, they are mutually constitutive. New forms of energy production have profoundly influenced many (if not all) aspects of society, such as the way we travel and what and how we eat.[10] The importance of this dialectical relationship between energy (security) and modernity cannot be overestimated.

Energy security has long included concerns over sustainability; in this historical sense sustainability implied that in order to maintain (and perhaps expand) energy systems, there needed to be sufficient supply of fuel and the system needed to be maintained for the duration of its use. In other words, successful energy systems have always been durable systems, which have operated or even expanded over the longer term.[11] In this sense, sustainability has always been closely aligned with the idea of energy security. Sovacool[12] provides a list of 45 definitions of energy security, most of which assume that it applies primarily to the nation-state; from a critical perspective it could be broadly defined as being achieved when there is sufficient energy available to satisfy the reasonable needs of the political community (the referent object) in an affordable, reliable and sustainable manner as long as pursuing it does not cause environmental insecurity to that or any other political community.[13]

This definition alludes to more recent interpretations of sustainability. In the last half century, the meaning of sustainability in relation to energy has slowly changed. Local pollution, oil crises, globalization and most recently climate change have revealed the many 'externalities' related to many energy production and consumption systems.[14] Nowadays, therefore, energy security and sustainability are no longer seen as aligned. In particular fossil fuel regimes are no longer considered to be sustainable, although they may very well be considered energy secure.[15] Some differentiation may occur between different types of fossil fuel: natural gas or 'clean coal' may be considered more sustainable than 'regular' coal, for example. The situation for renewable energy is more complex, as these can be seen as beneficial or detrimental for energy security. Opponents of renewable energy usually stress the unreliability of renewable energy sources, such as wind or solar. Yet others may point to the benefits of renewable energy to decrease their dependence on fossil-fuel producing countries.

The point is not to develop a definitive argument here whether renewable energy systems are more sustainable than fossil fuel systems. Indeed, the examples above show that this depends a lot on the type of energy system and the criteria used. However, the short narrative above does suggest that modernity is implicated in the history and debates about energy security and sustainability. In order to make these points more concrete, particularly in relation to Thailand, it is crucial to understand how energy security, modernity and sustainability have influenced Southeast Asia.

In Southeast Asia, the early developments of fossil fuel and electricity systems have often been strongly related to dependency, colonialism and state-building, aspects which often hang together.[16] Even in countries that were never colonized, such as Thailand, there is a strong link. The first electricity system in Bangkok, for example, was developed by a Danish company in 1884.[17] Thus, from very early on, the energy systems were following 'Western' models of modernity. Also in this case, the first electricity system and first power plant served the local elite – members of the Royal Family and wealthy areas in Bangkok.

From the late 19th century onwards, energy (and its related infrastructure) also played a critical role in the development of the nation-state in Southeast Asia. Roads and railroads were built to 'open up' the hinterlands of many countries in Southeast Asia and, as a consequence, increase their dependency on the capitals.[18] Parts of the country which previously were almost completely independent, gradually got closer (in terms of time) to each other.[19] As such, energy security was (and arguably still is) closely related to state-security and territorialization in Southeast Asia. While the process of expanding infrastructure is still ongoing, in countries such as Thailand, nearly all parts of the country have become connected to (rail)road and electricity networks (with some notable exceptions), while countries like Myanmar still have a long way to go.[20]

In some ways, the expansion of energy systems has always been a matter of energy security in Southeast Asia. This expansion was often limited by the resources and infrastructure available, as well as available expertise and technology. However, concerns about sustainability (in its contemporary meaning) also came into play at times, for example in the construction of hydropower. In many parts of Southeast Asia, early energy systems included large hydropower plants, such as Thailand, Laos and Vietnam.[21] These were not developed because the governments (and foreign investors and construction companies) were concerned about pollution or emissions, but rather because the resource was readily available. It is only more recently that sustainability, and particularly the threat of climate change, has been put forward as an additional concern.[22]

The concerns about sustainability have not only influenced ideas about energy security and modernity, but have also led to increasing contestation of certain energy provision models. While local groups have been resisting state-led energy developments in some parts of Southeast Asia for a long time, they usually have done so because of the local implications of such projects, such as the loss of land, increasing pollution, and influx of migrant workers.[23] Concerns about sustainability have been a more recent issue and have added a new dimension to environmental activism.[24] Some protest movements no longer contest projects for their local impacts only, but also because they continue to use fossil fuels or challenge other aspects of sustainability. Later in the chapter we examine two case studies of campaigns against fossil fuel projects but to contextualize this activism we first examine the climate risks that Thailand faces and its current energy politics and policies.

Climate security

While Thailand faces a plethora of environmental issues, in the long term the impacts of global climate change are likely to significantly exacerbate existing environmental insecurities. Climate change is likely to result in more extremes in cold, and particularly heat, affecting regional weather and climate patterns in Southeast Asia. Much of Thailand is low lying, coastal or otherwise susceptible to weather extremes such as cyclones that are likely to be exacerbated in frequency and intensity by climate change as water temperatures increase. Although too much water is often a significant contributor to insecurity drought is also becoming a problem due to increased monsoonal variability.

The most devastating potential impacts of global climate change will affect different geographical areas in different ways with Africa's mortality and economic loss risk largely due to drought while the impacts on Southeast Asia derive from a multitude of climate-related disasters including droughts, landslides, floods and tropical cyclones.[25] Thailand is particularly susceptible to flooding and is one of the top ten countries in the Global Climate Risk Index 1993–2012,[26] although the floods of 2011 caused 87 per cent of the total damage. The 2011 floods inundated most of central Thailand, and were the worst in 50 years; they also clearly demonstrated that some sectors and people are more important than others.

Climate change has resulted in substantially increased pre-monsoon rainfall in the Chao Phraya River Basin in recent decades and a significant sea level rise at the river outlet; both factors increased the severity of the 2011 floods, which resulted in more than 800 deaths and affected 13.6 million people.[27] Although many parts of Thailand experienced this flooding not all communities or people were affected or treated equally. The management of floods and other disasters in Thailand has been organized by elites and their bureaucracies to be deployed in ways that serve their interests and not those of more politically marginalized groups. This was particularly evident in the 2011 floods where privileged areas of the industrial sector and the associated Thai elites' assets were protected while other, less fortunate areas with fewer political connections, were sacrificed.[28]

The notion of climate security is therefore clearly dependent on the referent object of security. McDonald examines four climate security discourses focused on national security, human security, international security and ecological security and argues that the most powerful discourses of climate security are unlikely to inform a progressive or effective response to global climate change.[29] Similarly, inequalities within society may result in even human security approaches, if applied unevenly, resulting in unequal outcomes. Dominant sectors of society may be privileged in any policy response that attempts to either mitigate or adapt to climate insecurities. It is for this reason that it is crucial that environmental activists, and civil society more generally, are able to give voice to marginalized actors in society. Unfortunately, while Thailand has historically offered more political space for civil society than many of its Southeast Asian neighbours, political space has recently been constricted, with little end in sight.

Authoritarianism and environmental activism in Thailand

Thailand has long been afflicted by authoritarian governance, which has restricted the activities of civil society activists. The ability of social activists and movements to protest openly in Thailand has been largely determined by the nature of the contemporaneous political regime. The ability to substantially influence policy and political outcomes has been tenuous and, even under its most democratic governments, has tended to reflect the extent of accommodation by existing political power structures. These power structures, often allied to the monarchy and linked to structural inequality,[30] run deeply through Thai society and stretch back to its earliest history.

Thailand's nominally modern and democratic political era began in 1932 when a constitutional monarchy replaced absolute rule. For many of the subsequent years the military played a significant role in Thai politics with the prime minister a military officer for all but eight years over the period 1938–88.[31] This militaristic authoritarian rule generally constrained public dissent and criticism of the government. Following a military coup in 1991 and a violent crackdown on unarmed protesters, massive street demonstrations in May 1992, and a carefully orchestrated intervention by the king, squeezed the military from power. There followed a rapid expansion of social activism throughout the 1990s and 2000s in which there was a dramatic increase in NGO activism and increased public debate by academics and intellectuals, although many of the country's powerful and conservative bureaucratic and military structures remained. This increased activism created a new kind of responsive formal politics in Thailand, epitomized by the ascension of Prime Minister Thaksin Shinawatra, although his actions in government eventually cost him the support of many social and environmental activists.

Thaksin initially courted these activists, including those against the TTM Pipeline, but when in power his rhetoric against NGOs and activists provided cover for repressive crackdowns by the security services. Under his 'War on Drugs' between February and May 2003 approximately 2,500 alleged drug traffickers were killed.[32] UN Special Envoy for Human Rights, Hina Jilani, noted in the report of her mission to Thailand soon after that many Thai activists, including those opposing the TTM Pipeline, had reported that they were afraid to highlight human rights violations for fear of retaliation by local authorities, 'including possibly being killed under cover of the anti-drugs campaign.'[33] The threat to environmentalists at this time was highlighted by the murder of prominent activist against the Bo Nok power plant Charoen Wataksorn in June 2004.[34] Apart from the overt harassment of activists and NGOs by the military and police, there was also an increasing threat of violence perpetrated by non-uniformed assassins. Between 2001 and 2005 at least twenty environmentalists, human rights activists and community leaders, including monks, were killed in separate incidents, most of them shot.[35]

Despite these attacks on civil society, Thaksin's populist policies resulted in his parties dominating Thai electoral politics, with Thaksin-led parties winning comprehensive election victories in 2001 and 2005 and proxy parties winning in 2007 and, led by his sister Yingluck, 2011. Since the turn of the century, therefore, Thaksin or his proxies have won every national election.[36] Due to the undermining of democratic institutions and checks and balances under the 1997 Constitution the Thaksin government constituted a competitive authoritarian regime.[37] Nevertheless, this was far more democratic than military rule, which has repeatedly benighted Thailand both before and after Thaksin's government. Thaksin fled into exile following a royal-backed military coup in September 2006 and the military seized power once again in May 2014; it has since denied citizens basic rights including the right to assemble and freedom of expression, and political repression has been extensive.[38]

Soon after the 2014 coup its leader, General Prayut Chan-o-cha, established a military dominated national assembly, which elected him as prime minister. Having been demonstrably outmanoeuvred in every national election this century the military and its royal supporters have decided to avoid the inconvenience of democratic governance, with subsequent proposed draft constitutions effectively diminishing democratic rule in the country. This constriction of political space for activists, journalists and citizens has had broadly adverse impacts on the development of energy and environmental policies in the country, although improvements in some areas have occurred.

Thailand's energy policies

Thailand's history of democratic governance, although punctuated by authoritarian rule, has resulted in more progressive energy and environmental policies than many other countries in the Southeast Asian region.[39] Thailand's environment movement experienced some notable early successes with an official ban on logging in 1989 and the blocking of World Bank-backed Nam Choan Hydroelectric Dam in Kanchanaburi Province in 1988.[40] Unfortunately, illegal logging continued and the ban saw logging expand unchecked in Myanmar, Laos and Cambodia.[41] These impacts had parallels in the energy sector, particularly in the wake of the Nam Choan Dam cancellation and the long struggle around the Pak Mun dam,[42] with the state-owned Electricity Generating Authority of Thailand (EGAT) focusing on cross-border energy projects to import energy from its (at the time) more authoritarian neighbours through projects such as the Yadana, Yetagun and Zawtika Gas Pipelines and proposed Salween Dams in Myanmar and the completed Nam Theun 2 Dam and the Xayaburi and Don Sahong Dams currently under construction in Laos. These projects, nominally for the pursuit of national energy security, have had adverse impacts on the environmental security of local communities.[43] Despite some successes, however, the opportunities for activists to contribute to Thailand's energy and environmental policy development have been somewhat limited.[44]

The main developmental focus of most recent Thai governments has been on ensuring sufficient electricity for unrestricted domestic industrial development and acting as a regional hub of an ASEAN power grid.[45] About 70 per cent of Thailand's electricity is generated using natural gas with approximately one-third coming from Myanmar through the three aforementioned pipelines.[46] Approximately 5 per cent of electricity generation capacity is derived from large-scale hydropower, although if all the plans to import hydroelectric electricity from Myanmar and Laos came to fruition this figure would rise significantly. Dams and hydropower have been promoted by King Bhumibol throughout his long reign and, as with the king himself, have taken on almost mythological proportions.[47]

Despite assertions about the necessity of these projects, the actual electricity needs of Southeast Asian countries are often overstated, with Thailand's energy industry continually overestimating its projected electricity requirements. In 2004 the government's National Economic and Social Advisory Council examined projections by EGAT over the previous decade. It found that in the utility's previous eleven forecasts, ten had overestimated demand, sometimes by as much as 40 per cent. In addition Thailand's use of energy has been quite inefficient; it uses three times more energy per dollar of GDP than Japan.[48] Improved energy efficiency measures in conjunction with smaller scale decentralized renewable energy projects could have made some large-scale energy projects redundant.

The approach of Thai governments to energy production since 1992 has been broadly neo-liberal, particularly since Thaksin came to power and financed his promises to the poor by the privatization of state assets. Thaksin's first privatization was the sale in 2001 of around 32 per cent of the Petroleum Authority of Thailand (PTT), which was a partner in both the Yadana and TTM Pipeline projects. PTT was Asia's third largest oil and gas firm after China-based Sinpec and PetroChina. The sale, however, seemed to have been 'managed' as large holdings were issued to government ministers' families and friends. The issue price was also undervalued as it quintupled over two years. Five other smaller privatization projects followed the same pattern.[49] Privatization was clearly used to further enrich the already rich and powerful. As a result, a movement against privatization in Thailand formed with over one hundred civic organizations, development groups and trade unions joining together to oppose both privatization and the politics behind policy corruption.

Between 2004 and 2006 the government proposed twelve further privatization projects with the first and biggest being EGAT. When the privatization was announced in January 2004 it was strongly opposed by EGAT's union, its former governors and activist groups. Nevertheless, in 2005 the government enacted two Royal Decrees that dissolved the EGAT state enterprise and created the charter of EGAT Plc. The Supreme Administrative Court later revoked these decrees, however, and the 2006 coup effectively ended these privatization policies. Recent governments have embarked on a more subtle process of privatization of EGAT through the use of subsidiaries.

Despite the continued dominance of EGAT and associated state energy utilities in Thailand's electricity market, it has one of the most progressive renewable energy policies in the region, with reforms dating back to 1992 establishing markets with feed-in tariffs for Independent Power Producers (IPPs), Small Power Producers (SPPs) and Very Small Power Producers (VSPPs) (initially 1 MW). The SPP programme, launched in 1992, was designed for small plants, under 90 MW. The IPP programme was launched in 1994 and effectively liberalized the market for generating electricity that could be sold to EGAT. Thailand's VSPP regulations were approved by Cabinet under Thaksin in 2002. Since December 2006 VSPP generators have been allowed to export up to 10 MW to the grid with feed-in tariff subsidies for renewable electricity production, and also efficient fossil-fuel Combined Heat and Power (CHP). There are now a large number of small entrepreneurs active in this sector.[50]

Although it was largely government's neoliberal tendencies that launched the sector, NGOs, particularly Palang Thai,[51] which worked on the VSPP policies from 2001, have been influential in its development. Nevertheless, in recent years, particularly under Yingluck's government and the military regime, there have been significant backward steps in energy governance. Many of the effective energy and electricity governance structures that had been established over the last two decades have been undermined with the success of the renewable energy sector now creating fertile ground for well-connected corporations to extract rents. With high rents added onto costs, inevitable price rises are being associated with renewable energy in general, causing potentially long-lasting damage to community support for the sector as a whole.

This outcome seems consistent with Prime Minister Prayut Chan-o-cha's lauding of fossil fuels as an energy source in mid-2015, when he also instructed the Energy Ministry to boost 'public understanding' about the cost of producing electricity from renewable or alternative energy sources, which he argued would lead to higher power bills.[52] The prime minister has also used his absolute authority under Section 44 of the Interim Constitution to exempt all kinds of power plants, gas processing plants and other utility plants from regulations under the Town and City Planning Act.[53]

On the other hand, since the coup there has been some progress on climate change policy, at least on paper, with Thailand finally submitting its Intended Nationally Determined Contribution (INDC) to the UNFCCC in October 2015, stating that it intends to reduce its greenhouse gas emissions by 20 per cent from the projected business-as-usual (BAU) level by 2030.[54] The Climate Change Master Plan B.E. 2558–2593 (2015–2050) is still awaiting approval from Cabinet but in May 2015 the National Energy Policy Committee (NEPC) approved Thailand's Power Development Plan (PDP) 2015–2036.[55]

The first public hearing for the formulation of the PDP 2015 was held in August 2014,[56] three months after the coup which had banned protest and restricted political freedoms.[57] Journalists and activists had been arrested for voicing opposition to the military government. This was not the most conducive environment to encourage dissenting voices and the regime felt little need to listen if there were. An Alternative Energy Development Plan 2015–2036 exists under the PDP 2015 and is administered by the Department of Alternative Energy Development and Efficiency within the Ministry of Energy: it provides the framework for boosting renewable energy use in the country although as part of its remit it envisages 5 per cent of Thailand's electricity production coming from nuclear power by 2036 and, as discussed above, other government policies are acting to undermine the development of a sustainable renewable energy sector.

Activist groups have also critiqued the process of developing the PDP 2015 and its perceived focus on coal and large hydropower. The Network of People Affected by the Power Development Plan 2015, supported by the Thai Climate Justice Working Group, wrote an open letter to the Prime Minister and Minister of Energy requesting the cancellation of the PDP 2015 and the establishment of a more transparent and democratic process.[58] While Thailand's energy policies in the past have helped develop diverse and, for the region, progressive renewable energy policies, the current trajectory is clearly undermining earlier gains. While the space for environmental activists to influence policy has clearly narrowed it is still instructive for contemporary activists and policy makers to consider two contrasting campaigns against proposed fossil fuel projects.

Campaigns against fossil fuel energy projects

This section examines the strategies and tactics of two environmental campaigns against fossil fuel energy projects in Thailand. The first, ultimately unsuccessful, campaign was against the Trans-Thai Malaysian (TTM) Gas Pipeline Project in Songkhla Province. The second campaign, against the Bo Nok coal-fired power plant, was eventually successful with the power plants still not built on the site. Although both campaigns were against fossil fuel projects it is instructive to note that the gas project was completed while the coal project was cancelled. As noted earlier, emphases on climate security have made natural gas projects more desirable than coal projects. It should also be noted, however, that the gas project, which extracted the gas from the Gulf of Thailand, was geographically fixed while it was possible for the Bo Nok power plant to be moved to another province; this ultimately occurred, as well as shifting the fuel source from coal to gas. While both campaigns suffered from a variety of repressive tactics from government, the campaigns also took place in relatively democratic times, when elected governments could be held to account.

The campaign against the Trans-Thai Malaysian Gas Pipeline

The TTM Gas Pipeline Project through the largely Muslim Songkhla Province in the south of Thailand was punctuated by a variety of local protests following stymied attempts to participate in official public fora, including public hearings. The project required offshore drilling, the construction of two gas separation plants (GSPs) in Chana district on the east coast of Songkhla and the laying of a gas pipeline from the GSPs to the border with Malaysia in the west. The main project proponents were PTT and EGAT, which also planned to build a 700 MW gas-fired power station near the GSP. As the pipeline fed directly from Thailand into the pre-existing Malaysian pipeline network most of the protests and activism occurred on the Thai side of the border. The project was initiated during the Democrat Government of Chuan Leekpai but the project was also taken up by Thaksin Chinawatra when he came to power in January 2001.

Activists began the campaign against the TTM project following the signing of a Memorandum of Understanding between Chuan and Malaysian Prime Minister Mahathir Mohamad in April 1998 and the discovery of plans relating to the industrial development of Songkhla Province within the Indonesia-Malaysia-Thailand Growth Triangle. A broad coalition emerged comprised of environmental organizations, academics and local fisherfolk who argued that serious deleterious impacts upon local communities and their environments would occur for the duration of the project. Local academics at Prince of Songkla University in Had Yai, the capital of Songkhla Province, also questioned the need for the project pointing out that Thailand was importing gas from Myanmar through the Yadana pipeline while planning to export gas to Malaysia. This support from academics in Thailand was crucial to disseminating the campaign's messages throughout the country and was often sought out by the villagers themselves.

Local communities and ethnic minorities were not only most adversely affected by the project, but also the most voiceless communities in the decision-making process. Much of the local activism against the TTM project concerned the lack of genuine consultation and participation in the decision-making processes. The project was carried out under the now-superseded 1997 Constitution, which required greater public participation in development processes and improved checks and balances. The TTM Project was one of the first major tests for the public consultation processes in the Constitution. The Ministry of Industry set up a public hearing in Had Yai in July 2000 although the Environmental Impact Assessment (EIA) had already been completed and published four months earlier.[59] While the EIA process was underway four contracts had been signed by PTT and the Malaysian corporation Petronas on 30 October 1999. PTT argued that the contracts were non-binding and there would be no fine should PTT abandon the project on environmental grounds. This was at odds, however, with a special committee's investigation in January 2001, which claimed that severe penalty payments would arise over postponement due to the 'take-or-pay' nature of the contracts.[60] This therefore indicated to activists that the main decisions on the project had been decided upon prior to the results of the EIA and public hearings, with the result that public concerns were relegated to insignificance.

Compounding this impression was the manipulation of public events to avoid dissenting voices. Prior to the hearing military officers were employed as public relations officers and project opponents suggested these had been used to intimidate and harass them. The public hearing was held at the Municipality Hall in Had Yai on 29 July 2000 and academics and university students tried to broaden the discussion to consider the proposed Indonesia-Malaysia-Thailand Growth Triangle, which they considered was a critical issue to be discussed but the chairperson of the hearing, General Charan Kulavanija, rejected these attempts. Many of the villagers were also excluded from the hearing and violence finally erupted between the opponents of the project and its industry and military supporters as the hearing collapsed.[61] The Ministry of Industry arranged a second public hearing in Had Yai on 21 October 2000 that, again, led to clashes between project supporters and opponents causing the suspension of the meeting.

The new Thaksin government had promised to approach the activists differently to its predecessor and it commissioned reports from the Senate Committee on Environment, the National Human Rights Commission and Chulalongkorn University while Thaksin visited the protesters at Lan Hoy Siab on 4 January 2002 to promise a fair hearing. By May 2002 the reports had been submitted to Thaksin, all recommending that a final decision on the project be postponed until numerous issues related to human rights, the environment and the future energy needs of the region were resolved. Nevertheless on 11 May 2002 the government ignored this advice and announced that it had approved the pipeline.[62]

For the rest of 2002 local villagers in Chana district and environmental organizations, students and academics around the country took every available opportunity to lobby the government. On 20 December approximately 1,000 villagers accompanied by students and human rights activists travelled the 50 kilometres from their villages in Chana district to Had Yai to protest against the TTM project and to hand a petition to the Prime Minister who was meeting his Malaysian counterpart there.[63] Thaksin's aide told them to wait in a specified area and after they did so hundreds of policemen surrounded them and attacked them with batons. The resulting melee left 38 demonstrators and 15 policemen injured. Despite this authoritarian repression Thaksin's electoral popularity ensured that he was re-elected in 2005 with a large majority and the pipeline was eventually completed.

The campaign against the TTM Pipeline was ultimately unsuccessful, because it did not fulfil its aim of stopping the pipeline. It did, however, significantly increase the political and environmental awareness of local communities in Songkhla Province, ensuring that energy and the environment became important issues of political concern. History is, however, repeating itself in the neighbouring Tepha District where EGAT is planning to build a 2,000MW coal-fired power plant, with even worse outcomes expected for the livelihoods of local fishing communities and the Gulf of Thailand's ecosystem. The same tactics used by the government and proponents during the TTM campaign have been dusted off and re-used, with opponents of the project being excluded from public meetings on 27–8 July 2015, by order of the Songkhla governor.[64] This time, however, there are even fewer opportunities for dissent with a military government installed that is seemingly intent on pushing through fossil fuel projects while it can still rule by fiat.

At a similar time to the TTM campaign there was a more successful campaign against the coal-fired power station at Bo Nok. Despite the contemporary restrictions on political activities there are many lessons from the Bo Nok campaign that could be used to promote sustainability and energy and climate security in the campaign against the Tepha and other coal-fired power plants.

The campaign against the Bo Nok coal-fired power station

As demonstrated in the previous section, the conflicts around the siting of a coal-fired power plant in Bo Nok were significant for other environmental campaigns and more broadly for energy and climate transitions in Thailand. The Director-General of the Department of Alternative Energy Development and Energy Efficiency of the Ministry of Energy summarized it as follows:

> These conflicts were a turning point in Thai infrastructure planning, because they were in the time that the Thai society demanded more say in big projects. This was not the only site of protest, but it is the one that got stuck … and still has an impact, because many schemes are blocked. So planning has become more difficult, because Thailand is still growing, so there are less options.[65]

This section unpacks this statement by discussing 'Bo Nok' as a successful case of environmental activism.

The events in Bo Nok are directly related to the attempts of the Thai government to reform the energy sector, resulting from pressure from the World Bank and the IMF. As one of the first steps, Thailand introduced its IPP policy in 1992.[66] In the first round of bidding, two coal-fired power stations were proposed in the province Prachuab Khiri Khan, some 300 km south of Bangkok, one in Ban Krut (1400 MW) and one in Bo Nok (700 MW, to be extended to 1,400 MW later). Both projects were initially backed by international companies, Hin Krut by Japanese and Hong Kong investors and Bo Nok by the US-based Gulf company. The remainder of this case study will primarily focus on the situation in Bo Nok, although there are many parallels and indeed synergies with the situation in Ban Krut.

Soon after the IPP bidding was concluded, in 1995, people in Bo Nok noticed that a large amount of land was bought on the coast. This also marked the start of a long period of activism with varying degrees of intensity. An important characteristic was that the movement started and remained strongly driven by people in the local communities. When the company presented its plans to build the coal-fired power plant, there were some immediate disagreements and protests, as people were concerned about the consequences for the environment and their livelihoods. They found out that the coal-fired power station would damage their marine environment, and lead to local air, land and thermal pollution.[67] The 'movement' in Bo Nok started by contacting the company, the local government and organizing protest activities, such as sending 1,000 letters to the sub-district office and meet with the community from Ban Krut in 1997.

An important change in the movement and profile of the conflict came a few years later, in 1998, when the movement started to attract national, and later international, attention. The protest group in Bo Nok (and Ban Krut) started to team up with academics, NGOs, and other environmental movements in the country (TERRA, AEPS, Palang Thai), drawing on decades of experience and networking on environmental issues. Through these contacts, the movement was able to contest not just the project itself, but also the energy policy and EIA process in Thailand. For example, they proved that the EIA had left out critical information about the impacts on the coral reefs and sea animals. Moreover, they showed that the energy forecasts were too high, resulting in a reserve margin of more than 25 per cent. In addition, they managed to mobilize more than 10,000 people in front of the presidential office and block the Southern Highway, the main road connecting Bangkok with the provinces in the south. This activity – which ended violently – in particular featured in the national media and turned it into a 'national' problem and even internationally, through the involvement of Greenpeace and Probe International.

Importantly, while the movement was supported by a large number of people, it also divided the community in the area from the start. Some people believed that the power plant – and associated industrial development in the area – was a positive development. In addition, there were widespread rumours of the company trying to 'buy' the support of the community and its leaders. Proponents of the power plant in the local area say the same about the movement, pointing fingers at local and international NGOs. While there is no conclusive evidence either way, the siting of the Bo Nok coal-fired power station did lead to a strong divide in the community, which still remains.[68]

The advantage of the environmental movement in Bo Nok was time. By putting continuous pressure on the company and local, provincial and national governments, they managed to stall the development of the plant, leading to big revenue losses for the IPP company. In Bo Nok alone, 94 different letters were sent to various parties and at least 28 protest rallies organized. These ongoing activities and the changing political situation (Thaksin Shinawatra came to power in 2001), finally led the government to change the siting of the Bo Nok (and Ban Krut) power stations to Saraburi province (in the case of Bo Nok) and switch from coal- to gas-fired power stations.

The story of Bo Nok does not end here. During a series of court cases in the aftermath of the government's decision, the leader of the Bo Nok movement, Charoen Wataksorn, was killed by two hired gunmen in 2004. This killing has never been fully resolved, in part because both of them died shortly after in jail. Moreover, the community remains divided on the topic. As such, this violent history has turned the whole area into a 'place of concern'.[69]

To sum up, there are a number of important features that can be learned from this successful case of environmental activism. First, there are a number of rather unique features of this case, such as the fact that it all took place in a Central Thai province, in a well-off area in which the majority of people supports the conservative Democrat Party. This is in contrast to many environmental movements in North and North-eastern Thailand, where people have fewer monetary and political resources making it more difficult to organize prolonged protest. Moreover, in Bo Nok, the leaders and other key figures were able to challenge the company and the government on the actual content of the EIA and the Power Development Plans, through study of these documents and engagement with other critics of these policies. Many academics and policy makers agree that this was the first time that a 'local' protest movement managed to challenge, and eventually influence, national policy making.

Despite these features, the movement in Bo Nok was not a stand-alone activity, but rather part of a long history of social and environmental movements in Thailand. The people in Bo Nok were able to draw on support and experience from Thai NGOs, academics and also some international support. As the movement grew, the people in the movement also started to become an inspiration themselves for other movements in the other places, such as the controversial Mae Mo coal-fired power stations. Among the clearest features are the statue of the assassinated Charoen and a nearby training centre for environmental activists.

The final outcome of the activist movement was not only a result of the strength of the movement, but also a result of the changing political landscape and discourse about energy security in Thailand: the election of Thaksin Shinawatra as prime minister accelerated the decision not to build the coal-fired power stations. Moreover, there was an alternative solution in the form of a change in siting and fuel-switch from coal to gas. As such the company kept the contract and was even awarded a premium on their tariff as a result of the losses they suffered in the process.

Conclusion

During the writing of this chapter, the latest conflict around fossil fuel energy – an 800MW EGAT coal-fired power station in tourist destination Krabi – is becoming more prominent. This case has many similarities to the cases of Bo Nok and the TTM Pipeline and shows that, despite the non-democratic government, Thailand still has not found a way to transition peacefully to climate and energy security. On a more positive note, General Prayut Chan-o-cha put the Krabi coal-fired power plant project on hold in July 2015 and set up a joint committee that included all stakeholders to discuss, study and improve the plan, which brought an end to protests outside Government House.

This example makes clear that Thailand's elite is still trying to achieve modernity at the expense of less powerful actors in society, a process that is deeply engrained in its history. However, environmental activists have become more vocal and renewable energy and energy efficiency initiatives are now more seriously considered, even by some more conservative stakeholders. Moreover, in the quest for energy security, neoliberal tendencies have forced the Thai energy sector to open up to international finance. The downside of this process is that Thailand has increasingly 'outsourced' its energy externalities – pollution and protest – to neighbouring developing countries, such as Laos and Myanmar.

One of the case studies in this chapter, on the TTM pipeline, paints a grim picture of the future of energy security, modernity and sustainability in Thailand, namely more of the same centralized energy production without involvement of civil society. The other case study, of the coal-fired power plant in Bo Nok, shows that alternative scenarios are possible, and that the elite has to bend and perhaps at some point fully open up to the demands for more renewable, decentralized and cleaner energy production in Thailand and continue to be an example in the region.

Notes

1 S. Kreft, and D. Eckstein, *Global Climate Risk Index 2014* (Bonn, Germany: Germanwatch, 2013), https://germanwatch.org/en/download/8551.pdf.

2 A. Simpson, *Energy, Governance and Security in Thailand and Myanmar (Burma): A Critical Approach to Environmental Politics in the South* (New York: Routledge, 2014), 191–196.

3 Chuenchom Sangarasri Greacen, and C. Greacen, 'Thailand's Electricity Reforms: Privatization of Benefits and Socialization of Costs and Risks', *Pacific Affairs* 77, no. 3 (2004): 517–541.

4 A. Giddens, *The Consequences of Modernity* (Stanford: Stanford University Press, 1990); A. Martinelli, *Global Modernization: Rethinking the Project of Modernity* (London: Sage, 2005); U. Strohmayer, 'Modernity', in *Dictionary of Human Geography*, ed. D. Gregory et al. (Malden, MA: Blackwell, 2009), 471–474.

5 P. Wagner, *Theorizing Modernity: Inescapability and Attainability in Social Theory* (London: SAGE, 2001).

6 S.-J. Han, and Y.-H. Shim, 'Redefining Second Modernity for East Asia: A Critical Assessment', *The British Journal of Sociology* 61, no. 3 (2010): 465–488.

7 S. Jasanoff, and S.-H. Kim, 'Sociotechnical Imaginaries and National Energy Policies', *Science as Culture* 22, no. 2 (2013): 189–196.

8 G. Bridge, 'Past Peak Oil: Political Economy of Energy Crises', in *Global Political Ecology*, ed. R. Peet, P. Robbins, and M. Watts, 307–324 (New York: Routledge, 2011); V. Smil, *Energy at the Crossroads: Global Perspectives and Uncertainties* (Cambridge, MA: MIT Press, 2003).

9 A. Simpson, 'Challenging Inequality and Injustice: A Critical Approach to Energy Security', in *Environmental Security: Approaches and Issues*, ed. R. Floyd, and R. Matthew (New York: Routledge, 2013), 248–263.

10 D. Boyer, 'Energopolitics and the Anthropology of Energy', *Anthropology News* 52, no. 5 (2011): 5–7; V. Smil, *Energy in World History* (Boulder, CO: Westview Press, 1994); V. Smil, *Energy Transitions: History, Requirements, Prospects* (Santa Barbara, CA: Praeger Publishers, 2010).

11 F. W. Geels, 'Technological Transitions as Evolutionary Reconfiguration Processes: A Multi-level Perspective and a Case-study', *Research Policy* 31, no. 8–9 (2002): 1257–1274; G. Verbong, and F. Geels, 'The Ongoing Energy Transition: Lessons from a Socio-Technical, Multi-level Analysis of the Dutch Electricity System (1960–2004)', *Energy Policy* 35, no. 2 (2007): 1025–1037.

12 B. K. Sovacool, 'Introduction: Defining, Measuring, and Exploring Energy Security', in *The Routledge Handbook of Energy Security* (New York: Routledge, 2011), 3–6.

13 Simpson, 'Challenging Inequality and Injustice'.

14 D. Toke, and S.-E. Vezirgiannidou, 'The Relationship between Climate Change and Energy Security: Key Issues and Conclusions', *Environmental Politics* 22, no. 4 (2013): 537–552.

15 R. Hillebrand, 'Climate Protection, Energy Security, and Germany's Policy of Ecological Modernisation', *Environmental Politics* 22, no. 4 (2013): 664–682.
16 J. Rigg, *Southeast Asia: The Human Landscape of Modernization and Development*, 2nd edn. (London: Routledge, 2003).
17 C. Greacen, 'The Marginalization of "Small is Beautiful": Micro-hydroelectricity, Common Property and the Politics of Rural Electricity Provision in Thailand' (Ph.D. dissertation, University of California, Berkeley, 2004).
18 N. Fold, and P. Hirsch, 'Re-thinking Frontiers in Southeast Asia', *The Geographical Journal* 175, no. 2 (2009): 95–97.
19 Rigg, *Southeast Asia.*
20 R. Lee, 'Tools of Empire or Means of National Salvation? The Railway in the Imagination of Western Empire Builders and Their Enemies in Asia' (working paper, Institute of Railway Studies and Transport History, University of York, Heslington, UK, 2003); N. Starostina, 'Ambiguous Modernity: Representations of French Colonial Railways in the Third Republic', in *Proceedings of the Western Society for French History*, Vol. 38 (Ann Arbor, MI: University of Michigan Library, 2010).
21 P. Hirsch, 'Large Dams, Restructuring and Regional Integration in Southeast Asia', *Asia Pacific Viewpoint* 37 (1996): 1–20.
22 K. Bakker, 'The Politics of Hydropower: Developing the Mekong', *Political Geography*, 18, no. 2 (1999): 209–232.
23 B. D. Missingham, *The Assembly of the Poor in Thailand: From Local Struggles to National Protest Movement* (Chiang Mai, Thailand: Silkworm Books, 2003).
24 T. Forsyth, 'Social Movements and Environmental Democratization in Thailand', in *Earthly Politics: Local and Global in Environmental Governance*, ed. S. Jasanoff and M. L. Martello (Cambridge, MA: MIT Press, 2004).
25 C. Webersik, *Climate Change and Security: A Gathering Storm of Global Challenges* (Santa Barbara, CA: Praeger, 2010), 85.
26 Kreft, and Eckstein, *Global Climate Risk Index 2014.*
27 S.-Y. Parichart Promchote, S. Wang, and P. G. Johnson, 'The 2011 Great Flood in Thailand: Climate Diagnostics and Implications from Climate Change', *Journal of Climate* 29, no. 1 (2016): 367–79.
28 L. Lebel, J. B. Manuta, and P. Garden, 'Institutional Traps and Vulnerability to Changes in Climate and Flood Regimes in Thailand', *Regional Environmental Change* 11, no. 1 (2011): 45–58; D. Marks, 'The Urban Political Ecology of the 2011 Floods in Bangkok: The Creation of Uneven Vulnerabilities', *Pacific Affairs* 88, no. 3 (2015): 623–651; A. Salamanca, and J. Rigg, 'Adaptation to Climate Change in Southeast Asia: Developing a Relational Approach', in *Routledge Handbook of the Environment in Southeast Asia*, ed. P. Hirsch (New York: Routledge, 2016).
29 M. McDonald, 'Discourses of Climate Security', *Political Geography* 33 (March 2013): 42–51.
30 K. Hewison, 'Considerations on Inequality and Politics in Thailand', *Democratization* 21, no. 4 (2014): 846–866.
31 Pasuk Phongpaichit, and C. Baker, *Thaksin*, 2nd ed. (Chiang Mai, Thailand: Silkworm, 2009).
32 Amnesty International, *Thailand: Memorandum on Human Rights Concerns* (London: Amnesty International, 2004).
33 H. Jilani, *Report by the Special Representative of the Secretary-General on the Situation of Human Rights Defenders: Mission to Thailand* (United Nations Commission on Human Rights, Promotion and Protection of Human Rights: Human Rights Defenders, 2004), 18.
34 Somchai Phatharathananunth, *Civil Society and Democratization: Social Movements in Northeast Thailand* (Copenhagen: NIAS Press, 2006), 222.
35 Amnesty International, *Thailand.*
36 E. Biel, N. Hicks, and M. McClintock, eds, *Losing Ground: Human Rights Defenders and Counterterrorism in Thailand* (Washington, DC: Human Rights First, 2006), 22; Thitinan Pongsudhirak, 'Thailand's Uneasy Passage', *Journal of Democracy* 23, no. 2 (2012): 47–61.
37 Simpson, *Energy, Governance and Security.*
38 K. Hewison, 'Thailand: The Lessons of Protest', *Asian Studies: Journal of Critical Perspectives on Asia* 50, no. 1 (2014): 1–15; Pavin Chachavalpongpun, 'The Politics of International Sanctions: The 2014 Coup in Thailand', *Journal of International Affairs* 68, no. 1 (2014): 169–185.
39 A. Simpson, 'Challenging Inequality and Injustice: A Critical Approach to Energy Security', in *Environmental Security: Approaches and Issues*, ed. R. Floyd, and R. Matthew (New York: Routledge, 2013),

248–263; M. Smits, *Southeast Asian Energy Transitions: Between Modernity and Sustainability* (Farnham: Ashgate, 2015).

40 T. Forsyth, 'Environmental Social Movements in Thailand: How Important is Class?', *Asian Journal of Social Sciences* 29, no. 1 (2001): 5; J. Rigg, 'Thailand's Nam Choan Dam Project: A Case Study in the "Greening" of South-East Asia', *Global Ecology and Biogeography Letters* 1, no. 2 (1991): 46.

41 P. Hirsch, 'Globalisation, Regionalisation and Local Voices: The Asian Development Bank and Re-scaled Politics of Environment in the Mekong Region', *Singapore Journal of Tropical Geography* 22, no. 3 (2001): 241.

42 T. Foran, and K. Manorom, 'Pak Mun Dam: Perpetually Contested?', in *Contested Waterscapes in the Mekong Region: Hydropower, Livelihoods and Governance*, ed. F. Molle, T. Foran, and M. Käkönen (London: Earthscan, 2009), 55–80.

43 Piya Pangsapa, and M. J. Smith, 'Political Economy of Southeast Asian Borderlands: Migration, Environment, and Developing Country Firms', *Journal of Contemporary Asia* 38, no. 4 (2008): 485–514; A. Simpson, 'The Environment-energy Security Nexus: Critical Analysis of an Energy "Love Triangle" in Southeast Asia', *Third World Quarterly* 28, no. 3 (2007): 539–554; A. Simpson, 'Challenging Hydropower Development in Myanmar (Burma): Cross-border Activism under a Regime in Transition', *The Pacific Review* 26, no. 2 (2013): 129–152.

44 D. H. Ungera, and Patcharee Sirorosb, 'Trying to Make Decisions Stick: Natural Resource Policy Making in Thailand', *Journal of Contemporary Asia* 41, no. 2 (2011): 206–228.

45 Greacen, and Greacen, 'Thailand's Electricity Reforms', 538.

46 Simpson, *Energy, Governance and Security*; International Energy Agency, *Thailand: Statistics for This Country* (International Energy Agency, 2015), www.iea.org/countries/non-membercountries/thailand/statistics/.

47 D. J. H. Blake, 'King Bhumibol: The Symbolic "Father of Water Resources Management" and Hydraulic Development Discourse in Thailand', *Asian Studies Review* 39, no. 4 (2015): 649–668.

48 A. Imhof, 'Making Smart Choices for the Mekong', *World Rivers Review* 20, no. 5/6 (2005): 8–9.

49 Phongpaichit, and Baker, *Thaksin*, 13.

50 S. Tongsopit, and C. Greacen, 'An Assessment of Thailand's Feed-in Tariff Program', *Renewable Energy* 60 (2013): 439–445.

51 'Palang Thai', Palang Thai, 2015, www.palangthai.org/en/home.

52 *Bangkok Post*, 14 August 2015.

53 *Prachatai*, 22 January 2016.

54 Raweewan Bhuridej,*Submission by Thailand to UNFCCC: Intended Nationally Determined Contribution and Relevant Information* (Secretary General, Office of Natural Resources and Environmental Policy and Planning Bangkok, 2015), http://www4.unfccc.int/submissions/INDC/PublishedDocuments/Thailand/1/Thailand_INDC.pdf.

55 Chavalit Pichalai, *Thailand's Power Development Plan 2015 (PDP 2015)* (PowerPoint) (Bangkok: Ministry of Energy, 2015).

56 Ibid.

57 Hewison, 'Thailand: The Lessons of Protest'.

58 Network of People, *Open Letter: People's Demand to Cancel PDP2015 and Start a New Transparent Process* (Network of People, September 7, 2015), www.thaiclimatejustice.org/knowledge/view/126.

59 Warasak Phuangcharoen, 'The Failure of Public Participation in Developing Countries: Examples from the Yadana and JDA Pipeline Projects in Thailand', *Thai Khadi Journal* 2, no. 1 (2005): 27; Supara Janchitfah, *The Nets of Resistance* (Bangkok: Campaign for Alternative Industry Network, 2004).

60 Phuangcharoen, 'The Failure of Public Participation', 25–29.

61 Phuangcharoen, 'The Failure of Public Participation', 27–28; Janchitfah, *The Nets of Resistance*, 47.

62 Phuangcharoen, 'The Failure of Public Participation', 31–33.

63 Jilani, *Report by the Special Representative*, 15.

64 *Bangkok Post*, 30 October 2015.

65 Jo Garcia, interview with author, 15 September 2011.

66 Greacen, C. 'The Marginalization of "Small is Beautiful": Micro-hydroelectricity, Common Property and the Politics of Rural Electricity Provision in Thailand'. PhD. dissertation, University of California, Berkeley, 2004.

67 N. Kuze, *Multi-Organizational Relations in Social Movement: A Case Study of Anti-Power Plant Movements in Hinkrut and Bonok* (in Thai) (Bangkok: Chulalongkorn University, 2002).

68 J. J. Schatz, 'With Their Lives: Those Fighting Toxic Dumping and Coal-fired Power Plants Have an Unfortunate Tendency to Turn Up Dead', *Al Jazeera America*, 5 April 2014.
69 C. Schaeffer, and M. Smits, 'From Matters of Fact to Places of Concern? Energy, Environmental Movements and Place-making in Chile and Thailand', *Geoforum* 65 (2015): 146–157.

21

Managing energy and climate policy challenges in Pakistan

Modest progress, major problems

Michael Kugelman

Pakistan faces a conundrum. It is mired in an acute energy crisis, and at the same time it is highly vulnerable to the effects of climate change. And yet in its efforts to ease its energy woes, the country risks worsening its climate vulnerability in a big way.

The good news is that Pakistan recognizes the links between energy and climate change. It has taken some small but promising steps, both in its energy and climate change policies, to mitigate the effects of global warming. Ultimately, however, Pakistan faces tremendous obstacles in its efforts to get relief from an all-encompassing energy crisis, and from the increasingly alarming effects of climate change. For policymakers, the challenges are daunting.

The stakes are high. In volatile, nuclear-armed Pakistan, the security implications of failing to properly address these immense energy and climate challenges are immense.

A deep and destabilizing energy crisis[1]

Over the course of 2014 and 2015, Pakistan experienced energy deficits of 4,500 to 5,000 megawatts, or MW (they sometimes soared to 8,500 MW – more than 40% of national demand). These figures approximate those of similarly energy-insecure India, which saw shortfalls between 3,000 and 7,500 MW over the same period of time.[2] Pakistan's urban areas regularly experience several hours of daily outages, while in some rural regions residents are lucky to receive four hours of electricity per day. Consumption levels of Pakistan's two most heavily utilized sources of energy, oil and gas, are so high that Pakistan's national oil and gas company has predicted that indigenous oil reserves will be exhausted by 2025, and domestic natural gas reserves by 2030.

Pakistan's energy problems, however, are arguably rooted more in shortages of governance than of pure supply. The energy sector suffers from widespread inefficiencies, including transmission and distribution losses that exceed 20%, as well as from several billion dollars of debt. The losses are caused by bad equipment, poor maintenance, and energy theft. The debt is a consequence of cash flow problems: energy generators, distributors, and transmitters lack funds. This is due in part to a flawed pricing policy: the Pakistani government charges a pittance for energy, and yet few customers pay their bills. As a result, revenue is scarce, and the sector literally cannot afford to provide energy.

Pakistan's energy crisis has troubling implications for its fragile economy and volatile security situation. In recent years, power shortages have cost the country up to 4% of gross domestic product (GDP). Hundreds of factories (including more than 500 in the industrial hub city of Faisalabad alone) have been forced to close. According to the World Bank's 2016 Doing Business rankings, businesses in Pakistan have estimated losses from power outages at up to a whopping 34% of annual revenue.[3]

Meanwhile, the energy crisis has sparked demonstrations that sometimes turn violent. Additionally, militants are happy to exploit Pakistan's energy insecurity. In recent years, separatists in the insurgency-riven province of Baluchistan have targeted dozens of gas lines. In January 2015, insurgents in Baluchistan blew up two key towers near a major power station, tripping the national grid and plunging 80% of the country into darkness. In 2013, the Pakistani Taliban attacked a power station that cut off electricity throughout the city of Peshawar, which has a population nearly as large as Los Angeles.

Wide expanses of Pakistan's population are affected by this energy crisis. Shortages not only prevent people from working, but also from cooking and receiving proper medical care (in some hospitals, services have been curtailed). Not surprisingly, public opinion polls in Pakistan identify electricity shortages as one of the country's top problems.

Severe climate change vulnerability[4]

The low-lying, lower-riparian, flood-prone nation of Bangladesh is often held up as the poster child for climate change vulnerability in South Asia. In fact, Pakistan is right up there with it – and by some measures it actually qualifies as one of the most climate vulnerable nations in the world. Germanwatch's Global Climate Risk Index measures the extent to which nations are impacted by weather-related disasters (mainly using the criteria of death tolls and financial losses). The 2015 index ranks Pakistan as the 10th most impacted country in the world during the 1994–2013 period – though it was number three in 2012 and number six in 2013.[5] A comprehensive United Nations Development Programme study from 2015 reaches a similar conclusion: Pakistan, it contends, "is assessed to be one of the most vulnerable countries in the world to climate change." The study continues:

> Pakistan's extreme vulnerability ... is understandable owing to its geographic, demographic and diverse climatic conditions. Of particular concern are the CC [climate change] threats to water, energy and food security due to the inherent arid climate coupled with the high degree of reliance on water from glacial snowmelt.[6]

Pakistan's severe climate vulnerability is amplified by the staggering scale and array of natural disasters and extreme weather events that have hit the country in recent years. In the summer of 2015, a searing heat wave killed more than 1,200 people (most of them in the megacity of Karachi) over a one-week period – a death toll that some observers attributed in part to stifling homes suffering through hours-long power outages.[7] In 2014, severe drought in the bone-dry Thar desert region of Sindh province claimed the lives of dozens of people (many of them children) over a three-month period. A sobering UN assessment found that the drought "devastated" crops and livestock, and displaced hundreds of thousands of people.[8]

These scorching temperatures and drought conditions exacerbate Pakistan's already-severe water insecurity.[9] According to a 2015 International Monetary Fund report, Pakistan is the third-most water stressed country in the world, and its per capita annual water availability has plunged perilously close to the 1,000 cubic meter scarcity threshold.[10] In the coming years, Pakistan's water shortages could be further intensified by glacial melt. Pakistani officials have claimed that glacial recession on the country's mountains has increased by nearly 25% in recent years. This is all the more significant given that the Indus River Basin – Pakistan's chief water source – obtains its water stocks from the snows and rains of the western Himalayas. This majestic mountain region has experienced glacial thinning of up to a meter per year.

Pakistan also experiences torrential rains and damaging floods. In 2010, climatologists judged that the catastrophic deluge that convulsed Pakistan that year – submerging a fifth of the country and displacing millions constituted "the worst natural disaster to date attributable to climate change." The next year brought record-setting monsoon rains; monsoon amounts were a staggering 1,170% above normal in the country's south. The destructiveness of Pakistan's floods is exacerbated by rampant deforestation. The country suffers from the highest annual rate of deforestation in Asia (it lost a full third of its forest cover between 1990 and 2010), with barely 2% of its total area remaining forested today. One reason for this state of affairs is the illicit logging trade of the Pakistani Taliban. In 2009, when it briefly controlled the northern region of Swat, its timber business eliminated up to 15% of Swat's forest cover.

The uptake is troubling. Experts estimate that roughly a quarter of Pakistan's land area and half of its population of nearly 200 million are vulnerable to climate change-related disasters. Environmental officials have warned that in the southern province of Sindh alone, "millions of people" face "acute environmental threats." It is easy to understand why. A 2015 study of coastal communities in Sindh found that strong fluctuations in air temperatures, along with dry weather and warm surface sea temperatures, contributed to reductions in mangrove tree growth; decreases in agricultural growth; habitat losses for numerous types of species; and human displacement and migration.[11] Meanwhile, Baluchistan – a poor and dry, though resource-rich, province – is also highly vulnerable to climate change. A 2012 assessment concluded that climate change poses multi-dimensional risks to many critical sectors, including water, agriculture, horticulture, forestry, and livestock. "If unmanaged," the study warns, climate impacts could pose "a severe development hazard."[12]

Environmentally damaging energy fixes

There is reason to fear that in its zeal to tackle the energy crisis, Pakistan could worsen its climate vulnerability. The national government, led by Prime Minister Nawaz Sharif, has announced its intention to revamp the country's energy mix by making a big push for coal. (Pakistan's energy mix has been dominated by oil and gas in recent years, with coal making a more modest contribution.) Islamabad vows to exploit vast untapped coal reserves in Thar, and to develop coal projects in Baluchistan (the government also hopes to ramp up imports of coal while these indigenous extractive endeavors are underway). The Thar reserves are estimated to hold a total of 175 million tons of coal, and Pakistan hopes to use them to generate 100,000 MW of power.[13]

Proponents of this policy contend that coal boasts three key qualities in Pakistan that other energy sources, including renewables, do not: immediate availability, low cost, and capacity to provide uninterrupted power supply. Top energy officials in Pakistan acknowledge coal's harmful climate impacts, but they contend that economic considerations must trump environmental ones. "We are a poor country," Sharif's energy adviser said in 2014. "We have to create a portfolio that is affordable."[14] In effect, Pakistan intends to ease its energy crisis by embracing one of the world's dirtiest fuels.

To be sure, Pakistan's coal policy is fraught with challenges. The country lacks much of the capital and technology necessary for intensive mining, and its roads and railways are poorly equipped to transport extracted coal across the country. Help is on the way, however, thanks to a mammoth Chinese infrastructure investment, announced in 2015, that will bring an estimated $35 billion worth of energy projects to Pakistan. These include, notably, several coal-fired power plants.

In addition, renewables are given relatively short shrift in the country's National Power Policy of 2013 (one of Pakistan's most recent, and only, comprehensive energy policies). The policy contains only a few mentions of wind and solar projects, most of which are described as aspirational more than operational (there is also a vague reference to subsidizing low-cost renewable energy for consumers). The term "climate change" is not mentioned at all.[15]

Furthermore, Pakistan's 2015–16 annual budget allocated very few resources to climate-related issues. In fact, of the 40 federal ministries or divisions awarded funds from Pakistan's Public Sector Development Program, the Ministry of Climate Change received the fourth-lowest allotment – an infinitesimal 0.016%. The ministry's allotment of 40 million rupees, roughly $380,000, was actually higher than the 25 million rupees awarded to it in the previous two budgets.[16] On the whole, federal climate-related expenditures have totaled about 6 to 8% of all federal expenditures in recent years.[17]

Small but encouraging steps for climate change mitigation

All of the above suggests that Pakistan does not accord much if any priority to climate change mitigation, and strengthens the assertion that corrective measures for its energy crisis are doomed to exacerbate climate vulnerability instead of ease it.

This assertion, however, is belied by the reality on the ground. Pakistan has in fact expressed considerable concern about its climate vulnerability and taken concrete measures to address it. As far back as 2002, the country formed a Global Change Impact Study Center to conduct climate change research and to advise policymakers and planners about climate issues. In 2005, the government established a Committee on Climate Change, overseen by the prime minister. In 2010, Pakistan's Planning Commission – a government research and advising body – published a report on climate change impacts in Pakistan. This report inspired the Ministry of National Disaster Management to develop a National Climate Change Policy and Action Plan.

New policies

Three years later, in 2013, Pakistan's Ministry of Climate Change launched a formal National Climate Change Policy (NCCP) (it was drafted in 2012).[18] The policy acknowledges that Pakistan's energy sector is the country's largest source of greenhouse gas emissions, and highlights the need for an integration of climate change and energy policy. While admitting that Pakistan's future embrace of coal is all but inevitable, it makes a strong pitch for the acquisition of clean coal technologies and for coal production techniques that minimize environmental damage. The policy also encourages the development of renewable energy resources, and advocates for new building designs that incorporate solar panels and other earth-friendly measures (when the NCCP was launched, there were already plans for some buildings in Karachi to use stormwater harvesting for plant-watering, and wastewater for fountains, fire control, and restrooms). Additionally, the NCCP proposes the imposition of a carbon tax to discourage environmentally damaging energy generation; the development of indigenous technologies for carbon dioxide capture and storage; and the slow introduction of "green fiscal reforms" in different sectors (including water and energy) to reduce carbon emissions.

Significantly, the NCCP also links environmentally friendly energy policies to broader national benefits. More energy efficiency and conservation, it argues, constitute "excellent and cost-effective ways" to ensure sufficient energy supplies and the achievement of economic development goals.

Later in 2013, an implementing framework was published for the NCCP. It outlines climate change adaptation actions across a variety of sectors, including energy, water, and agriculture, and assigns explicit timeframes for the completion of such actions.[19]

Pakistani officials have remained acutely aware of this matter in the few years since the NCCP was launched. In late 2014, the Pakistan Agricultural Research Council urged agricultural scientists, policymakers, and environmental experts to develop comprehensive policies to mitigate climate change impacts on the agricultural sector (which is the largest contributor to Pakistan's economy).[20] In August 2015, Pakistan's Senate Standing Committee on Climate Change described climate change as a "big threat" and called on all government ministries to help mitigate it – in ways that include launching public awareness campaigns and reducing the amount of illegally held forestland.[21] Several weeks later, top authorities at the Ministry of Climate Change and the Parliamentary Task Force on Millennium Development Goals joined forces to announce the need for a greater focus on climate change and its threats.[22]

Meanwhile, the donor community is doing its part to help. After a meeting between U.S. President Barack Obama and Prime Minister Sharif in October 2015, their two countries announced the formation of a U.S.-Pakistan Clean Energy Partnership to help attract investment for renewable energy projects.[23] Earlier in 2015, Pakistan agreed to arrangements with two European companies (Vestas of Denmark and PROPARCO of France) to develop wind energy projects with a combined capacity of 350 MW.[24] Additionally, the aforementioned $35 billion Chinese investment package for Pakistan's energy sector does involve several renewable projects, including a solar energy park. In May 2015, Pakistan launched its first solar power plant – a project funded by the Chinese – which initially had a modest 100-MW generation capacity but was expected to ramp up to 1,000 MW in 2016.[25]

Producer and consumer incentives

Pakistan is also taking concrete steps to incentivize producers and consumers to embrace renewable energy – a smart move in a country with plenty of sunshine and wind. On the producer side, Pakistan in 2015 announced generous upfront tariffs – fixed costs paid out as a lump sum – to solar and small hydro power producers (this encouraging development, however, was somewhat dampened by the announcement several months later that upfront tariffs would be reduced for wind energy producers).[26]

Meanwhile, on the consumer side, government authorities have approved new measures that facilitate the installation of rooftop solar panels for private use, and enable solar-powered homeowners to receive credits on future energy bills if they allow their excess solar power to be supplied to the national grid. In early 2015, the State Bank of Pakistan and the Alternative Energy Development Board announced a new mortgage financing option that enables homeowners to borrow up to $50,000 against their mortgage to pay for the installation of rooftop solar panels.[27] The provincial government of Khyber-Pakhtunkhwa has gone even further. In February 2015, it announced that it would provide solar power to nearly 6,000 off-grid households in 200 villages. Provincial authorities agreed to foot 90% of the bill – leaving only 10% to be paid by households in one of Pakistan's poorest regions.[28]

Climate change justice: a case study

Perhaps the most striking illustration of the seriousness with which Pakistan regards climate change comes not from the government, but from the courts – and specifically the High Court of Lahore, the capital of Pakistan's most populous province of Punjab. In September 2015, in a ruling with few precedents anywhere in the world, Judge Syed Mansoor Ali Shah ordered that the Pakistani government do more to enforce the climate change adaptation measures articulated in the NCCP several years earlier. The ruling not only obliges the government to carry out its climate change commitments, but also establishes a new climate change commission – and names 21 officials that must sit on it – to oversee the process. Judge Shah's ruling came after a farmer filed a public interest litigation case against the government, alleging that climate change effects were causing him undue hardship and that officials' inaction had violated his "fundamental rights."[29]

In a country where justice often moves at a glacial pace, the alacrity with which the court's ruling took effect is remarkable. On October 1, 2015, just days after the verdict, the new climate change commission held its first meeting. Members have been asked to identify achievable items from the NCCP's implementing framework, and to discuss the process for implementation at future meetings. Subsequent orders issued by Judge Shah have laid out a detailed timetable for meetings of the climate change commission and its expected deliverables.

Environmental lawyers around the world were stunned by the case; one admitted that "Pakistan was nowhere in the list of my countries where I would have expected to see this kind of a ruling."[30] This is because, in the blunt words of one Pakistani environmental specialist, "Pakistan is one of the countries most vulnerable to climate change but, despite having a national climate policy, no one appears to care."[31]

The fact that the case did occur in Pakistan can perhaps be attributed to Judge Shah himself, a fervent advocate for more robust climate change mitigation policies. He saw firsthand the extent of Pakistan's climate vulnerability when he served on a commission that surveyed flood damage. That experience served as a "rude awakening," and helped crystallize what he describes as the "grave risk" posed to Pakistan by climate change. He soon discovered that federal and provincial government officials, including senior-level government ministers, did not appreciate the gravity of the situation and did not even know what was meant by climate change. All of this helped inspire his ruling.[32]

In the days after the ruling, Judge Shah expressed confidence that his orders would be carried out successfully, even though several key members of the climate change commission did not show up for initial meetings. He pointed to several early successes, including a request from forestry department officials for training and assistance.[33] Parvez Hassan, an environmental lawyer serving as the commission chair, was similarly sanguine. He said that the commission's work was "progressing rapidly and effectively," and that the government was "fully cooperating and supporting" the commission.[34]

Major obstacles to overcome

Unfortunately, impressive as these efforts have been to enhance resilience to climate change through clean energy projects and other mitigation measures, Pakistan faces a long and hard road ahead. Progress, while impressive, could ultimately prove limited. Seven factors help explain why.

Competing priorities

Every government faces multiple demands and lacks the resources and capacity to tackle each one equally. Policy planners tend to emphasize issues that they perceive as the most pressing at the present moment. Despite evidence that climate change has become a real and current threat, many countries continue to shrug it off as an abstraction – a notable yet far-off threat that can be saved for another day.

This is particularly the case in Pakistan, where the government faces so many immediate challenges. These range from terrorism and malnutrition to out-of-school children and water-borne disease. Additionally, given the volatility of Pakistani politics, governments are often preoccupied with political survival issues. The current government, for example, nearly did not survive an anti-government protest campaign spearheaded by the political opposition in the summer of 2014. Renewable energy and climate change mitigation simply do not stack up with these immediate issues in terms of priority; government officials do not have the luxury of steadfastly pursuing the implementation and enforcement of climate-related policies – despite the best efforts of individuals such as Judge Shah.

Energy in general, however, is very much a top priority for the government. It figured prominently in Sharif's 2013 election campaign (energy took up more pages than any other issue in his party's election manifesto document), and he took office that year with a strong mandate to fix the crisis. Still, for political reasons, Pakistani officials are constantly under pressure to boost generation on a broad scale, and fast, in order to narrow large supply-demand gaps. Renewables currently do not enjoy sufficient scale to be very helpful in this regard. Importing more foreign hydrocarbons – particularly in an era of relatively cheap global oil prices – can address demand more widely and rapidly, and is therefore a politically safer move.

The military

Pakistan's military has ruled the country for nearly half its existence, and even at times when it has not – such as the current era – it has exerted heavy influence over the policy sphere. Defense spending habitually hogs the annual budget, which helps explain why areas such as education and healthcare have historically received such small allocations – and why subjects perceived as peripheral, such as climate change, receive even smaller allotments. The Pakistani security establishment is not known for taking an interest in climate change issues. The military enjoys a vast economic empire, valued at about $20 billion, with assets that range from farms to construction facilities and cereal companies.[35] One can comfortably assume that this empire leaves a deep carbon footprint and does not embrace earth-friendly tactics and technologies.

One can also reasonably argue that in the interest of national stability, the military has an interest in resolving the energy crisis in the fastest way possible. However, this entails a preference for rapid-generation, hydrocarbon-focused, supply-side-centered policies implemented during past eras of military rule (and in the current era of civilian rule), and not for renewables or other greener options, or for energy efficiency or other demand-side policies. The military, for example, strongly supports China's $35 billion energy investments – several of which are focused on clean energy, but most of which are focused on coal. Without strong levels of buy-in from the all-powerful Pakistani military, it is unrealistic to believe that the government can make a case for climate policy to become a front-burner issue anytime soon.

High costs

It is perhaps no coincidence that Pakistan's recent renewable power generation success stories – such as the launch of its first solar power plant – have largely been one-off projects that enjoy the largesse of foreign donors. Pakistan, quite simply, cannot afford the sky-high costs necessary to support renewable energy projects on a wide scale, and in a way that meets baseline demand.

This is not to say that renewable energy is simply too expensive to touch. On the contrary, costs have fallen in recent years. Additionally, Pakistan's annual budget for 2015–16 features a number of cost-friendly measures for producers and consumers. These include tax exemptions for solar imports and solar- and wind-energy producing industries; government financing of mark-ups on loans for solar-based tubewells; and interest-free loans for small farmers using solar tubewells.

Nevertheless, solar and wind energy costs – particularly for technology, storage, and overall production – are quite high for producers. This can be attributed in part to Pakistan's paucity of energy infrastructure. If potential solar and wind producers want to build grid-connected power plants, then they need to work the steep costs of building transmission lines into their financial calculations – because these transmission lines are lacking in Pakistan.

Renewable energy costs are also high for consumers. A major reason why is the unregulated market for renewable energy, and particularly for solar. Islamabad has offered tax exemptions and other discounts on imported solar panels and related products, but the positive benefits of such policies are negated by middle men who sell solar materials (often subpar ones) at jacked-up prices.

The cost of renewable energy is also high for Pakistan's government. This is because of project financing plans that must offer generous returns to investors, in order to ease concerns about risk in an unstable country where the renewable energy portfolio remains an unproven frontier.[36] In effect, the government must absorb high capital and start-up costs to attract nervous investors. Ultimately, with the energy sector already handicapped by crippling levels of debt that have approached $5 billion, it is unrealistic to believe that the high costs associated with solar and wind power can easily be absorbed or overcome.

Decentralization

In 2010, Pakistan passed its 18th constitutional amendment, which reduced the power of the presidency and devolved many federal-level government functions, resources, and responsibilities to the provinces. Theoretically, this was a policy of democratization in that it empowered provincial governments, which have traditionally had to defer to often-abused central power sources such as the presidency and military. In practice, however, the 18th amendment has overburdened already capacity-constrained and frequently corrupt provincial authorities.[37] This has troubling implications for policy development and implementation.

Climate issues have been directly and deleteriously affected by the 18th amendment. Pakistan's environment ministry was one of the 17 that were eliminated after the amendment's passage, meaning that key policy areas such as environmental pollution and ecology became the full responsibility of the provinces – even though, as the environmental expert Ahmad Rafay Alam has written, provinces had little experience with environmental regulation. Even today, he says, climate change, mitigation, and adaptation "all remain foreign terms to the provinces."[38] There is also no longer a clear national-level coordination framework for environment and climate change policy. Additionally, the devolution process for environmental management has been implemented in a disorganized and inconsistent fashion. A 2015 World

Bank study found that these responsibilities have been devolved more fully in the province of Punjab than in the province of Sindh.[39]

Furthermore, the 18th amendment has also obliged beleaguered climate-focused provincial authorities to operate against a confusing and chaotic institutional backdrop. The abolished environment ministry was replaced with a federal climate change ministry, though in 2014 the latter was downgraded to a division due to budget cuts – before once again being restored to a ministry in 2015 (curiously, however, in announcing its 2015–16 budget allocations, Pakistan still refers to it as a division, not a ministry). Critics describe the current climate ministry as ineffective and cash-starved. Alam has quipped that its current budget is "less than the cost of a Toyota."[40]

In effect, well-intentioned political decentralization has produced a regrettable result for climate policy: the onus has been put on overwhelmed provincial governments to manage climate portfolios with which they have little if any prior experience, while federal-level climate authorities lack the resources to provide meaningful support.

Institutional dysfunction

It is not just climate change-focused institutions that are troubled in Pakistan; those associated with energy are as well. This bodes ill for the country's efforts to successfully manage policy for these two interrelated sectors. Pakistan has no overarching energy ministry; rather, over a dozen different entities are charged with energy-related functions. Different government units manage different energy resources – and in the case of biomass, different units manage the same resource. Additionally, each major energy resource has its own regulating entity. There is little to no coordination (though plenty of competition) between these various energy-focused actors.[41]

Such a chaotic institutional structure constrains effective policy planning, development, and implementation. It also complicates any potential effort to fashion a revised energy mix that draws more on renewables – a major policy shift that would need to win over key stakeholders in the vast energy bureaucracy and in state-owned hydrocarbon companies. It is hard to imagine that such a chaotic institutional environment – and the rivalries and turf wars that it engenders – could support such delicate exchanges and negotiations, much less the development and execution of a more renewables-focused energy policy on the whole.

Urbanization

Like many developing countries, Pakistan is experiencing rapid urbanization – and its 3% annual urbanization rate is the fastest in South Asia. Karachi's population grew 80% between 2000 and 2010 – the largest increase of any city in the world. Most estimates contend that Pakistan's urban-based denizens, presently a third of the total population, will be about 50% by 2025. However, according to density-based definitions of urbanization – which classify urban space as any area with 1,000 people per square mile, regardless of whether these areas are administratively classified as cities – Pakistan's population could already be up to 65% urban today.[42]

Many manifestations of urbanization – including heavy industrialization and exhaust-belching automobiles – drive up carbon emissions. Pakistani cities do not do much on a large scale to mitigate environmental damage. For example, there are few wastewater treatment facilities in Pakistani cities, despite the widespread contamination of urban water supplies due to largely unregulated industrialization.

To be sure, some small-scale pro-environment measures are in place. A number of urban buildings, for example, have incorporated stormwater harvesting into their water usage plans. Additionally, some would argue that bringing more people on to the electricity grid (which happens when people settle in urban spaces) can reduce climate change effects, because more people on the grid means fewer people using traditional, off-grid – and heavily polluting – energy sources such as firewood and biomass. This may be true to an extent, but ultimately such an argument is flawed. Many poor urban residents refuse to go on the grid because they much prefer to keep using cheap or free (and polluting) off-grid options instead of opting for more expensive (and cleaner) on-grid options. Additionally, many of those who go on the grid in Pakistani cities use dirty fuels. This will certainly continue to be the case if Pakistan succeeds in integrating coal more fully into the country's energy mix. The country intends to serve urban demand not only through Thar-based coal reserves, but also by converting furnace-oil based power plants to coal.

In sum, urbanization deepens environmental risks more than it mitigates them. It also complicates the ability of Pakistani officials to fashion a greener and more sustainable energy mix. The financial and public health consequences are quite serious: environmental degradation damages in cities are estimated at several billion dollars per year, and air pollution kills nearly 25,000 urban residents annually.[43]

Lack of policy implementation

This is arguably the most significant factor of all, and one that ails Pakistan's broader policy environment. Many promising policies across the board are conceptualized and developed – but never implemented, much less enforced. There are numerous reasons why. One is technical. In Pakistan, policy documents are often written in a way that precludes any chance of implementation. There is often little explicit information or instructions as to how a particular policy is to be carried out. Another reason is purely political: a lack of political will to push through major reforms or policies, even if they could have long-term positive effects for the country. Pakistan – along with many other countries, to be sure – tends to have leaders who focus more on their political fortunes than on the broader needs of their country.

Another reason that may help explain Pakistan's implementation problems is one of political economy. One may argue that a rentier state mentality compels Pakistan's political class and other powerbrokers (especially the military) to assume that the outside world will provide assistance to help keep the country afloat, including measures to keep the energy situation from spiraling out of control. The fact that Pakistan has placed so much hope in China's $35 billion energy investment package is illustrative. When a state can assume that it will get bailed out by external support, then it has fewer incentives to take ownership of policy challenges and hunker down to address them. To this end, while recent European, American, and Asian investments in Pakistan's fledgling renewable energy sector are encouraging, they may also exacerbate the complacency that prevents Pakistan from taking bigger steps on its own.

The stakes of inaction: trigger for destabilization

The potential consequences of failing to take sufficient action to mitigate climate change effects have been well chronicled. These include the destruction of agriculture and ecosystems, widespread water scarcity, and waves of climate refugees. Sadly, these are all very real possibilities for Pakistan in the coming decades. The steep price tag of climate change is another frequently discussed consequence. In Pakistan, a former environmental minister has projected an eventual figure of $14 billion per year – even as current annual climate change needs in the country cost anywhere from $13 billion to $32 billion. These are expenses that the country is presently unable to finance.[44]

Equally important, though less discussed, are the stability implications of an angry and abused environment unleashing its full wrath, and particularly on a nation such as Pakistan that is deeply fractured, militancy-riven, and nuclear-armed. On a general level, there is the possibility of intensified privation – spawned by livelihoods losses, acute water and food insecurity, mass displacements, and other climate change effects – heightening the risks of radicalization. Pakistan's young masses – about two-thirds of the country's population is under the age of 30 – could be more likely to succumb to the blandishments of militants, particularly given the state's repeated failure to meet the general population's basic needs even under ordinary circumstances.

More specifically, there are three troubling climate-change-related scenarios that could imperil Pakistan's fragile stability.[45]

First, climate change effects could inflame relations with India, which in 2015 were already suffering from one of their most difficult periods in years. Pakistani hardliners, including the Lashkar-e-Taiba terror group, routinely accuse upper riparian India of contributing to, if not outright precipitating, Pakistan's floods and droughts. India, they allege, manipulates Indus Basin river flows so that water either gushes downstream or is diverted upstream (there is little evidence that India does either, at least not intentionally). They contend that "liberating" India-held Kashmir, the point of origin for many of the rivers and tributaries flowing into Pakistan, is the only way to stop India's hydro machinations.

Increasingly hot temperatures, glacial melt, and other climate events could exacerbate Pakistan's flooding and droughts – thereby supplying Pakistani militants with further ammunition for their accusations and even pretexts for attacks on India. New Delhi, meanwhile, suffers from its own dry spells and water shortages. It could well decide, for the sake of the national interest, to do the very thing that Pakistani militants falsely accuse it of doing now: diverting river flows to provide more water and hydro power for Indians. The Indus Waters Treaty, which has helped keep India and Pakistan from fighting a war over water, could be put to the test – and so could the bilateral relationship itself.

Second, environmental stress could deepen Pakistan's urban violence. Karachi is often convulsed by such strife, and much of it arises from fierce competition over precious land. Yet Karachi – a coastal, low-lying metropolis – is vulnerable to flooding, cyclones, and other climate-related phenomena that could easily wipe out vast swaths of the city's heavily contested real estate. This means the land that remains could become even more precious, thereby raising the stakes for the city's fighting factions and likely increasing violence.

Third, and perhaps most troubling, Pakistan's environmental insecurity imperils nuclear security. The concern here is not of militants seizing nuclear weapons, but of the country experiencing a catastrophic nuclear accident similar to that of Japan's Fukushima nuclear plant in 2011. A key Pakistani nuclear facility, the Karachi Nuclear Power Plant (KANUPP), sits not only in a flood- and storm-prone area, but also in one of the most densely populated parts of the country. Back in 2012, a study released jointly by the journal *Nature* and Columbia University found that more than 8 million people live within 30 kilometers of KANUPP – the largest figure for any nuclear facility in the world.[46] KANUPP, which is nearly 45 years old, has been described by Pervez Hoodbhoy, a prominent Pakistani nuclear physicist, as a "chronically incontinent" reactor that frequently leaks heavy water. Given the combination of a fragile old plant, a large nearby population, and Pakistan's poor emergency-response capabilities, the consequences of a tsunami or cyclone strike on or near KANUPP could be truly catastrophic. According to Hoodbhoy, the release of deadly radioactivity would be only one of several threats. Others would include clogged roads, a collapse of vital services, and Karachi (the country's largest city and financial capital) succumbing to the predations of looters, criminals, and, perhaps, militants.[47]

Over the last few years, China has committed to constructing several new nuclear reactors in Pakistan. However, this has not led to the shuttering of KANUPP. On the contrary, it remains operational. Even more concerning is that the Chinese are building two new reactors on the very grounds of KANUPP. One of these new reactors was inaugurated in 2015. Pakistani officials have defended these new investments and their locations, contending that the integrity of KANUPP's structure is sound and that it has proper safety measures in place – thereby ruling out doomsday scenarios.[48] Nonetheless, given KANUPP's rickety state, the construction of two new reactors right next to it is quite worrisome.

Recommendations and conclusions

Pakistan faces a conundrum of serious energy problems coupled with severe climate vulnerability. The government is aware of each dilemma and has taken encouraging steps to address both of them. However, the challenges are daunting. Fortunately, Pakistan, with assistance from international donors, can execute several policy interventions that would better equip the country to overcome them, or at least to manage them.

First, to help get climate change off the policy back burner, Pakistan's vibrant civil society should lead awareness-building campaigns about the imminent threat it poses. These campaigns should target Pakistan's influential and far-reaching mass media – specifically radio in rural areas and private television channels in urban areas. The spokespersons for such campaigns should include prominent personalities who can shape public sentiment. These might include television news anchors, religious leaders, and Bollywood and athletic stars. If civil society can build awareness in this way, and put pressure on the government to act, then policymakers are likely to take climate issues more seriously.

Second, the powerful Pakistani military, which exerts strong influence over the policy environment but has little awareness of or understanding of climate change, must better understand the connections between environmental security and national security. Specifically, efforts should be made to impress upon the security establishment the links between climate change effects and stability.

Third, to help bring the costs of renewable energy down to more reasonable levels, Pakistan's government should formulate policies that encourage the most cost-efficient production of as much energy as possible in order to bring down the per unit price as low as possible. Quite simply, policymakers should strive for policies that encourage investment in energy generation at a price that the country can afford.[49] In all reality, for costs to come down in this way, major correctives will be necessary. These include the construction of more transmission lines and other infrastructure, more regulation of renewable energy markets, and across-the-board anti-militancy and anti-extremism measures that usher in a calmer long-term security situation.

Fourth, with decentralization reforms saddling overburdened provincial governments with difficult new climate change responsibilities, international donors should sponsor capacity-building training programs to help these officials better learn how to manage their resources and to carry out their mandates.

Fifth, Pakistan should take steps to bring some order to a chaotic energy sector. This should entail scaling back the number of government entities charged with energy responsibilities, and above all establishing an overarching body – such as a new energy ministry – to coordinate energy policy. This would admittedly be a hard sell, given the deep inertia of Pakistan's energy bureaucracy and the strong resistance of various vested interests within the energy sector. And yet the benefits would be great. More institutional coherence would facilitate a more effective and organized process of energy planning, development, and execution.

Sixth, Pakistan should acknowledge that while it – like any country – is in no position to halt climate change outright, it can nonetheless institute a number of correctives to blunt its effects and reduce the scope of potential damage. These measures include passing stringent laws against Pakistan's deforestation, which happens at alarming rates, and establishing more robust disaster risk reduction mechanisms.

To be sure, even if all these recommendations were to be successfully implemented – and that is a mammoth "if" – Pakistan would still face major energy challenges and climate change vulnerability. In the coming years, Pakistani policymakers will need to confront these problems, whether or not they are ready to do so. Even if Pakistan's more well-known and immediate challenges were to disappear – if militancy were to be completely stamped out, if the military were to permanently withdraw to the barracks and usher in a golden age of civilian democracy, if the economy were to enjoy spectacular growth – then energy and environmental matters will still loom large.

This is a reality that cannot be wished away. Devising more effective – and, in time, integrated – energy and climate policies is a worthwhile way to start preparing for this future and inevitable state of affairs.[50]

Notes

1 This section is drawn in part from Michael Kugelman, ed., *Pakistan's Interminable Energy Crisis: Is There Any Way Out?* (Washington, DC: Woodrow Wilson Center, 2015), www.wilsoncenter.org/sites/default/files/ASIA_150521_Pakistan's%20Interminable%20Energy%20Crisis%20rpt_0629.pdf. See also Wendy Culp, "Pakistan's Energy Insecurity: Anatomy of a Crisis and How to Move Forward – An Interview with Michael Kugelman," National Bureau of Asian Research, July 22, 2015, www.nbr.org/downloads/pdfs/eta/kugelman_interview_072215.pdf; Michael Kugelman, "Could Pakistan's Energy Crisis Bring Down the Government?" War on the Rocks, February 10, 2015, http://warontherocks.com/2015/02/could-pakistans-energy-crisis-bring-down-the-government/.

2 Michael Kugelman, "Powerless in Pakistan," *Foreign Policy*, June 30, 2015, http://foreignpolicy.com/2015/06/30/powerless-in-pakistan/.

3 World Bank Group, *Doing Business 2016: Measuring Regulatory Quality and Efficiency* (Washington, DC: World Bank Group, 2015), www.doingbusiness.org/~/media/GIAWB/Doing%20Business/Documents/Annual-Reports/English/DB16-Full-Report.pdf, 70.

4 This section is drawn in part from Michael Kugelman, "Pakistan's Climate Change Challenge," *Foreign Policy*, May 9, 2012, http://foreignpolicy.com/2012/05/09/pakistans-climate-change-challenge/.

5 Sonke Kreft, David Eckstein, Lisa Junghans et al., *Global Climate Risk Index 2015* (Germanwatch, 2014), https://germanwatch.org/en/download/10333.pdf.

6 United Nations Development Programme (UNDP) in Pakistan, *Pakistan: Climate Public Expenditure and Institutional Review (CPEIR)* (United Nations Development Programme in Pakistan, 2015), www.pk.undp.org/content/dam/pakistan/docs/Environment%20&%20Climate%20Change/UNDP%20Climate%20Report%20V10.pdf, 11.

7 Michael Kugelman, "Powerless in Pakistan." *Foreign Policy*, June 30, 2015. http://foreignpolicy.com/2015/06/30/powerless-in-pakistan/.

8 Michael Kugelman, "Pakistan's Impending Famine," *The Diplomat*, March 14, 2014, http://thediplomat.com/2014/03/famine-threatens-pakistan/; United Nations International Children's Emergency Fund, "Drought Disasters: Asia" (United Nations International Children's Emergency Fund, 2014), www.unicef.org/drought/asia.htm.

9 For background on Pakistan's water crisis, see Michael Kugelman, ed., *Running on Empty: Pakistan's Water Crisis* (Washington, DC: Woodrow Wilson Center, 2009), www.wilsoncenter.org/sites/default/files/ASIA_090422_Running%20on%20Empty_web.pdf.

10 Kalpana Kochhar, Catherine Pattillo, Yan Sun et al., *Is the Glass Half Empty or Full? Issues in Managing Water Challenges and Policy Instruments*, International Monetary Fund, June 2015, www.imf.org/external/pubs/ft/sdn/2015/sdn1511.pdf.

11 Kashif Majeed Salik, Sehrish Jahangar, Waheed ul Zafar Zahdi, et al., "Climate Change Vulnerability and Adaptation Options for the Coastal Communities of Pakistan," *Ocean & Coastal Management* 112 (2015): 61–73. www.sdpi.org/publications/externalpublications/climate-change-vulnerability-and-adaptation-options-for-the-coastal-communities-of-Pakistan.pdf.
12 Saadullah Ayaz, *Climate Change and Coastal Districts of Baluchistan Communities – Situation Analysis, Implications, and Recommendations* (International Union for the Conservation of Nature-Pakistan, Embassy of the Netherlands, and Government of Baluchistan, 2012), https://cmsdata.iucn.org/downloads/cc_and_coastal_districts_analysis_study.pdf. While much of the literature on Pakistan's climate risks focuses on Sindh and Baluchistan, the provinces of Khyber-Pakhtunkhwa and Punjab are quite vulnerable as well. In fact, some studies rank Punjab as Pakistan's second-most vulnerable province after Baluchistan. See Sadia Mariam Malik, Haroon Awan, and Nizullah Khan, "Mapping Vulnerability to Climate Change and its Repercussions on Human Health in Pakistan," *Globalization and Health* 8, no. 31 (2012), www.globalizationandhealth.com/content/pdf/1744-8603-8-31.pdf.
13 Khalid Mansoor, "How Coal Can Help Address Pakistan's Energy Crisis," in *Pakistan's Interminable Energy Crisis*, ed. Michael Kugelman (Washington, DC: Woodrow Wilson Center, 2015), 40–52.
14 Musadik Malik, "Pakistan's Energy Crisis: Challenges, Principles, and Strategies," in *Pakistan's Interminable Energy Crisis*, ed. Michael Kugelman (Washington, DC: Woodrow Wilson Center, 2015), 23–39.
15 Ministry of Water and Power (Pakistan), *National Power Policy 2013* (Ministry of Water and Power (Pakistan), 2013), www.ppib.gov.pk/National%20Power%20Policy%202013.pdf.
16 Finance Division (Pakistan), *Federal Budget 2015–16: Budget in Brief* (Finance Division (Pakistan), 2015), www.finance.gov.pk/budget/Budget_in_Brief_2015_16.pdf.
17 UNDP in Pakistan, *Pakistan.*
18 Ministry of Climate Change (Pakistan), *National Climate Change Policy* (Ministry of Climate Change (Pakistan), 2012), www.mocc.gov.pk/gop/index.php?q=aHR0cDovLzE5Mi4xNjguNzAuMTM2L21vY2xjL3VzZXJmaWxlczEvZmlsZS9Nb2NsYy9Qb2xpY3kvTmF0aW9uYWwlMjBDbGltYXRlJTIwQ2hhbmdlJTIwUG9saWN5JTIwb2YlMjBQYWtpc3RhbiUyMCgyKS5wZGY%3D.
19 Climate Change Division (Pakistan), *Framework for Implementation of Climate Change Policy (2014–2030)*, Climate Change Division (Pakistan), 2013, www.pk.undp.org/content/dam/pakistan/docs/Environment%20&%20Climate%20Change/Framework%20for%20Implementation%20of%20CC%20Policy.pdf.
20 "Pakistan Faces Major Risks from Climate Change: PARC," *Daily Times*, November 11, 2014, www.dailytimes.com.pk/islamabad/11-Nov-2014/pakistan-faces-major-risks-from-climate-change-parc.
21 Associated Press of Pakistan, "Senate Body Wants Solid Steps to Mitigate Climate Change Threats," *Daily Times*, August 19, 2015, www.dailytimes.com.pk/national/19-Aug-2015/senate-body-wants-solid-steps-to-mitigate-climate-change-threats.
22 Associated Press of Pakistan, "Need for Climate Change Awareness Plan Stressed," *Daily Times*, September 15, 2015, www.dailytimes.com.pk/islamabad/15-Sep-2015/need-for-climate-change-awareness-plan-stressed.
23 White House, Office of the Press Secretary, "2015 Joint Statement by President Barack Obama and Prime Minister Nawaz Sharif," White House, Office of the Press Secretary, October 22, 2015, www.whitehouse.gov/the-press-office/2015/10/22/2015-joint-statement-president-barack-obama-and-prime-minister-nawaz.
24 Greentech Lead, "Vestas Signs MOU to Develop Pilot Wind Projects in Pakistan," Greentech Lead, February 20, 2015, www.greentechlead.com/wind/vestas-wind-signs-mou-develop-pilot-wind-projects-pakistan-21833; "France PROPARCO Invests 20m to Support Pakistan Wind Energy Projects," *Daily Times*, March 11, 2015, www.dailytimes.com.pk/business/11-Mar-2015/france-proparco-invests-20m-to-support-pak-wind-energy-projects.
25 Anam Zehra, "Pakistan Opens First Solar Power Plant, Built with Chinese Investment," Reuters, May 5, 2015, http://in.reuters.com/article/2015/05/05/pakistan-solar-idINL4N0XW2VP20150505.
26 "Discovering New Resources: NEPRA Approves Upfront Tariff for Solar Power," Associated Press of Pakistan, January 29, 2015, http://tribune.com.pk/story/829221/discovering-new-resources-nepra-approves-upfront-tariff-for-solar-power/; Khaleeq Kiani, "Upfront Tariff for Small Hydro Plants Okayed," *Dawn*, April 3, 2015, www.dawn.com/news/1173562; Mushtaq Ghumman, "Wind Power Projects: Upfront Tariff of Rs. 10.6048/kwh," *Business Recorder*, June 26, 2015, www.brecorder.com/fuel-a-energy/193/1199787/.
27 Ian Clover, "Pakistan Overhauls its Solar Industry for the Better," *PV Magazine*, January 2015, www.pv-magazine.com/news/details/beitrag/pakistan-overhauls-its-solar-industry-for-the-better-_100017687/#axzz3n9bpnDEG.

28 Aamir Saeed, "Solar to Power Thousands of Off-Grid Homes in North Pakistan," Reuters, February 18, 2015, www.reuters.com/article/2015/02/18/us-pakistan-energy-solar-idUSKBN0LM0OR20150218.
29 Ahmed Rafay Alam, "Pakistan Court Orders Government to Enforce Climate Law," *The Third Pole*, September 24, 2015, www.thethirdpole.net/2015/09/24/pakistan-court-orders-government-to-enforce-climate-law/; Natasha Geiling, "Pakistan Now has a 21-Person 'Climate Council,' Thanks to a Judge's Ruling," *Think Progress*, October 7, 2015, http://thinkprogress.org/climate/2015/10/07/3710023/pakistan-court-climate-change/.
30 Raveena Aulakh, "Pakistan Orders State to Enforce Climate Policies," *The Star*, October 3, 2015, www.thestar.com/news/world/2015/10/03/pakistan-judge-orders-state-to-enforce-climate-policies.html.
31 Alam, "Pakistan Court Orders Government to Enforce Climate Law."
32 Judge Syed Mansoor Ali Shah, phone interview with the author, October 7, 2015.
33 Ibid.
34 Parvez Hassan, email interview with the author, October 31, 2015.
35 Very little work has been done on this opaque subject, though the most definitive account is Ayesha Siddiqa, *Military Inc.: Inside Pakistan's Military Economy* (Karachi, Pakistan: Oxford University Press, 2007).
36 For a more detailed discussion, see Musa Khan Durrani, "Is Solar Power Really the Best Solution for Pakistan?" *Dawn*, October 2, 2015, www.dawn.com/news/1209933.
37 See Michael Kugelman, *Decentralization in Pakistan: The Lost Opportunity of the 18th Amendment* (Norwegian Peacebuilding Resource Center, 2012), www.peacebuilding.no/var/ezflow_site/storage/original/application/fcfee16739837e8125703a4ae5e750df.pdf; Colin Cookman, "The 18th Amendment and Pakistan's Political Transitions," Center for American Progress, April 19, 2010, www.americanprogress.org/issues/security/news/2010/04/19/7587/the-18th-amendment-and-pakistans-political-transitions/.
38 Alam, Ahmed Rafay, "Pakistan Court Orders Government to Enforce Climate Law," The Third Pole, September 24, 2015. www.thethirdpole.net/2015/09/24/pakistan-court-orders-government-to-enforce-climate-law/.
39 Ernesto Sanchez-Triana, Santiago Enriquez, Bjorn Larsen, et al., *Sustainability and Poverty Alleviation: Confronting Environmental Threats in Sindh, Pakistan* (Washington, DC: World Bank Group, 2015), 132.
40 Zofeen T. Ebrahim, "Pakistan's New Climate Change Ministry Merely 'Cosmetic,'" *Dawn*, February 6, 2015, www.dawn.com/news/1161895.
41 For a more detailed discussion of the institutional dysfunction within Pakistan's energy sector, see Ziad Alahdad, "Pakistan's Energy Sector: Putting It All Together," in *Pakistan's Interminable Energy Crisis*, ed. Michael Kugelman (Washington, DC: Woodrow Wilson Center, 2015), 134–151.
42 Michael Kugelman, ed., *Pakistan's Runaway Urbanization: What Can Be Done?* (Washington, DC: Woodrow Wilson Center, 2014), www.wilsoncenter.org/sites/default/files/ASIA_140502_Pakistan's%20Runaway%20Urbanization%20rpt_0530.pdf.
43 Sania Nishtar, Farrukh Chishtie, and Jawad Chishtie, "Pakistan's Urbanization Challenges: Health," in *Pakistan's Runaway Urbanization*, ed. Michael Kugelman (Washington, DC: Woodrow Wilson Center, 2014), www.wilsoncenter.org/sites/default/files/ASIA_140502_Pakistan's%20Runaway%20Urbanization%20rpt_0530.pdf, 107–125.
44 Lisa Friedman, "Can Pakistan Survive Climate Change?" *E&E News*, May 28, 2013, www.eenews.net/stories/1059981842.
45 These three scenarios are adapted from Kugelman, "Pakistan's Climate Change Challenge."
46 Declan Butler, "Reactors, Residents, and Risk," *Nature*, April 21, 2011, www.nature.com/news/2011/110421/full/472400a.html.
47 Pervez Hoodbhoy, "Some Learned from Fukushima. Did We?" *Express Tribune*, March 11, 2012, http://tribune.com.pk/story/348572/some-learned-from-fukushima-did-we/.
48 Imtiaz Ali, and Shahid Ghazali, "PM Nawaz Inaugurates K-2 Powerplant at Kanupp," *Dawn*, August 21, 2015, www.dawn.com/news/1201662; Abdul Hafeez, "Kanupp Defends its Ambitions, Rules out Doomsday Scenario," *Geo*, May 25, 2015, www.geo.tv/article-185958-Kanupp-defends-its-ambitions-rules-out-doomsday-scenario.
49 For a detailed discussion of appropriate pricing policies, see Shannon Grewer, "How to Incentivize Energy Innovation and Efficiency, and Encourage the Rapid Deployment of Affordable Solutions," in *Pakistan's Interminable Energy Crisis*, ed. Michael Kugelman (Washington, DC: Woodrow Wilson Center, 2015), 102–119.
50 Michael Kugelman is grateful for the invaluable research assistance provided by Aleena Ali and Nausheen Rajan.

22

Energy transition in a carbon consuming country

India

Lydia Powell

Introduction

The energy transition that many desire to see in India is one that will transform the coal dependent country into one that is less so. While this transition is important, what is equally important if not more is a transition that will increase access to modern energy sources to millions of people who depend on traditional sources of energy such as kerosene, firewood and dried animal dung for lighting and heating.

There are fears among some that the pursuit of the former would necessarily mean a compromise on the latter. The argument is that the transition to a low carbon economy could potentially limit the use of coal which is currently the cheapest source of electricity generation in India.[1] Cheap electricity is seen as the only source that can be scaled up to replace kerosene lamps used for lighting in more than a quarter of Indian households.

'Sustainable growth', a narrative that assures that there is no trade off between the low carbon transition and the energy access transition, is used to mediate these fears. But close observation of trends in India indicates that the two transitions are being pursued simultaneously but independently. The low carbon transition is showcased to the international audience concerned over carbon emissions through the increasing share of renewable energy in India's energy basket. The energy access transition is showcased to the domestic audience through growing access to affordable petroleum fuels and cheap electricity generated using coal.

Some see the combination of the two independent transitions as a transition towards sustainability. Others see it as a wise strategy that hedges the low carbon transition with the carbon intensive energy access transition. The impact of one transition on the other is likely to unravel slowly but the current environment is filled with optimism that the low carbon transition will prevail in the longer term.

The low carbon transition

India's current energy basket is dominated by fossil fuels. If the share of fossil fuels in India's energy basket is taken as an indicator of its transition towards a low carbon economy there is insufficient evidence to show that India is firmly on a low carbon growth path. If the share of traditional energy sources such as fire-wood and dried animal dung which are labelled 'non-commercial energy sources' are excluded, the share of fossil fuels in India's commercial energy basket was about 95 per cent in 2011 according to Indian sources[2] or 96 per cent in 2012 according to international sources.[3]

This share falls to about 71–73 per cent if non-commercial energy sources are included.[4] In a business as usual scenario this share of fossil fuels in commercial energy is expected to decrease marginally to 93 per cent by 2022 with nuclear power and renewable energy taking a share from fossil fuels.[5] However the revised targets for renewable energy set by the new government that came to power in 2014 are expected to lower the share of fossil fuels considerably in the next two decades.

The budget for 2015–16 announced by the new government revised the target for renewable energy capacity to 175 GW by 2022 that included 100 GW of solar (from about 4 GW today), 60 GW of wind (from about 23 GW today), 10 GW of biomass and 5 GW of small hydro power.[6] The government also increased the target for solar water pumps to over 100,000 from fewer than a thousand today. To reach the overall target set for renewable energy, India would have to develop, in just seven years, renewable energy capacity that exceeds coal based generation capacity that India developed in the last sixty years.

The largest increase is planned for solar energy that is expected to increase by a factor of 25 in the next seven years with a compound annual growth rate (CAGR) of over 55 per cent.[7] Wind power is expected to triple in capacity in the next seven years with a CAGR of about 14 per cent. This unprecedented growth in renewable energy capacity is to be achieved by state led incentive schemes offered to renewable energy projects.[8]

The previous government offered incentives such as preferential credit terms, capital subsidies and duty exemptions on import of technology to renewable energy projects estimated to cost about $20 billion over a ten year period to renewable energy projects. The new government has announced additional incentives and these include but are not limited to a support of over $150 million to central government owned companies to set up 1,000 MW grid connected solar photovoltaic (PV) projects, $610 million for setting up 25 solar parks each with a capacity of 500 MW and $113 million for defence and paramilitary establishments to set up 300 MW capacity solar plants.[9]

The new government has also increased the budget of the Ministry of New and Renewable Energy (MNRE) by over 65 per cent.[10] Accelerated depreciation benefit to wind projects has been restored by the new government.[11] This incentive was withdrawn by the previous government as it was seen to be used by investors as a financial instrument to improve their balance sheets rather than as an incentive to generate wind power as intended by the policy.[12]

The current government also doubled the clean energy tax on coal (cess) from about one dollar per tonne of coal to about two dollars per tonne of coal to finance renewable energy projects and in 2016 it was increased to 5 dollars per tonne. In addition the government has advised banks to encourage rooftop PV installations by extending affordable loans to private households and other buildings. In order to speed up clearances for renewable energy projects, solar, wind and small hydro projects have been moved from the red (higher environmental risk) to the green (lower environmental risk) category.

The government has also secured a line of credit of €100 million to the Indian Renewable Energy Development Agency (IREDA) from Agence Française de Développement (AFD) of

France and about ¥30 billion (equivalent to about $250 million) from the Japan International Co-operation Agency (JICA) for 30 years. In the words of the government, 'with these decisions, India will emerge as a major solar power producing country.'[13]

Optimism over meeting the targets set for renewable energy is sustained by regular news on the flow of funds for investment in renewable energy projects. Most recently, the minister in charge for power and renewable energy announced that banks in India including a few private sector banks had renewable energy financing commitments of about $52 billion until 2021–22.[14] The minister also announced that 387 companies including global firms had submitted green energy certificates for renewable energy power generation worth about 272 GW in the next five years.[15]

Two issues may be raised at this point. The first is the extent to which commitment to invest will translate into actual investments. If the past is any guide to the future, the probability of commitments translating into investments is not very high. For example, private investors who had committed to invest in the semiconductor material processing capacity to cater to both the information technology and solar energy industries in 2009 when the government announced attractive policies for processing and refining semiconductor materials in India have not materialized so far.[16]

India continues to be a net importer of silicon ingots, wafers and modules rather than a primary producer of semiconductor materials as intended by the policy.[17] India's commitment to develop 20 GW of nuclear energy by 2020 when it signed a historic nuclear deal with the United States in 2005 has also not materialized and the target of 20 GW is now seen as unachievable.[18]

Renewable energy commitments may have greater probability of materializing into actual investment on account of international pressure on India to take on a responsible role in addressing climate change but it is too early to be optimistic on this front. The eventual outcome depends on decisions of a number of domestic and foreign private investors with uncertain incentives for keeping commitments.

The second is the extent to which renewable energy based energy generation (primarily in the form of electricity) will actually contribute to total energy generation assuming all commitments to invest materialize and all renewable energy capacity does come on-stream.

Based on optimistic assumptions on capacity utilization for all renewable energy capacity (say 17 per cent of installed capacity),[19] renewable energy could potentially meet roughly 16 per cent of India's electricity demand in 2022.[20] India's 36 GW of renewable energy capacity currently accounts for 4–5 per cent of electricity generation.[21]

As per India's 12th plan that was prepared by the previous government, renewable energy was expected to meet 16 per cent share of electricity generated only by 2030.[22] According to the projections in the plan, if the shares of nuclear energy and hydro-power are included as components of low carbon energy sources, roughly 36 per cent of India's electricity generation is expected to come from low carbon sources by 2030.[23]

India's Intended Nationally Determined Contributions (INDCs) towards addressing climate change submitted to the 21st Conference of Parties (COP) in Paris in December 2015 indirectly indicated a target of 350 GW of renewable energy by 2030.[24] The exact commitment in the INDC was that the share of non-fossil fuel in electricity generation capacity will increase to 40 per cent by 2030 provided technology and funding is available. As long as renewable energy replaces fossil fuel based energy generation capacity as opposed to merely supplementing fossil fuel based energy generation, the increase in renewable energy generation capacity may lead to a reduction in carbon emissions. This does not appear to be the case.

The revised target for coal production and coal based electricity generation announced by the new government suggests that renewable energy will supplement rather than displace fossil fuel based energy generation. If coal production and consumption doubles as proposed by the revised targets, the share of low carbon energy sources in India's commercial energy basket may actually fall below 30 per cent.

India's coal demand in 2012–13 was about 710 million tonnes (mt) including 145 mt of imported coal and coal demand in 2013–14 is expected to increase to 740 mt.[25] By 2016–17 India's total coal demand is expected to touch 980 mt.[26] This is close to coal demand projected for 2035 under the new policies scenario of the International Energy Agency (IEA).[27] The new government has revised the target for coal production to 1 billion tonnes (bt) by 2019. The objective of the new government is to limit coal imports and increase the supply of cheap electricity to provide affordable electricity access to all by 2019.[28]

The '450' scenario of the IEA (under which India transforms into a low carbon country) requires India's coal demand to decrease to 354 million tonnes of oil equivalent (mtoe) or 531 mt of hard coal by 2040.[29] This means that India's coal demand in 2040 must return to the level of coal demand in 2012 after peaking at 636 mt in 2020. India's current coal demand exceeds the peak levels prescribed for 2020 by the 450 scenario of the IEA.

Most projections by international institutions such as the IEA say that India is unlikely to demonstrate a marked shift away from fossil fuels even if India manages to implement policies to increase the share of renewable energy and improve the efficiency of energy use. In the 'current policies' scenario of the IEA, which takes into account policies put in place until 2013 such as India's national action plan for climate change (NAPCC) that include policies to introduce tradable renewable energy certificates, a national solar mission to increase the share of solar energy to 20 GW by 2022 and policies to enhance energy efficiency,[30] the share of fossil fuels in India's commercial (excluding traditional fuels such as firewood, dried animal dung, etc.) energy basket is expected to fall only by 2 per cent to 94 per cent by 2040[31] and the share of renewable energy is expected to increase to 1.2 per cent.[32]

Under the IEA's 'new policies' scenario which assumes that India will manage to achieve a 20 per cent reduction in carbon dioxide (CO_2) intensity[33] by 2020 compared with CO_2 intensity levels in 2005, the share of fossil fuels in India's commercial energy basket is expected to fall by 7 per cent to 89 per cent by 2040 and the share of renewable energy increase to 3 per cent.[34] Under the IEA's '450' scenario that is expected to contain the concentration of greenhouse gases (GHGs) in the atmosphere within the limit of 450 parts per million (ppm), the share of fossil fuels in India's energy basket is expected to fall by 21 per cent to 75 per cent by 2040 and the share of renewable energy is expected to increase to 8 per cent.[35]

Projections for energy demand in India in the future by domestic agencies concur with this conclusion.[36] The base line scenarios in all projections expect fossil fuel use to grow through 2030 and coal use to double.[37] Even the most optimistic scenario which assumes that all potential renewable energy resources will be exploited sees coal demand growing 1.5 times by 2030.[38] Projections by the Government for 2022 released after the revision of renewable energy targets in 2015 also show continued dominance of coal.

The energy access transition

According to household level surveys, 67 per cent of the households in India used electricity as their primary source of lighting and 71 per cent of the households used traditional fuels such as firewood, dried animal dung, coal and kerosene as the primary fuel for cooking in 2011.[39] The share of households using modern energy sources for lighting and cooking may have improved marginally since then but the fact that one-third of the population lacks access to electricity and two-thirds lack access to modern cooking fuels such as natural gas or liquid petroleum gas

(LPG) despite several decades of policy push to increase access is not only an aberration in India's image as a rising economic power but also a reflection of the state's inability to facilitate change.

Access to electricity for lighting

Electrification of households in rural India accelerated when India made a transition from a predominantly hydro-power based electricity generation system to a coal based system in the early decades as an independent country. Hydro-power shared the dominance in power generation with coal in India until the 1950s but coal based power generation outpaced growth of hydro-power generation by more than an order of magnitude in the subsequent periods.[40] Coal's emergence as the dominant source of power generation in India is underpinned by its abundant availability and its affordability.[41]

India has the fifth largest coal reserves in the world after the United States, Russia, China and Australia and it is currently the world's third largest producer of coal after China and the United States.[42] Coal extraction in India is largely dependent on low skilled labour, once again a plentiful, though not desirable resource in India which has ensured stable and predictable prices for coal over the last five decades. As the cheapest source of base load power, coal currently accounts for over 70 per cent of power generation[43] and coal based power generation capacity is projected to grow at a CAGR of 10 per cent for economic growth rates of around 9 per cent a year.[44]

Increase in coal based power generation capacity in India corresponds to a substantial increase in access to electricity as well as per person electricity consumption.[45] In the six decades following independence in 1947, power generation dominated by coal based generation coincided with expansion of village electrification by a factor of over 150.[46] At the time of independence only 0.5 per cent of Indian villages where more than 90 per cent of the population lived, had access to electricity; today more than 94 per cent of the villages where 66 per cent of the population lives have access to electricity.[47]

Though access to electricity only represents the availability of infrastructure to transmit electricity from the grid to the village and not necessarily a connection to individual households or the supply of electricity, it represents a realistic potential for households to use electricity. In the last six decades, electricity consumption per person has grown at roughly 6 per cent annually and increased from 16 kWh in 1942 to about 1,010 kWh in 2015, a sixty-fold increase.[48]

Despite this, average per person electricity consumption in India at 1,010 kWh in 2015 is roughly half of the world average and far below that in large developing countries such as those in the BRICS[49] group.[50] The low per person consumption figures partly reflects the fact that those who have access to electricity supply infrastructure are not always supplied with electricity.[51]

Moreover, average per person electricity consumption figures do not reflect household level reality in electricity consumption as they simply divide available electricity equally among the entire population of the country. According to household expenditure surveys, average electricity consumption per person in low income households in India is estimated to be as low as 150 kWh per person per year.[52] This is a tenth of what is considered the norm for meeting basic human needs. Average consumption among all households in India estimated at about 400 kWh per person per year is also lower than the threshold of 1,000 kWh required for decent living.[53]

Even when household level averages are used, 'the average' rarely constitutes the majority given the significant differences in incomes and access to resources between the affluent households and poor households. Poor households exceed the number of affluent households by an order of magnitude.[54] Out of rural households that constitute 66 per cent of Indian households, three-fourths have incomes well below national average; only half have electricity as primary source of lighting; less than a tenth have modern cooking stoves or refrigerators.[55]

The inadequate consumption of electricity in India is generally interpreted as a problem of inadequate supply and recommendations are made to augment the supply of electricity with abundant renewable sources such as solar power. Availability does facilitate consumption as the large difference in energy consumption patterns between households in rural and urban areas from the same income quartile indicates.[56] But it is also true that relatively affluent households with electrical goods consume more electricity than households without such goods.[57] Energy consumption is thus both the cause and consequence of material wealth. Without effective demand for electricity that is backed by purchasing power the supply of electricity will not materialize in a system that is increasingly influenced by market fundamentals.

In 2013 India had roughly 240 GW of power generating capacity including renewable power.[58] Theoretically, this is sufficient to supply over 1,200 kWh to each citizen for a year utilizing 60 per cent of generating capacity. While un-electrified households may desire to consume 1,200 kWh of electricity, their 'desire' to consume electricity does not constitute effective demand. To address this complex socio-economic challenge, a parliamentary panel has asked the government to estimate the extent of latent demand (un-served desire) for electricity in India and find ways to serve latent demand for electricity.[59]

Without general growth in material circumstances of households, energy consumption and consequently the quality of life of people is unlikely to change dramatically merely through an increase in supply of energy, be it from coal or from solar energy.

Solar energy in context of energy access

Many within and outside India believe that millions of Indian households without access to electricity constitute a huge opportunity for 'leapfrogging' to electricity provided by renewable energy sources such as solar energy.[60] The high cost of setting up transmission and distribution networks for grid based electricity supply in remote villages that are sparsely populated and the negative impact of coal mining and coal combustion on the health of the society are among arguments offered in favour of off-grid, stand-alone solar energy solutions for increasing household access to electricity.

These arguments are valid but close observation of ground realities in India shows that solar energy projects are moving towards the grid and not away from it as they should if replacing the grid with decentralized energy solutions to poor rural households is the objective. In other words solar power generators prefer to supply to the grid rather than to poor households. In 2014, installed capacity of grid connected solar projects in India was 30 times the installed capacity of stand-alone off-grid projects.[61] As observed earlier, the new government has scaled up the target for solar energy from 20 GW to 100 GW in 2015 and expanded the range of state assistance extended to solar energy projects.[62] Most of the solar energy projects planned after the revision of targets are grid connected projects.[63]

The superior economics of distributed grid connected PV systems over off-grid systems[64] may be among many reasons why grid connected PV systems dominate the sector in India. The economic case for grid connected PV systems is based on figures for feed in tariffs.[65] Feed in tariffs of grid-connected PV systems in India have achieved what has come to be labelled 'grid-parity'.[66] This has given rise to the notion that electricity generated by solar PV systems is now cheaper than grid based electricity largely generated using coal. This is not necessarily true.

The levelized cost of electricity (LCOE) captured in the feed in tariff for solar electricity implicitly values all kilowatt hours of electricity generated to be the same regardless of when they are generated. Electricity generated during the day is not valuable for a rural household that only uses electricity for lighting at night. LCOE also does not reflect the project's ability to provide capacity to meet uncertain demand (or ramping up capability). Even if solar capacity increases dramatically in India the need for dispatchable capacity will not be reduced significantly.

Capacity credit defined as the solar energy capacity that can be confidently relied upon at times of high demand is low for solar energy. Grid connected PV systems shift the cost of backup/storage to other non-solar grid users and hence receive an implicit subsidy from them.[67] In off-grid decentralized applications that seek to meet the need for lighting in poor rural households, this subsidy is unavailable and hence they have higher costs. When the cost of the battery/storage incorporated in off-grid systems is counted, solar power is neither competitive with grid based electricity nor is it valued as the same as grid based electricity.[68]

Most rural households in India that are 'electrified' (have access to infrastructure to receive electricity supply) but do not necessarily have a regular supply of electricity use kerosene as a substitute for electricity in lighting.[69] When the grid eventually reaches villages that have been electrified earlier with solar panels, households tend to abandon solar electricity in favour of cheaper grid electricity.[70]

The cross subsidy that solar PV users get from grid based users has raised the issue of equity in countries such as Germany and the United States which have substantial grid connected distributed PV systems.[71] In India this cost is invisible as yet because the share of electricity from grid connected PV systems is infinitesimally small compared to total generation based on fossil fuels.[72]

There are success stories in off-grid electrification in rural areas, some of which are underwritten by external grants or state subsidies or by pioneering business models promoted by private entrepreneurs.[73] Some of the successful business models are based on the willingness of poor households to pay a price for a unit of electricity that is significantly higher than what affluent households pay for the same electricity in urban areas. From a public policy perspective this would amount to a subsidy from the poor to the rich in making a contribution towards reducing carbon emissions. Whether this is desirable or sustainable in the longer term in a country with widespread poverty is a question that is yet to be debated seriously in India.

On the whole, the ability of off-grid PV systems to scale-up and provide electricity access to millions of households will depend on a massive internally or externally funded subsidy programme. Given that the elimination of existing subsidies including those for energy is now seen as a milestone in India's extended economic reform process, such a large subsidy is unlikely to materialize. In this light, large scale increase in access to electricity in the next two decades is likely to be driven by continued expansion of India's grid based rural electrification programme rather than through stand-alone solar and other renewable based systems.

Though the rural electrification programme is also driven by a cross subsidy derived from industrial electricity consumers, it is now an entrenched part of the system. Furthermore, the state capacity for electrification through grid expansion has deep and established roots. Since independence successive governments in India at the state and central level have implemented policies to increase grid access. These policies include but are not limited to financial support for the installation of infrastructure for transmitting and distributing electricity in villages and the cross-subsidization of electricity tariff for household and agricultural electricity consumption by industrial and commercial electricity consumption.

The 'Deen Dayal Upadhyaya Gram Jyoti Yojana' programme launched by the central government in July 2015 is among the newest of grid expansion programmes. It assures round the clock power to rural households and adequate power for irrigation pump sets in the next five years.[74] This will require substantial increase in base load power generation. Apart from this the 'make in India' programme which is one of the flag-ship programmes of the government that seeks to increase manufacturing sector employment generation will also require substantial increase in uninterrupted base load power generation. This will inevitably mean the continued dominance of coal in power generation.

Rural areas in India are not static entities socio-economically or demographically. Urbanization is changing the profile of villages in India as it pushes the younger generation in search of better economic prospects into cities and towns. Effectively this may move people to the grid much faster than the government can move the grid to them.

Access to modern cooking fuels

India's push to increase access to modern cooking fuels such as bottled liquid petroleum gas (LPG) and natural gas is more recent compared to its pursuit of providing access to electricity for lighting. Until the late 1980s, the government presumed that commercially procured fuel-wood will remain the dominant fuel for cooking. In the late 1970s and 1980s government reports consistently projected an increase in demand for fuel-wood and warned of the possibility of a supply shortage.[75]

The anticipated shortage of fuel-wood did not materialize as hydrocarbon based liquid and gaseous fuels for cooking rapidly displaced fuel-wood in urban households in the 1980s. The stabilization of oil prices in this period enabled the government to make kerosene and LPG available at a steep discount to market prices. Kerosene and LPG became fuels of choice even in middle class households in urban areas by the 1990s. Between 1970 and 2011 kerosene use increased by 50 per cent and LPG by 2,000 per cent albeit from a small base.[76] However the adoption of these modern cooking fuels remains an urban phenomenon even today.

In 1987, most of the urban and rural households used biomass for cooking.[77] By 2010, 60 per cent of urban households switched to LPG but only 10 per cent of rural households did.[78] According to the most recent census only 13 per cent of rural households and 74 per cent of urban households have access to LPG or natural gas.[79] Overall only about a third of Indian households use modern cooking fuels while the majority of the households depend on non-commercial biomass (twigs and farm residue collected informally by households) and dried animal dung along with limited quantities of kerosene available at a subsidized rates for poor households.

The persistence of biomass as dominant fuel for cooking in Indian households is a widely studied issue. One of the intriguing conclusions from these studies is that increase in household income does not automatically result in a switch to modern cooking fuels. Complex socio-economic circumstances of many poor rural households such as low opportunity cost of collecting biomass, low levels of female literacy and marginalization of women in the household that gives them little or no bargaining power also play a key role in the choice of fuels for cooking.[80]

Women were likely to adopt efficient cook stoves using LPG or other fuels if they were free but less likely to do so if there was a charge, demonstrating the fact that women bear the responsibility of cooking food but have no control over resources of the household.[81] Perversely even the gender of the first child appeared to have an impact on choice of fuel used for cooking. Urban households in India with a male first child were 2 per cent more likely to adopt efficient cooking fuels than comparable households with a female first child, illustrating a temporary rise in the woman's status within the household when she bears a son.[82] A policy that offers free LPG cooking stoves with free or subsidized supply of LPG when the first child of a poor household is female may facilitate change but this is not an area that has attracted as much attention as that of household electrification from the government.

The marginalized Indian woman's burden does not end there. Generally she has to collect and 'consume' (burn) more energy than her wealthier counterparts in urban areas to cook the same quantity of food because over a fourth of the energy content of biomass is dissipated in inefficient open cooking stoves.

For the nation as a whole, the opportunity cost of women using their time in collecting and using firewood has been estimated to be more than $6 billion a year even if the wage rate is assumed to be just US$1.33 per day per person.[83] The energy expended in collecting and processing biomass primarily by women is a hidden subsidy that millions of households utilize but do not acknowledge. The adverse health outcomes of using biomass for cooking caused by inhalation of fine particles are also largely borne by women.[84]

The transition of Indian households from biomass to modern cooking is an extremely slow one that reflects the slow transformation in the gender politics of India. Unlike the transition to electricity even urbanization may not accelerate it as long as benefits of such a transition are seen to accrue primarily to women.

In all scenarios for the future (low as well as high carbon growth scenarios) the share of traditional sources of energy in total primary energy is expected to fall from the current level of 23 per cent, which is the second largest share after coal.[85] The share of traditional sources of energy in total energy use that may be taken as an indicator of the extent of lack of access to modern fuels, particularly modern cooking fuels, is the largest in the '450' scenario at 19 per cent.[86] Loosely this could be interpreted as the trade-off between the transition to a lower carbon energy basket (lowest shares of fossil fuels) and the transition to universal access to energy (lowest share of traditional energy sources).

Even by 2030, 20 per cent of the Indian population is expected to have no access to electricity and 46 per cent of the population is expected to remain dependent on traditional fuels for cooking.[87] This projection for those with lack of access to electricity could be proved wrong if the government successfully implements its policy of providing electricity access to all through its 'Deen Dayal Upadhyaya Gram Jyoti Yojana' programme introduced in 2015.[88] As for cooking fuels the projection may be closer to reality as it requires more than just a subsidy on fuels such as kerosene and LPG that is available now. Innovative development policies that increase the opportunity cost of women as fuel collectors through subsidized and mandated access to education could accelerate the shift towards clean fuels but it requires long-term commitment and vision from the government.

Conclusions

India's transition towards a low carbon economy as well as its transition towards the provision of universal access to modern energy sources raise some intriguing issues.

The first is that both transitions are state led. Though the private sector is the dominant investor in the manufacture and production of low carbon energy such as solar and wind they are driven by state policy. The dominance of the state is in line with the structure of the Indian energy sector which has been state led for most of its history as an independent country. The irony which also constitutes India's energy transition challenge is that these two state led transitions are occurring at a time when India is beginning a transition away from a state dominant energy governance model towards a private sector led (and to a lesser extent market led) model to correct 'state failure' in increasing the supply and consumption of energy. Whether or not the state with a poor record as an agent of change will orchestrate the low carbon and energy access transitions successfully is a matter of conjecture at this point in time.

The Indian energy sector is a product of independent India's industrial policy. India's industrial policy resolution of 1948 and 1956 stated the government's aspiration and future plans for core industries like petroleum and reserved all future development for public sector undertakings. The government at the federal and regional level now controls vital segments of the energy industry such as production, transport (or transmission in the case of electricity), distribution and pricing in the coal, power, oil and gas sectors.

The 1970s was the decade when India nationalized most of the companies operating in the energy sector. The nationalization of the coal industry that began in 1973 was seen as the solution to poor labour conditions, absence of economically and geologically sound scientific methods for mining and low production volumes of coal.[89] The coal sector continues to be dominated by the state even today and the state is assigned almost all the blame for the inefficiency in the sector.

The nationalization of Anglo-American international oil companies (IOCs) operating in India companies was the result of government anger over their refusal to comply with its request to control the price of oil products during oil embargoes of the 1970s.[90] Though the sector is now open to the private sector, private participation is limited as they see the presence of the state in pricing decisions of key petroleum products such as LPG and kerosene as well as the pricing of natural gas as a source of distortion in the sector.

The worst performance of the state is reserved for the power sector where the governments at the central and regional level share responsibility. State (regional) electricity boards (SEBs) that were created after the enactment of the Electricity Supply Act of 1948 to increase access to electricity are now facing acute liquidity problems as government influenced low tariff rates for electricity consumption by households and agriculture pump sets have ruined their balance sheets.

In the light of the long history of 'state-failure' in increasing access to energy the question remains as to whether the state can successfully orchestrate the transition towards a low carbon energy system. Furthermore, given that the private sector (on which state policies for low carbon energy investment depend) sees the continued presence of the state as the biggest barrier to investment, the challenge before the state becomes more acute.

The second issue is over the adequacy of policies to facilitate the low carbon transition. Globally the longer term shift towards low carbon energy sources will be engineered primarily through technologies. Countries have the choice to become consumers or suppliers of these technologies. Current policies of the Indian government indicate that India is seeking to become both a consumer and supplier of new low carbon technologies.

The former is driven by its energy policy and the latter by India's industrial policy. On both there is lack of clarity in India's long-term objectives and the means to achieve them. Policies for the energy sector seem to be targeting primarily a quantitative target of installing 175 GW of renewable energy capacity. Even if this quantitative capacity target is achieved it may not automatically translate into generation and consumption of renewable energy.

India's industrial policy seeks to mandate domestic content in renewable energy equipment installed in India. This seems to signal India's desire to become a supplier of low carbon technologies. While both positions are desirable, neither policy is qualified with the necessary capabilities such as (a) financial incentives for utilities that are the largest buyers of electricity to opt for relatively expensive electricity when their balance sheets are severely constrained, (b) investment in infrastructure that will accommodate intermittent renewable energy into the grid and (c) investment in fundamental material science that will facilitate the emergence of India as a net supplier of cutting edge low carbon technologies.

The energy competition of the future will involve intangibles such as property rights and carbon rights (to emit and to clean it up). Mere addition of renewable energy capacity is unlikely to equip India with the desired rights and make it a serious player in the future.

The third is the trade off between the low carbon transition and the energy access transition. Many see coal as the main adjustment variable between the 'business as usual' and the 'low carbon' energy paths for India. In reality, people awaiting their first taste of carbon (electricity) and hydrocarbon (LPG and natural gas) based fuels in millions of Indian households are the main adjustment variable between the two paths.

If they are provided with these fuels that are readily available commodities today their quality of life will improve but it will add to the volume of carbon content in India's energy basket. However if they are asked to use or wait for yet to be commercialized technologies it would add to the heavy burden of social and economic injustice they already bear.

Many expect India to make a significant contribution towards saving the global commons independent of its socio-economic context. For them the persistence of fossil fuel use in India is a deliberate sub-optimal choice that would increase CO_2 emissions globally and expose millions of lives to climate risks domestically. Others think that India's socio-economic context limits its ability to make a contribution towards saving the global commons. From their perspective, the persistence of fossil fuels in India is not a deliberate sub-optimal choice but the only choice as it tries to address mass material deprivation in an unforgiving and predatory international economic environment.

There is evidence that a transition to a low carbon energy basket would impose considerable costs on developing nations.[91] The transformation in the energy systems of advanced countries such as Germany is a post-industrial transformation that is not necessarily a model for India. As long as the strong correlation between economic growth energy related carbon emissions in the early stages of industrialization of a country remains intact, containing CO_2 emissions though a priority is likely to remain subordinate to India's primary pursuit of material wealth.

In the next two decades, both of India's energy transitions, one towards becoming a low carbon country and the other towards increasing access to modern energy sources are likely to record substantial progress. However the probability that they will remain incomplete transitions that are independent of each other is high. Despite a phenomenal increase in renewable energy capacity, India may remain a fossil fuel dominant country even by 2040 and neither the market nor the state may provide universal access to modern energy sources.

India may also remain an energy poor country with low levels of energy consumption per person compared to global benchmarks and perversely, this will limit the perceived threat posed by the scale of India's energy transitions on global boundary conditions such as the global carbon budget. Ironically the continued co-existence of both high and low carbon growth and the co-existence of energy poverty and energy affluence may confirm, yet again the truism that for everything that is true in India, the opposite is also true.

Notes

1 Jitendra Singh, *Comparative Cost of Generation of Electricity* (New Delhi: Lok Sabha, 2015).
2 Planning Commission, *Twelfth Five Year Plan* (New Delhi: Government of India, Sage, 2013).
3 International Energy Agency (IEA), *World Energy Outlook* (Paris: IEA, 2014).
4 Planning Commission, *Twelfth Five Year Plan*; IEA, *World Energy Outlook* (2014).
5 Planning Commission, *Twelfth Five Year Plan*.
6 Ministry of Finance, *Key Features of the Budget 2015–16* (New Delhi: Government of India, 2015).
7 Ministry of New and Renewable Resources, 31 July 2015, http://mnre.gov.in/mission-and-vision-2/achievements/.

8 Press Information Bureau, *Revision of Cumulative Target under National Solar Mission*. June 17, 2015. http://pib.nic.in/newsite/PrintRelease.aspx?relid=122566. Accessed September 7, 2015.

9 Press Information Bureau, press release, December 2014, http://pib.nic.in/newsite/pmreleases.aspx?mincode=28.

10 Ibid.

11 Ibid.

12 Natalie Obiko Pearson, 'India's $3 Billion Wind Market to Slump as Tax Break May End', *Bloomberg Business*, 12 July 2011, www.bloomberg.com/news/articles/2011-07-12/india-s-3-billion-wind-market-to-slump-as-tax-break-nears-end.

13 Press Information Bureau, 'Renewable Energy Programmes Gets a New Impetus; Focus on Development of Energy Infrastructure', December 2014, http://pib.nic.in/newsite/pmreleases.aspx?mincode=28.

14 Press Information Bureau, 'Investment in Renewable Energy [illegible]', [illegible] 2015, http://pib.nic.in/newsite/pmreleases.aspx?mincode=28.

15 Press Information Bureau, 'Share of Solar Energy', March 2015, http://pib.nic.in/newsite/PrintRelease.aspx?relid=116567.

16 *EFY Times*, February 19, 2000, [illegible] Billion Semiconductor-Facility-In-Gujarat.

17 D. N. Bose, 'Polisilicon Production in India', *Current Science* 107, no. 1 (2014): 20–21.

18 Carl Paddock, *India-US Nuclear Deal: Prospects and Implications* (New Delhi: Epitome Books, 2009).

19 Renewable energy sources are intermittent and so utilization of capacity created for generation of electricity is much lower than utilization of capacity from conventional thermal projects. According to the IEA (International Energy Agency, *World Energy Outlook*, Paris: International Energy Agency, 2013), the capacity credit (dispatchable capacity that can be confidently relied upon during periods of high demand) for wind projects is 0–10 per cent for wind and 0–5 per cent for solar.

20 On the basis of the capacity credit in note 19, 17 per cent is an optimistic assumption but it is the average of the capacity utilization for renewable energy projects assumed by the Ministry of New and Renewable Energy of the Government of India. The share of renewable energy based electricity generation is calculated using demand projected by the 18th power survey of India for 2022.

21 Central Electricity Authority, *All India Installed Capacity of Power Stations as on 31.07.2015* (New Delhi: Government of India, 2015).

22 Planning Commission, *Twelfth Five Year Plan*.

23 Ibid.

24 Government of India, 'Intended Nationally Determined Contribution', New Delhi, 2015.

25 Central Statistical Office, *Energy Statistics 2015* (New Delhi: Central Statistical Office, Government of India, 2015).

26 Planning Commission, *Sectoral Demand for Coal Demand/Off-take for Annual Plan, 2012–13*, Government of India, https://data.gov.in/catalog/sectoral-coal-demandoff-take-annual-plan#web_catalog_tabs_block_10.

27 IEA, *World Energy Outlook* (2014).

28 Press Information Bureau, 'Modi Government is Committed to Provide Affordable and Clean Power 24x7 for all', January 2015, http://pib.nic.in/newsite/PrintRelease.aspx?relid=114557.

29 1 million tonnes of oil equivalent of coal is equal to 1.5 million tonnes of hard coal.

30 Ministry of Environment and Forests, *National Action Plan for Climate Change* (New Delhi: Government of India, 2008).

31 IEA, *World Energy Outlook* (2014).

32 Ibid.

33 CO_2 intensity is the amount of CO_2 emitted per unit of GDP.

34 IEA, *World Energy Outlook* (2014).

35 Ibid.

36 Navroz K. Dubash, Radhika Khosla, Narasimha D. Rao, et al., *Informing India's Energy and Climate Debate: Policy Lessons from Modelling Studies* (New Delhi: Centre for Policy Research, 2015).

37 Ibid.

38 Ibid.

39 Census of India, *Households, Household Amenities and Assets Data* (New Delhi: Government of India, 2011).

40 Central Electricity Authority, *Growth of Electricity Sector in India from 1947 to 2013* (New Delhi: Government of India, 2013).
41 Planning Commission, *Fourth Five Year Plan* (New Delhi: Government of India, 1970).
42 BP, *BP Statistical Review of World Energy* (London: BP, 2015), www.bp.com/content/dam/bp/pdf/energy-economics/statistical-review-2015/bp-statistical-review-of-world-energy-2015-full-report.pdf.
43 Planning Commission, *Twelfth Five Year Plan.*
44 Ministry of Coal, *Report of the Working Group on Coal & Lignite for the Formulation of the 12th Plan* (New Delhi: Government of India, 2011).
45 Central Electricity Authority, *Growth of Electricity Sector in India from 1947 to 2013.*
46 Ibid.
47 Central Electricity Authority, *Growth of Electricity Sector in India from 1947 to 2013*; Planning Commission, *First Five Year Plan* (New Delhi: Government of India, 1951).
48 Central Electricity Authority, *Growth of Electricity Sector in India from 1947 to 2015* (New Delhi: Government of India, 2015).
49 BRICS: acronym of Brazil, Russia, India, China and South Africa that constituted relatively countries with high economic growth rates in the last decade.
50 Central Electricity Authority, *All India Installed Capacity of Power Stations as on 31.07.2015* (Government of India, 2015).
51 Ministry of Home Affairs, 'Census of India,' 2011, www.censusindia.gov.in/2011census/hlo/House listing-housing-PCA.html.
52 National Sample Survey Organisation, *66th Round Consumption Expenditure, Survey 2009–10, 65th Round Housing Stock Survey 2008–09* (New Delhi: Government of India, 2010).
53 National Sample Survey Organisation, *66th Round Consumption Survey.*
54 Census of India. *Socio-Economic and Caste Census 2011* (New Delhi: Government of India, 2015).
55 Census of India, *Households, Household Amenities and Assets Data*; Census of India. *Socio-Economic and Caste Census 2011.*
56 National Sample Survey Organisation, *66th Round Consumption Survey.*
57 Ibid.
58 Central Electricity Authority, *All India Installed Capacity of Power Stations as on 31.07.2015* (Government of India, 2015).
59 'Power Ministry Should Find Out Latent Demand of Electricity: Parliamentary Panel', *DNA*, August 10, 2015.
60 Sunita Dubey, Siddharth Chatpalliwar, and Srinivas Krishnaswamy, *Electricity for All in India: Why Coal is Not King* (New Delhi: Vasudha Foundation, 2014).
61 Central Statistical Office, *Energy Statistics 2015.*
62 Press Information Bureau, 'Revision of cumulative targets under National Solar Mission from 20,000 MW by 2021-22 to 1,00,000 MW', June 2015, http://pib.nic.in/newsite/PrintRelease.aspx?relid=122566.
63 Anilesh S. Mahajan, 'The 100 GW Headache', *Business Today*, May 10, 2015.
64 Massachusetts Institute of Technology (MIT), *The Future of Solar Energy* (Cambridge MA: Massachusetts Institute of Technology, 2015).
65 The compensation required by the investor for each kWh of solar electricity supplied.
66 Tariff that is comparable or lower than the tariff of electricity supplied by conventional networks. Deutsche Bank, *2015 Solar Outlook* (Frankfurt am Main: Deutsche Bank, 2015).
67 Schalk Cloete, 'The Effect of Intermittent Renewables on Electricity Prices in Germany', Energy Collective, January 9, 2014, www.theenergycollective.com/schalk-cloete/324836/effect-intermitten t-renewables-electricity-prices-germany.
68 Amy Rose, Andrew Campanella, Reja Amatya, et al., 'Solar Energy Applications in Developing Countries' (working paper, Massachusetts Institute of Technology, Cambridge, MA, 2015).
69 Chao-yo Cheng, and Johannes Urpelainen. 'Fuel Stacking in India: Changes in Cooking and Lighting Mix 1987–2010', *Energy* 76, no. C (2014): 306–317.
70 'Dimming Glory of Sundarbans Solar Power Projects', *GFY News Service*, 7 January 2015, www.geo graphyandyou.com/32-featured-stories/200-dimming-glory-of-sunderban-solar-power-projects.html.
71 MIT, *The Future of Solar Energy.*
72 Central Statistical Office, *Energy Statistics 2015.*

73 Global Network on Energy for Sustainable Development, *Renewable Energy-based Rural Electrification: The Mini-Grid Experience from India* (New Delhi: Global Network on Energy for Sustainable Development, 2014).
74 Press Information Bureau, *387 Companies committed to build renewable energy capacity of 2.7 lakh MW in 5 years.* April 23, 2015. http://pib.nic.in/newsite/pmreleases.aspx?mincode=28. Accessed September 7, 2015.
75 N. C. Saxena, *Wood Fuel Scenario and Policy in India* (Bangkok: Food and Agriculture Organisation of the United Nations, 1997).
76 Central Statistical Office, *Energy Statistics 2015*; Planning Commission, *First Five Year Plan.*
77 Cheng, and Urpelainen, 'Fuel Stacking in India.'
78 Ibid.
79 Census of India, *Households, Household Amenities and Assets Data.*
80 Jessika J. Lewis, and K. Subhrendu Pattanayak, 'Who Adopts Improved Cook Stoves? A Systematic Review,' *Environmental Health Perspectives* 120, no. 5 (2012): 637–645.
81 Avinash Kishore, and Dean Spears, 'Having a Son Promotes Clean Cooking Fuel Use in Urban India: Women's Status and Son Preference,' *Economic Development and Cultural Change* 62, no. 4 (2014): 637–699.
82 Kishore, and Spears, 'Having a Son Promotes Clean Cooking.'
83 Planning Commission, *Integrated Energy Policy: Report of the Expert Committee* (New Delhi: Government of India, 2006).
84 Lewis, and Pattanayak, 'Who Adopts Improved Cook Stoves?'
85 IEA, *World Energy Outlook* (2014).
86 Ibid.
87 International Energy Agency (IEA), *World Energy Outlook* (Paris: IEA, 2012), 250–253.
88 *DNA*, 'PM Modi to Launch Deendayal Upadhyaya Gram Jyoti Yojana in Patna,' July 25, 2015.
89 Chakrabarti, Prabhas Kumar, *Investment Decisions in the Indian Public Sector* (New Delhi: Northern Book Centre, 2002).
90 Batra, C. D. *Oil Industry in India* (New Delhi: Mital Publishers, 2004).
91 Michael Jakob, Jan Christoph Steckell, Stephan Klasen, et al., 'Feasible Mitigation Action in Developing Countries,' *Nature Climate Change* 4, no. 11 (2014): 961–968.

23

Jordan's response to acute energy insecurity

Searching for a winning combination

John Calabrese

On 10 May 2015, the Government of Jordan (GOJ) launched 'Jordan 2025 Vision and National Strategy', a 10-year blueprint for economic and social development that will guide the country's economic, fiscal and social policies over the next decade. The document lays out a set of ambitious targets that include increasing the gross domestic product (GDP) growth rate from 3.1% in 2014 to 7.5% by 2025.[1] Reaching these targets will be difficult given the host of challenges Jordan faces – energy insecurity arguably being the most important and the most vexing of them.[2]

Impending crisis can supply the impetus for bold policy initiatives of potentially momentous consequence. In the case of Jordan, mounting domestic economic pressures coupled with the spillover effects of political upheaval and conflict in the surrounding region have spurred determined efforts aimed at rescuing the country from its precarious, and indeed, unsustainable energy situation. The past decade has been marked by the unfurling of a comprehensive energy strategy and the enactment of legislation and policy reforms designed to facilitate its implementation, followed by a flurry of deals and Memoranda of Understanding (MOUs), as well as the launch of new energy projects and energy-related initiatives.

This chapter explores the relationship between Jordan's dire energy situation and the country's economic performance and prospects by addressing three main questions: What are the sources and severity of Jordan's energy insecurity? How have Jordan's energy-related problems undermined economic progress and, as a consequence, placed the country's very stability at risk? What steps have Jordanian authorities taken in order to extricate the country from its energy predicament and unleash its economic potential?

The structure and struggles of the Jordanian economy

In the early decades following Jordan's independence in 1946, the country's economy was characterized by a small industrial base, limited agricultural production, and the dominance of the services sector.[3] As Figure 23.1 shows, the services sector continues to contribute the largest share by far to Jordan's GDP.

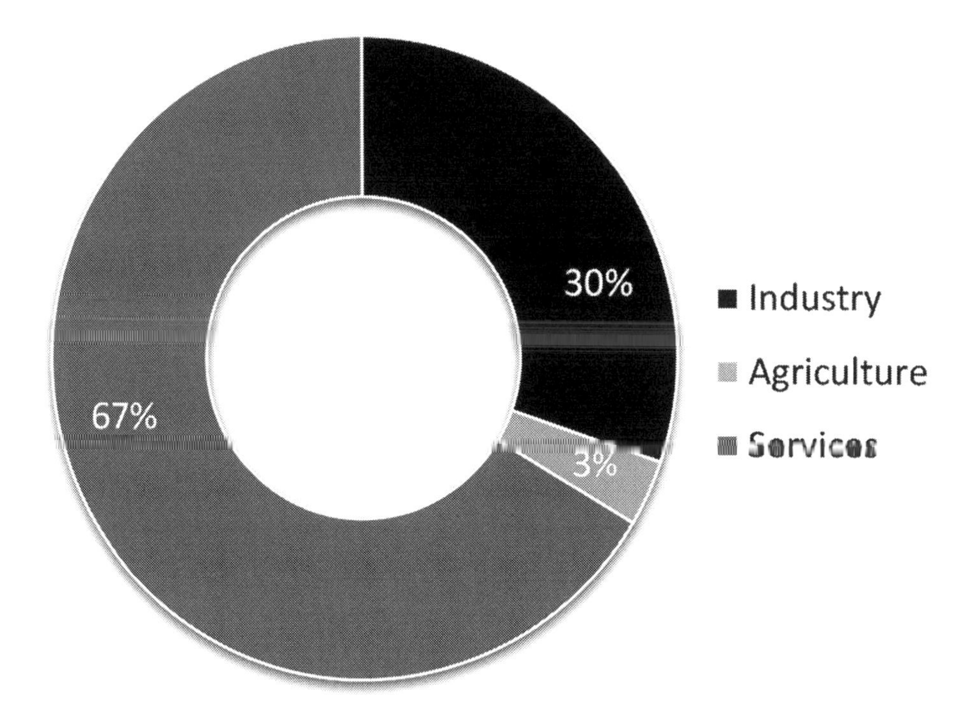

Figure 23.1 GDP composition by sector of origin, 2014
Note: Includes services, commerce, education, and the military.
Source: World Bank, Jordan; see http://data.worldbank.org/country/jordan.

Since the early 1990s, there has been greater emphasis on the need to expand industrial production.[4] However, the further development of Jordan's industrial sector requires the long-term availability of inexpensive and stable energy supplies. After achieving an average growth rate of 6.5% in 2000–09, the economy slowed to an average of 2.6% between 2010 and 2013 – buffeted by the effects of the global financial crisis and gripped by a crippling energy crisis. In fact, Jordan was then, and remains, a country struggling to overcome acute energy insecurity.

Energy security is the continuous availability of energy in varied forms, in adequate quantities, and at reasonable prices.[5] Many factors can determine the degree to which a country enjoys 'energy security': its vulnerability to transient or longer disruptions of imported supplies; the availability of local and foreign sources of supply to fulfill, over time and at reasonable prices, growing demand requirements; environmental challenges; policy and regulatory frameworks; and global energy market forces. Jordan confronts a complex set of energy-related challenges, beginning with a widening gap between rising consumption requirements and domestic production capabilities.

Jordan's energy imbalances[6]

Jordan's energy situation is characterized by 1) strong energy demand growth, particularly in the electricity sector, 2) a primary energy supply that is overwhelmingly tilted toward fossil fuels, primarily oil and gas, and 3) limited indigenous primary energy resources, and thus heavy reliance on foreign sources of supply.

Strong demand growth

In recent years, primary energy consumption has risen rapidly. As shown in Figure 23.2, surging consumption of crude oil and oil products has accounted for much of that increase.

Jordan's primary energy demand growth has been dominated by the transport, household and industrial sectors (Figure 23.3).

Jordan's electricity consumption per caput – primarily responsible for driving primary energy demand – has increased dramatically in recent years,[7] spurred mainly by demand growth in the

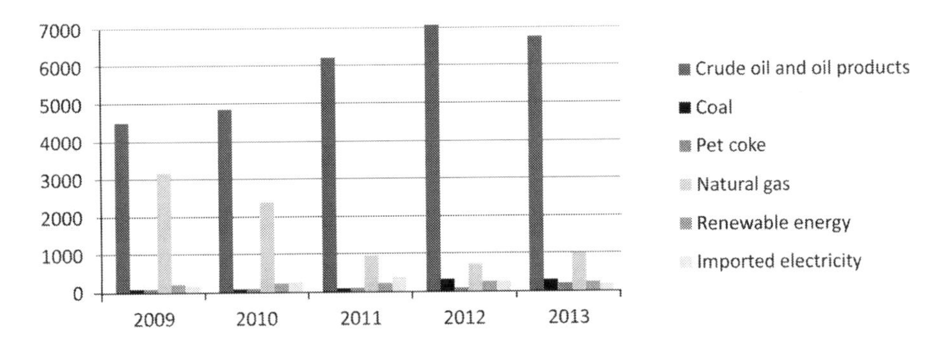

Figure 23.2 Primary energy consumption, 2009–13 (thousand toe)
Source: Jordanian Ministry of Energy and Mineral Resources (MEMR), MEMR, http://www.memr.gov.jo/.

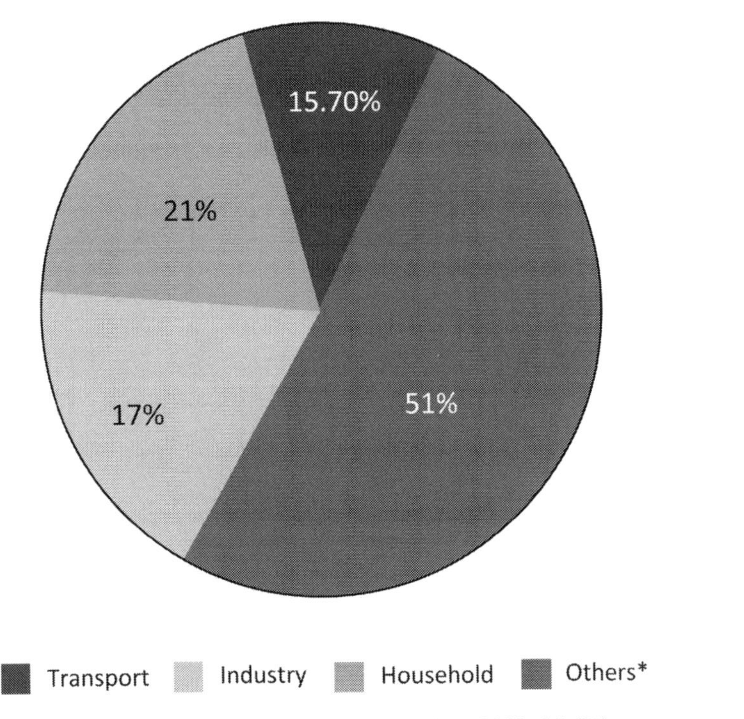

Figure 23.3 Sectoral distribution of final energy consumption, 2009–13 (%)
Source: Jordanian Ministry of Energy and Mineral Resources (MEMR), MEMR, http://www.memr.gov.jo/.

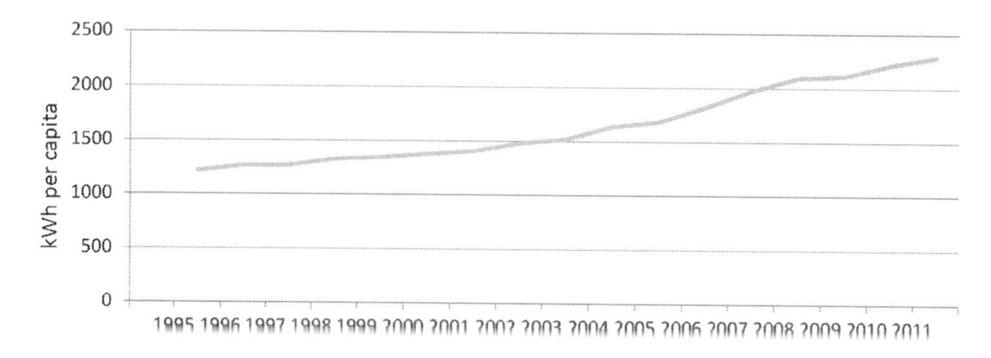

Figure 23.4 Electricity consumption (kWh per capita)
Source: World Bank, http://databank.worldbank.org/data/home.aspx.

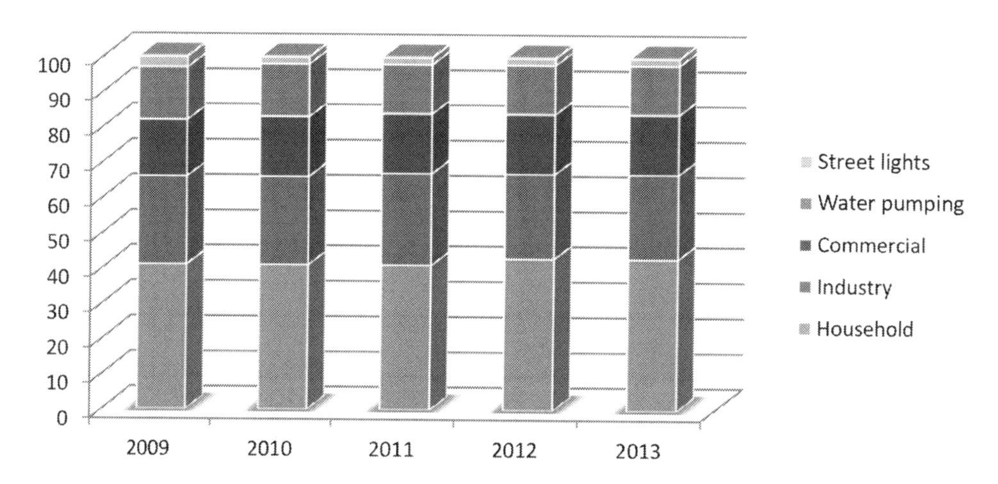

Figure 23.5 Sectoral consumption of electricity, 2009–13 (%)

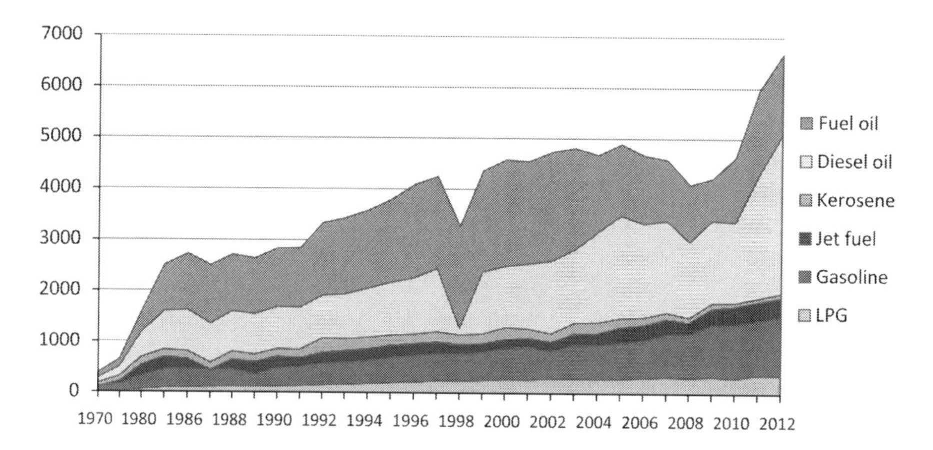

Figure 23.6 Petroleum products consumption (000 tons)
Source: World Bank.

residential and industrial sectors.[8] (See Figures 23.4 and 23.5.) According to Energy Minister Mohammad Hamed, electricity requirements are expected to triple by 2030.[9]

Historically, Jordan has relied very heavily on diesel and fuel oil to generate electricity, which is reflected in their disproportionate share of petroleum products consumption.[10] (See Figure 23.6.)

Drivers of energy/electricity demand growth

The main drivers of energy demand growth in Jordan are population growth, rapid urbanization, increased per capita consumption due to rising standards of living and the adoption of new social practices, and large government subsidies.

Jordan's demographics are changing dramatically. The average annual population growth rate for the period 2010–15 was estimated at 3.5%.[11] Even at the present, lower growth rate of 2.2%, Jordan's population is expected to double in about 30 years, from 6.3m to 13.2m in 2040.[12] The massive influx of refugees, whose total the UN High Commissioner of Refugees (UNHCR) forecasts could climb to 1m by the end of 2015,[13] has placed an additional strain not just on Jordan's infrastructure but on its energy resources as well.

Urbanization is a major demographic driver of energy demand. In recent years, the pace of urbanization in Jordan has accelerated. The annual rate of change for the period 2010–15 is estimated at 3.79%. As of 2014, urban residents accounted for 83.4% of the country's population.[14]

Severe domestic energy supply constraints

Jordan is an extremely resource poor state with limited renewable freshwater supplies, no reserves of crude oil and little natural gas. While Jordan is believed to possess large shale oil deposits, they are currently not recoverable in large quantities at viable prices.[15] Thus it is unsurprising that local production of energy (i.e. crude oil, natural gas and renewable energy) in 2013 amounted to a mere 3% of Jordan's total energy needs.[16] In the period between 2009 and 2013, as depicted in Figure 23.7, domestic production of crude oil and natural gas steadily declined.

Jordan's ability to meet its energy consumption requirements has also been hampered by energy and electricity infrastructure constraints. Plans to upgrade and expand the supply capacity of the country's sole oil refinery at Zarqa – long overdue – have been repeatedly postponed as Jordan has struggled, among other things, to assemble the financing needed to undertake the project.[17] As in the case of the refinery, Jordan's electricity generation fleet is relatively old, with most existing power plants scheduled for decommissioning by 2020. Projects are planned or underway to address capacity shortages. Some have already materialized. The Qatrana Power Station began operation in 2011, the Manakher Power Station opened in 2014, the Amman Asia Power Plant was completed in 2015, and the multi-phase plan to boost capacity at the Al-Samra Power Generating Station has advanced – all significant steps forward in addressing the country's acute electricity shortages though likely to provide just temporary relief.[18]

Heavy energy import dependence and exogenous shocks

Government statistics indicate that in 2012 the use of crude oil and oil products accounted for approximately 88% of Jordan's total primary energy demand.[19] For the reasons discussed above, Jordan has had to rely heavily on imports of these resources as well as of natural gas to meet its mounting energy requirements. (See Figure 23.8.)

Jordan's heavy dependence on ever-increasing volumes of energy imports has resulted in acute vulnerability to supply disruptions. The year 2003 marked the beginning of the end of an era of privileged access and cheap energy supplies from Arab neighbours. By 2005, the Jordanian authorities had acknowledged that the country faced an 'energy insecurity scenario'.[20]

The first exogenous shock came with the change of regime in Baghdad in 2003 and the subsequent cancellation of the existing arrangement whereby Jordan had obtained half of its supply of Iraqi oil at no charge and the rest at steeply discounted prices. Initially, Kuwait, Saudi Arabia and the UAE agreed to offset the loss of Iraqi oil and maintain low prices. However, these agreements expired in 2006, forcing Jordan to enter the international oil market and compete for supplies.

Jordan revised its energy strategy in 2003, switching most of its power plants from oil to natural gas (80% of electricity generation is now based on gas) and entering a long-term, fixed price supply agreement with Egypt, at a cost estimated at one-third of the market price.[21] This preferential arrangement came under considerable scrutiny following the fall of the Mubarak regime in February 2011. Five months later, the Jordanian and Egyptian (transitional) governments signed a new 12-year agreement but on terms requiring Jordan to pay much closer to the international market price.[22]

However, as it happened, Jordan not only had to pay higher prices for Egyptian gas, but also had to contend with unstable supplies. Egypt struggled to meet its export commitments, due to high domestic demand and falling production, resulting in the redirection of gas from overseas sales to the local market. In addition, the Arab Gas Pipeline (AGP) – which runs from Egypt through Jordan and had been the latter's principal source of natural gas imports – was repeatedly targeted by saboteurs. As a result of these attacks, the pipeline's throughput dropped precipitously in 2011 and 2012. Citing continuing attacks on the pipeline, the Egyptian gas holding company EGAS announced in April 2014 the indefinite suspension of natural gas supplies to Jordan.[23]

The Syrian civil war and the emergence of the Islamic State in the Levant (ISIL) have adversely affected Jordan's energy situation as well. The massive influx of Syrian refugees has resulted in additional power demand in cities and new electricity power allocations for the Zaatari refugee camp, partly in order to expand access to water. As a consequence, Jordan has incurred additional capital and operational costs. The rise of ISIL also has had a negative impact on Jordan's energy situation. Oil exports from Iraq to Jordan were halted in early 2014 due to deteriorating security in Anbar province. Insecurity has also delayed construction of an oil pipeline – a project agreed in March 2013 and approved by the Iraqi parliament the next year – that will extend from the oilfields of Basra to the Haditha region in the far west of Anbar to the Jordanian port of Aqaba.[24]

In sum, for years Jordan had enjoyed lower than international market prices at the cost of dependency on one or two regional suppliers and at the risk of supply disruptions. Over the past decade, Jordan has had to switch suppliers and fuel inputs due to disruptions in production and transportation that it could neither foresee nor prevent, thereby incurring costs that have placed a severe strain on the economy.

Jordan's soaring energy import bill has led to mounting fiscal and external imbalances. The cost of imported crude oil, oil products and natural gas in 2013 amounted to a whopping 17% of GDP.[25] In recent years, the public sector has registered record deficits, reaching 16.4% of GDP in 2012. Aside from the central government, the main contributor to these deficits has been the National Electric Power Company (NEPCO), whose losses climbed to 5% of GDP in 2011 and 2012.[26] Rising deficits led Jordan to enter into an IMF Stand-By Arrangement (SBA) in August 2012, whose aim was to bring NEPCO back to profitability by gradually raising electricity tariffs, the first of which took effect at the beginning of 2014.[27] Plunging oil prices, which dropped by more than 50% from June 2014 to March 2015, have provided some relief for the economy and eased the strain on the government's finances – though only temporarily.

Jordan's energy problems and the burden they have placed on the economy have been aggravated by the country's fragile water situation. The provision of water and energy is interdependent. Increasing the supply of one is largely contingent on the availability of the other. When water scarcity increases, additional energy is required to pump water over longer distances and lift it over differential elevations, or to produce it through alternative means, such as desalination.[28]

Jordan is in a situation of water deficit, with demand higher than the country can sustainably supply. The country's water resources are dwindling. Its groundwater is overexploited. The Dead Sea provides a substantial portion, but its level is falling fast. Furthermore, the availability of surface water and groundwater is uncertain since a large proportion originates from transboundary, shared aquifers.[29]

The Water Authority of Jordan (WAJ) is already the country's largest electricity consumer, using about 15% of Jordan's entire electricity production. Included among the key elements of *Jordan's Water Strategy 2008–2020* is substantially increased dependence on desalination.[30] However, constructing and installing the infrastructure, as well as powering booster pump stations and desalination facilities will require a massive amount of electricity. Moreover, continued reliance on imported fuel to power Jordan's water sector will also expose water prices to volatilities in the oil and gas markets.

The search for a winning combination

Against the backdrop of the trends and developments discussed above, the GOJ has had to confront two main challenges: 1) how to keep pace with surging electricity demand and 2) how to lessen the country's dependence on imported fossil fuels and the associated risks of price volatility and unreliable supplies.

It is generally understood that the most effective means of ensuring energy security include managing domestic demand (e.g. by increasing efficiency), improving the reliability of external sources through a more diversified supply structure, and/or increasing domestic energy supplies. The Royal Energy Strategy issued in 2005 and updated in 2007 envisages meeting Jordan's growing energy demand, particularly in the electricity domain, by raising the contribution of indigenous energy sources in the primary energy mix by 2020 to 39%.[31] To reach this target, the strategy places emphasis on investing in exploiting unconventional resources (mainly oil shale), boosting domestic gas production, increasing the utilization of renewable energy (primarily wind and solar), and developing nuclear power.

Joining the shale revolution

Oil shale is Jordan's only major proven indigenous energy source.[32] In fact, Jordan possesses the world's fourth-largest reserves of oil shale after the United States, China, and Russia.[33] Vast deposits of near-surface oil shale are widely distributed throughout the country.[34]

Over the past two decades, the GOJ has commissioned numerous feasibility studies and test programmes; and Natural Resources Authority (NRA) has conducted extensive geological surveys. Although moves to develop shale deposits date back to 2006, the high price of extraction was a major hindrance in doing so. To be sure, improvements in shale technologies have lowered the cost of extracting and processing oil shale so as to make it more competitive with conventional crude oil. However, arguably the strongest impetus for Jordan to forge ahead was the interruption of Egyptian gas supplies and the related financial drain caused by the need to switch to more expensive fuel inputs for power generation.[35] As presented in Table 23.1, Jordan has signed a number of deals aimed at developing shale over the past several years.

Despite this progress, Jordan faces headwinds. For one thing, production of oil from oil shale in Jordan is not expected to start in 2018. At first, this will be in small amounts, though it is expected to rise gradually to reach significant quantities by early 2020. The venture with Sacos is not expected to yield results until 2019. It is likely to take more than a decade before JOSCO is able to produce oil from oil shale in commercial quantities. Thus, even if expectations regarding oil shale are eventually met, they will not address Jordan's immediate needs.

A second and related issue is that of securing funding. Attracting private sector investment is difficult because oil shale projects generally require large initial capital outlays and involve comparatively long lead times before facilities become productive and profitable.

Yet another issue is the environmental risk associated with producing oil from shale. Processing oil shale is water-intensive, and thus a serious problem in itself for water-poor Jordan. Furthermore, producing oil from shale could contaminate and/or lower the level of groundwater as well as result in mine water leakage into surface water bodies.

Developing domestic gas production – grasping the LNG and Israeli gas lifelines

A key objective of Jordan's energy strategy is to augment the share of gas in the primary energy mix by developing domestic sources, the most promising of which was thought to be from British Petroleum's (BP) activities at the Al Risha field.[42] Indeed, Jordanian officials were hopeful that intensive exploration and drilling at Al Risha would lead to the discovery of extensive recoverable gas reserves.[43] However, the decision by BP in January 2014 to end its drilling operation due to unsuccessful appraisal work[44] dealt a sharp blow to these expectations, leaving Jordan scrambling for alternatives. One such option was speeding up efforts aimed at importing liquefied natural gas (LNG). Another was trying to cement a long-term gas supply agreement with Israel. Jordanian authorities have assiduously pursued both.

Shortly after the withdrawal of BP from Al Risha, Jordan launched a tender for the supply of LNG to feed a planned floating storage and regasification unit (FSRU) in the southern port of Aqaba.[45] In fact, from as early as 2012, Jordanian officials had declared they were committed to securing direct, as opposed to transit supply of LNG, as part of a comprehensive scheme to develop the Aqaba Special Economic Zone Authority (ASEZA), including the enhancement of Jordan's energy intake capabilities.[46] Concern about the unreliability of Egyptian gas supplies coupled with the disappointment of BP's termination of its operations at Al Risha are likely to have further spurred these efforts. In January 2015 Royal Dutch Shell signed a five-year agreement to supply LNG to Jordan's NEPCO, with LNG amounting to a quarter of the company's daily needs for power generation.[47]

Israel could provide Jordan with an inexpensive and reliable means of meeting most, if not all of its gas requirements given the size and potential of the Tamar and Leviathan offshore gas fields. In light of this possibility, beginning in 2012, American officials reportedly encouraged the US firm Noble Energy to approach Jordanian authorities and supported formal negotiations. But since then the prospective partners have had to overcome or find ways around numerous domestic political, regulatory and other hurdles. In September 2014, a group of Israeli energy companies signed a letter of intent to supply gas from the Leviathan field to Jordan.[48] However, several months later Israel's Antitrust Authority decided to reconsider the status of the project's main investors (Noble Energy and the Dalek Group), effectively freezing the deal.[49] Meanwhile, in Jordan, over two dozen organizations, including trade unions and political parties, waged a campaign to scuttle the deal.[50] Eventually, an arrangement was crafted authorizing

Table 23.1 Jordan's oil shale deals

Year	*Developer*	*Project description*
2009	Royal Dutch Shell PLC (RDSA)	Deal to explore for oil shale[36]
2010	Enefit [Estonia]	Concession to explore for oil shale and gas in Attarat area, southern Jordan
2011	Karak International Oil, a subsidiary of Jordan Energy & Mining Ltd. (JEML) [UK]	Deal to explore the Al Lajjun area in Karak governorate[37]
2012	Global Oil Shale Holdings (GOSH) [Canada]	Agreement to conduct feasibility studies and drilling operations in the central region of al-Attarat-Um el-Ghudran area[38]
2013	Shandong Electric Power Construction Corporation (SEPCO III) and HTJ Group [China] with Al-Lajjun Oil Shale Company [Jordan]	MOU to build a $2.5b. shale-fired (900 MW) power plant in southern city of Karak[39]
2014	Saudi Arabian Corporation for Oil Shale (Sacos)	A $2b. deal for the right to extract and develop oil shale in al-Attarat Um al-Ghudran region
2014	Attarat Power Company (APCO), a subsidiary of Enefit [Estonia] and YTL International Berhad [Malaysia]	Final agreement reached on $2.2b. deal to build oil shale-fueled power plant with a capacity of 470 MW.[40]
2015	Questerre Energy Corporation [Canada]	Deal to conduct initial technical and geological studies in the south of Jordan[41]

separate supply deals with the Jordanian mineral companies Arab Potash and Jordan Bromine (with gas to be drawn from the Tamar, not the Leviathan field). Nevertheless, the status of Noble Energy and the Dalek Group in the Israeli gas sector has not yet been resolved while Jordanian access to Israel's Leviathan reservoir remains uncertain.[51]

'Jordan's hope': boosting renewables

Addressing attendees at the Second Jordan Energy Investment and Projects Summit, Energy Minister Hamed stated, 'Jordan is facing an unprecedented increase in electricity demand, and the only way we can meet this is by investing and utilizing local energy resources, particularly solar and wind power.'[52] In remarks to Parliament, Prime Minister Abdullah Ensour referred to renewable energy as 'Jordan's hope'.[53]

Large-scale deployment of renewable energy (RE) offers Jordan three benefits: 1) reduced dependence on fossil fuel imports and, therefore, lower risk of fuel shortages due to supply disruptions from foreign sources; 2) absorption of rising costs of future electricity generation and thus insurance against increasing fossil fuel prices; and 3) job creation, thereby contributing to economic development.

As laid out in the 2013 National Energy Efficiency Action Plan (NEEAP), Jordan seeks to expand renewable energy from 1% of the energy mix in 2010 to 10% by 2020.[54] The Action Plan envisages all projects being linked to the grid by 2018. Jordan's quest to boost the share of RE in the primary energy mix is focused mainly on solar and wind energy. That is not surprising. Jordan, whose solar irradiance is one of the highest in the world (ranging from four to seven kilowatt hours per square metre) has enormous solar energy potential. However, solar, like all renewable energy, has thus far been underutilized. The GOJ is determined to change that – setting a target to install 600 megawatts (MW) solar photovoltaic (PV) capacity and 1,200 MW wind energy capacity by 2020.

The passage of the Renewable Energy and Energy Efficiency Law (REEL) in April 2012 was a major boost to the RE sector. The REEL put in place a number of measures designed to facilitate investment in RE, including a system whereby private local and international companies can bypass a competitive bidding process and negotiate directly with the Ministry of Energy and Mineral Resources to implement new projects; and the provision for the establishment of a 'Renewable Energy and Energy Efficiency Fund' dedicated to supporting projects that aim to reduce energy consumption and/or utilize RE sources.[55]

As originally crafted, the plan to increase locally produced RE input is to proceed in three phases, the first of which was launched in 2011 and the second in 2013. Both of these invitations for private sector involvement in the expansion of Jordan's RE sector have borne fruit. Speaking at the opening of the Powering Middle East Summit in September 2014, Energy Minister Hamed noted that the Kingdom had signed 12 power purchase agreements (PPAs) to develop solar projects – mainly in the southern Ma'an Governorate – with a total capacity of 200MW.[56] Apart from a total of 12 solar PV projects, two wind energy projects were also allocated through the first tender. The second round of Jordan's solar independent power producer (IPP) tender, totaling 200MW, drew exceptionally low bids that could bode well for the future of the RE sector in Jordan, and indeed of other Middle Eastern countries.[57]

The Jordanian Hashemite Royal Court has sought to lead by example, connecting its grid to a 5.6MW solar power plant established inside the compound.[58] King Abdullah II personally has been proactive in garnering support, domestically and internationally, for Jordan's RE sector. On the sidelines of the World Economic Forum on the Middle East and North Africa (WEF-MENA), the King announced project launches and new agreements totalling $6.9b., including a letter of intent between Hanergy (a Chinese firm) and MEMR through a 20-year PPA to upgrade the local grid capacity (valued at $1.5bn.) and a 20-year PPA between Green Watts and the NEPCO for an 82 MW wind farm in Al Rajef (valued at $174m).[59]

In undertaking these efforts, Jordan has benefited from the strong support of international partners. In 2013, a particularly important accord was struck between the GOJ and Masdar (Abu Dhabi), which allows for direct consultation between the two parties – drawing on the latter's expertise – in order to determine the viability and facilitate the implementation of RE projects in Jordan.

Several financing programmes also have been launched to support the development of the RE sector and to promote energy efficiency:

- KfW (German) Development Bank: The establishment, together with the GOJ, of a public-sector energy efficiency fund.[60]
- Gulf Cooperation Council (GCC): The allocation of approximately $300m to support renewable energy projects.[61]
- European Bank for Reconstruction and Development (EBRD) and the French Development Finance Institution PROPARCO: The provision of loans for solar projects totalling $100m.[62]
- International Finance Corporation (IFC) (The World Bank): The provision of a $207.5m loan package to fund the construction of seven solar plants.[63]

- Nippon Export and Investment Insurance, Mizuho Bank, Japan for International Cooperation, and Standard Chartered Bank: Provision of debt financing for the Shams Maan project.[64]
- Kuwait Fund for Economic Development: Award of $215m for energy projects, with $150m earmarked for building the Maan wind farm.[65]

In addition, numerous foreign companies are engaged in implementing these projects, including Trina Solar Ltd. (China),[66] Qatar Nebras, Diamond Generating Europe (a subsidiary of Japan's Mitsubishi Corporation), and First Solar (US)[67] – to name only a few.

These encouraging results notwithstanding, there are a number of obstacles in the way of achieving the ambitious targets that the GOJ has set for the RE sector. The Renewable Energy Law – enabling MEMR to enter direct negotiations with firms and providing tax incentives for development of renewable energy – though drafted in 2009 was not enacted by parliament until 2012. There were significant delays in the first and second rounds of the direct proposals process for renewable energy, as companies that had signed the PPAs worked with lenders to bring the projects to financial closure. The third procurement round for solar projects was cancelled, reportedly because of failure to secure funding for expanding grid capacity in order to accommodate the projects.[68] An additional problem is that RE technology has not yet advanced to the point of enabling large-scale energy storage. As a result, Jordan cannot rely upon RE – at least not in the near term – as the base load source for power.

Nuclear energy – 'a strategic choice'

The Jordanian Atomic Energy Commission (JAEC) describes the country's pursuit of nuclear power as 'a strategic choice'. It should come as no surprise that Jordan has taken nuclear energy seriously in light of its energy predicament. Indeed, as far back as April 2007, then Energy Minister Khaled Sharida announced that Jordan intended to build a nuclear power plant to produce electricity by 2015.[69] Four months later, King Abdullah II remarked that the country hopes to obtain 30% of its electricity from nuclear power by 2030.[70] Clearly, the 2015 date proved overly ambitious – so too might the 30% target. Nevertheless, proponents of Jordan's civilian nuclear programme have remained steadfastly committed to it. In fact, over the past several years the GOJ has signed a number of bilateral agreements – including with Argentina, Canada, the Czech Republic, China, France, Japan, Romania, Russia, Saudi Arabia, Spain, South Korea, Turkey and the United Kingdom – aimed at fostering cooperation in the nuclear energy sphere.

But progress in the quest for nuclear power has been slow. Negotiations with the United States on a '123' agreement have been suspended since early 2011.[71] The JAEC's position that nuclear power can provide the pathway for the attainment of energy independence – based partly on the argument that the country possesses substantial quantities of commercially exploitable uranium – has yet to be validated. On the contrary, Jordan declined to renew the licence of the AREVA-backed Jordanian French Uranium Mining Company, ending a once-promising partnership.[72] A year earlier, Rio Tinto, whose uranium exploration activities had focused on Wadi Sahab/Sahra Abiad, close to the Saudi Arabian border, withdrew. China National Nuclear Corporation (CNNC) has continued searching for uranium at Hamra-Hausha in the north, and Wadi Baheyya in the south.[73] All the while, in the face of these setbacks and uncertainties, Jordanian officials have remained steadfast in their commitment and optimistic in their assessment of the prospects for developing large quantities of commercially viable grades of uranium.

Similarly, they have remained resolute in the face of concerns and criticism at home and abroad regarding the nuclear programme as a whole. Jordan's plan to develop civilian nuclear power has evoked apprehension in Israel[74] and sparked domestic political opposition – from environmentalists and activists, and even from lawmakers. A May 2012 vote of the lower house of parliament ordered the suspension of nuclear activities, including uranium exploration, though the JAEC forged ahead anyway, contending that their activities were compatible with lawmakers' demands.

Since then, opponents of the nuclear programme have continued to raise a variety of objections.[75] Some insist that, far from assuring Jordan's energy independence, the quest for nuclear power will make the country dependent on foreign companies.[76] There have also been questions about the financial feasibility of Jordan's pursuit of civilian nuclear energy.[77] In March 2015, it was reported that the JAEC and Rosatom (of Russia) had reached agreement to proceed with the plan first announced in 2013 to build two 1,000MW reactors in the northern part of the country.[78] The partners expect to be able to attract international funding for the project. However, at least one possible source of such funding, the IFC – the World Bank's lending arm – ordinarily does not finance nuclear projects.

Yet, Jordanian authorities seem undeterred by the setbacks, roadblocks and uncertainties they have encountered. Though forced to consider, and then reject several possible sites on which to build the country's first nuclear power plant, they none the less persevered, homing in on the Amra desert area north of Amman. However, this latest choice too is problematic, as the proposed site of the plant lies in close proximity to the Azra Wetland Reserve and would sit atop the Azra aquifer.[79] It remains to be seen whether Jordanian authorities will be able to assuage the concerns of local farmers, tribal leaders and others – or ultimately decide simply to proceed despite them.

But pinning its hopes on the nuclear sector could be a risky option for Jordan's political establishment given the susceptibility of such projects to early cancellation, questions about Jordan's ability to mobilize sufficient capital to cover its own obligations, as well as doubts about its ability to implement the requisite upgrades to the power grid, scale up the necessary technical skill base, and avoid potential cost overruns – not to mention the opportunity cost of diverting financial resources away from renewables.

Conclusion

Jordan's economy grew steadily between 1999 and 2008, bolstered by the implementation of reforms. The economy then entered a three-year period in which growth dramatically declined and the country's fiscal situation sharply deteriorated, buffeted by the ill effects of the global financial crisis. Since then, the economy has been struggling to regain its stride. Regional instability has hampered the recovery: the persistence of conflict in Syria and the escalation of violence in Iraq over the past two years have caused Jordan's trade with those countries to plummet[80] while the deluge of Syrian refugees has placed an additional burden on the economy.

However, weak oil prices have partially offset these difficulties – narrowing the current account and helping boost construction activity and mining exports. Meanwhile, foreign reserves, estimated at $14b, have shown resilience. These bright spots have raised expectations that Jordan's economy is on track to grow at a rate exceeding 3% in 2015.[81] Nevertheless, Jordan's acute energy insecurity, which has undermined its economic progress, remains a major risk factor. Whether, when and how Jordan's complex energy challenges are addressed will profoundly affect the Jordanian economy's structure, stability and further development.

In the face of the country's chronic energy insecurity, Jordanian authorities cannot justifiably be accused of inertia. Particularly over the past four years, seized by the gravity of the dire economic situation in the late 2000s, the GOJ has demonstrated resourcefulness, resilience and determination in its efforts to execute an energy strategy that is comprehensive and forward-looking. Though some of the headline goals and targets may prove unattainable – at least within the originally conceived timeframe – there is clear evidence of progress on a number of fronts, including in energy infrastructure (especially LNG) and renewable energy (solar and wind) development.

Jordan's geopolitical centrality within the Middle East region has endowed it with a strategic significance far greater than its small size might otherwise warrant. It is therefore not surprising that the involvement of regional and international partners has been – and will remain – of critical importance to Jordan's efforts in addressing the country's energy challenges, whether in furnishing technical advice and support, competing for tenders, assisting with financing or providing investment capital.

Yet, at the same time, Jordan's efforts to tackle energy insecurity have been plagued by delays, setbacks and uncertainties. Plans to exploit the country's untapped uranium ore deposits have not yet materialized. Nor have efforts to expand domestic natural gas production. Turmoil in Iraq makes it impossible to gauge when construction of the dual pipeline (gas and oil) project agreed in 2013 – promising a lower import bill and substantial transit fees – might begin, much less when it might be completed.

Furthermore, Jordan remains at the mercy of the market even in areas where it has made significant headway, such as entry into the LNG industry. LNG will replace diesel and heavy fuel oil, which are more harmful to the environment and generally more expensive. However, savings on electricity generation costs will depend on the co-evolution of oil and gas prices. Jordan remains at the mercy of regional geopolitics and domestic politics as well. Additional gas imports may become available from the Eastern Mediterranean basin but territorial disputes and public opposition have thus far stood in the way.

Otherwise 'successful' efforts can sometimes engender new problems or complications. If, for example, security conditions were to improve, thereby facilitating the construction of the Iraq-Jordan 'Aqaba pipeline', the Kingdom would presumably have to absorb at least some operations and maintenance costs and would need to expand its refinery capacity to accommodate augmented supplies. The further development of renewable energy, too, could pose continuing challenges for Jordan, necessitating, for example, the upgrading of the country's grid in order to cope with the additional generation capacity provided by new RE power plants.

The allure of a big payoff could come at great risk and/or cost. Such might be the case with Jordan's shale development plans and nuclear power programme – both of which will require large amounts of water. In the case of the proposed nuclear power project, Jordanian authorities are planning to cool the plants with waste water to minimize the impact on scarce fresh-water deposits. However, such a 'solution' will increase costs.

There is no panacea for Jordan's energy insecurity. Nevertheless, the country's complex energy challenges are surmountable. In order to tackle them effectively, Jordanian authorities will have to strike a balance between the need to take urgent action and the need to ensure that the actions taken are sustainable in the longer term. This, in turn, will require a combination of flexibility and steely resolve, as well as a deft hand in harnessing the continued strong support of the international community, securing financing and mobilizing private sector investment, and marshaling broad public acceptance for its initiatives.

Even if the twin goals of energy self-sufficiency and of Jordan becoming a net energy exporter and regional energy transit hub prove not to be far-fetched, it will be far into the future before they are realized. In the meantime, Jordan will have to struggle to manage energy insecurity by seeking to mitigate the risks and costs associated with it. Some of the steps taken so far have produced encouraging results – and may well constitute, or at least help construct a winning combination.

Notes

1 Omar Obeidat, 'Gov't Launches "Jordan 2025" Development Blueprint', *The Jordan Times*, 11 May 2015, http://jordantimes.com/govt-launches-jordan-2025-development-blueprint.

2 According to Dr Nemat Shafik, deputy head of the International Monetary Fund, 'Energy is the [illegible] Al Khalidi, 'IMF Says Jordan Energy Crisis Poses Toughest Reform Challenge', *Reuters*, 6 March 2013, http://english.alarabiya.net/en/business/2013/03/06/IMF-says-Jordan-s-energy-crisis-poses-toughest-reform-challenge.html.

3 The trade and services sector contributes to nearly two-thirds of Jordan's GDP and has been steadily [illegible]

4 Warwick Knowles, *Jordan since 1989: A Study in Political Economy* (London: I. B. Tauris, 2005).

5 Deese defines energy security as 'a condition in which a nation perceives a high probability that it will have adequate energy supplies at affordable prices'. See David Deese, 'Energy: Economics, Politics and Security', *International Security* 4, no. 3 (1979): 140.

6 Jordanian Ministry of Energy and Mineral Resources, *Jordan's Energy Balance 2012* (Ministry of Energy and Mineral Resources, 2012), www.memr.gov.jo/Portals/12/statistics/Energy%20Balance.htm.

7 World Bank, 'Electric Power Consumption (kWh per capita)', World Bank, http://data.worldbank.org/indicator/EG.USE.ELEC.KH.PC?page=3&display=default.

8 Jordanian Ministry of Energy and Mineral Resources (MEMR), www.memr.gov.jo/Portals/12/statistics/Electricity%20Consumption.htm.

9 Mohammad Ghazal, 'Electricity Demand Expected to Triple by 2030 – Minister', *The Jordan Times*, June 11, 2014, http://jordantimes.com/electricity-demand-expected-to-triple-by-2030——minister.

10 MEMR.

11 United Nations Department of Economic and Social Affairs, *World Statistics Pocketbook 2014*, United Nations Department of Economic and Social Affairs, 2014, http://unstats.un.org/unsd/pocketbook/WSPB2014.pdf, 102.

12 US Agency for International Development, *Jordan Country Development Cooperation Strategy (CDCS)* (Washington, DC: US Agency for International Development, 2012), 8.

13 United Nations High Commissioner for Refugees, '2015 UNHCR Country Operations Profile – Jordan', United Nations High Commissioner for Refugees, 2015, www.unhcr.org/pages/49e486566.html.

14 Central Intelligence Agency; see also World Bank, http://data.worldbank.org/indicator/.

15 Jordan National Energy Research Center, 'Oil Shale in Jordan', Jordan National Energy Research Center, www.nerc.gov.jo/OilShale/OilShaleInJordan.html.

16 Jordanian Ministry of Energy and Mineral Resources (MEMR), *2013 Report* (MEMR, 2013), www.memr.gov.jo/LinkClick.aspx?fileticket=B495BBqcNs4%3d&tabid=111, 20.

17 See Mohammad Ghazal, 'Work on Refinery Expansion to Start in 2015', *The Jordan Times*, 26 November 2014, http://jordantimes.com/work-on-refinery-expansion-to-start-in-2015; Omar Obeidat, 'Jordan Showcases Investment Opportunities Worth $20b in Seven Sectors', *The Jordan Times*, 23 May 2015, http://jordantimes.com/jordan-showcases-investment-opportunities-worth-20b-in-seven-sectors.

18 Jordan Atomic Energy Commission, *White Paper on Nuclear Energy in Jordan* (Jordan Atomic Energy Commission, 2011), www.jaec.gov.jo/cms/uploadedfiles/c18cbcac-92e9-481b-a781-498ca0bf7e9c.pdf; 'KEPCO Opens Jordan Power Plant', *The Korea Herald*, April 30, 2015, www.koreaherald.com/view.php?ud=20150430000793, 13–14.

19 Jordanian Ministry of Energy and Mineral Resources (MEMR), www.memr.gov.jo/Portals/12/statistics/Import%20of%20Crude%20oil%20and%20Petroleum%20Products.htm.

20 M. Mason, M. A. Al-Muhtaseb, and M. Al-Widyan, 'The Energy Sector in Jordan: Current Trends and the Potential for Renewable Energy', in *Renewable Energy in the Middle East*, ed. M. Mason and A. Mor (Dordrecht: Springer, 2009), 41–54.

21 Jordanian Electricity Regulation Committee, 'Electricity, Energy and National Economy', Jordanian Electricity Regulation Committee, www.erc.gov.jo/English/ElectricityNationalGrowth/Pages/default.aspx.
22 'Jordan and Egypt Finalize Amended Gas Deal', *Al Arabiya News*, 22 December 2011, http://english.alarabiya.net/articles/2011/12/22/183967.htm.
23 Mohamed Adel, 'Egypt Ceases Supplying Gas to Jordan Indefinitely: EGAS', *Egypt News*, 4 August 2014, www.dailynewsegypt.com/2014/08/04/egypt-ceases-supplying-gas-jordan-indefinitely/.
24 'Iraq Agrees to Extend Oil Pipeline to Jordan', *Al Arabiya News*, 24 December 2012, http://english.alarabiya.net/articles/2012/12/24/256896.html; John Lee, 'Companies Sought for Iraq-Jordan Pipeline', *Iraq-Business News*, 5 March 2015, www.iraq-businessnews.com/tag/iraq-jordan-pipeline/; Omar al-Shahr, 'Iraqi Oil to Reach Jordan, Egypt in Pipeline through Aqaba', *Al Monitor*, 12 March 2013, www.al-monitor.com/pulse/originals/2013/03/iraq-jordan-egypt-oil.html#.
25 See Abdel-Motaleb Al-Nugrush, *Jordan Energy Statistics Compilation Methods* (presented at the International Energy Agency Energy Statistics and Indicators, Cairo, October 2014), www.iea.org/media/training/presentations/egypttrainingoct2014/pdf/JordanPresentationcairo2014.pdf.
26 World Bank, 'Maintaining Stability and Fostering Shared Prosperity amid Regional Turmoil', *Jordan Economic Monitor* (Spring 2013), www.worldbank.org/content/dam/Worldbank/document/MNA/Jordan_EM_Spring_2013.pdf, 8, 11–12.
27 Jordan removed fuel subsidies in 2013 and replaced them with cash payments for lower income households. See 'New Electricity Tariff Effective at Year Turn', *The Jordan Times*, 31 December 2013, http://jordantimes.com/new-electricity-tariff-effective-at-year-turn.
28 Jakob Granit, 'Elaborating on the Nexus between Energy and Water', *Journal of Energy Security* (23 March 2010), www.ensec.org/index.php?option=com_content&view=article&id=238:elaborating-on-the-nexus-betweenenergy-and-water&catid=103:energysecurityissuecontent&Itemid=358.
29 Valerie Yorke, 'Politics Matter: Jordan's Path to Water Security Lies through Political Reforms and Regional Cooperation' (working Paper No. 2013/19, NCCR, 2013), www.nccr-trade.org/fileadmin/user_upload/nccr-trade.ch/wp5/5.5a/Valerie_Yorke_NCCR_WP_2013_19_v3.pdf, 15. See also Stephen Mcilwaine, 'Managing Jordan's Water Budget: Providing for Past, Present and Future Needs' in *The Jordan River and Dead Sea Basin: Cooperation amid Conflict*, eds C. Lipchin, D. Sandler and E. Cushman, NATO Science for Peace and Security Series (New York: Springer, 2009), 61–73.
30 Jordanian Ministry of Water and Irrigation, *Water for Life: Jordan's Water Strategy, 2010–2012* (Jordanian Ministry of Water and Irrigation, 2009), http://web.idrc.ca/uploads/user-S/12431464431JO_Water-Strategy09.pdf.
31 Jordanian Ministry of Energy and Mineral Resources, *Summary of the Updated Master Strategy of Energy Sector in Jordan for the Period (2007–2020)* (Ministry of Energy and Mineral Resources, 2007), www.memr.gov.jo/Portals/0/energystrategy.pdf.
32 For a discussion of Jordan's oil shale potential, see M. S. Bsieso, 'Jordan's Experience in Oil Shale Studies Employing Different Technologies', *Oil Shale* 23, no. 3 (2003): 360–370.
33 Oil shale, which is different from shale oil, can be heated to produce oil, either below ground or after extraction. See Colorado Oil & Gas Association, 'Oil Shale vs Shale Oil', Colorado Oil & Gas Association, 18 June 2013, www.coga.org/pdf_Basics/Basics_OilShale.pdf.
34 See Jordan National Energy Research Center, 'Oil Shale in Jordan'.
35 Suleiman Al-Khalidi, 'Lights Go Out in Jordan as Energy Crisis Bites', Reuters, 3 April 2013, http://uk.reuters.com/article/2013/04/03/uk-jordan-energy-idUKBRE9320NG20130403.
36 Taylor Luck, 'Jordan Energy Demand "Outpacing" Supply, Officials Warn', *The Jordan Times*, 25 November 2015, http://jordantimes.com/jordan-energy-demand-outpacing-supply-officials-warn.
37 'Investment Boost for Jordan Oil Shale', *Petroleum Economist*, 5 October 2011, www.petroleum-economist.com/Article/2912542/Investment-boost-for-Jordan-oil-shale.html.
38 Mohammad Tayseer, 'Jordan, Canada's GOSH Agree to Assess Shale Oil, Petra Says', *Bloomberg Business*, 18 September 2012, www.bloomberg.com/news/articles/2012-09-18/jordan-canada-s-gosh-agree-to-assess-shale-oil-petra-says.
39 'Jordan, China Ink $2.5 Bn Deal to Build Oil Shale-fired Plant', Agence France Presse, 19 September 2013.
40 'Jordan to Build Oil Shale-fueled Power Plant', *Oil Review*, 6 October 2014, www.oilreview.me/industry/jordan-signs-us-2-2bn-deal-to-build-oil-shale-fueled-power-plant.
41 *Big News Network*, 18 May 2015, www.bignewsnetwork.com/index.php/sid/232946499.

42 Karen Ayat, 'Jordan's Efforts towards Energy Security', Natural Gas Europe, 8 May 2013, www.naturalgaseurope.com/jordan-energy-security-interview-dr-khaled-toukan. Interview with Dr. Khaled Toukan, Chair of the Jordan Atomic Energy Commission and Chair of Higher Ministerial Nuclear Steering Committee.
43 'BP Begins Drilling in Risha's Gas Field', Reuters, 18 June 2012, www.reuters.com/article/2012/06/18/us-jordan-gas-drilling-idUSBRE85H0JK20120618.
44 Summer Said, 'BP's Risha Exit Means Slim Pickings for Jordan's Energy Needs', *The Wall Street Journal* (blog), 29 January 2014, http://blogs.wsj.com/middleeast/2014/01/29/bps-risha-exit-means-slim-pickings-for-jordans-energy-needs/.
45 'BP Steps Back from Jordan's Risha Gas Field', *Interfax Natural Gas Daily*, 30 January 2014, http://interfaxenergy.com/gasdaily/article/6281/bp-steps-back-from-jordans-risha-gas-field.
46 Mahmoud Al Abed, 'Aqaba to Open LNG Terminal on Independence Day', *The Jordan Times*, 15 May 2015, http://jordantimes.com/aqaba-to-open-lng-terminal-on-independence-day.
47 Natural Gas Asia, 'Shell, Jordan Close LNG Supply Deal', Natural Gas Asia, 21 January 2015, www.naturalgasasia.com/jordan-shell-sign-gas-deal-14554.
48 Sara Toth Stub, and Sarah Kent, 'Israel Plans to Sell Natural Gas to Jordan', *The Wall Street Journal*, 3 September 2014; Sharon Udasin, 'Israel and Jordan Sign 15-year Gas Supply Deal Worth $15 Billion', *The Jerusalem Post*, 3 September 2014, www.jpost.com/Middle-East/Israel-and-Jordan-sign-15-year-gas-supply-deal-worth-potentially-15-billion-374332.
49 Sharon Udasin, 'Jordanian Parliament Member Announces Pause in Gas Negotiations with Israel', *The Jerusalem Post*, 4 January 2015, www.jpost.com/Middle-East/Jordan-suspends-talks-with-Israel-over-15-billion-natural-gas-deal-386600.
50 Patrick O. Strickland, 'Gas Deal with Israel is "Knife at Jordan's Throat", Say Campaigners', *The Electronic Intifada*, 5 March 2015, http://electronicintifada.net/content/gas-deal-israel-knife-jordans-throat-say-campaigners/14325.
51 Sharon Udasin, 'Israel Approves Gas Export Deal with Jordan', *The Jerusalem Post*, 2 April 2015, www.jpost.com/Israel-News/Israel-approves-gas-export-deal-with-Jordan-396038.
52 Luck, 'Jordan Energy Demand "Outpacing" Supply, Officials Warn'.
53 'Kingdom to Become Renewable Energy Regional Leader – PM', *The Jordan Times*, 13 March 2015, http://m.jordantimes.com/article/kingdom-to-become-renewable-energy-regional-leader——pm.
54 Regional Center for Renewable Energy and Energy Efficiency, 'Summary: The National Energy Efficiency Action Plan of Jordan (NEEAP)', Regional Center for Renewable Energy and Energy Efficiency, 2013, www.rcreee.org/content/summary-national-energy-efficiency-action-plan-jordan-neeap.
55 Oxford Business Group, 'Legislative Reform: Moves to Strengthen the Mandate for Renewable Energy Usage', Oxford Business Group, www.oxfordbusinessgroup.com/analysis/legislative-reform-moves-strengthen-mandate-renewable-energy-usage.
56 Mohammad Ghazal, '1800 Megawatts of Renewable Energy Projects to Be Connected to Grid by 2018', *The Jordan Times*, 17 September 2014, http://jordantimes.com/1800-megawatts-of-renewable-energy-projects-to-be-connected-to-grid-by-end-of-2018.
57 LeAnne Graves, 'Low Bids on Project in Jordan Likely to Trigger Solar Energy Boom', *The National*, 21 May 2015, www.thenational.ae/business/energy/low-bids-on-project-in-jordan-likely-to-trigger-solar-energy-boom.
58 Anthony McAuley, and Frank Kane, 'Jordan Unveils Dh48 Billion of New Deals', *The National*, 22 May 2015, www.thenational.ae/business/economy/jordan-unveils-dh48-billion-of-new-deals; 'Jordan Targets Energy Diversification with Bumper Investment Deals', *Arabian Business*, 23 May 2015, http://m.arabianbusiness.com/jordan-targets-energy-diversification-with-bumper-investment-deals-593610.html.
59 Walid Khoudouri, 'Jordan Turns to Wind Power in Search of Renewable Energy', *Al Monitor*, 5 August 2014, www.al-monitor.com/pulse/business/2014/08/jordan-wind-power-project-energy-consumption.html#/; 'Jordan's Competitiveness Body Sets Up Deals', *Zawya*, 24 May 2015, www.zawya.com/story/Jordan_Competitiveness_body_set_up_deals_signed_at_WEF-ZAWYA20150524062847/.
60 'Jordan Signs 30m Euros Loan Deal with KfW for Water Project', *The Jordan Times*, 10 July 2014, http://jordantimes.com/jordan-signs-30m-euros-loan-deal-with-kfw-for-water-project.
61 Taylor Luck, 'Gulf Funds to Support Solar, Wind Projects in South', *The Jordan Times*, 29 May 2015, http://jordantimes.com/article/gulf-funds-to-support-solar-wind-projects-in-south.
62 Nibal Zgheib, 'EBRD Finances Solar Power Plants in Jordan', *EBRD News*, 10 November 2014, www.ebrd.com/news/2014/ebrd-finances-solar-power-plants-in-jordan.html.

63 See International Finance Corporation, 'IFC Finalizes $207.5 Million Financing for Ground-breaking Solar Program in Jordan', press release, International Finance Corporation, 15 October 2014, http://ifcext.ifc.org/IFCExt/pressroom/IFCPressRoom.nsf/0/8C749563CC07E1B285257D72004766D3; Mohammad Ghazal, 'International Finance Corporation to Fund Construction of Seven Solar Power Plants in Jordan', *The Jordan Times*, 16 October 2014.
64 'Jordan, Japan Sign deals to Expand Cooperation', *The Jordan Times*, 18 January 2015.
65 Justin Doom, 'Kuwait Fund Gives $215 Million for Jordan Energy Projects', *Bloomberg Business*, 6 May 2015.
66 Ehren Goossens, 'Trina to Build 10 Megawatt Solar Power Plant for Shamsuna', *Bloomberg Business*, 19 November 2014, www.bloomberg.com/news/articles/2014-11-19/trina-to-build-10-megawatt-jordan-solar-power-plant-for-shamsuna.
67 Mohammad Ghazal, 'US Company to Help Construct Maan Solar Power Plant', *The Jordan Times*, 5 August 2014.
68 For description and status of RE projects, see Squire Sanders, *The Future for Renewable in the MENA Region* (Cleaner Energy Pipeline, 2014), www.cleanenergypipeline.com/Resources/CE/ResearchReports/The%20Future%20for%20Renewable%20Energy%20in%20the%20MENA%20Region.pdf, 8–9; and Eversheds, *Developing Renewable Energy: A Guide to Achieving Success* (PwC, 2015), www.pwc.com/en_M1/m1/publications/documents/eversheds-pwc-developing-renewable-energy-projects.pdf, 9–10. For the announcement of the cancellation of the third round of procurements, see Jordanian Ministry of Energy and Mineral Resources, www.memr.gov.jo/LinkClick.aspx?fileticket=KvRV6fuTyC8%3d&tabid=36 and for the reasons related thereto, see Mohammad Ghazal, 'Gov't Scraps Plans for Four Renewable Energy Power Plants', *The Jordan Times*, 6 August 2014, http://jordantimes.com/govt-scraps-plans-for-four-renewable-energy-power-plants.
69 Yoav Stern, 'Jordan Announces Plans to Build Nuclear Power Plant by 2015', *Haaretz*, 2 April 2007, www.haaretz.com/news/jordan-announces-plans-to-build-nuclear-power-plant-by-2015-1.217260.
70 'Jordan's King Urges Speeding Up the Nation's Nuclear Program', Associated Press, 26 August 2007, www.haaretz.com/news/jordan-s-king-urges-speeding-up-the-nation-s-nuclear-program-1.228200.
71 The so-called '123' agreements allow for the exchange of technology, expertise and research between the signatories. The sticking point appears to be insistence by the United States, for which proliferation concerns are paramount, that Jordan foregoes the uranium enrichment and the development of other sensitive technologies such as the production of heavy water. Absent such an agreement, US corporations would be barred from involvement in the nuclear sector.
72 Taylor Luck, 'Jordan, AREVA Part Ways over Uranium Mining', *The Jordan Times*, 23 October 2012, http://jordantimes.com/jordan-areva-part-ways-over-uranium-mining; Elizabeth Whitman, 'Jordan Scrambles to Avoid Energy Crisis amid Regional Turmoil', *Middle East Eye* 24 July 2014, www.middleeasteye.net/news/jordan-scrambles-avoid-energy-crisis-amid-regional-turmoil-986749875.
73 World Nuclear Association, 'Nuclear Power in Jordan', World Nuclear Association, last modified October 2015, www.world-nuclear.org/info/Country-Profiles/Countries-G-N/Jordan/.
74 See, for example, 'Jordan Briefs Israel on Nuclear Plans', *WikiLeaks*, 22 June 2009, www.wikileaks.org/plusd/cables/09AMMAN1394_a.html.
75 Alice Su, 'Jordan Faces No-nukes Campaign', *Al Monitor*, 12 November 2013, www.al-monitor.com/pulse/originals/2013/11/jordan-nuclear-rosatom-environment-energy.html#.
76 Raed Omari, 'Deputies Vote to Suspend Nuclear Project', *The Jordan Times*, 30 May 2012, http://jordantimes.com/Deputies+vote+to+suspend+nuclear+project-48497.
77 Claire-Louise Isted, 'Rosatom Expects Funding for Foreign Projects', *Nucleonics Week*, 21 November 2013; Chen Kane, 'Are Jordan's Nuclear Ambitions a Mirage?' *Bulletin of Atomic Scientists*, 15 December 2013.
78 Sam McNeil, 'Jordan, Russia Sign $12 Billion Deal on Nuclear Power Plant', PennEnergy, 6 April 2015, www.pennenergy.com/articles/pennenergy/2015/04/jordan-russia-sign-10-billion-deal-on-nuclear-power-plant.html.http://www.jica.go.jp/jordan/english/office/topics/c8h0vm00008wsa3l-att/press141002_01.pdf.
79 Areej Abuqudairi, 'Jordan Nuclear Battle Heats Up', *Al Jazeera*, 14 April 2014, www.aljazeera.com/news/middleeast/2014/02/battle-heats-up-over-jordanian-nuclear-power-201422685957126736.html.
80 John Reed, 'Closure of Syria's Last Border Crossing Hits Jordanian Economy', *The Financial Times*, 8 April 2015, www.ft.com/cms/s/0/c0df376a-dd27-11e4-a772-00144feab7de.html#axzz3bTBBAChp.
81 'Lower Energy Prices to Increase Competitiveness', *Arab News*, 21 May 2015, www.arabnews.com/economy/news/749676; Suleiman Al-Khalidi, 'Jordan Seeks 3.8 pct Growth Despite Regional Wars',

Reuters, 22 May 2015, https://en-maktoob.news.yahoo.com/jordan-sees-3-8-pct-gdp-growth-despite-150459848–sector.html.

24

Analyzing Turkey's energy transition

Challenges and opportunities

Mehmet Efe Biresselioglu

In the aftermath of the 1970s energy crises, the governments around the globe, especially western governments, have realized the importance of security of supply. It is difficult to clearly state that "energy security" as a concept has emerged as a priority; however the security of supply aspect has become prominent. In classical terms, security of supply simply refers to the threat of disruption in energy supply, rather than to a broader concept.[1] Moreover, it was assumed to be directly related with national security, rather than a distinct issue in its own right.[2] Therefore, in early writings, energy security was usually considered a part of defense policy, focusing on the availability of energy resources. In this mentality, producing countries were seen as key actors due to their reserves, while there was only a limited recognition of interdependency among the producer and consumer countries.[3] Accordingly, diversification appeared as a key issue for security of supply.

During the 1970s and 1980s, the energy security concept focused on the impact of oil prices on supply, rather than the broader issues. As I argued in previous writings, the energy security literature became concerned with the connections among energy, conflict behavior and political change, considering mainly military aspects and to a lesser extent economic and political aspects.[4] As mentioned, the high oil prices triggered the interest in the energy security concept. Hence, an era of low oil prices from the late Cold War period until the end of the 1990s deprioritized energy security in governmental agendas. However, this allows us to see the importance of consumer countries to producing countries as well, who are highly dependent on revenue from energy exports even when prices are low. Therefore, *security of demand* has also became a part of the energy security matrix, in addition to the *security of supply,* and *availability of supply.*

One other factor that has become an important part of the energy security concept was the *security of infrastructure*, which includes the terrorism threat, geopolitical rivalries and instability in producing regions.[5] There have been more than 4,650 terrorist attacks observed in 95 countries, since 1970.[6] This data underlines the crucial role of security of infrastructure, since attacks may result in a serious disruption of the supply.

Therefore, as explained by Barton et al., energy security comprises these four above categories, focusing on the economic, political and security aspects.[7] However, these ignore the important issue of the environment. Due to increasing awareness of environmental issues, after 1990, it is inconsistent to define energy security without acknowledging the environment. Realization of climate change and the Kyoto Protocol were key consequences of the growing awareness towards environmental issues resulting from increasing fossil fuel consumption. Therefore, a contemporary approach to energy security needs multifaceted trade-offs between economic, political, security and environmental goals.

A further issue that needs to be understood is that the global geopolitics is constantly changing, and directly impacting on our contemporary approach to energy security. It is certain that geopolitical risks have a central role in the energy security matrix. Political and economic crises, conflicts, price fluctuations, and power politics are among issues that have a direct impact on the geopolitics of energy.

Thus, energy security issues that need to be considered by governments are global energy consumption, energy prices, global geopolitics, availability of supply, security of supply, diversification, security of energy infrastructure, and cost of energy supplies for economies, reliability of supply, interdependency, climate change and environmental issues.

It is argued by many that there is no consensus on the definition of energy security.[8] Hence, contemporary energy security should not only simply be defined as "the uninterrupted availability of energy sources at an affordable price" as described by the International Energy Agency (IEA), i.e. focusing only on the security of supply and economic side. A better definition in line with the European Commission is as a "strategy for energy supply security [which] must be geared to ensuring, for the well-being of its citizens and the proper functioning of the economy, the uninterrupted physical availability of energy products on the market, at a price which is affordable for all consumers, while respecting environmental concerns and looking towards sustainable development."[9] This definition more comprehensively covers the aspects proposed above. Accordingly, Yergin's short definition of energy security, "the protection of the entire energy supply chain and infrastructure"[10] could be considered as a more appropriate definition.

Recently, the World Energy Council coined the "energy trilemma" term in order to cover all aspects that possibly affect the entire energy supply chain and infrastructure.[11] The Energy trilemma is based on the dimensions of energy security, energy equity and environmental sustainability. Moreover, countries are also analyzed through their political, societal and economic strength. Such aspects are considered to contribute to securing energy supply, despite the fact that energy prices are increasingly unaffordable due to conflicts, and demand is growing at a time when reducing carbon emissions is crucial. Such constraints make the target difficult to attain. Therefore, not only developing countries, but also the developed ones face energy security problems.

This makes formulizing an energy policy a great challenge. Also, it is a known fact that no single energy resource is optimal; therefore, formulating a balanced energy policy is a key issue. Clear criteria are needed to formulate an energy policy. At this point, European Union (EU) energy policy is relevant with its three dimensions, namely security of supply, competitiveness and sustainability.[12] These three core areas are the key to a successful policy formulation. Here, it is important to understand the weaknesses of the implementation of the so-called common European energy policy since it is really difficult to combine all 28 member countries' interests in a single approach. However, this study is not the appropriate place to discuss this particular issue. Together with WEO's energy trilemma approach, the EU's three core dimensions are nevertheless important in the analysis of energy policy in a carbon constrained world.

Therefore, this chapter aims to analyze and evaluate Turkish energy policy making from an energy security perspective, considering the policy trilemma that governments are currently facing. WEO's energy trilemma index will be used as a basis for analyzing Turkey's energy transition process, together with references to the EU's energy policy triangle. Thus, it will be examined from the dimensions of energy security, energy equity and environmental sustainability. Hence, priorities will be discussed, drawn from the constraints of key documents related to Turkish energy policy.

Priorities in Turkish energy policy making

Energy security, currently at the top of the Turkish government's agenda, is a key component in the formulation of energy policy. Security of supply, alternative energy resources, diversification, and utilization of indigenous resources, sustainability, liberalization and energy efficiency are priorities for Turkish energy policy making.

In order to determine Turkey's fundamental goals and targets, in this study, key energy related documents are reviewed. These documents which have important roles in evaluating Turkish energy policy making, consist of the Ministry of Energy and Natural Resources' (MENR) Strategic Plan for the period of 2015–2019, Ministry of Development's 10th Development Plan, Turkish General Directorate of Renewable Energy's (GDRE) Energy Efficiency Strategy Document, Privatization Administration of the Turkish Prime Ministry's Electricity Sector Reform and Privatization Strategy Document, and Ministry of Energy and Natural Resources' Electricity Market and Supply Security Strategy Document.

According to the above mentioned top energy related policy documents, the main aims and targets could be listed as:

- Increasing the level of energy efficiency.
- Increasing the level of diversification through diversifying suppliers, routes and resources.
- Utilizing indigenous energy resources, including fossil fuels and renewables.
- Increasing the share of renewables in the energy mix.
- Liberalizing the energy market.
- Increasing economic competitiveness.
- Decreasing the level of carbon emission generation.
- Increasing the level of awareness towards environmental issues.
- Increasing the investment level of energy infrastructure.
- Becoming an energy transit country and a potential energy hub.
- Becoming more active in energy diplomacy, especially in pipeline politics.

Each of these above listed priorities will be considered in the following analysis of Turkey's energy profile and transition. There are also specific targets, such as an increase in installed capacities based on different fuels, reduction for carbon emissions, and promotion of energy efficiency. These will be further discussed in the following parts.

Turkey's performance in WEC's Energy Trilemma Index

After, the main aims and targets have been identified, it is important to analyze Turkey's performance in WEC's Energy Trilemma Index in order to demonstrate its position compared to other countries in the world. This index has been widely used to analyze countries' individual performance and compare energy indices and performances across countries.[13] Therefore, it will support the current study's aim of evaluating Turkey's energy policy, focusing on the tradeoffs between the above mentioned pillars and demonstrate the country's capacity for implementing a balanced policy making approach.

The 2015 WEC Energy Trilemma Index is the 5th edition, the first of which was published in 2011, covering 130 countries. The index measures the energy systems by evaluating the data under three energy system dimensions:

1 Energy security: the level of security of supply, the reliability of infrastructure and the ability to meetithe current and future demand.
2 Energy equity: the affordability and accessibility of energy supply.
3 Environmental sustainability: the levels of energy efficiency, carbon emissions level and contribution to a low carbon economy.

Energy performance is also supported by the rankings of contextual performance, namely (1) political strength, (2) societal strength and (3) economic strength.

Turkey ranked as 76th out of 130 countries for balanced score among three pillars in 2015 in the WEC's Energy Trilemma Index, moving up six places since 2011, following a slight increase in its scores, as shown in Figure 24.1. This score designated Turkey as among the two lowest performing European countries compared to EU member countries alongside Bulgaria (81), and among the seven lowest performing countries in Europe, alongside Montenegro (98), Macedonia (106), Ukraine (110), Serbia (112) and Moldova (127).

According to the index, Turkey ranked 71st under energy security behind Paraguay, Pakistan and Algeria, moving up 22 places compared to 2011. It ranked 72nd under energy equity, moving down 11 places compared to 2011, behind Georgia, Armenia and Montenegro. However, the poorest performance was under environmental sustainability, where it was ranked 79th behind Malaysia, Bolivia and Mauritania, moving down 11 places compared to 2014, as shown in Figure 24.1.

Turkey has been seen to sustain a steady position in the index over the last five years, successfully balancing the three pillars. However, it is not possible to say that Turkey has been successful in achieving high average rankings, compared to other countries. Performance in environmental sustainability is lower than the other two pillars, a key issue in the current

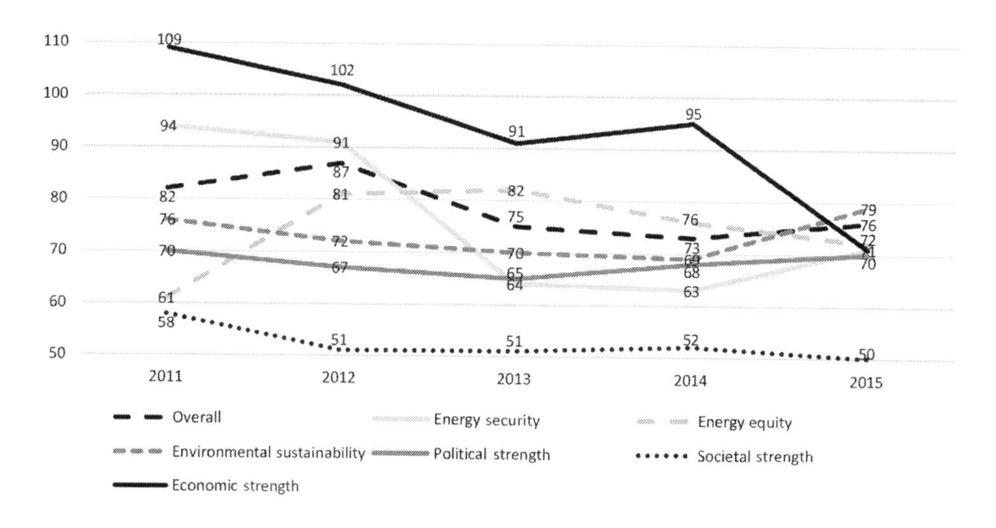

Figure 24.1 Turkey's ranking in WEC's Energy Trilemma Index (2011–2015)

context. Similar to its energy performance, Turkey has only managed to sustain a steady position in its contextual performance, despite a significant improvement in economic strength.

Together with the highlighted priorities in Turkey's energy policy making, its performance in the energy trilemma index is a key factor in the current evaluation of Turkey's energy profile and transition.

Analyzing Turkey's energy profile: opportunities and challenges

Turkey's energy consumption is rapidly increasing, corresponding to economic growth and increasing population levels. Turkish energy consumption has tripled in the last 25 years, increasing from 46 million tonnes oil equivalent (mtoe) in 1990 to 74 mtoe in 2000 and to 125 mtoe by the end of 2014.[14] In the same period, Turkey's carbon dioxide emissions increased 2.5-fold, from 138 million tonnes (mt) to 350 mt, as shown in Figure 24.2. [15] This is the result of rapid industrialization based on fossil fuels, similar to many developing countries, such as Malaysia, Indonesia and Mexico. This trend in Turkey is anticipated to continue; it is expected energy consumption will increase by 60%, reaching 218 mtoe by 2023.[16]

Oil is currently the primary source in global energy consumption, with 33% share, followed by coal with 30%, natural gas with 24%, hydro with 7%, nuclear with 4%, and renewables with 2%.[17] Accordingly, fossil fuels account for 87% of the total world energy consumption.

Likewise, fossil fuels have the highest share, 91%, in Turkey. However, there is a significant difference from global trends in terms of ranking of the consumption, as the share of natural gas is highest with 35%, followed by coal with 29% and oil with 27%, as shown in Figure 24.3. Natural gas usage in Turkey began to be prioritized in the 1980s, due to high levels of air pollution resulting from increasing consumption of coal for both electricity generation and domestic heating, together with fuel oil, as demonstrated in Figure 24.6. Although this shift in energy consumption has had a clear and direct effect in reducing air pollution, the decision to increase the share of natural gas has also had negative consequences for Turkey, particularly an increase in energy dependency, and decrease in the security of supply.[18]

The domination of fossil fuel consumption means that the share of renewables is only 9%, among which, 7% belongs to hydro and the remaining 2% belongs to other renewables such as wind and solar, as seen in Figure 24.3.

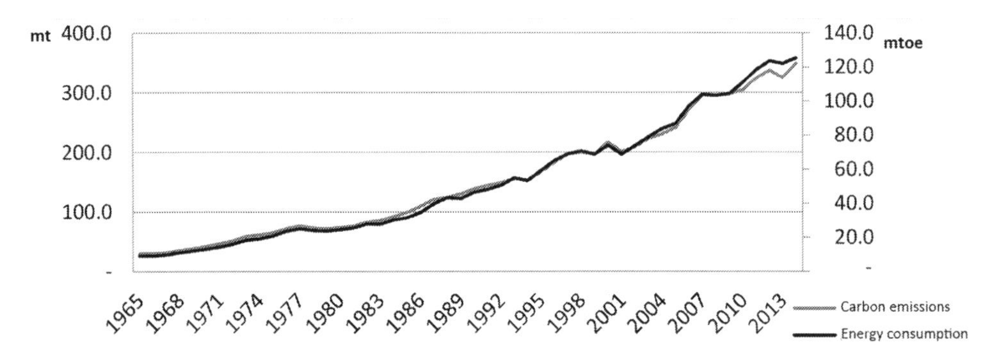

Figure 24.2 Turkey's carbon emission level vs. energy consumption (1965–2014)
Source: BP, *Statistical Review of World Energy*, London: BP, 2015. http://www.bp.com/en/global/corporate/energy-economics/statistical-review-of-world-energy.html.

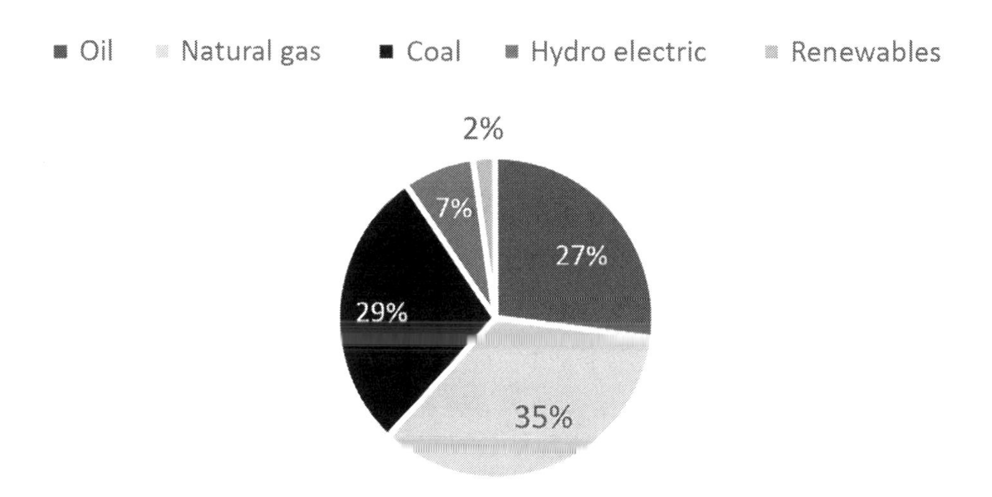

Figure 24.3 Turkey's final energy consumption by fuel type (2014)
Source: BP, *Statistical Review of World Energy*, London: BP, 2015. http://www.bp.com/en/global/corporate/energy-economics/statistical-review-of-world-energy.html.

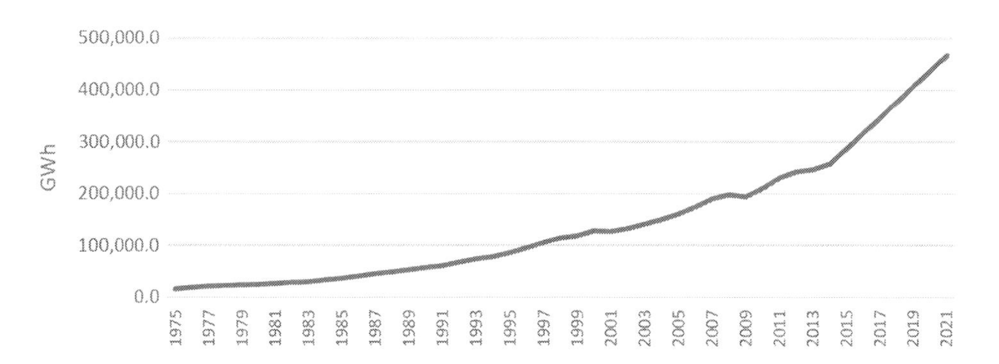

Figure 24.4 Turkey's electricity consumption (1975–2014) and projection for 2014–2021
Source: TEIAS, *10 Years Projection*; Turkish Electricity Transmission Company (TEIAS), "Statistics for Electricity," 2015, http://www.teias.gov.tr/TurkiyeElektrikIstatistikleri.aspx.

Similar to Turkey's final energy consumption, Turkey's electricity consumption has been steadily increasing since 1975, as shown in Figure 24.4. The increase was almost four-fold between 1990 and 2014, from 56,000 gigawatt hour (GWh) in 1990 to 128,275 GWh in 2000, and to 257,220 GWh in 2014, as shown in Figure 24.4. Moreover, the Turkish Electricity Transmission Company[19] has projected that Turkish electricity consumption will exceed the level of 450,000 GWh by the year 2021. Comparable to final energy consumption, natural gas has the highest share in electricity generation. Currently, Turkey's installed capacity is more than 70 GW. Hydro has the highest installed capacity followed by natural gas and coal as shown in Figure 24.8. In 2014, natural gas accounted for almost half of the electricity generation, as shown in Figure 24.5. Hard coal and lignite combined have the highest share in the remaining part of the generation, with more than 30%, followed by hydro with 15% and renewables with 5%. Fossil fuels dominate in Turkey's final energy consumption, hydro and other renewables' share is around 20%, in electricity generation.

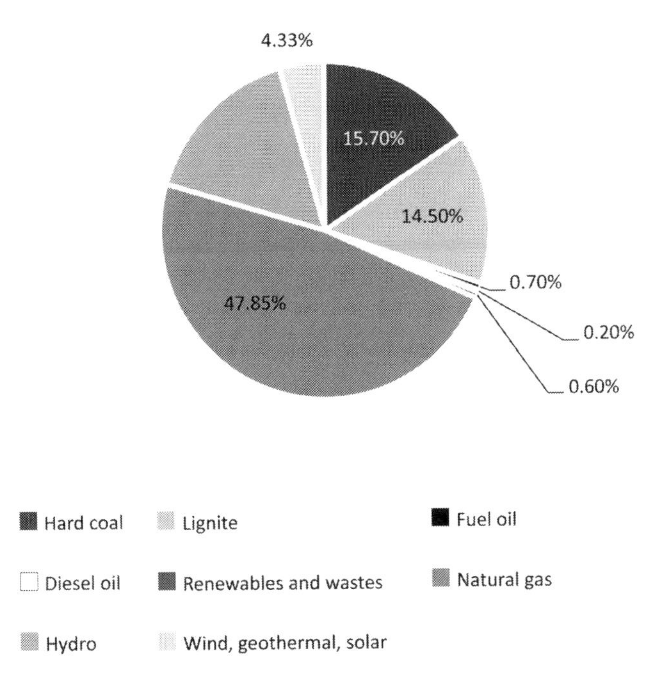

Figure 24.5 Share of fuel types in Turkish electricity generation (2014)
Source: TEIAS, *10 Years Projection*; Turkish Electricity Transmission Company (TEIAS), "Statistics for Electricity," 2015, http://www.teias.gov.tr/TurkiyeElektrikIstatistikleri.aspx.

In spite of efforts to promote renewables in recent years, it is not yet possible to say that these have been successful. However, apart from hydro, wind can be distinguished from the other renewables, as it has experienced a solid growth in the last five years, tripling its installed capacity, as shown in Figure 24.8.

Figure 24.4 demonstrates the energy consumption pattern of Turkey, from 1965 to the present. During the last 50 years, consumption has experienced a 5.5% annual average growth rate, in line with economic growth. In this period, similar to almost all countries, oil has remained indispensible, due to its vital role in the transportation sector. In addition, fuel oil was used extensively for electricity generation until the mid-1980s, when its consumption started to decline with the advent of natural gas.

Natural gas

As shown in Figure 24.6, natural gas was first introduced in 1986 as a result of the first contract with the Soviet Union, followed by contracts with Russia, Iran and Azerbaijan as piped gas, and Algeria and Nigeria as liquefied natural gas (LNG). It increased its importance, becoming the primary source of energy, and today it has a share of more than 35%. Here, it is crucial to mention that the average annual natural gas consumption growth rate over the last 30 years is 18%, more than three times the total energy consumption growth rate in the same period. Since the 1990s, natural gas has been clearly favored by the Turkish state to reduce air pollution, as explained, and also because of its flexibility of use in both electricity generation and heating.

The decision to increase the use of natural gas has been seen as controversial for a country with negligible reserves. However, in the 1990s, natural gas was reasonably priced compared to

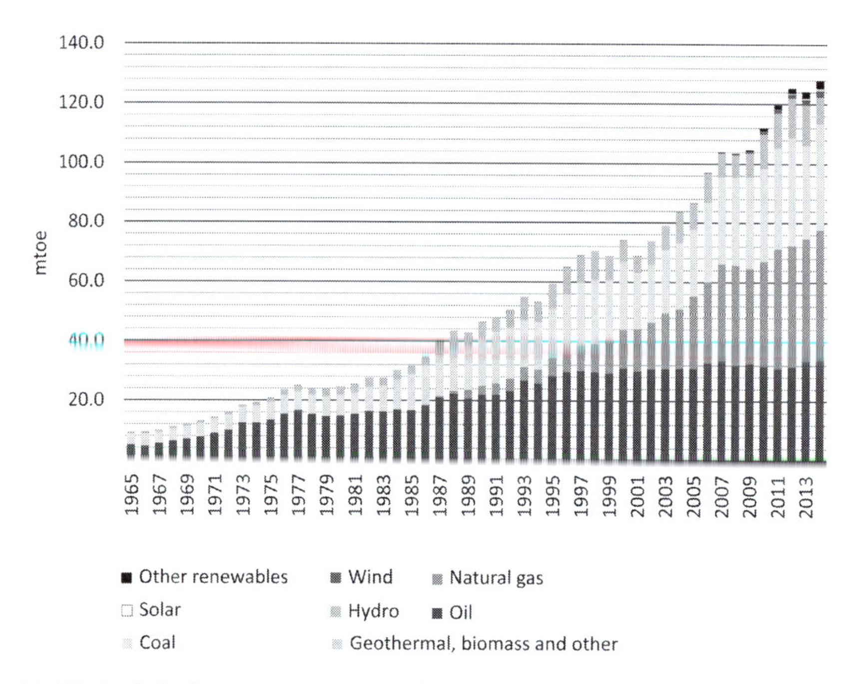

Figure 24.6 Turkey's final energy consumption by fuel type (1965–2014)
Source: BP, *Statistical Review of World Energy*, London: BP, 2015 (http://www.bp.com/en/global/corporate/energy-economics/statistical-review-of-world-energy.html).

the period after 2000. Therefore, natural gas fired power plants were cost effective. Furthermore, natural gas fired power plants have capability to respond to urgent energy needs, and are more flexible than other power plant types. It is also important to mention that renewable energy technology other than hydro has not developed according to its full potential, because focusing on renewable energy has not been the primary aim of Turkey. It is certain that the decision to increase the share of natural gas had a direct impact on Turkey's final energy consumption. However, its greater impact on Turkish electricity generation can be seen in Figure 24.7, which shows almost 50% of the electricity generation originated from natural gas fired power plants.

Turkey's natural gas consumption passed 50 billion cubic meters (bcm) in 2014, doubling the levels of 2004.[20] Seventy-three out of 81 provinces are currently supplied; the remaining eight provinces are in the process of building the infrastructure.[21] This rapid increase in natural gas consumption has been translated into high import dependency, currently more than 98%.[22] This high level of dependency emerged as a threat to energy security. Accordingly, this issue has been significantly discussed in energy related strategy papers and is at the top of the Turkish government's agenda. Although natural gas consumption is increasing rapidly, according to the Ministry of Energy and Natural Resources' Strategic Plan for the period of 2015–2019, a key objective is to decrease the share of natural gas in electricity generation to 38% by the end of 2019.[23] In order to achieve this, the government is prioritizing the utilization of indigenous coal reserves for electricity generation, and the installation of additional capacity for renewables, potentially wind and solar.

The Turkish government has not only prioritized to diversify energy resources, but also suppliers. In 2014, Turkey imported 76% of its natural gas needs from pipelines and the remaining in LNG format. Russia is the leading supplier with 55%, followed by Iran with 18%,

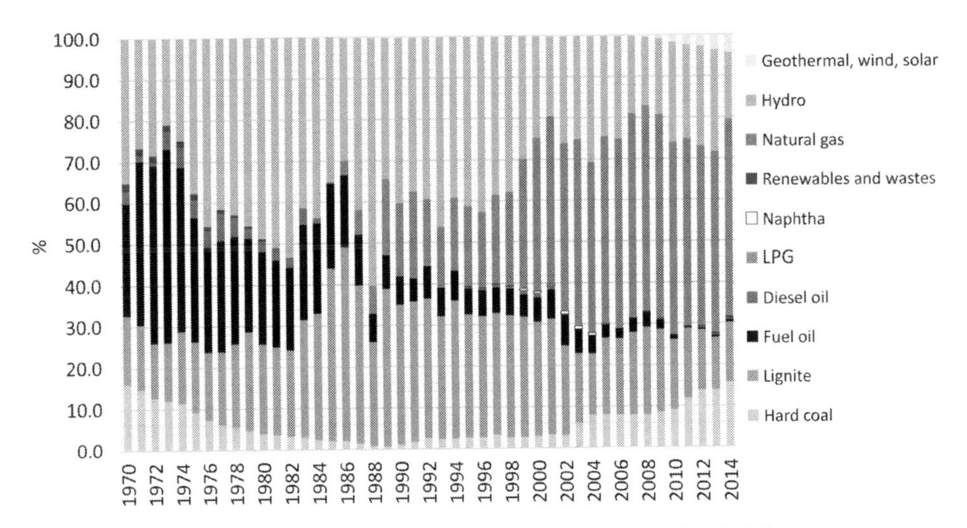

Figure 24.7 Share of fuel types in Turkish electricity generation (1970–2014)
Source: Turkish Electricity Transmission Company (TEIAS).

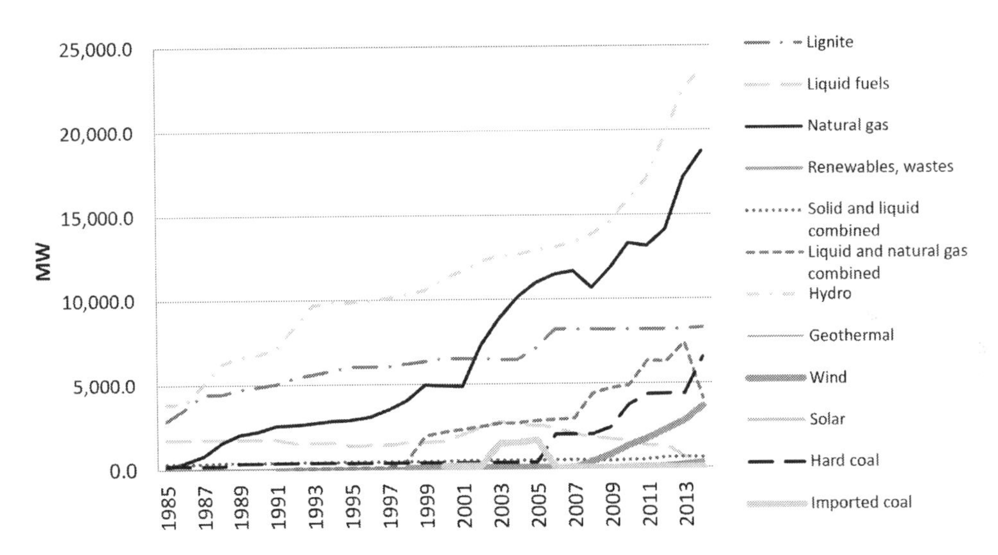

Figure 24.8 Annual development of Turkey's installed capacity by fuel types (1985–2014)
Source: Turkish Electricity Transmission Company (TEIAS).

Azerbaijan with 12.5%, Algeria with 8.5% (LNG), Nigeria with 3% (LNG) and others (spot LNG) with 3.5%.[24] One government priority is to decrease the dependency on Russian natural gas. The main aim is to decrease the level of dependency on a single supplier country to a maximum level of 50%.[25] In order to achieve this goal, new pipeline projects have been developed, such as the Trans Anatolian Pipeline under construction and the proposed Iraq-Turkey pipeline. These projects, together with the proposed Turkish Stream, will also support the aim of becoming an energy transit country, and potentially, an energy hub, thus leading to the capacity to influence price formation in its region. Moreover, the Turkish government is promoting Turkish private sector involvement in the natural gas market. By liberalizing the natural gas market, private companies have been encouraged to conduct studies for procuring natural gas in several countries in the region, such as Iraq and Turkmenistan.[26] In addition, a new program encouraging the exploration of domestic natural gas reserves has been implemented, with an emphasis on shale gas. If successful, these targets and aims will significantly enhance Turkey's energy security level, especially from the supply security perspective, as well as increasing the economic competitiveness of Turkey, especially in the industrial sector, potentially decreasing the costs.

Coal

Even though the natural gas consumption of Turkey started to increase sharply from the 1990s, this had no major impact on the coal consumption. As shown in Figure 24.6, natural gas has been used to meet the shortfall caused by increasing energy demand in Turkey, together with hydro, rather than replacing coal whose share in final energy consumption has never fallen under the level of 25%, as demonstrated in Figure 24.6. According to BP Statistical Review of World Energy, Turkey's coal consumption has more than doubled, compared to the levels of the 1990s.[27] Similar to consumption, production increased by 50% in the same period. Even though coal production has increased with consumption, Turkey is currently 50% dependent on imports. Turkey currently has 8,702 mt of coal reserves, of which 95% are sub-bituminous and lignite.[28] Consequently, Turkish coal reserves are characterized by relatively low heat content, leading to the import of hard coal which has the highest carbon content. Similar to final energy consumption trends including domestic heating, coal has a 30% share in electricity production. It is also important to note that the share of hard coal has been higher than lignite, starting from 2013.[29] This is an emerging burden for Turkey's already high level of energy dependency. Hence, according to Ministry of Energy and Natural Resources' Strategic Plan for the period of 2015–2019, utilization of domestic coal resources is currently a major policy making priority. The goal is to reach a level of 600,000,000 kilowatt hours per year (kWh/year) by the end of 2019, almost doubling current capacity, from known indigenous coal reserves combined with efforts to discover new resources.[30]

It is a known fact that burning coal produces carbon dioxide, and a significant amount is produced by electricity generation and heating. Accordingly, there has been significant growth in the development of clean coal technologies (CCT) in IEA member countries. Currently, the main aim is to reduce the greenhouse gas emissions, mostly carbon dioxide, through CCT such as carbon dioxide capture and storage.[31] As an IEA member, Turkey has also set CCT as a priority in its coal fired power plant development. It is recognized that Turkish carbon emission generation is significantly increasing in line with energy consumption, reaching to almost 350 mt, as shown in Figure 24.2. Turkish energy consumption has increased by as much as 12 times since 1965.[32] This sharp increase has been based on fossil fuels, mainly coal, as discussed and this level of coal consumption is unsurprising, even though carbon emission generation per capita is much less than developed countries. According to the Turkish Statistical Institute (Turkstat), the share of the energy sector in carbon emission generation is 82.2%, proving this significant contribution.[33] Therefore, a tradeoff is required between supply security and environment due to ambitious current plans to increase the share of coal in electricity generation. However, as stated above, there is doubt over Turkey's performance under environmental sustainability of WEC's index (2015).[34] Therefore, it could be better for Turkey to consider the role of renewable energy while balancing its energy supply with coal. However, coal could emerge as a suitable option, as long as CCT is used. Otherwise, increasing the share of coal in electricity generation could generate another burden on environmental sustainability while increasing the level of energy security and energy equity.

Oil

Unlike natural gas and coal, oil is primarily used for transportation in Turkey, rather than electricity generation or domestic heating, similar to other countries with very limited oil reserves. According to the United States' Energy Information Administration (EIA), Turkey has 296,000,000 barrels of oil reserves,[35] with a production level of 50,000 barrels of oil per day and consumption level of 720,000 barrels of oil per day, translating into almost 90% import dependency.[36] Compared to Turkey's natural gas import mix, crude oil imports are much more diversified. Iraq is the leading supplier with 31%, followed by Iran with 30%, Saudi Arabia with 11%, Nigeria with 10%, Kazakhstan with 8.5%, the remainder supplied by Russia, Colombia, Egypt, Libya and Yemen.[37] Together with natural gas import dependency, oil import dependency is a priority in the Turkish energy security agenda, as discussed in the Strategic Document of Electrical Energy Market and Supply Security.[38] Moreover, exploration of indigenous oil reserves is prioritized. The main aim is to increase the share of domestic oil to 13.6% by the end of 2019, including potential shale oil reserves.[39] Similar to the natural gas market, domestic private and public companies are currently encouraged to acquire fields overseas.

Oil dependency is a more substantial challenge compared to natural gas dependency due to the indispensable role of oil in the transportation sector, similar to many other countries. Like many developed and developing countries, Turkey is considering other options, such as the development of domestically produced electricity powered cars for the transportation sector; however, these have not yet proven to be widely available against the existing efforts. Hybrid electric vehicles would be an alternative for Turkey, but due to high tax rates for imported cars, these are similarly not widely available yet. In addition, one should recognize that Turkey is a highly energy dependent country, with 74% overall.[40] Therefore, more than 60% of its electricity generation is also based on imported sources, namely natural gas and hard coal – almost completely imported sources, as shown in Figure 24.5. As a result, hybrid or electricity powered cars would emerge as another environmentally positive dependence, unlike the current dependency on fossil fuels. Vehicle ownership is rapidly increasing together with economic growth.

The number of vehicles per 1,000 people has recently passed 250.[41] When compared to developed countries, the current level is relatively low. Therefore, the rapidly growing economy will certainly translate into an increase in the number of vehicles per 1,000 people. Turkey should include creating mass transportation alternatives, together with promoting energy efficient vehicles in its energy policy making in order to combat environmental challenges. It is also important to state here that the number of air passengers carried has doubled in the last four years, reaching 100,000,000 passengers.[42] Therefore, policies regarding the transportation sector should be revised in order to successfully meet with the challenges regarding security of supply and environmental sustainability.

Hydro

Additional installed capacity has been fueled by natural gas, together with hydro, starting from the mid-1990s, as shown in Figure 24.6. Hydro has always been an important dimension of the Turkish energy mix, from the establishment of the Turkish Republic. Hydro was the primary resource for electricity generation in Turkey between 1975 and 1995, having a greater share than fossil fuels, especially coal, as shown in Figure 24.7. Only in 1985 and 1989, the share of hydro decreased due to low rainfall and dry winters.[43] The following years were dominated by natural gas as the main source for electricity generation. Compared to Turkey's increasing energy consumption, the rate of hydro power electricity generation is lower than natural gas. Hence, considering the development rate of hydro power installed capacity, it is much higher than natural gas installed capacity, as shown in Figure 24.8. Due to the higher capacity factor of natural gas, and hydro's dependence on rainfall and weather conditions, natural gas has become by far the leading source of electricity generation.

It is a known fact that hydro has played an important role in electricity generation for countries with high energy dependence, such as Turkey. The Turkish government also has prioritized hydro in its energy policy, similar to other renewable energy sources. As with its utilization of indigenous coal reserves, hydro has been prioritized in diversification efforts and the struggle against increasing energy dependency. One of the strategies in the Ministry of Energy and Natural Resources' Strategic Plan for the period of 2015–2019 is to construct and operate pumped-storage hydroelectric power stations.[44] According to MENR, Turkey's economical hydropower potential is 140,000 GWh/year, equal to 16% of the Europe's total economical hydropower potential, however only 40% of this potential has been exploited and 27% is under construction.[45] By the end of October 2015, Turkey's hydro power installed capacity is 26,000 megawatts (MW).[46] The aim is to increase the installed capacity to 36,000 MW in line with the aim of increasing the share of renewables in Turkish electricity generation by 2023, the centenary of the Turkish Republic.

Renewables

Similar to hydro, Turkey has prioritized other renewable sources, such as wind, solar and geothermal in its energy generation. There are two priorities related to renewables in MENR's Strategic Plan for the period of 2015–2019: (1) to increase the share of renewables in electricity generation, and (2) to utilize them as a heat source.[47] Currently, renewables excluding hydro have a really minor share both in Turkey's final energy consumption and in electricity generation, 2% and 4.5% respectively.

Since Turkey has no significant fossil fuel reserves, together with concerns about air pollution, renewable energy has played an important role in its energy policy. Renewable energy has mainly been promoted via feed-in tariff systems and other incentives since 2010 via the recent Law No. 6,094 – Amendment of the Law on Utilization of Renewable Energy Resources for the Purpose of Generating Electrical Energy in Turkey. Together with the Turkish government's increasing awareness related to security of supply, factors including environment, energy scarcity, high energy dependency on imports and fluctuating prices have all resulted as an increase in renewables investment.

Compared to renewable sources other than hydro, over the last decade, wind and geothermal have experienced significant growth in Turkey. Wind energy consumption has the biggest growth, as demonstrated in Figure 24.7. The installed capacity has increased from 51 MW in 2006 to 4,200 MW in July 2015, 80-fold.[48] The aim is to increase the installed capacity to 20,000 MW by 2023, the centenary of the Turkish Republic.[49] However, under current conditions, it is not expected that this target is in fact achievable.

Similar to wind, geothermal has experienced significant growth, as shown in Figure 24.6. The installed capacity has increased from 28 MW in 2006 to 614 MW in October 2015.[50] The aim is to increase the installed capacity to 1,000 MW by the centenary of the Republic.[51]

Solar energy is a different case for Turkey. The installed capacity of solar energy was 45 MW by the end of 2014, despite the fact that there is a noteworthy potential, as shown in Figure 24.8.

The Turkish government received submissions for the first tender round of licenses worth 600 MW in 2013. It received applications for licenses worth 8.9 GW, showing a significant interest from investors.[52] Nevertheless, the licenses have not yet been issued due to the complex bureaucracy and financing requirements. One other important issue in the Turkish solar market is the unlicensed segment. It is important to state that the additional installed capacity is increasing due to recently widely spreading unlicensed installations, accounting for less than 1 MW each. However, there are still complex bureaucracy and financing requirements for these installations. As of October 2015, there was 119 MW of installed capacity from the unlicensed segment.[53] However, there are ambitious targets related to solar power. According to the strategy plan, the installed capacity of solar power is expected to increase to 3,000 MW.[54] In addition to electricity generation, solar energy is expected to be used for heating, the pre-heating process in thermal power plants, and irrigation systems, in addition to electricity generation.

Nuclear

At the moment, Turkey has no installed capacity of nuclear power, although the first discussions related to nuclear power took in 1955, resulting Turkey being a signatory to the Atoms for Peace agreement. Following that, there have been several attempts to build nuclear power plants, starting from 1965.[55] The main reason behind this rationale has been related to security of supply concerns and diversification efforts.

Hence, despite these efforts by the Turkish government, no progress was made until an intergovernmental agreement with Russia in 2010 for the construction of the first nuclear power plant in Akkuyu, on the Mediterranean coast of Turkey. Russian Rosatom will finance the power plant under the build-own-operate model. Subsequently five years later in 2015, the Energy Market Regulatory Authority (EMRA) issued a preliminary license for electricity generation. The construction is expected to start in early 2016.[56] There will be four reactors, totaling 4.8 GW, equal to almost 7% of Turkey's current total installed capacity.

The Turkish government has accepted the bid of the consortium led by Mitsubishi and Areva for building a second nuclear power plant in Sinop, on the Black Sea coast, with 4.8 GW of installed capacity. However, there is an important difference in this tender. Unlike Akkuyu, Turkish state owned Electricity Generation Company, EUAS will be involved, and it will be financed under build-operate-transfer model.[57] Recently, another site close to the Bulgarian border has been announced as the site of a possible third nuclear power plant.

Thus, nuclear power is one of the important pillars of MENR's Strategic Plan for the period of 2015–2019. It is stated that nuclear energy will be included in the Turkish electricity generation mix, and is one of the main priorities for Turkey.[58] The aim is to have two nuclear power plants online, and the third under construction by 2023, the centenary year.[59] It is not expected that Akkuyu will be operational before 2022, and Sinop, sometime after.

Energy efficiency

Energy efficiency has emerged as a "hidden fuel" for governments around the globe due to increasing global energy consumption and energy scarcity. Since energy efficiency refers to the capability of delivering the same service with less energy input,[60] or simply extracting more from existing resources (Sustainable Energy for All), it has become an important pillar of the contemporary energy security mentality. Turkish energy intensity has decreased by only 7% from 2000 to 2013, less than the EU-28's average of 20%.[61] Consequently, efforts have been increased to implement measures related to energy efficiency, recognizing its importance. In

2007, Law No. 5,627 – Energy Efficiency was published in order "to increase efficiency in using energy sources and energy in order to use energy effectively, avoid waste, ease the burden of energy costs on the economy and protect environment."[62]

Later, the Turkish government published Energy Efficiency Strategy Paper 2012–2013 in 2012, determining the route map for possible actions, together with targets; the main aim is to decrease energy intensity by 20% by 2023.[63] This paper has been quoted in the other major energy related documents mentioned above. The Turkish government has also realized that one of the main policy tools is to increase public awareness in order to improve energy efficiency. Therefore, MENR is implementing awareness campaigns aimed at the public and the consumer. In addition, Turkey has implemented several measures, such as Improving Energy Efficiency in All Sectors, Regulating and Promoting Energy Efficiency, Support Scheme for Energy Efficiency in Industry, Energy Labels for Household Appliances and Monitoring Energy Efficiency in Sectors.[64] Increasing levels of energy efficiency will have a direct positive impact on three pillars of the energy trilemma.

Liberalization of Turkish energy markets

Liberalization of energy markets began with Law No. 4,628 – Electricity Market Law and Law No. 4,646 – Natural Gas Market Law in 2001. In the same year, the Turkey Energy Market Regulatory Authority was established, responsible for enacting key regulations for both the energy market and its liberalization process. This was followed by a number of new laws: Law No. 5,015 – Petroleum Market Law in 2003, Law No. 5,346 – Renewable Energy Law and Law No. 5,307 – Liquefied Petroleum Gas Market in 2005; Law No. 5,627 – Energy Efficiency Law, Law No. 5,686 – Geothermal Law and Law No. 5,654 – Nuclear Energy Law in 2007; Law No. 5,784 – Amendment to Electricity Market Law in 2008, and Law No. 6,446 – New Electricity Market Law. Together with the new laws, production and distribution began to be privatized; only transmission remained state-owned.

Law No. 6,446 – New Electricity Market Law is critical for the liberalization process of Turkish energy markets. This law encouraged the establishment of EPIAS – the Turkish energy stock exchange, which was intended to bring transparency into Turkish energy markets, thus stabilizing prices and increasing reliability. It will also operate day ahead markets and daily markets, as well as participating in international energy markets. First, it will operate in the electricity market, followed by the natural gas market, and eventually, other energy markets. EPIAS will increase the energy equity level of Turkey by fully liberalizing the energy markets.

Moreover, due to growing energy demand in Turkey, more than $1,200,000,000 of investment will be required by 2023.[65] Therefore, the Turkish government has been encouraging private sector investments, by for instance, increasing the share of private sector ownership in the installed capacity of electricity from 32% in 2002 to 65% in 2014.[66] Liberalization of the markets will also assure private sector participation.

Conclusion

Turkish energy demand is increasing due to rapid economic growth and population increase. It is expected that energy needs will increase by 80% by 2023. Hence, additional energy supply and major investment in the energy sector are needed.

Currently, the overall Turkish energy dependency level is 74%, and dependency on natural gas and oil imports are 98% and 90% respectively. Considering the high share of natural gas in electricity generation and in final energy consumption, and the high share of oil in the transportation sector, energy security, particularly, security of supply, has emerged as a primary concern for Turkish energy policy makers. Therefore, diversification has been at the center of policy calculations as described. However, diversification efforts have not only focused on suppliers, but also on resources. Hence, there are ambitious targets to decrease the share of natural gas in electricity generation. There are two main policy options for this decrease. The first is to increase the share of renewables and the second is to increase the share of coal in electricity generation. In fact, it has been stated in the policy documents that these two policy options will be implemented together. However, both options have different tradeoffs. On the one hand, selecting renewables will result in a decrease in carbon emissions generation, and an increase in security of supply, but will create another type of dependency, a technological one. On the other hand, selecting coal will result in a direct increase in the level of economic competitiveness and the level of security of supply, but will create much greater burden on environmental sustainability, unless CCT is implemented. Therefore, there needs to be a balance between these resources in order to improve Turkey's position in terms of the energy trilemma. These policy options need to be supported by the development of domestic resources, as argued. Inclusion of nuclear energy in Turkey's energy mix will also have a positive impact on its energy security calculations if well managed and maintained.

Moreover, efforts related to liberalization of energy markets have become crucial elements of energy policy making. Turkey has already taken important steps, starting from the early 21st century and is currently implementing new ones to reach a fully competitive market with transparency and a reliable legal and regulatory framework. If these steps can be effectively implemented, it will encourage private sector participation, not only in the electricity market, but also the others. This will have a positive impact on the required energy investment. However, it is important that these reforms should promote the transition to an energy market with a higher level of renewable energy.

At the moment, fossil fuels continue to have a 91% share in Turkey's final energy consumption. This has resulted as an increase in carbon emissions in line with increasing energy consumption. Although Turkey has made efforts to reduce the influence of fossil fuel consumption on the environment, there has only been a slight decrease in energy and carbon intensity. These issues are highlighted in the key policy documents, as well as in climate action plans and strategies. A number of targets, such as increasing share of lignite will make no positive contribution to reducing carbon emission or increasing the level of environmental sustainability. However, there is clearly concern that an ambitious climate policy could have a possible negative impact on economic competitiveness.

Reviewing Turkish energy policy documents, one can also see in energy policy calculations the importance of energy efficiency which has emerged as an important policy tool for the Turkish government, and is being strongly promoted. Especially, energy efficiency improvement opportunities in the industry sector have been realized.

As a result, Turkey currently appears to favor energy security and economic competitiveness over environmental sustainability, despite the fact there is a clear increase in overall environmental awareness.

Notes

1 D. A. Deese, and Joseph S. Nye, eds, *Energy and Security* (Cambridge, MA: Ballinger Publishing Company, 1981), 489.
2 Deese, and Nye, *Energy and Security*; H. Bucknell III, *Energy and the National Defense* (Lexington, KY: The University Press of Kentucky, 1981).
3 Robert Keohane, and Joseph S. Nye, *Power and Independence* (Boston, MA: Harper & Collins, 1977).

4 M. E. Biresselioglu, "The Contribution of Renewables in Turkish Energy Security," *Turkish Studies* 13 no. 4 (2011): 615–632.
5 D. Yergin, "Ensuring Energy Security," *Foreign Affairs* 85 no. 2 (2006): 69–82.
6 M. E. Biresselioglu, and I. O. Yumurtaci, "Evaluating the Nature of Terrorist Attacks on the Energy Infrastructure: The Periodical Study for 1970–2011," *International Journal of Oil, Gas and Coal Technology* 10, no. 3 (2015): 325–341.
7 B. Barton, C. Redgewell, A. Ronne et al. (eds), *Energy Security: Managing Risk in a Dynamic Legal and Regulatory Environment* (Oxford: Oxford University Press, 2005).
8 V. Vivoda, "Evaluating Energy Security in the Asia-Pacific Region: A Novel Methodological Approach," *Energy Policy* 38, no. 9 (2010); C. Winzer, "Conceptualizing Energy Security," *Energy Policy* 46 (July 2012): 36–48; L. Chester, "Conceptualising Energy Security and Making Explicit its Polysemic Nature," *Energy Policy* 38, no. 2 (2009): 887–895.
9 European Commission, *Green Paper-Towards a European Strategy for the Security of Energy Supply* (European Commission, 2000), http://eurlex.europa.eu/smartapi/cgi/sga_doc?smartapi!celexapi!prod!DocNumber &lg=EN&type_doc=COMfinal&an_doc=2000&nu_doc=0769&model=guicheti.
10 Yergin, "Ensuring Energy Security."
11 World Energy Council, "World Energy Trilemma," World Energy Council, 2010, www.worldenergy.org/work-programme/strategic-insight/assessment-of-energy-climate-change-policy/.
12 European Commission, *Green Paper.*
13 B. W. Ang, W. L. Choong, T. S. Ng, "Energy Security: Definitions, Dimensions and Indexes," *Renewable and Sustainable Energy Reviews* 42 (February 2015): 1077–1093; M. Sencar, V. Pozeb, and T. Krope "Development of EU (European Union) Energy Market Agenda and Security of Supply," *Energy* 77 (2014): 117–124; K. Narula, and B. S. Reddy, "Three Blind Men and an Elephant: The Case of Energy Indices to Measure Energy Security and Energy Sustainability," *Energy* 80 (February 2015): 148–158.
14 BP, *Statistical Review of World Energy* (London: BP, 2015), www.bp.com/en/global/corporate/energy-economics/statistical-review-of-world-energy.html.
15 Ibid.
16 T. Yıldız, *His Presentation to Turkish Grand National Assembly's Commission on Planning and Budgeting* (Turkish Minister of Energy and Natural Resources, 2014), www.enerji.gov.tr/File/?path=ROOT%2F1%2FDocuments%2FB%C3%BCt%C3%A7e+Konu%C5%9Fmas%C4%B1%2F2015+Y%C4%B1l%C4%B1+Plan+B%C3%BCt%C3%A7e+Komisyonu+Konu%C5%9Fmas%C4%B1.pdf.
17 Ibid.
18 Biresselioglu, and Yumurtaci, "Evaluating the Nature of Terrorist Attacks."
19 Turkish Electricity Transmission Company (TEIAS), *10 Years Projection for Turkish Electricity Generation (2012–2021)* (Turkish Electricity Transmission Company, 2012), www.teias.gov.tr/KAPASITEPROJEKSIYONU2012.pdf.
20 BP, *Statistical Review of World Energy*.
21 Turkish Statistical Institute, "Road Motor Vehicles Statistics," 2015, Turkish Statistical Institute, www.tuik.gov.tr/PreHaberBultenleri.do?id=18768.
22 International Energy Agency, "Emergency Response Systems of Individual IEA Countries," in *Energy Supply Security: The Emergency Response of IEA Countries* (International Energy Agency, 2014), www.iea.org/media/freepublications/security/EnergySupplySecurity2014_Turkey.pdf.
23 Turkish Ministry of Energy and Natural Resources (MENR), *Strategic Plan 2015–2019* (MENR, 2014), www.enerji.gov.tr/File/?path=ROOT%2F1%2FDocuments%2FStrategic+Plan%2FStrategicPlan2015-2019.pdf.
24 Republic of Turkey Energy Market Regulatory Authority (EMRA), *Natural Gas Market Report 2014* (EMRA, 2015), www3.epdk.org.tr/documents/dogalgaz/rapor_yayin/DPD_RaporYayin2014.pdf.
25 Yıldız, *His Presentation to Turkish Grand National.*
26 MENR, *Strategic Plan 2015–2019.*
27 BP, *Statistical Review of World Energy*.
28 Ibid.
29 TEIAS, "Statistics for Electricity."
30 MENR, *Strategic Plan 2015–2019.*
31 International Energy Agency, *Clean Coal Technologies: Accelerating Commercial and Policy Drivers for Deployment* (International Energy Agency, 2008), www.iea.org/publications/freepublications/publication/Clean_Coal_CIAB_2008.pdf.

32 BP, *Statistical Review of World Energy*.
33 Turkish Statistical Institute, "Turkish Greenhouse Gas Emissions Inventory" (Turkish Statistical Institute, 2015), www.turkstat.gov.tr/PreHaberBultenleri.do?id=18744.
34 World Energy Council. "World Energy Trilemma."
35 US Energy Information Administration, "International Energy Statistics," US Energy Information Administration, 2015, http://199.36.140.204/cfapps/ipdbproject/IEDIndex3.cfm.
36 US Energy Information Administration, *Turkey Report* (US Energy Information Administration, 2015), www.eia.gov/beta/international/analysis_includes/countries_long/Turkey/turkey.pdf.
37 Republic of Turkey Energy Market Regulatory Authority, *Oil Market Report 2014* (Republic of Turkey Energy Market Regulatory Authority, 2015), http://www3.epdk.org.tr//documents/petrol/rapor_yayin/PPD_RaporYayin20141.pdf.
38 Turkish Ministry of Energy and Natural Resources, *The Strategic Document of Electrical Energy Market and Supply Security* (Turkish Ministry of Energy and Natural Resources, 2009), www.enerji.gov.tr/File/?path=ROOT%2F1%2FDocuments%2FBelge%2FArz_Guvenligi_Strateji_Belgesi.pdf.
39 MENR, *Strategic Plan 2015–2019*.
40 World Bank, "Energy Imports, Net (% of Energy Use)," World Development Indicators, 2014, http://data.worldbank.org/indicator/EG.IMP.CONS.ZS.
41 Turkish Statistical Institute. "Road Motor Vehicles Statistics," 2015, Turkish Statistical Institute, www.tuik.gov.tr/PreHaberBultenleri.do?id=18768.
42 World Bank. "Air Transport, Passengers Carried," World Development Indicators, 2014, http://data.worldbank.org/indicator/IS.AIR.PSGR.
43 Turkish Electricity Transmission Company (TEIAS).
44 MENR, *Strategic Plan 2015–2019*.
45 Turkish Ministry of Energy and Natural Resources, "Hydro," Turkish Ministry of Energy and Natural Resources, 2015, http://enerji.gov.tr/tr-TR/Sayfalar/Hidrolik.
46 TEIAS, "Statistics for Electricity."
47 MENR, *Strategic Plan 2015–2019*.
48 Turkish Wind Energy Association, *Turkish Wind Energy Statistics Report* (Turkish Wind Energy Association, 2015), www.tureb.com.tr/attachments/article/542/%C4%B0statistik%20Raporu%20%20Temmuz%202015%20High.pdf.
49 Yıldız, *His Presentation to Turkish Grand National*.
50 BP, *Statistical Review of World Energy*; TEIAS, "Statistics for Electricity."
51 Yıldız, *His Presentation to Turkish Grand National*.
52 I. Clover, "Unlicensed PV sector could propel Turkey to 10 GW by 2023," *PV Magazine*, 2014, www.pv-magazine.com/news/details/beitrag/unlicensed-pv-sector-could-propel-turkey-to-10-gw-by-2023_100016076/#ixzz4CEhVyoUV.
53 Unlicensed Electricity Generation Association, "List of Accepted Unlicensed Projects," Unlicensed Electricity Generation Association, 2015, www.lisanssizelektrik.org/?p=Basvuruprojeonayvekabulbilgileri.
54 MENR, *Strategic Plan 2015–2019*.
55 M. E. Biresselioglu, "The Prospective Position of Nuclear Power in Turkish Energy Policy," *Mediterranean Journal of Social Sciences* 3, no. 9 (2012): 207–214.
56 World Nuclear Association, "Nuclear Power in Turkey," World Nuclear Association, 2015, www.world-nuclear.org/info/Country-Profiles/Countries-T-Z/Turkey/.
57 World Nuclear Association, "Nuclear Power in Turkey."
58 MENR, *Strategic Plan 2015–2019*.
59 Yıldız, *His Presentation to Turkish Grand National*.
60 International Energy Agency, "Energy Efficiency," (International Energy Agency, 2015). www.iea.org/topics/energyefficiency/.
61 European Environment Agency, "Total Energy Intensity, Relative Energy Intensity and per capita Consumption," European Environment Agency, 2015, www.eea.europa.eu/data-and-maps/daviz/total-energy-intensity-relative-energy-2#tab-chart_1.
62 Turkish Official Gazette, "Energy Efficiency Law – Law No. 5627," 2007, http://faolex.fao.org/cgi-bin/faolex.exe?rec_id=120779&database=faolex&search_type=link&table=result&lang=eng&format_name=@ERALL.

63 Turkish General Directorate of Renewable Energy, *Energy Efficiency Strategy Paper 2012–2023* (Turkish General Directorate of Renewable Energy, 2012), www.eie.gov.tr/verimlilik/document/Energy_Efficiency_Strategy_Paper.pdf.

64 International Energy Agency, "Energy Efficiency: Turkey Statistics," (International Energy Agency, 2015). www.iea.org/policiesandmeasures/energyefficiency/?country=Turkey.

65 Yıldız, *His Presentation to Turkish Grand National.*

66 Ibid.

Part V

Energy transitions in the carbon reduction countries

25

France and the energy trilemma

How the Fifth Republic has sought to balance energy security, affordability and environmental sustainability

John S. Duffield

Although not the leading power it once was, France continues to play an important role in world energy affairs. It remains among the 10 largest energy consumers. And for many years, it has been the second largest producer of nuclear power, after the United States.

Since the 1970s, France has had a relatively easy time balancing the often competing imperatives of energy security in the form of reliable access to adequate supplies of energy, reasonably priced energy sources to promote economic growth and competitiveness, and environmental sustainability. This happy state of affairs is largely attributable to the country's heavy investment in nuclear power following the first oil shock in 1973. The primary purpose of the nuclear program was to reduce France's reliance on what had come to be unreliable and expensive energy imports, especially petroleum. But the program also resulted in relatively inexpensive electricity supplies and low CO_2 emissions.[1]

Nevertheless, France's emphasis on nuclear power has not been an unmitigated blessing. As a result of nuclear bounty, France paid relatively little attention to renewable sources of energy. And energy efficiency took a back seat to finding ways to consume the prodigious amounts of electricity produced by France's nuclear power system.

In more recent years, the tradeoffs between the horns of the energy trilemma have grown sharper for France. One reason has been heightened concern about the effects of greenhouse gas emissions, especially climate change, despite France's relatively low per head emissions levels. Another has been new worries about the safety of nuclear power, especially following the 2011 disaster at Fukushima Daiichi in Japan. And the cost of building new nuclear reactors has escalated far beyond initial expectations. As a result, the perpetuation of nuclear power as the centerpiece of French energy policy has become less certain, and French energy policy is now in flux. Increased attention is being belatedly paid to energy savings and renewable sources of energy.

This chapter takes a chronological approach, examining how France has addressed the energy trilemma over time. It begins with a review of the situation before the oil shocks, when France was becoming steadily more dependent on imports of oil. It then describes how France responded to the oil shocks, devoting particular attention to the rapid buildup of the country's nuclear power capacity. A third section covers the review of energy policy that took place in the first years of the new millennium, which culminated in a renewal of France's commitment to nuclear power. The following section examines the developments of the past several years. These have called into question more than ever before France's long-standing reliance on nuclear power and resulted in a much ballyhooed "energy transition." What difference this latest policy initiative will make, however, remains to be seen.[2]

French energy priorities and policies before 1973

During much of the postwar era, France faced not an energy trilemma but a dilemma: balancing the security of energy supplies against their costs. Environmental concerns ranked well behind these first two considerations. During the decade and a half prior to the first oil shock, moreover, it appeared that there were few or no tradeoffs to be had. It appeared that France could have virtually all the low cost energy, principally in the form of imported oil, it needed at little or no cost in terms of energy security. In fact, however, the costs and risks of foreign oil dependence were mounting beneath the surface, waiting for the right combination of geopolitical conditions to reveal them.

During the first decade and a half after World War II, the French economy relied primarily on coal, much of which was produced at home. As oil production exploded in the Middle East, however, the government began to promote the use of oil over coal. The former had become less expensive, and its abundance led many to believe that prices would remain low indefinitely. At the same time, the government withdrew support for the coal industry, offering less protection than any other country in Western Europe.

As a result of these actions and market forces, French petroleum consumption soared during the following decade, increasing at an average annual rate of 12% between 1960 and 1973. Over the same period, oil's share of total primary energy consumption (PEC) jumped from just 30% to more than two-thirds. By 1973, 60% of energy demand in the critical industrial and power sectors was met by oil. Meanwhile, coal production peaked at 60 million tonnes in 1960, when coal still accounted for more than half of PEC, and then steadily declined. By 1973, just one-sixth of PEC was provided by coal.

Because France had little oil of its own, it became increasingly dependent on energy imports, especially oil. In 1960, France produced 62% of the energy it consumed. By 1973, that figure had dropped to less than 25%, and nearly 70% of France's energy needs was met by foreign oil alone. Consumption of natural gas also grew substantially 1960 to 1973, to 8.5% of PEC, but a growing share, more than half, was also imported.

To be sure, the French government took several precautions against this growing dependence on imported oil. As early as 1958, it required that importers maintain a stockpile equivalent to three months of domestic sales. It promoted the formation of national oil companies that could compete with the so-called "majors" for control of overseas oil resources. It sought to diversify the sources of France's imports away from the Persian Gulf, with particular emphasis on Algeria, which soon provided one-third of France's crude oil imports. And after the 1967 Middle East war, it adopted a pro-Arab foreign policy

Finally, France initiated a nuclear energy program of significant size. By 1973, it had built 11 experimental and small power reactors, and it had begun construction on or had ordered six 900 megawatt (MW) pressurized water reactors (PWRs). The plan was to order a total of 10 commercial reactors between 1971 and 1975 with a total generating capacity of just under 10 gigawatts (GW).

In the end, however, these measures did little to buffer France against the shock that was to come. Nuclear power accounted for just 2% of PEC in 1973. The best that can be said is that they did lay the groundwork for a rapid shift in energy policy when that became necessary.

Priorities and policies after the oil shocks

The 1970s and 1980s did indeed see a dramatic shift in French energy policy. The previous emphasis on low energy prices, with its concomitant open door to imported oil, was hastily replaced by an emphasis on energy security through self-sufficiency. This goal was to be achieved primarily through the rapid buildup of France's nuclear power generating capacity. It turned out, though, that the ambitious nuclear program yielded the additional benefits of keeping electricity prices low while reducing France's CO_2 emissions.

Impact of the oil shocks

The primary impetus for the shift in French policy was the first oil shock of 1973. In fact, however, France had experienced a setback of its own just two years previously. At that time, Algeria nationalized a controlling share of the French oil concessions in that country, causing most of the French oil companies to leave.

In fact, France was not directly targeted by the Arab oil embargo that began in 1973, thanks to its pro-Arab policy in the Middle East. Nevertheless, the French economy was hit hard by the Arab oil production cuts and the quadrupling of world oil prices that followed. Although France suffered a reduction in oil supplies of just 7%, and then for a duration of just four months, the economic consequences of the oil price rises were severe. French GDP growth was cut in half, while inflation and unemployment more than doubled. Despite efforts to limit oil imports, their cost tripled from 1973 to 1974, from 14.6 billion to 43 billion French francs. Even after the production cuts ended, it was widely assumed in France that oil scarcity would eventually return.[3]

Policy actions

In response to these events, France adopted what French energy policy expert Robert Lieber has called "the most coherent and vigorous energy program of any principal oil consuming state."[4] Abroad, France engaged in intense energy diplomacy aimed at securing access to foreign sources of oil. At home, it sought to reduce oil consumption while developing substitutes for oil, especially in the form of nuclear power for electricity production.

Following the Arab oil production cuts and selective embargo, France moved quickly to lock up oil supplies from foreign producers. As early as December 1973, it struck a deal with Saudi Arabia to provide 27 million tonnes of oil over three years. This initial agreement was followed in early 1974 by a more substantial one promising 800 million tonnes over 20 years, or enough to meet a third of French oil demand at the time. That same year, France reached an agreement with Iraq for 20 – later increased to 30 – million tonnes per year. These efforts continued into the early 1980s, when France signed a 5 million tonnes per year contract with Mexico, which had just emerged as a major producer.

In the end, however, France's energy diplomacy yielded only very limited benefits. France received little or no break on the prices that it paid for oil. And its bilateral deals with Persian Gulf states did nothing to protect France from the oil supply disruptions that subsequently resulted from the Iranian revolution and the Iran-Iraq war.

Arguably more important, then, were the efforts France made at home to reduce its dependence on imported oil. On the demand side, it increased the prices that could be charged for petroleum products while maintaining gasoline taxes that were among the highest in Europe. The government also increased insulation standards for buildings and provided subsidies for investments in energy savings.

Most notable of all were France's efforts to develop alternatives to oil. After pushing the substitution of oil for coal during the previous decade and a half, it now encouraged the use of coal in industry and power generation. Nevertheless, coal consumption increased only modestly before resuming its previous slide in 1979.

Instead, the biggest gains were made in the promotion of nuclear power, which constituted "the most ambitious program in any Western country during the 1970s."[5] The center-right governments in power between 1973 and 1981 ordered more than 40 additional reactors, with the standard reactor size increasing from 900 MW to 1.3 GW. Although the Socialist government that took office in 1981 initially cancelled one reactor and suspended work on 18 others, it eventually raised the goal of nuclear capacity to 56 GW by 1990 and ordered six more units in the early 1980s. Overall, construction began on no fewer than 55 reactors between late 1971 and 1984.[6]

Outcomes

What were the fruits of these efforts? Collectively, they resulted in a significant increase in France's energy security without compromising the other goals of affordability and environmental sustainability, at least in the short to medium term. In March 1974, the French government had set a number of ambitious energy goals for 1985.[7] These included:

1. cutting PEC from a projected 285 million tonnes of oil equivalent (mtoe) to just 240 mtoe (although this would represent a 30% increase from the actual PEC of 185 mtoe in 1973);
2. reducing the share of PEC obtaining from imported energy from 76% to just 55–60%, and reducing oil's share of PEC from 70 to just 40%;
3. installing more than 40 GW in new nuclear capacity and raising the nuclear share of PEC from less than 2 to 25%; and
4. limiting the amount of oil imported from any single country to 15% of the total.

By 1985, these goals had been largely met and in some cases even exceeded. PEC was up only 6% from 1973 and 18% lower than the target of 240 mtoe. Oil's share of PEC was down to 43% and still dropping. Fifty-four nuclear reactors had either entered commercial operation since 1978 or were under construction. The share of PEC coming from nuclear power had already reached 25% and was still rapidly rising, on its way to more than 35%.

Meanwhile, the substitution of nuclear power for oil and eventually coal in power generation had had some sizeable side benefits. Thanks to a high degree of standardization in reactor design, the cost of the nuclear program had been kept in check, resulting in low electricity prices. The program had also contributed to a sharp decline in CO_2 emissions, which had fallen from 538 million tonnes in 1973 to just 410 million tonnes in 1985, approximately where they would remain for the next 25 years.

If anything, the rapid buildup of nuclear generating capacity had been too successful. By the mid-1980s, it had become clear that France would be able to produce more electricity than it needed. In response, the government heavily promoted electricity use by industry and in the residential sector.

The 2000s: renewal of the nuclear commitment

For the next two decades, from the mid-1980s to the early 2000s, France enjoyed the security, economic, and environmental fruits of its post-oil shock energy investments, especially those in nuclear power. In the early to mid-2000s, however, a new set of concerns prompted France to undertake a comprehensive review of its energy policy. The principal result was a renewed commitment to nuclear power. At the same time, however, the country adopted ambitious energy efficiency measures and, for the first time, began to promote renewable energy to a significant extent.

Motives for the review of energy policy

The review of French energy policy was prompted by growing concerns in each of the three areas captured by the concept of the energy trilemma. One was the rising cost of oil and other energy imports. Beginning in 2000, the price of oil, France's principal energy import, rose above $20 a barrel for only the second time in a decade and continued to climb. Between 2003 and 2006 alone, the cost of France's energy imports doubled, nearly reaching the all-time high set a quarter century before during the peak of the second oil shock.

As for energy security, it was not too soon to start thinking about the future of the nuclear program. Although the first reactors would not reach the end of their expected 40 year lifespans until late the following decade, planning for what would come next would have to begin well in advance, given the long lead times for reactor design and construction. And because all but four of France's 58 commercial reactors had commenced operation between the late 1970s and 1990s, large numbers of replacements might have to start coming on line by 2020.

Finally, with regard to environmental sustainability, France had begun to take seriously the problem of climate change. In 2000, it adopted a comprehensive National Program to Combat Climate Change, which set a goal of reducing greenhouse gas emissions in 2010 by about 10% below projected levels. The program specified some 100 low cost measures for achieving that goal, though it relied primarily on voluntary agreements rather than taxation and regulation. Then in 2002, France ratified the Kyoto Protocol, whereby it pledged to bring emissions back down to 1990 levels by the end of the decade.[8]

Policy responses

These challenges were spelled out in a 2003 white paper on energy, which set the stage for the adoption of a broad framework law two years later.[9] The resulting energy law, adopted in 2005, reflected the changing constellation of concerns, putting the fight against climate change on par with the long-standing goals of guaranteeing the security of energy supplies and assuring competitive energy prices (as well as ensuring energy access to all). It established ambitious long-term targets for cutting CO_2 emissions by 75% by 2050 and reducing France's energy intensity on an annual basis by 2% through 2015 and 2.5% thereafter. In the short term, the law called for reducing energy imports by 10 mtoe, or nearly 10%, by 2010 and for increasing the share of renewables in France's PEC and electricity supply to 10% and 21%, respectively, over the same time period.[10]

To achieve these goals, the government took a number of actions. Most important was a renewal of France's commitment to nuclear power, which was now justified in part as a contribution to the fight against climate change. The first step would be to extend where possible the lifetimes of the existing reactors beyond 40 years, although upgrades costing on the order of 400–600 million euros per unit would be required. The national power company, Électricité de

France (EDF), estimated that service periods of up to 60 years would be possible. This meant that even the oldest reactors might operate for another two decades.

At the same time, the government laid the groundwork for the construction of a new generation of reactors, the so-called European Pressurized Reactor (EPR). Work on the first EPR began in 2007 and was expected to be completed by 2012. Two years later, the site for a second EPR was chosen, with construction expected to begin in 2012.

Not least important, the government took steps to build and maintain broad public support for nuclear power. One of these was the creation in 2006 of an independent Nuclear Safety Authority (Autorité de Sûreté Nucléaire, or ASN), which would have to approve all lifetime extensions and sign off on new reactors before they could begin commercial operation.

On the demand side, the government established an ambitious "White Certificate" program, which placed an obligation on energy suppliers to generate energy savings by their customers. For the first three-year phase (2006–2009), the goal was extremely modest. But it was expected to rise over time, eventually resulting in significant energy savings. The government also adopted new regulations to reduce energy use in both new construction and existing structures, eventually setting a target of cutting consumption by 38%. And it took steps to limit oil consumption in the transportation sector.

Finally, France made its first major foray into the promotion of renewable energy sources. It had introduced a feed-in tariff for electricity in 2000, but this program was limited to plants with a capacity of less than 12 MW and had not resulted in much new generating capacity. So the tariffs were made more favorable in 2006, with particular emphasis placed on wind power. At that time, a goal was set for 2015 of 17 GW of wind capacity, later raised to 25 GW by 2020. And when in 2011 the desired level of investment continued to fall short, the government announced tenders for 3 GW of offshore wind power. Meanwhile, in the transportation sector, the government sought to increase the share of biofuels in the fuel supply by further cutting taxes. It set goals of 7% of the fuel supply by 2010 and 10% by 2015.

One potential measure that France did not pursue during this period was the use of hydraulic fracturing to increase domestic production of natural gas. The U.S. Energy Information Administration estimated that France might possess on the order of 4 trillion cubic meters of technically recoverable shale gas, which would put it near the top of European gas reserves.[11] But strong public concerns about the potential environmental impact of fracking led the Parliament to prohibit the practice in 2011.

Outcomes

During the years leading up to the presidential election of 2012, France's energy situation showed notable improvements. PEC experienced a gradual decline, from 262 mtoe in 2005 to 244.5 in 2012. Oil consumption dropped even more dramatically over the same time period, from 93.1 million tonnes in 2005 to just 80.3 million in 2012. And partly as a result, CO_2 emissions fell steadily, from 433.4 in 2005 to 379.5 in 2012. Meanwhile, wind-generated electric power rose rapidly, from just 1 terawatt-hour in 2005 to more than 14 in 2012. Biofuel production jumped as well, although it plateaued in 2009 at just 2.3 mtoe, or the equivalent of about 3% of France's oil consumption.[12]

2010s: the energy transition

Hardly had the policies adopted in the second half of the 2000s gone into effect, however, than did the new Socialist government of François Hollande, elected in 2012, undertake a further sweeping review of energy policy. During the campaign, Hollande had pledged to cut the share of France's electricity generated by nuclear power by one-third, to just 50%, by 2020 and to shut down France's oldest nuclear power station before the next presidential election in 2017. According to David Buchan, Hollande's pledge was the result of electoral politics in the wake of Germany's 2011 decision to accelerate its long-planned nuclear phaseout.[13] Early in the campaign, another Socialist candidate had called for eliminating nuclear power altogether, albeit over 25–30 years, and some significant gesture in that direction was needed if the Socialists were to establish a desired electoral alliance with the French Green Party. Indeed, Hollande's pledge corresponded roughly with the degree to which Germany was planning to reduce its reliance on nuclear power.[14] But the so-called "energy transition" was not simply the outcome of short-term political calculations; deeper forces were at work.

Motives for the adoption of a new energy law

One impetus for the new review was problems encountered with the nuclear reactor program. The construction of the first EPR was subject to delays and cost overruns. As a result, the price more than doubled and the date for commercial operation was pushed back until at least 2017, some five years later than originally expected. Then, in 2015, tests found potential flaws in the reactor pressure vessel, which led to a further delay in the expected start-up date, to late 2018, and yet another price increase, to nearly $12 billion.[15] A further consequence of these problems is that EDF decided to put the start of the second EPR on hold indefinitely.

Then, in 2011, the nuclear disaster at Fukushima Daiichi in Japan raised searching questions about the safety of nuclear power. In response, the ASN imposed a demanding new set of safety standards in early 2012. EDF would have to quickly make a number of physical and organizational improvements at each of its facilities in order to increase their robustness to extreme situations, and it would be required to deploy a national nuclear rapid response force that could deal with such contingencies. The new measures were expected to cost at least 13 billion euros.

A third factor was developments within the European Union in response to growing concerns about climate change. These developments culminated in the adoption in 2009 of the Climate and Energy Package, which set ambitious goals for energy savings, renewable energy, and greenhouse gas emissions for 2020. Under the Effort Sharing Decision that followed, France agreed to cut its 2020 greenhouse gas emissions by 14% below the 2005 level in sectors not covered by the EU Emissions Trading System. And under the subsequent Renewable Energy Directive and National Renewable Energy Action Plan, France pledged to increase the share of renewables in final energy consumption to 23% and in electric power generation to 27%, both also by 2020.

Yet in 2012, France still had a long way to go to reach these targets. Hydroelectricity still accounted for the lion's share of renewable power; other forms of renewables were providing only about 5% of total electricity production.[16] Solar power capacity stood at less than 4 GW. Instead, the greatest hopes were placed on wind power, which had grown in capacity by roughly 1 GW per year since 2005, to some 7.7 GW in 2012. In order to achieve the government's goal of 25 GW, including 6 GW offshore, by 2020, however, it would have to grow even faster during the remainder of the decade.[17]

As the International Energy Agency (IEA) noted in 2010, however, "wind farm developers face significant challenges related to the complexity of the permitting process and to public opposition."[18] Wind turbines could be installed only in specified zones that were a minimum distance from areas designated for housing. And in 2011, the government created additional obstacles by establishing more cumbersome review procedures for wind turbines over 50 meters tall and requiring that wind farms have a minimum of at least five turbines. As a result, there were good grounds to expect that because of environmental, aesthetic, or other concerns, the development of wind power would proceed more slowly than the government had hoped.[19] Thus to achieve the targets for renewable energy, policies more aggressive than those adopted by the previous conservative governments might well be required.

Development of the new energy law

Despite the prominence given to energy in the 2012 presidential campaign, the adoption of the energy law took more than three years, or nearly two years longer than originally expected.[20] The process began with the government conducting another national "debate" during late 2012 and much of 2013. The debate was largely inconclusive, especially with regard to the future of nuclear power. According to Buchan, "all participants agreed on the need … to define a strategy for the evolution of the nuclear fleet … but not much more."

The government did not present a draft energy law to the French Parliament until June 2014. By then, the pressure was growing to take action prior to the UN Climate Change Conference scheduled to be held in Paris in late 2015.[22] Although the proposal received swift passage in the lower house, the French Parliament as a whole did not adopt the final legislation until July 2015.[23] The delays reflected continuing divisions within the government, the legislature, and French society regarding the appropriate direction for energy policy, and especially the role of nuclear power. Indeed, the French senate offered some 1,000 amendments to the original proposal, although most of these were removed from the version ultimately adopted.[24] The new energy law, which contained 215 articles, was formally published on August 18, 2015.[25]

Provisions of the new energy law

The new energy law reflected the escalating nature of targets for reductions in greenhouse gas emissions, energy savings, and the penetration of renewable energy. Most importantly, France would:

1. reduce greenhouse gas emissions over the 1990 baseline by 40% in 2030 and by 75% in 2050;
2. reduce final energy consumption from the 2012 baseline by 20% in 2030 and by 50% in 2050;
3. reduce the consumption of fossil fuels by 30% by 2030 in comparison with 2012; and
4. increase the share of renewables in final energy consumption to 23% by 2020, consistent with the EU's Renewables Directive, and to 32% by 2030 – by that same year, renewables should account for 40% of electricity production, 30% of heat, 15% of liquid fuel, and 5% of gas.

The law also contained some short-term goals, especially in the area of transport. Greenhouse gases from transportation would fall by 10% by 2020 and 20% by 2025. Meanwhile, the share of renewable energy in the transport sector would rise to 10% by 2020.

The reductions in energy use and greenhouse gas emissions would be achieved, moreover, while seemingly reducing France's reliance on nuclear power. As Hollande had previously pledged, the share of electricity provided by nuclear power would be reduced to 50% around 2025, although no precise target date was included. In addition, France's nuclear generating capacity would be capped at the current level of 63.2 GW. This meant that EDF would have to close older nuclear reactors in order to bring any new ones on line, and, according to some estimates, France would have to close as many as 23–25 of its 58 reactors to meet the 50% target.[26] In addition, the law created a stricter regulatory framework for the operation of facilities more than 40 years old. Nevertheless, the Minister of Ecology, Sustainable Development and Energy, Ségolène Royal, who had spearheaded the effort, declared that nuclear power would remain the "linchpin" of France's energy mix.[27]

How, then, was France to achieve the goals of energy policy – security, affordability, and environmental sustainability – while reducing reliance on nuclear power, which had previously been the secret of France's success? The new energy law was generally short on detailed policies for achieving these targets, although it did offer some specifics.[28]

With regard to reducing energy consumption, the primary focus was the building sector, which accounted for 44% of energy use in 2012. One tactic would be to accelerate the renovation of the existing housing stock, increasing the number of units renovated to 500,000 per year. To this end, the government would offer tax credits and interest free loans. A second approach would be to improve the performance of new construction by instituting even higher standards. And greater attention would be paid to the management of consumption through the deployment of smart meters and the tracking of individual energy use in apartment buildings.

Other measures would target the transportation sector, which was responsible for about 27% of France's greenhouse gas emissions in 2011. Particular emphasis was placed on promoting the use of electric and other "clean" vehicles through the use of rebates and the proliferation of charging stations. In addition, the governmental and other public bodies would set an example by buying "clean" vehicles for their fleets.

With regard to renewables, which accounted for just 14% of energy consumption in 2012, the law would simplify the procedures for establishing wind parks as well as the environmental permitting process for both wind and biogas installations. The government also established a new support scheme for all types of renewable electricity and called for the creation of 1,500 biogas projects in rural areas over three years.

Perhaps the biggest surprise was the last-minute inclusion of a progressive rise in the carbon tax, which had been introduced in 2014. The tax would now rise from 14.5 euros per tonne of carbon in 2015 to 22 euros in 2016, and then to 56 euros in 2020 and 100 euros in 2030.[29]

Overall, though, the new law was relatively silent on the specific programs that would be employed to achieve the headline goals. There was no mention, for example, of how the share of electricity provided by nuclear power might be reduced, even as overall energy consumption declined, and there were few provisions for promoting the massive buildup of renewable energy sources envisioned by the law. Likewise, how the various long-term targets for reductions in greenhouse gas emissions, final energy consumption, and fossil fuel use would be met was left largely unclear. Most of the concrete measures that were adopted by the government were very short-term in nature. Instead, the determination of many details and difficult decisions, such as which nuclear plants to close, were deferred to a later date. The law merely called upon the government to open a discussion by the end of the year on the development of a multi-year implementation strategy.

Analysis

The new energy law represented potentially the greatest shift in French energy policy since France's response to the oil shocks some four decades ago. For the first time in many years, France would significantly reduce its reliance on nuclear power. At the same time, it would bring down substantially both its energy consumption and greenhouse gas emissions, in part by expanding the share of energy coming from renewable sources.

Nevertheless, or perhaps because of its ambitious nature, the new energy law left some important questions unanswered. For example, it appeared that the buildup of renewable power generation would progress more slowly than the decline in the nuclear share of electricity production. If so, what sources of power, other than fossil fuels, would make up the difference? Likewise, the reduction in greenhouse gas emissions by 2030 would proceed more rapidly than that of fossil fuel consumption. What other measures would contribute to achieving the former target?

In addition, little information has yet emerged regarding the many considerations that went into determining the details of the energy law. The limited number of financial provisions reflected the fact that government resources remained limited, as France continued to struggle to reduce a persistent budget deficit on the order of 4% of gross domestic product. At the same time, the nuclear industry remained a very powerful force in French politics, representing some 200,000 generally well-paying jobs,[30] and localities that hosted nuclear power plants were generally opposed to shuttering them.[31] Hence it is not surprising that the law contained no concrete timetable or provisions for meeting the planned reduction of nuclear power as a share of total electricity consumption.

Other recent developments in French energy policy

Before concluding, at least two other developments in French energy policy in the 2010s merit mention. One concerned the ban on hydraulic fracturing that the Parliament had adopted in 2011. The ban was quickly challenged in court, but it was upheld by a ruling of the French Constitutional Council in 2013.[32]

Another was the establishment of a capacity market for electric power generation that would enhance the security of electricity supplies. This market was created in response to a steady growth in peak electricity demand. Between 2002 and 2012, the peak load increased by about 30%, even as total electricity consumption increased by just 10%.[33] As a result, insufficient generating capacity to meet peak demand was expected as early as 2015, and the problem would only become more acute thereafter with the closing of coal-fired power plants under the EU's Large Combustion Plant Directive.

The underlying cause of the problem was the sensitivity of French electricity consumption to temperature changes because of the country's widespread use of electricity for heating. According to one estimate, a 1 degree drop in temperature created an extra 2.3 GW of electricity demand.[34] The problem was expected to be exacerbated by the growing penetration of intermittent renewable generating capacity.

The overall goal of the new capacity market was to keep flexible but otherwise unprofitable power generation available to meet spikes in demand. The mechanism, which was established in early 2015, placed responsibility on each electricity supplier to balance supply and demand. The supplier could meet this reliability obligation by acquiring sufficient capacity certificates from power generators. Suppliers could also pay customers to reduce their demand in the event of a supply shortage. The first certificates were issued in 2015 with the expectation that the system would go into operation in 2017.

Conclusion

Since the 1970s, France has had a relatively easy time of balancing the often competing goals of energy security, energy affordability, and environmental stability. In addressing the energy trilemma, France got considerable mileage out of its heavy investment in nuclear power in the 1970s and 1980s. Nuclear power helped greatly to reduce both France's imports and consumption of fossil fuels, which in turn did much to lower its CO_2 emissions. In addition, the substantial economies of scale made possible by such a large program and a high degree of standardization meant that the cost of generating electricity from nuclear plants was kept relatively low in comparison with other fuels and other countries.

Nuclear power has not been an unqualified blessing, however. Its virtues also served to inhibit the development and introduction of new renewable sources of power. Likewise, the low electricity prices it made possible and an excess of generating capacity acted as a brake on efforts to improve energy efficiency.[35] And the potential long-term costs of decommissioning reactors and nuclear waste storage were amplified by the large size of the nuclear program.

Recent developments, moreover, have caused France to reconsider its substantial reliance on nuclear power. Particularly consequential was the 2011 nuclear accident at Fukushima Daiichi, which raised sweeping questions about the safety of nuclear facilities. The potential economic benefits of nuclear power also appear to have evaporated with the repeated delays in and the escalating costs of the construction of the first EPR in France as well as the expensive safety measures for nuclear facilities that have been ordered by the ASN. Thus, although most of the existing reactors may be able to operate for many more years, it seems unlikely that they will ever be replaced on something like a one-for-one basis.[36] The prospects for an eventual significant decline in nuclear power's contribution to France's energy mix are now greater than ever.

Instead, France will eventually have to develop a new formula for managing the energy trilemma. The outlines of such a formula have been suggested by the energy law adopted in mid-2015. As in many other developed countries, greater emphasis will have to be placed on renewable sources of energy as well as advances in energy conservation and efficiency. These alternatives are still relatively expensive in France, however, and how quickly they can compensate for reductions in the use of fossil fuels and nuclear energy production remains to be seen. Many renewable sources, moreover, still suffer from the disadvantages of intermittency. France can only hope that its large nuclear legacy will tide it over until the costs of these alternatives, including large-scale power storage, can be brought down to prevailing energy price levels.

Notes

1 According to a former CEO of Électricité de France, Henri Proglio, French electricity prices are 35% lower than the European average and French per head CO_2 emissions are one-third lower than in Germany. "France Needs More Electricity, Not Less Nuclear," *World Nuclear News*, October, 15, 2012, www.world-nuclear-news.org/EE_France_needs_more_electricity_not_less_nuclear_1510121.html.

2 This chapter draws on John S. Duffield, *Fuels Paradise: Seeking Energy Security in Europe, Japan, and the United States* (Baltimore, MD: Johns Hopkins University Press, 2015).

3 Duffield, *Fuels Paradise*, 122.

4 Robert Lieber, *The Oil Decade: Conflict and Cooperation in the West* (New York: Praeger, 1983), 76.

5 Ibid., 81.

6 David Buchan, *The French Disconnection: Reducing the Nuclear Share in France's Energy Mix* (Oxford: Oxford Institute for Energy Studies, 2014), www.oxfordenergy.org/wpcms/wp-content/uploads/2014/01/SP-32.pdf, 3.

7 Duffield, *Fuels Paradise*, 123.

8 International Energy Agency, *Energy Policies of IEA Countries: France 2000 Review* (Paris: International Energy Agency, 2000), 49, 546; International Energy Agency, *Energy Policies of IEA Countries: France 2004 Review* (Paris: International Energy Agency, 2004), 62.

9 Nicole Fontaine, *Livre blanc sur les énergies*, 2003, www.ladocumentationfrancaise.fr/var/storage/rapports-publics/034000650.pdf.

10 "Loi n° 2005–781, 13 juillet 2005 de programmation fixant les orientations de la politique énergétique," (LPOPE), 2005, www.legifrance.gouv.fr/affichTexte.do?cidTexte=JORFTEXT000000813253.
11 Energy Information Administration, *Technically Recoverable Shale Oil and Shale Gas Resources: An Assessment of 137 Shale Formations in 41 Countries Outside the United States* (Washington, DC: Energy Information Administration, 2013), 6.
12 BP, "Statistical Review of World Energy 2015: Data Workbook," BP, 2015, www.bp.com/content/dam/bp/excel/energy-economics/statistical-review-2015/bp-statistical-review-of-world-energy-2015-workbook.xlsx.
13 Buchan, *The French Disconnection*.
14 Ibid., 1.
15 "Flamanville EPR Timetable and Costs Revised," *World Nuclear News*, September 3, 2015, www.world-nuclear-news.org/RS-French-regulator-expects-to-report-soon-on-EPR-anomaly-2101501.html.
16 Buchan, *The French Disconnection*, 7.
17 Réseau de transport d'électricité (RTE), *2014 Annual Electricity Report* (Réseau de transport d'électricité, 2015), www.rte-france.com/sites/default/files/bilan_electrique_2014_en.pdf.
18 International Energy Agency (IEA), *Energy Policies of IEA Countries: France 2010 Review* (Paris: International Energy Agency, 2010), 121.
19 IEA, *Energy Policies of IEA Countries*, 98, 121; Alexander Ochs, and Serre Camille, "An Analysis of France's Climate Bill: Green Deal or Great Disillusion?" World Watch, 2013, www.worldwatch.org/analysis-france%E2%80%99s-climate-bill-green-deal-or-great-disillusion.
20 Buchan, *The French Disconnection*, 2.
21 Ibid., 5.
22 Marion Bitoune, *The German and French Energy Transitions: Have the Two Changed European Energy Policy?* (Washington, DC: Heinrich Böll Stiftung, 2015).
23 Michel Rose, "French Energy Law Dodges Decisions on Nuclear Cuts," *Reuters*, July 22, 2015, www.reuters.com/article/2015/07/22/france-energy-bill-idUSL5N10242K20150722.
24 Pierre Le Hir and Laetitia Van Eeckhout, "Transition énergétique: comment le Sénat a changé la loi," *Le Monde*, February 19, 2015, www.lemonde.fr/planete/article/2015/02/19/transition-energetique-comment-le-senat-a-change-la-loi_4580129_3244.html.
25 The full text of the law, *LOI n° 2015–992 du 17 août 2015 relative à la transition énergétique pour la croissance verte*, is available at www.legifrance.gouv.fr/eli/loi/2015/8/17/2015-992/jo/texte.
26 Geraldine Amiel, "France to Dim its Reliance on Nuclear Power," *The Wall Street Journal*, June 18, 2014, www.wsj.com/articles/france-to-dim-its-reliance-on-nuclear-power-1403113287.
27 Cécile Barbière, "Nuclear Remains Linchpin of French Energy Transition," Euractiv, August 1, 2014, www.euractiv.com/sections/energy/nuclear-remains-linchpin-french-energy-transition-303832.
28 Ségolène Royal, *La Transition Énergétique pour la Croissance Verte*, August 18, 2015, www.developpement-durable.gouv.fr/IMG/pdf/CP_Loi_TECV_publication_citoyens.pdf. See also Ministry of Ecology, Sustainable Development, and Energy, *The Energy Transition: A User's Guide* (Ministry of Ecology, Sustainable Development, and Energy, 2014), www.developpement-durable.gouv.fr/IMG/pdf/14123-2_plaq-NMTE-parlementaires_GB_DEF_Light-2.pdf.
29 Tara Patel, "France Passes New Energy Law Quadruples Carbon Price," *Bloomberg Business*, July 23, 2015, www.bloomberg.com/news/articles/2015-07-23/france-passes-new-energy-law-quadruples-carbon-price.
30 Amiel, "France to Dim its Reliance on Nuclear Power."
31 Rose, "French Energy Law Dodges Decisions."
32 David Jolly, "France Upholds Ban on Hydraulic Fracturing," *International New York Times*, October 11, 2013, www.nytimes.com/2013/10/12/business/international/france-upholds-fracking-ban.html?_r=0.
33 RTE, *2014 Annual Electricity Report*.
34 Buchan, *The French Disconnection*, 9.
35 Ibid., 13.
36 Ibid., 60.

26

Struggles in Denmark's transition towards a low carbon future

Shifts in the energy technology assemblage

Peter Karnøe and Jens Stissing Jensen

Introduction[1]

In 2015, Denmark set a world record when wind power generated 42.1% of the nation's total electricity.[2] For 16% of the time, wind power alone generated more than 100% of the electricity demand in western Denmark, and excess electricity could be sold and transmitted to neighbouring countries like Norway, Germany and Sweden. This high level of wind power penetration has enabled the country to shift away from the use of fossil fuel-based power plants to generate electricity, and is the main reason that Denmark is the top-ranked country based on the Climate Change Performance Index created by Climate Action Network Europe and Germanwatch for 2015.[3] The ranking is based on factors such as overall development of a CO_2 emissions strategy, electricity generation from renewable sources, energy efficiency, and an ambitious and consistent climate policy. The recent ranking was based on 2013 statistics, which reflected the cumulative effects of past policies driven by social democratic governments, sometimes with pressure from left wing and green parties. The assessment also highlighted Denmark's 2012 Energy Agreement, which cemented a broad political compromise for reducing CO_2 emissions through energy efficiency by combining electricity and heat generation and making wind power the core energy technology in the future low carbon energy system.

However, with the election of a new centre-right minority government in June 2015, the ambitions of Denmark's previous climate policies were not to be realized. This government has introduced the notion of 'green realism' into its climate policy, aimed at tempering the transition process towards a fossil-free society by 2050, similar to the centre-right government of 2001–2008 that almost brought climate policies to a full stop. Green realism is based on the argument that given its top rank worldwide, Denmark can relax its policies and reduce its ambitions due to the material and economic costs of integrating increasing amounts of wind-generated electricity into the electricity/energy system.

The policy change was noticed immediately, and at the Conference of Parties (COP 21) in Paris in December 2015, Denmark was given the distinct 'honour' of receiving the 'fossil-of-the-day' award, which is given to the country that has abandoned former climate policies with no broader concerns for the greater good. This distinction was bestowed by the Climate Action Network, which earlier that year had ranked Denmark highest based on its Climate Change Performance Index mentioned earlier. The fossil-of-the-day award may be a surprise to many outsiders, but it is a reminder that societal transitions cannot be reduced to technological innovation alone. Whenever a centre-right government has come into power in Denmark, climate policies have been almost completely thwarted. However, the context for stopping or relaxing climate policies in 2015 is very different from 2001, given increasing momentum of the transition to a low carbon energy system outside Parliament.

First, the large-scale penetration of wind power (42.1% in 2015, up from 10% in 2001) represents a new material reality. Wind-generated electricity thus functions as a 'living material thing' that threatens to destabilize the frequency in the Danish electricity grid, which must be balanced at 50 Hz at all times to prevent blackouts. Furthermore, the market design of the Nordic electricity market appears to be incapable of generating prices that provide adequate financial incentives for 'system electricity services' (e.g., different plant types and load services associated with spot balancing load capacity). Current prices are too low, and as a consequence, 35% of central power plant capacity was shut down between 2008 and 2013. In this way, high wind power penetration represents a new reality that has disrupted the intertwined material, regulatory, organizational and market arrangements that served the fossil based electricity system quite well.[4] The increasing amount of wind-generated electricity is thus turning into a fundamental systemic problem for the energy system as a whole.

Second, while the centre-right minority government advocates green realism, many actors in the energy system and its suppliers already act on the basis of a new world view focused exclusively on making a full transition to an energy system based on wind power. Green realism thus faces opposition not only from NGOs, but increasingly from established organizations in the energy and industrial sectors, and prominent CEOs from new green, low carbon, clean-tech companies. While the priorities of the right wing minority government are fairly predictable, it is somewhat surprising to witness this outspoken resistance to the green realism policy from many organizations, industries, municipalities and experts. Indeed, the Danish energy and climate policies have created new business opportunities and new industrial clusters and strongholds for wind power, district heating, and energy efficiency-related goods and services, which have given rise to shifting interests.

Indeed, this kaleidoscopic picture of the current diverse shifts in the Danish transition to a low carbon energy system reveals conflicting political positions[5] yet, at the same time there is a strengthening of new material and discursive realities that constitute a new normal. As such, climate policy must address not only the systemic problems of wind power integration, but also the new business opportunities and industrial transformations associated with investing in the goods and services that will comprise the future low carbon energy system. In this chapter, we describe the surprising outcome when green realism meets the new climate policy normal, in which both material problems and industrial cluster business opportunities must be considered. We use the concept of sociotechnical assemblage to understand the conflict between political-ideological realities and the effects of increasing momentum and material and structural demands for continued change, as the concept allows for the co-existence of differentiated ambitions and fragmentation while maintaining momentum.

Shifting sociotechnical assemblages in the energy system transition

A salient concern in many contemporary climate policies is how to de-carbonize large-scale societal systems in domains such as water, energy, transportation and food. While technological components and infrastructures constitute a large part of such systems, empirical analyses in the tradition of science, technology and society studies (STS) stress that transitioning from high carbon to low carbon energy systems cannot be reduced to managing technical dimensions.[6] From an STS perspective, there is no such thing as pure technology[7] just as there is no such thing as a 'natural and pure market'.[8] Scholars in this tradition typically perceive societal systems as 'seamless webs' in order to underscore how such systems only become 'real' through multiple relations and reciprocities among people, politics, regulation, technical design and production, science and the economic market.[9]

Specifically, we draw on the STS-inspired concept of socio-technical assemblages in order to tell the story of the transition of the Danish energy system. This concept is particularly well-suited to understanding both *persistence and change* as phenomena that are endogenous to the established structures, building blocks and processes of societal systems. The assemblage concept presents a relational way to understand 'systems', and is a new way of examining the part-whole connection without resorting to analysing completely ordered wholes using one script or translator, as is the presumption in systems theories.[10] Instead, assemblages consist of networked actors made from combinations of various intermediaries such as technical, scientific and legal texts, imaginaries and narratives, artefacts, money, skills and competences as well as criteria used to justify and legitimize actions.[11]

Importantly, however, a socio-technical assemblage is not a fixed entity, but an 'open building site under constant construction' comprised of groupings of heterogeneous actors and intermediaries that shape and define each other's technologies, cognitions, interests and identities.[12] This mutual shaping may create enough convergence to generate momentum[13] that harnesses the inputs of distributed actors.[14] For example, the fossil fuel-based centralized electricity system is an example of an assemblage that grew from nothing in the 1880s to complete dominance in the 1980s through translation of the heterogeneous elements that were shaped and connected to constitute the emergent 'system', and thereby make coordinated actions possible.[15] Although the centralized fossil fuel-based electricity system may appear to be relatively fixed, it is comprised of stabilized relations between many elements that may be challenged and reorganized. Since the 1980s, the particular relations of this configuration have been increasingly challenged by concerns over material effects on the environment and climate that eventually were addressed in COP meetings in strategic battles of transition between resistant 'incumbents' and 'challengers'.[16] While such declarations are important, COP21 also illustrates that systems should not be mistaken for homogenous wholes. The groupings of 'parts' (such as nations, citizens' groups, corporations, municipalities or 'municipalities for climate change') comprise particular and often precarious networked constellations of contextually-situated action points. The degree of convergence, stability and sharedness that characterizes such a constellation is an empirical matter.

Despite the possibility of convergence, the assemblage perspective differs from more static and structural views of societal systems (e.g., system concepts such as socio-technical regimes or techno-institutional complexes). Static concepts tend to represent systems as if there is a common overarching whole that 'acts' on all involved actors homogenously. Static-structural approaches thus filter away the heterogeneity of situated practices and 'frames of reference' of those involved in a societal system. This makes it analytically difficult to accommodate both developmental stability and the potentially transformative dynamics of agents as they engage in new concerns, struggles and negotiations about 'next steps'.

Unlike the static structural views of societal systems, the concept of socio-technical assemblage embraces the incompleteness, multiplicity and precariousness of the composite relations that make up so-called systemic wholes or totalities. Whatever coherence and convergence exists is constituted by the *situated actions* of the participating actors, who ascribe sense and meaning to their actions based on particular situated framings of their entanglement.[17] A situated entanglement is associated with particular frames of reference, which are rooted in concrete social and material practices in such domains as design and production, science and research, politics and regulation, media, NGOs, markets and users.

Since a socio-technical assemblage is an 'open building site under constant construction' comprised of groupings of heterogeneous actors and intermediaries that shape and define each other, it is also important to make an analytical distinction between an assemblage and the different networked agencies that comprise it. Depending on the state of the assemblage (i.e., the degree of convergence, stability and sharedness) these agencies may have different and competing agendas or interests, and consequently may want to take the unfolding of the assemblage in different directions. Networked agencies are integral to assemblages, but may actually compete with each other.

A methodological strategy by which to embrace both the static nature of a structure and the processual dynamics of a societal system is to apply the notion of socio-technical assemblage both in a 'noun version' and in a 'verb version' as suggested by Law: 'If "assemblage" is to do the work [of replacing views of systems] that is needed then it needs to be understood as a tentative and hesitant unfolding, that is at most very partially under any form of deliberate control. It needs to be understood as a verb as well as a noun.'[18]

The 'verb perspective' considers how associations recursively unfold in localized processes by which networked agencies typically manoeuvre, which are neither static nor given, but constantly in-the-making and re-making. We argue that this relational understanding of a 'system' is particularly sensitive to the co-existence of multiple agencies. Agencies are networked constellations of regulations, resources and calculation located within the broader assemblage that provides actors with locally-situated 'frames' for how to make sense of their concrete socio-material engagements in the operation, reproduction or transformation of the energy system. Even from the verb perspective, assemblage activities may nevertheless stabilize, as did the networked agencies associated with the centralized electricity system, and become somewhat predictable as they become increasingly irreversible and gain momentum.

The 'noun perspective' shifts attention to how relations among the socio-technical elements of the assemblage have stabilized and make up a relatively fixed structure (i.e., Callon's 'Techno-economic Networks and Irreversibility'). From the noun perspective, it is possible to make partial summaries (representations) of the state of an assemblage in terms of its politics, policy plans, and scientific and technological accomplishments (e.g., 42.1% wind power penetration), current public sentiment, or the effectiveness of current market arrangements. Various reports and rankings are examples of such summaries of the state of an assemblage; however, the particular 'reality made visible' from these reports depends on the context of the observer and the concepts, categories, etc. used to make the measurement.[19]

In this chapter, we draw on both the verb and noun perspectives in describing the Danish transition away from fossil fuels by following some elements and agencies involved in shifting the assemblage. From a processual view, transitions are not the result of outside structures 'acting upon' existing actors. Assemblages shift as locally-situated networked agencies in the domains of politics and regulation, design and production, science and technology research, markets and users, and media and public debate take new actions and attempt to interest, provoke or coercively force other actors to re-frame themselves and their actions and contribute to shifting the socio-technical assemblage through recursive processes.[20] Any networked agency needs legitimate referents for maintaining or shifting action, and materially, these may be highly diverse. The reference points may include demonstrations that provoke new policy actions, watershed events like the Fukushima accident, the gradual acceptance of IPCC climate science as a reference point for new actions, or new discursive (i.e., the ideological doctrine of 'free markets') and calculative standards and valuation frames used to justify investment decisions in companies or national policy actions. The shifting of assemblages is not like linear rational policymaking, with clear steps for getting from A to B; nor is it guaranteed to be successful, as such shifts happen through distributed agencies, struggles, skills and negotiations that temporarily mobilize and enrol new allies. We use the concept of socio-technical assemblage to follow the 40-year transition of the Danish energy system, and show how it has been modulated in a series of contradicting and piecemeal movements whereby actors have struggled to develop and promote new framings of the energy system that ascribe new roles to individual components.

Towards a model of shifting technologies and practices in the energy assemblage, 1976–2001

In 1973, before the first energy crisis, Denmark had created a highly reliable energy supply to provide both electricity and heating to homes, industry and the public sector. The availability of light and the so-called '21-degree' cultural standard were now defining the new normal of comfort in modern society.[21] These services were provided by an energy assemblage of centralized electricity and heat comprised of oil boilers, district heating systems, and co-generation. Fuel-wise, this assemblage produced about 95% of its energy by burning cheap and abundant oil, which had replaced city gas, coal and coke. The biggest challenge seemed to be when to add nuclear power to the energy assemblage built around the centralized electricity system. A strong coalition of major groups from the political, science, industrial and electricity domains were behind nuclear power as the next natural step.

Denmark's oil dependency was revealed during the 1973 oil crisis. When Arab oil producers enacted an embargo on Denmark (after it became known that the prime minister supported Israel during a closed meeting), the price of oil quadrupled overnight.[22] Denmark had taken great pride in its energy assemblage that provided a reliable supply of centralized generation of electricity and heat; yet, it had no contingency energy policy.

The first political response was to create a series of energy policy reports on possible energy futures and provide some techno-economic calculations so as to compare various scenarios. Policy priorities outlined in the first Energy Plan created in 1976 were:

1. reduce oil dependency and improve energy supply security (coal and nuclear power);
2. build a multi-directional energy supply that included domestic Danish energy sources;
3. promote energy savings and energy efficiency; and
4. establish a national heat plan that prioritized how to heat buildings most effectively in different districts.

This framing of the official energy policy did not go uncontested, however. A new agency network comprised of NGOs, scientists and civil society actors published the so-called Alternative Energy Plan (AE'76). It suggested a policy based on energy efficiency by advocating more co-generation of power and heat, and envisioned that wind power parks could contribute 12% of domestic electricity production by 1995. The AE'76 was a central calculative device with different calculations and scenarios for energy growth, and contributed to an alternative socio-technical vision that helped to broaden the situated and public mind-set of what could be a possible future energy system.[23] The role of wind power was, however, regarded as 'white noise' or 'meaningless words/sounds' (similar to reactions when the electronic synthesizer was introduced)[24] in relation to the dominant and normalized framing of the future energy system. For example, the Danish Electricity Association (DEA) ridiculed the idea of using wind power from the very beginning in the 1970s.[25] In a brochure, the DEA claimed that the electricity generated from wind power would not be sufficient to heat the waterbeds in which the (hippie) windmill owners were sleeping.[26] However, this renewable energy network continued to shape the public debate by advocating and demonstrating alternative futures by offering reports such as the Alternative Energy Plan of 1983, which criticized the official Energy Plan of 1981. Their texts materialized alternative visions of a renewable energy system based on wind power, that deviated completely from the official energy plans and DEA's nuclear future, and were, by the majority of actors, seen as 'white noise' and not immediately incorporated into the official and dominant energy policy discourse.

The official 1976 and 1981 Energy Plans did, however, form the basis for two relatively separate policy approaches that shifted agencies in the energy technology assemblage. The main approach consisted of a reorganization of the established energy system involving three major changes. First, power plants were converted from oil to coal and natural gas. Second, in order to build domestic energy capacity, a piped heating infrastructure for district heating and natural gas was constructed in the Danish part of the North Sea with the intent of fulfilling the heat demands of more than 50% of all dwellings. Third, to save energy, new (and later upgraded) insulation standards were incorporated into building regulations, and taxes on fossil fuels were raised to incentivize energy efficiency in industry.

These initiatives triggered new agencies to act and collaborate. For example, district heating infrastructures were developed by municipalities in collaboration with user-owned local district heating plants. Further, municipal gas companies were added to this infrastructure along with incineration plants that produced heat from waste. Energy taxes were structured to generate economic incentives to act in accordance with policy goals and regulations. During the 1990s, a more fundamental reorganization took place as many centralized power plants were replaced by hundreds of private, decentralized production units located closer to consumers with heat demands. New formalized agencies with new competencies, roles and interests were being developed in the energy technology assemblage.

These policies of the 1980s and 1990s were highly successful, and energy consumption stabilized around 800 PJ despite a doubling of national economic output from 1985 to 2010. In addition, the energy tax was generating about €7 billion annually for the state budget by 2013. While this was good for state finances, it also made shifting assemblages difficult because replacement of tax revenues would come at the expense of private actors.

Relatively isolated from this large-scale reconfiguration of the oil-dependent energy technology assemblage, the development of wind turbines took off in the early 1970s. Grassroots organizations and entrepreneurs engaged in early experimentation independent from the national electricity grid, while others focused directly on developing grid-connected wind power and set in motion shifting assemblages related to regulations for the ownership and connection of wind turbines to the grid[27] (multiple accounts from grassroots found in Beuse et al., 2000). These efforts were complemented by the development of a research programme and policy approval of economic subsidies for wind turbines. There was, however, no a priori functional plan for these subsidies; rather, wind power policy was assembled as a new policy object in a 'garbage can-type non-linear process'.[28] The convergence was, however, weak and precarious as critics and sceptics strongly contested wind power in order to influence adaptive expectations about its role in the electricity system.

For instance, one debate in the early 1980s centred on the size of 22–30–55 kW wind turbines, deemed 'too small to matter energy production wise', compared to the large-scale wind power programme based on 600 kW wind turbines. However, during the mid-1980s, the wind power network was strengthened by a surprising set of relations to new issues: export income and jobs in Denmark. Exports of small (55–75 kW) Danish wind turbines increased from 30 to 2,000 per annum between 1982 and 1985, making the wind turbine more popular and legitimate in Denmark.[29] The booming export market had a positive effect on the domestic market, as it shaped new expectations, through which wind power policies were increasingly justified by reference to the effects on employment and export benefits. This happened at a point in time when the existence of wind power in Denmark was fragile and contested for being subsidized, expensive and irrelevant to base-load electricity production. As part of the 1985 Energy Act, energy authorities made an agreement with the electric utilities that they would install 100 MW of wind power energy between 1986 and 1990, on the condition that private installations would be restricted. This represented a substantial increase in the home market, as private installations from 1978 to 1985 totalled only 75 MW. Wind power's share of Danish electricity generation only amounted to 2% by 1990, however, and the material presence of wind power in the grid was not problematic.

By the year 2000, the energy assemblage thus consisted of an energy system organized around piped heating infrastructures (natural gas and district heating infrastructures) and a high level of co-generation of heat and power. Relatively separated from this, the wind power assemblage was materially gaining momentum, accounting for about 10% of electricity generation. In the following three sections, we outline the recent transition processes of the Danish energy system by analysing how these two energy technology assemblages were increasingly interweaved after 2000. The story focuses on how the shifting of agencies created various material effects as wind power and other low-carbon technologies shifted the energy assemblage.

Assemblage shift 1: Struggles related to framing the wind power assemblage as a policy object

A quick glance at the development of the Danish energy system assemblage since 2000 reveals that wind power has gained an increasingly important position. Figure 26.1 shows the growth in total wind power installations in Denmark from 1990 to 2020 (projected). In 2015, installed wind power accounted for approximately 5 GW of the 13 GW total electricity capacity generated by wind power, small and large scale power plants, and independent producers. Wind power is seen as the new core element in the future energy system, and many actors expect an installed base of 11 GW by 2030.[30] In 2015, wind-generated electricity accounted for more than 42% of total electricity production.

The process by which wind power came to play an increasingly central role was not a smooth one. While growth was high (25%) between 1994 and 2003, development was flat from 2004 to 2010/2011, before it started growing again through the beginning of 2016 with expectations for continued growth until 2020. These shifts in growth rates show the immediate effects of shifts in governments. The social-democratic coalition governments have supported wind power, whereas the centre-right governments have sought to stop or reduce wind power development immediately upon taking office.

In this section, we demonstrate how the shifts in political agencies (i.e., social-democratic governments versus centre-right governments) have been associated with the promotion of different calculative devices and valuation frames for wind power. These valuation frames have been used to assign different industrial and economic values to wind power vis-à-vis other energy technologies.[31]

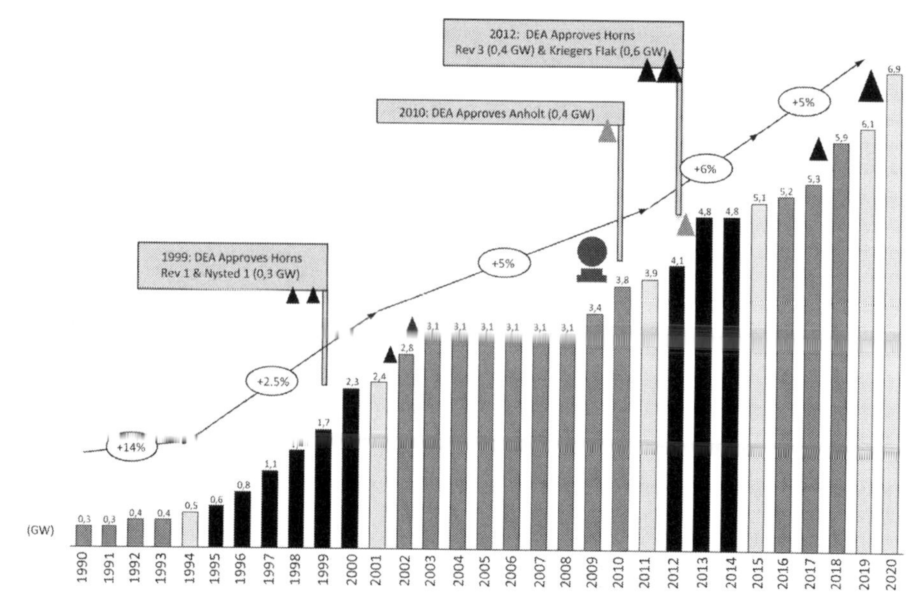

Figure 26.1 The increased installation of wind power in Denmark (1990–2015) and projected installations (2016–2020).

Note: The bar colours indicate a centre-right (in grey) or social democratic (darkest) government in office in a given year. The lightest bars represent election years. Black lines show the compound annual growth rate for given government term. The triangles show connections between the approval and installation years for Denmark's three important offshore wind farms (Horns Rev, Nysted, and Anholt) and those expected to be installed by 2020 (Horns Rev 3 and Kriegers Flak). The colours indicate under which government they were approved to be built. The grey circle represents COP15 held in Copenhagen in 2009.

A critical component in the strong (25%) growth of the wind power assemblage from 1994–2002 was a climate policy rooted in a new valuation frame promoted by the new social democratic Minister of Energy and Environmental Affairs, Svend Auken. Coming into government after the centre-right government (1991–1993) unsuccessfully sought to stop wind power development in Denmark, Auken's response was to create a 'super ministry' by combining the strong Ministry of Environment with the Ministry of Energy. This reorganization enabled biological and technical expertise in environment and energy to be combined and serve as a policy platform. For example, in 1988, the Ministry of Environment had already presented a calculation-based report on the consequences of CO_2 emissions that was used in the 1990 Energy Plan. The Auken regime successfully forged strong new linkages between renewable energy (most notably wind power development), CO_2 reductions and industrial policy cluster effects. The new Energy Plan of 1996, EP'21, built upon an explicit commitment to outputs of the 1995 UN Climate Convention (specifically, binding CO_2 reduction targets for industrialized nations) and further specified the development of industry and exports as key policy objectives. Auken was also a strong national and international promoter of the 1997 Kyoto agreement.

The valuation frame backing the policy for the wind power assemblage now consisted of two components. First, complex calculations in EP'21 produced a reality claim that wind power was valuable in relation to CO_2 reductions. Second, wind power was framed as being valuable as a strategic industrial cluster with high future employment, growth and export potentials that deserved to be protected and cultivated. This valuation frame, however, relied on a precarious network. First, it was dependent upon the highly charismatic and influential minister, Sven Auken, who successfully increased the budget of the 'super ministry' and prioritized climate and environmental concerns within the government. These policies enabled Danish wind power installations to grow from 3% in 1994 to about 13% by 2000. The Auken-regime, however, was increasingly criticized by the agency-network of the centre-right opposition in Parliament, which argued that expenditures on environmental policy were growing disproportionately.

The shift to a centre-right government in 2001 paved the way for a dramatic shift in the valuation frame of wind power. The centre-right coalition government stopped the development of two planned off-shore wind farms, and enacted savings of more than 2.5B DKK (2015 currency) in the 2002 state budget. In addition, the farmer-friendly government had an explicit policy to reduce governmental involvement in environment and climate issues and terminated more than 400 jobs related to those activities in 2002 alone.[32] The Danish Centre for Alternative Societal Analysis (CASA) later released a report based on proposed environmental cuts from the financial ministry, indicating that budget cuts to environmental activities would total between 21B and 26B DKK (2015 currency) from 2002 through 2005.[33]

The shift in government enabled the increasingly powerful Ministry of Finance to replace the 'hands-on' industrial policies based on strategic support to prioritized industries with more neo-liberal, 'free-market', 'hands-off' policies. Energy sources would now be judged on their ability to 'compete' in what was perceived as a 'subsidy-free market'. The hands-off policy approach penetrated the environment and energy domain under the guise of socio-economic analysis, which constituted a new calculative device for producing 'realities' that enabled environmental policies to be evaluated.[34] This socio-economic calculative device was not used to address ecological or resource-related issues, but to compare and contrast the economic efficiencies of various policies, and the extent to which they constituted a liability to the broader national economy. Further, in order to ensure that environmental activities would not once again spin out of control, the government established an environmental assessment institute in 2002. Climate sceptic Bjørn Lomborg, with the declared goal to evaluate the socio-economic cost-effectiveness of environmental policies using socio-economic analysis, headed the institute. Lomborg resigned in 2004 after strong critique of the quality of the reports, but created the 'Copenhagen Consensus Centre' think tank in 2006, which would receive funding from the annual budget of 2007 and onwards. In total, the centre-right wing government supported Lomborg and his affiliated institutions with more than 138 m DKK from 2002–2011.[35] It would be the new environmental minister of the Social-Democratic-led 2011 government Ida Auken, Svend Auken's niece, who would lead the way in removing what she categorized as 'ideologically-based funding' to Lomborg.[36]

This dominant new valuation frame based on the optimal allocation of resources given the assumption of a perfect market provided a new reference point for governmental policies related to the wind power assemblage. Yet it also paved the way for new struggles and contestation among various economic experts over the basic elements in any valuation process (i.e., what counts and how it counts).[37]

For example, the socio-economic analysis produced by the Ministry of Finance (2001) showed that investments in wind turbines were, in fact, associated with considerable positive socio-economic effects, mainly because the value of reductions in local air pollution was included in the calculation. Members of the independent and prestigious Danish Economic Council were critical and concluded that the analysis suffered from a series of methodological flaws.[38] First, they argued that the amount of pollution associated with electricity production from traditional thermal power plants had been overestimated, thus the environmental benefit of wind power had been overestimated as well. Second, they argued that the initial analysis had ignored so-called market distortion effects associated with tax-based financing of subsidies for wind power production. This suggested that the societal cost of wind power production had been underestimated. By re-calibrating the calculative device, the council concluded that wind power had negative socio-economic effects and thus constituted a societal liability. Their analysis further indicated that national climate policy goals could be more cost-effectively addressed by taxing fossil fuels and reducing CO_2 emissions abroad. Drawing on neoclassical supply-side economics, the analysis 'naturally' questioned the industrial policy cluster framing of the wind power industry that had been a key component of the governmental valuation frame during the 1990s.

In this new valuation frame, wind power was valued as an object in a socio-economic efficiency contest rather than in an industrial cluster driven by CO_2-reducing climate policies. This change reframed wind power from a potential future industrial stronghold into a societal burden. This conclusion can be seen as inevitable or tautological in the sense that all industrial policy is inappropriate, because the market is assumed to be perfect.

Competing valuation frames and policies in Denmark

The valuation frame of the Auken regime turned the IPCC report and CO_2 emissions into salient reference points for energy/climate policy and industrial cluster policy, and resulted in strong growth in wind power. The centre-right government shifted the valuation frame in 2001, and the socio-economic calculation approach contributed to legitimizing a drastic reduction in the influence and budget of the Ministry of Environment and Energy. Also, the new government cancelled the construction of two new offshore wind farms. However, due to pressure from social democrats, the socialist people's party, and the social liberal party, the centre-right government enacted a new political agreement in 2004 and resumed development of the two offshore wind farms that had been cancelled in 2001.[39] The centre-right accepted this for pragmatic reasons, namely, the continued importance of environmental concerns among voters and the importance of exporting wind turbines and other energy and environmental technologies.

The centre-right policy was criticized by the wind turbine industry and other spokespersons from the pro-climate policy network of actors, but the majority of industrial actors and related interest organizations did not criticize the drastically reduced policy ambitions and the diminishing expertise and power of the Ministries of Environment and Energy (which were separate once again). However, members of the centre-right began to change their position around 2006 when Denmark was chosen to host the COP15 meeting in Copenhagen in 2009. For example, the conservatives shifted their position in 2006 to embrace more decarbonization climate policies, highly driven by Connie Hedegaard, the new Minister for Climate Affairs (she later became the EU Commissioner for Energy and Climate, 2010–14). The year before COP15, in the autumn of 2008, the strong-minded anti-climate change Prime Minister Fogh Rasmussen announced at a public party meeting that 'I have very long belonged to those who were a little in doubt about climate change' and he went on to say 'that was maybe wrong … many of us have been cautious if not to say a little foot-dragging in all of this … We (Venstre party) have not been the climate policy avant-garde'.[40] As part of this shift in interest and orientation the government created a Climate Commission (2007–10) to establish whether it was possible for Denmark to have a 100% renewable energy system by 2050. However, except for some 'paper tigers' about future energy strategies, there were no further shifts in policies before the new social-democratic government took office in 2011.

The new charismatic Minister for Climate and Energy, Martin Lidegaard (who had been director of the green think tank CONCITO) followed Auken's policy approach and used new knowledge from the Climate Commission (and CEESA) to support the further transition of the energy system. The important 2012 Energy Act specified the overall energy policy until 2020, and represented a continuation of the valuation frame of the 1990s. The act promoted increasing energy efficiency, incorporating more renewables (biomass, biogas and wind power) into the energy supply, and funding energy research. There was no direct link between new calculative devices and shifts in policies, as little attention was paid to socio-economic calculations in the execution of the agreement, and green energy technology was once again framed as an important industrial cluster. The communication thus emphasized: 'The initiatives in the agreement generate green growth until 2020 and take the competitiveness of businesses into account.'[41] Longsighted 'points of orientation' were further defined as part of the agreement in an attempt to stabilize expectations and policy commitment to the energy system transition. These stipulated that 50% of electricity should be generated by wind power by 2020, that electricity and heat should be CO_2-neutral by 2035, and that the entire energy system (including transportation) should be CO_2-neutral by 2050.

In summer 2015, however, the new centre-right government took office and implemented their green realism policy approach, questioning the 2012 re-establishment of associations between these 'points of reference for action'. Denmark experienced a déjà-vu of 2002, as the annual state budget for 2016 included cuts to environmental activities both in Denmark and in the poorest areas of the world. The cuts amounted to more than 1B DKK (2015 currency) and were combined with the elimination of numerous environmental taxes and regulations, in particular to help the agricultural sector.[42] Further, green realism re-politicized the cost of transitioning towards a renewable energy system.

Splitting the bill in 2015: the PSO controversy and re-politicization of the transition cost

Right-wing politicians and some members of industry have once again framed the development of wind power as a burdensome societal cost. Wind power has been framed as a burden, exemplified by an extra cost component introduced on the electricity bills received by all consumers in Denmark. This so-called PSO (Public Service Obligation) is a tax that finances the feed-in tariff that ensures investors and owners of renewable energy sources (wind turbines, solar panels, biogas, etc.) receive a minimum price for electricity.[43] The PSO covers the gap between the feed-in tariff and the electricity market price, and consequently, the higher the market price on electricity, the lower the PSO and vice versa. The Merit-Order-effect and the low CO_2 prices (adding to the cost of fossil fuels) reduce electricity market prices, and this tax has increased in recent years. Critics have argued that the tax constitutes a critical economic burden to electricity-intensive industries such as horticulture and cement production.

This argument has not gone unchallenged, however. One counterargument has been that even when the energy tax is included, the Danish electricity price is still highly competitive in the European context. Another counterargument is that the PSO serves as an economic market incentive to become more efficient (i.e., as a green energy technology facilitator) because the PSO is part of a set of regulations for a favourable and predictable domestic market for new products and services.

To eliminate the argument that the PSO is a 'cost burden' for industry, some green realism advocates seek to shift the PSO from consumers' electricity bills to the state budget. Some smart energy system advocates see this as a smart way to 'constantly politicize' the funding of the transition to wind power and other renewables like solar power and heat pumps.[44] If the PSO is part of the state budget, the transition funding will have to compete with other issues such as social welfare benefits, investments in education and research, and tax cuts for low or high-income groups.

Since the EU is scrutinizing all national subsidy programmes related to energy technologies for treaty violations, there is considerable 'interpretative flexibility' for these subsidy programmes.[45] The green realism coalition has used this review to claim that the Danish PSO model violates the EU treaty. However, the EU has not confirmed this, and since most programmes can be adjusted to conform to the EU treaty, this is an example of the politicization of an important policy instrument.[46]

Summary of assemblage shift 1: valuation frames and policy objects

Competing valuation frames have constituted organizing nodes in the configuration of the networked agencies that make up the Danish wind power assemblage over the past four decades. These different valuation frames have turned wind power into different policy objects. One valuation frame driven by social-democratic governments has facilitated networked agencies configured around the climate change agenda and a clustered approach to the wind power industry. Since the early 1990s their tactic has been to enrol members of the green tech industry and energy utilities in this valuation frame.

The competing valuation frame of the centre-right governments has facilitated networked agencies based on socio-economic efficiency criteria, downplayed the importance of responding to climate change, and represented the wind power assemblage as a liability to economic growth and industry competitiveness. This valuation frame has been supported by the new socio-economic calculation method promoted by the powerful Ministry of Finance, and has attracted the support of agriculture and electricity-intensive industries. From 2001 to 2012 or so there was very little critique of this less ambitious, rolling-back centre-right climate policy from industry-based actors, except those in the wind turbine industry.

These struggles over which valuation frame to use as the basis for policy formulation illustrate that wind power has been subject to continuous political contestation. However, as we shall see in the next sections, there are two materially-based changes in the 2015 sociotechnical energy assemblage that constitute a very different context for this ongoing policy valuation struggle compared to 15 years ago.

Assemblage shift 2: wind power as a materially-based systemic concern in the electricity assemblage

The centre-right government reframed the wind power assemblage as a socio-economic 'efficiency' policy object just as an important material shift in the energy system assemblage became manifest. While electricity from wind power contributed less than 0.2% of Danish electricity production in 1981, this share had increased to 2% in 1990 and 13% in 2000. When electricity from wind turbines only contributed a marginal share of total electricity production, wind power electricity was considered to be of limited importance to the broader architecture of the electricity system as well as the energy system as a whole before 2000. The share of electricity produced by wind power, however, was rapidly increasing, and the average size of installed wind turbines increased from 150 kW in 1990 to more than 600 kW in 2000 (and subsequently increased to about 1.5 MW in 2010). Wind power was indeed becoming a 'large-scale' material penetration, and a new phenomenon emerged, as documented by articles in the Danish media and in the journal *The Engineer* stating, 'Denmark is flooded with excess electricity'. Only transmission lines to Norway and Sweden assure that someone used the power – but at a very low price.

Thus, by 2000, main actors within the energy system began to recognize that electricity from wind turbines was no longer an innocent source of energy that could be readily absorbed by the existing electricity system. Some had previously warned of the potentially disruptive consequences of wind power. When the 1981 Energy Plan noted the possibility of achieving 10% wind power by 2000, the electric utilities voiced concerns. In a research report published in 1983 based on a simulation of 600 MW of wind power in the electricity system, the utilities concluded that the unpredictability of wind power made it an unreliable source of energy, and further stated that wind power would result in 'certain operational disturbances, the solution to which requires extra expense' (our translation).[47] The report thus warned:

> The production from a wind power generation system will be characterized by limited predictability (contrary to the electricity load [i.e., demand]), and frequent and often strong variations. When the regulation of power plants is characterized by inertia, there can be difficulties in balancing the production of the power plants in relation to the wind power production and the electricity demand. (our translation)[48]

The challenge of integrating 'unpredictable' wind power into an electricity system designed to be run by 'predictable' power generation, however, was only realized and articulated as a pressing material concern after 2000. This articulation was the result of a report commissioned by the DEA in 2001 from involved stakeholders – including wind turbine owners, electric utilities, TSOs, the Organization for Renewable Energy, and experts – to analyse the system problem and come up with solutions to excess electricity.[49] The experts invented the new term *critical excess electricity* to describe new situations in which the electrical load cannot be consumed within a given area and other power plants have difficulty in reducing production further, making export the only option to avoid critical system failure.

Excess electricity is linked to a phenomenon called *technically bounded electricity generation*, which occurs when power plants produce electricity when there is demand for heat. While this electricity generation could compensate for a lack of wind, both heat demand and wind speeds are high during the winter months, which has the net effect of putting too much electricity into the grid. Power plants cannot regulate their power generation down, and since wind turbine owners have priority rights to generate electricity, grid load balance is threatened. There are two types of excess electricity: exportable and critical. While exportable excess electricity can be sold if there is capacity in transmission lines out of the load balancing area, critical excess electricity would cause a system breakdown 'if an actor did not step in and regulate the generation or demand'.[50]

Critical excess electricity defined a completely new situation, since the category did not exist in the regulatory framework of electricity supply law. Consequently, there were no legal provisions that allowed the transmission system operator (TSO) to regulate critical excess electricity.[51] The new systemic concerns were materially based, and provoked different programmes aimed at optimizing the energy system for wind power. For example, research by wind power integration experts demonstrated that the technical problems and integration costs depended on the complementarity of existing technologies in the energy system.[52]

The share of wind-generated electricity continued to increase, reaching 18% in 2005 and 22% in 2010. Consequently, the electricity market for traditional thermal power plants decreased. Power plants that continued to operate improved their technical flexibility by building heat storage capacity and electrical heaters, which enabled them to considerably regulate power generation on short notice in order to more efficiently respond to fluctuating pricing in the electricity market. Wind power was materially becoming an increasingly critical component of the overall energy system assemblage; as a result, pressure was mounting to redesign other components of the broader energy system.

New system concepts and new agencies: from a smart grid to smart energy systems

These material and economic challenges pertaining to wind power integration fostered new agencies centred on the search for concepts to address system problems. The first generation of system concepts, the *smart grid*, began to be promoted in mid-2000 by different consortia of researchers, energy distribution companies and companies in the IT field. Within the smart grid framework, wind power was viewed as an integration challenge pertaining to the electricity sector alone. The main strategy was to solve the integration puzzle by developing a 'smarter' demand side that could 'shave the peak loads' and shift electricity consumption from high to low price periods. Advocates argued that a 'smart' demand side response could be facilitated by an 'intelligent' electricity infrastructure aided by modern information and communication technologies that combine price signals with data from smart meters and intelligent electric appliances. However, Danish research projects in the smart grid tradition (I-Power, Eco-Grid, CITIES, Smart Energy in Cities) concluded that the material effects of the demand side responses were much more limited than anticipated, as they had little economic value. This was partly due to 'over-designed distribution cables' and failed attempts to push for increased electrification of transportation (electric vehicles) and heating (heat pumps) infrastructures.

To address the perceived shortcomings of the smart grid approach, the Energy System Research group at Aalborg University promoted a broader *smart energy system* approach[53] based upon interdisciplinary studies in Coherent Energy and Environmental System Analysis.[54] Instead of viewing the electricity system in isolation, the smart energy system framework redefines the boundaries of the energy system to include infrastructures related to electricity, heating and gas, and end user domains of energy services related to transportation, heating and industry. Only such a radical shift can facilitate the development of sufficiently responsive complementary technologies in a low carbon energy system based on fluctuating wind power. This requires that the TSO strategy must strike a new balance between investing in transmission lines to neighbouring countries (increasingly also with wind power) and investing in integration of the national energy system.

Disruption and redesign of the electricity market

Another critical system component that came under pressure due to the increasing penetration of wind power was the Nordpool electricity market platform that has been operated jointly by the Nordic TSOs since 2001. In spring 2014, the Danish TSO organized a meeting with its stakeholders to discuss serious concerns that the traditional market model was breaking down.[55] Centralized power plants were struggling; between 2008 and 2015, 35% of traditional thermal power plants ceased operating.

The critical problem was that the high share of wind power (with zero marginal cost) drove down prices by shifting the supply curve relative to a given demand, which lowered the prices for all owners of power plants. This phenomenon, the *merit order effect*,[56] occurred whenever large amounts of low marginal cost wind power were added to the electricity market exchange, which had been designed to set prices in a market based on the marginal cost of fossil fuels. As a result, Nordpool set prices that were too low to support the business models of the power plant owners. The Danish TSO was especially concerned about the serious reduction in controllable capacity, as Danish power plants had closed with record speed due to these inadequate prices. A related issue was an expected increase in wind power in neighbouring countries, which was anticipated to drive prices even lower. Increased wind power in neighbouring countries would make it increasingly difficult to export excess wind power, since these areas also would be flooded by excess electricity from their own wind turbines.

In response to these market issues, a working group of stakeholders was assembled to outline a 'Market Model 2.0'. This working group concluded that the traditional market model was not well-suited to securing the flexibility and capacity of services in the new electricity system due to its inability to set accurate prices. The group thus stated a need to 'identify new "system-services" that could be priced so as to make them attractive for market participants'.

This new (revolutionary) 'understanding' spread rapidly throughout the electricity field. For example, at the EWEA (European Wind Energy Association) in Paris in November 2015, a professional from the IEA stated that building a plant to provide base-load energy is an old-fashioned way of thinking about load balancing problems and the economic value of electricity services, since future energy systems will need flexibility more than base-load.[57] Service flexibility was thus being framed as integral to the new approach to load balancing in the smart energy system.

The new climate policy normal: smart energy systems thinking

The smart energy systems concept constitutes a second generation framework that has become the dominant system framework among key actors within the energy system. The stabilization of this new system framing was supported by the 2012 Energy Act, which stipulated that the Danish energy system should become CO_2-neutral by 2050. Achieving this goal was expected to require a drastic increase in wind power capacity, which would make the integration challenge even more critical. Presently, the smart energy system philosophy forms the backbone of energy system concepts and scenarios promoted by actors such as the Danish TSO, the Danish Energy Agency, The Danish Energy Association and the Danish Society of Engineers. A climate council established to advise the government has further advocated the smart energy system approach as a cost-effective way of transitioning to a CO_2-neutral energy system.[58] Other analysts even argue that the transition to a smart energy system is socio-economically competitive with a continuation of the established energy system design. In recent years, the smart energy system approach has thus successfully cultivated a shared system framing that promotes fundamental transitions across the sectors of electricity, heating, gas and transportation with wind power as the core energy production technology.

No implementation plan for increased electrification

However, this increasingly shared system frame does not appear capable of ensuring a strongly coordinated transition. For example, inherent in the smart energy system approach is the need to increase electrification, especially in flexible consumption areas that are able to decrease consumption when wind speed is low and increase consumption when power production is high. An area with high potential for increased electrification is district heating, where large heat pumps are able to convert electricity to hot water very efficiently. Current investments are nevertheless directed towards biomass boilers, given tax advantages (zero taxation) over electrification (high taxation). The present Minister of Energy has refused to change the tax structure because during the election campaign those currently in office had promised not to increase taxes; moreover, reducing the electricity tax would negatively impact the public budget. This political stalemate in relation to tax reforms has been strongly criticized by the Danish district heating association, which has already perceived future changes in district heating due to the integration of wind power:

> We destroy our opportunity to take advantage of this situation and seize the production of large heat pumps because of the electricity tax. It is pure nonsense, says Kim Mortensen from the Danish District Heating Association. He can only say that rather than the efficient use of the state-sponsored wind energy in large heat pumps, then for example, Norwegians welcome the fact that the average export price of wind power is down to about 157 kr. per MWh. On the other hand, the Danes have had to import at a cost of nearly 212 kr. per MWh when the wind turbines have stood still, and the Norwegians have turned on the water power.[59]

This lack of coordination has been criticized by Kathrine Richardson, former chairman for the Danish Climate Commission (2007–10) who strongly warned the centre-right government not to reduce ambitions related to increased electrification of the demand side where technologies already exist (heat pumps, electric vehicles).[60] Despite the lack of coordination in some current investment patterns, the diffusion of the smart energy system concept as a shared system frame illustrates that wind-generated electricity is defining a new material reality for the system as whole. This new reality has now been established as a 'hard fact' among key actors of the energy system.

Summary of assemblage shift 2: materiality strikes back and intensifies the need to reconfigure the energy assemblage

In 2000, the stronger material presence of wind power began to have increasingly obvious repercussions on the hitherto well-functioning – and since 2001, market-based – electricity system. The fluctuating nature of wind power generated critical mismatches between technological, market and regulatory elements linked to the established electricity generation assemblage. The material presence of wind power fractured old understandings of the system and created a need to frame new (ontological) realities. New agencies were cultivated within the energy system assemblage and they began to construct new system vocabularies.

Now, it is no longer considered possible to discuss wind power integration without viewing the energy system as a whole. From this new material reality, the smart energy system concept has emerged as one response, and wind power has been framed in such a way that wind power development and energy system development have become inseparable. This new framing has shifted perceptions of wind power mainly as a system problem to being framed as the central organizing component in the future low carbon energy system. For example, the existing infrastructures in district heating and gas have been creatively assigned new roles, and are to be reconfigured as assets for a wind-powered low carbon energy system. Denmark did not originally invest in gas infrastructure and co-generated district heating to facilitate the integration of wind power, but has benefited from lucky timing.[61] Countries such as Germany, the UK and China do not have existing assets like district heating infrastructures that can become relatively cheap assets for integrating large shares of wind power.

The smart energy systems framing suggests the need for a fundamental rebuilding of the energy system assemblage, not only with new technologies but also with new regulatory strategies and market arrangements able to facilitate a commercially based, efficient low carbon energy system.

Assemblage shift 3: shifting political coalitions and the new material grounding of the industrial cluster in policy

Under the green realism campaign, the right wing government elected in 2015 problematized the ambitions, speed, and directions of the energy system transition outlined in the 2012 Energy Act with reference to socio-economic considerations and the burdens of higher energy prices on industrial competitiveness. Danish energy and climate policies have never resulted from ambitions linked to a stable and strong national consensus in Parliament. While there have been compromises with broad political support, the ambitions in these compromises have always been driven by a coalition of the left wing, social liberals, and the social democrats. The cocktail of renewable energy and new jobs has been important for the left wing and social liberals since the 1970s, and the social democrats joined wholeheartedly in the mid-1980s. Since then, the coalition has strongly supported renewable policies and has had high ambitions to challenge the existing energy sector and use strong policy instruments (e.g., the energy policy goals from 1981, wind power market stimulation, and coercive regulation of electric utilities). In opposition to this political framing of the energy system assemblage, a centre-right coalition comprised of liberals, conservatives and the Danish people's party, has always been sceptical about the need for radical new energy policies, the promotion of renewable energy, and coercive regulation of the energy sector.

New allies and a new material grounding of the industrial cluster in policy

The relative influence of these competing framings of the energy system assemblage, however, is also rooted in the extra-parliamentary socio-technical collective of actors from grassroots and industry, science, and the media. Ever since the 1970s, NGOs and grassroots movements with direct connections to left wing parties have been active in the public arena. They have made allies with prominent Danish researchers who have used their expertise and academic positions at universities to strengthen and legitimize their claims. Similarly the pro-nuclear and anti-wind power movements formed a socio-technical collective by joining forces with prominent researchers in physics, engineering and economics, and were supported by the media in newspaper editorials. In the 1970s, the Danish Industry Association supported nuclear power, not only because it could produce cheap energy, but also because of the potential for Danish industry to help build nuclear power plants and thus upgrade the scientific and technological capacities of companies and research institutions.[62] Political framings of the energy system assemblage are thus highly dependent on strong alliances with extra-parliamentary agency networks.

An outcome of the ongoing reconfiguration of the energy system assemblage has been the development of a new and increasingly active collective of business and industry actors involved in the production of green energy technologies such as wind power, biogas and district heating. Energy system-related green tech companies focused on wind power, district heating and automated regulation systems have been growing rapidly and now constitute a high value sector of the economy. Since 2005, the collective formed by these green tech industries has been increasingly promoting itself using industrial cluster arguments cultivated by left wing governments during the 1980s and 1990s concerning the future employment and export potential of a wind-based transition of the energy system.

This collective has made it increasingly difficult for right wing governments to speak 'on behalf of industry', when criticizing governmental support of the energy system transition. Today, the tables have turned, and green realism is facing criticism from industry. This criticism is not only voiced by the usual suspects from the wind turbine industry, but from an increasingly broad range of industry actors who were initially sceptical about the low carbon energy system, but now see it as a fundamental part of the future based on revenues from new goods and services. These new critics exemplify a tipping point-type shift in the energy assemblage, the cumulative effect of many small and gradual changes that have strengthened business engagement in and commitment to the energy system transition. Many business actors seem to have completely shifted position in favour of the transition.

This new collective of industrial actors argues that numerous new business opportunities are associated with the transition to a new low carbon energy system, but that these business opportunities rely on stable boundary conditions for the market arrangements that are going to finance them. Especially after the 2012 Energy Act, business actors shifted their framing of future investment opportunities towards the green, low carbon renewable energy system. For instance, commenting on green realism, the CEO of Grundfos, the world's largest energy-efficient water circulation company, stated, 'Too bad the ambitions have been lowered on environment and climate – and the timing is so bad before COP 21'.[63] Likewise, the CEO of Siemens Wind, which has 10,000 Danish employees, argued that Denmark should not reduce its offshore wind power ambitions: 'We would like to have R&D and production jobs in Denmark, but Denmark must think green and global.'[64] Former EU Climate Commissioner Connie Hedegaard joined this public debate, stating, 'We are repeating Fogh's mistakes', and pointing out that Denmark is losing industrial leadership positions associated with the transition to the green economy while other countries are accelerating their investments.[65]

The strategic industrial cluster thinking is also supported by the labour union. Their newly established green think tank reported, 'There are up to 55,000 new jobs in green transition'.[66] The green think tank is headed by a former head of the Ministry of Finance during the 1990s who has been a key spokesperson for green transition policies. He later became CEO of the state-owned energy company DONG, where in 2006 he announced a bold initiative: by 2030, 85% of DONG's power generation will be from renewable resources (most notably wind), and 15% from fossil fuels, a dramatic shift from the 2006 proportion of 85% from fossil fuels and 15% from renewable resources. DONG has since become a world leading developer and operator of offshore wind power, thereby becoming a new powerful spokesperson for the continuation and acceleration of changes in the assemblage to accommodate the low carbon energy system. Importantly, the new engineering and business competencies in offshore wind power stem from the series of coercive political acts since the first 100 MW agreement in 1985 that have incentivized Danish electrical utilities to invest in on- and offshore wind power. Coercive regulations forced them to do things that they did not like, but the competences they gained in the process now form the basis for their leading global position.

Summary of assemblage shift 3: homogenous industrial opposition to green realism

The gradual materialization of clean-tech based revenue streams among many industrial companies has fostered rather homogenous industrial opposition to green realism. This new clean-tech voice is materially anchored in interests, competences, and expectations related to new future business opportunities for clean-tech products and services.[67] In short, the identities of many agencies in the Danish industry have been transformed due to their increased involvement with the gradual transformation of the energy technology assemblage.

Conclusion: shifting assemblages where the present government is out of sync with materially-anchored concerns and interests in the present state of the assemblage

In this chapter, we have used the relational assemblage perspective to provide a kaleidoscopic picture of some diverse shifts in the Danish transition towards a low carbon energy system. The socio-technical assemblage is not a fixed entity, but an 'open building site under constant construction' made from *new relational constellations* of heterogeneous actors and intermediaries that shape, contest, and define each other's technologies, cognitions, interests, and identities.

An important thread in the Danish transition has been the competing networks of the assemblage associated with the social-democratic and centre-right dominated governments, which have used competing valuation frames as reference points to legitimize their energy and climate policies. Another important thread has been the material dimension associated with the penetration of wind power in the electricity system and revenue streams from clean-tech products and services.

The materiality dimension has enabled us to see that the green realism approach promoted by the centre-right government has been confronted with new agencies that respond to new material and discursive realities. These realities involve both a new system concept (i.e., the smart energy system) aimed at solving the material systemic problems of wind power integration, and include new business opportunities and industrial transformation associated with investing in the goods and services that will comprise the future low carbon energy system. The agencies involved in the reconfiguration of the Danish energy system assemblage did not come from outside contexts or structures. New agencies were assembled as embedded actors struggled to cope with difficulties and tensions associated with their own socio-material engagements in the system. Since 2000, this interrelatedness has been transformed from being framed mainly as a 'system problem' or 'system cost' to being framed as a source of 'industrial innovation' and as a driver of a societally desirable system transition.

In addition, the smart energy system concept seems robust, as it is associated with a new climate policy normal. Indeed, major actors in the energy assemblage (TSO, IDA, DEA, energy analysts, engineering consultancies) are converging on a framework of the future low carbon energy system with wind power as the central technology. Even if the Danish centre-right government with its focus on green realism once again seeks to roll back or even stop climate policies, the increasing socio-material momentum in the assemblage is making it more and more difficult for them to do so.

Wind power has been transformed from 'white noise' with no meaning and relevance for the broader configuration of the system assemblage into the centrepiece of the new climate policy normal, understood as broad acceptance of a transition towards a low carbon energy system by 2050 – a renewable energy system based on wind power. Wind power is no longer a hippie thing or a relic from the past, but has become a valued technology that can save society from the damage caused by CO_2 emissions from the fossil fuel-based energy system. Wind power has also become a 'material irreversibility' that has triggered new concerns and provoked actors to construct new agencies. Controversies over next steps characterize the current situation as Denmark heads toward achieving a low carbon energy system by 2050. The smart energy system will become a battlefield as diverse actors seek to defend their existing positions; but more than likely, green realism will delay rather than derail the further transition towards a wind-oriented reconfiguration of the energy system assemblage.

The key objective of our analysis has been to understand how the Danish energy technology assemblage was reconfigured through the construction of new agencies. In particular, we have shown how the process of transformation has been modulated through relational interactions and tensions among agencies and their valuation frames, and how these processes have enabled *actors to re-frame themselves and their actions* through new socio-material concepts and narratives that ascribe new roles to individual components of the system. The assemblage perspective has enabled us to give this account from an 'open building site' by capturing the incompleteness, multiplicity and struggle involved when emergent networks of distributed agencies seek to mobilize and enrol new allies in order to stabilize their sociotechnical worlds.

From an agency-based assemblage perspective, renewable energy transitions are no different from former historical societal transitions associated with centralized electricity generation, sanitation, and mobility realized with a particular transportation system based upon fossil fuels in individual automobiles.

Notes

1 We thank Henrik Bach Mortensen, Peter Karnøe's industrial PhD student, for creating Figure 26.1, finding important data, and providing helpful comments on this manuscript.

2 http://energinet.dk/DA/El/Nyheder/Sider/Dansk-vindstroem-slaar-igen-rekord-42-procent.aspx.

3 German Watch, https://germanwatch.org/en/download/10409.pdf.

4 Peter Karnøe, 'Large-Scale Wind Power Penetration – Breaking Up and Re-mixing Politics, Technologies, and Markets', *La Revue de l'Énergie* 64, no. 611 (2013).

5 A. Smith, and A. Stirling, 'Moving Outside or Inside? Objectification and Reflexivity in the Governance of Socio-Technical Systems', *Journal of Environmental Policy and Planning* 9, no. 3–4 (2007): 351–373; J. S. Jensen, 'Framing of Regimes and Transition Strategies: An Application to Housing Construction in Denmark', *Environmental Innovation and Societal Transition* 4 (2012): 51–62.

6 T. Mitchell, *Carbon Democracy: Political Power in the Age of Oil* (London and New York: Verso, 2011); Catherine Mitchell, *The Political Economy of Sustainable Energy* (London: Palgrave MacMillan, 2010); F. W. Geels, 'Technological Transitions as Evolutionary Reconfiguration Processes: A Multi-level Perspective and a Case-study', *Research Policy* 31, no. 8–9 (2002): 1257–1274; G. C. Unruh, 'Understanding Carbon Lock-in', *Energy Policy* 28, no. 12 (2000): 817–830.

7 W. E. Bijker, T. P. Hughes, and T. J. Pinch, *The Social Construction of Technological Systems* (London: MIT Press, 1987); M. Callon, 'Some Elements of a Sociology of Translation: Domestication of the Scallops and the Fishermen of St Brieuc Bay', in *Power, Action and Belief: A New Sociology of Knowledge?*, ed. J. Law (London: Routledge, 1986), 196–223; W. Bijker, and J. Law, *Shaping Technology/Building Society: Studies in Sociotechnical Change* (London: MIT Press, 1992).

8 M. Callon, *The Laws of the Markets* (Oxford: Wiley-Blackwell, 1998); M. Callon, F. Muniesa, and Y. Milo, eds, 'Introduction', in *Market Devices* (Oxford: Wiley-Blackwell, 2007); N. Fligstein, *The Architecture of Markets* (Princeton, NJ: Princeton University Press, 2001).

9 R. Garud, and P. Karnøe. 'Bricolage versus Breakthrough: Distributed and Embedded Agency in Technology Entrepreneurship', *Research Policy* 32 no. 2 (2003): 277–300.

10 M. Callon, 'Techno-economic Networks and Irreversibility', in *A Sociology of Monsters? Essays on Power, Technology, and Domination*, ed. J. Law (London: Routledge 1991), 132–161; J. Law, *After Method: Mess in Social Science Research* (London: Routledge, 2004); B. Latour, *Reassembling the Social: An Introduction to Actor-Network-Theory* (Oxford: Oxford University Press, 2005).

11 Callon, 'Techno-economic Networks and Irreversibility'; J. Markard, S. Wirth, and B. Truffer, 'Institutional Dynamics and Technology Legitimacy: A Framework and a Case Study on Biogas Technology', *Research Policy* 45, no. 1 (2016): 330–344; S. H. Kim, and S. Jasanoff, 'Containing the Atom: Sociotechnical Imaginaries and Nuclear Power in the United States and South Korea', *Minerva* 47 no. 2 (2009): 119–146.

12 M. Callon, 'What Does it Mean to Say that Economics is Performative', in *Do Economists Make Markets?*, ed. D. MacKenzie, F. Muniesa, and L. Siu (Princeton, NJ: Princeton University Press, 2008).

13 T. P. Hughes, *Networks of Power: Electrification in Western Society 1880–1930* (Washington, DC: Hopkins University Press, 1983).

14 R. Garud, and P. Karnøe. 'Bricolage versus Breakthrough: Distributed and Embedded Agency in Technology Entrepreneurship', *Research Policy* 32 no. 2 (2003): 277–300.

15 Unruh, 'Understanding Carbon Lock-in'; Hughes, *Networks of Power*.
16 N. Fligstein, and D. McAdam, 'Toward a General Theory of Strategic Action Fields', *Sociological Theory* 29, no. 1 (2011): 1–26.
17 L. Suchman, *Plans and Situated Action: The Problem of Human-machine Communication* (Cambridge: Cambridge University Press, 1987).
18 Law, *After Method*, 41–42.
19 Law, *After Method*; Latour, *Pandoras Hope – Essays on the Reality of Science Studies* (Cambridge MA: Harvard University Press).
20 Law, *After Method*; R. Garud, A. Kumaraswamy, and P. Karnøe, 'Path Creation or Path Dependence', *Journal of Management Studies* 47, no. 4 (2010): 760–774; Peter Karnøe, and R. Garud, 'Path Creation: Co-creation of Heterogeneous Resources in the Emergence of the Danish Wind Turbine Cluster', *European Planning Studies* 20, no. 5 (2012): 733–752.
21 M. Rüdiger, *Energi i forandring* (London: DONG Energy, 2011), 43.
22 Rüdiger, *Energi i forandring*, 45.
23 Kim, and Jasanoff, 'Containing the Atom'; Hvidtfelt K. Nielsen, 'Tilting at Windmills: On Actor-Worlds, Socio-Logics, and Techno-Economic Networks of Wind Power, 1974–1999' (PhD diss., Aarhus University, 2001).
24 T. Pinch, 'Moments in the Valuation of Sound: The Early History of Synthesizers', in *Moments of Valuation – Exploring Sites of Dissonance*, A. Antal, M. Hutter, and D. Stark (eds), (Oxford: Oxford University Press, 2015).
25 N. I. Meyer, 'Politik og Vedvarende Energi', in *Vedvarende Energi i Danmark 1975–2000* (Renewable Energy in Denmark 1975–2000), ed. E. Beuse et al. (Aarhus, Denmark: Organization for Renewable Energy, 2000), 75–110.
26 Konrad Jensen, *Mænd i Modvind* (Copenhagen: Børsens Forlag, 2003).
27 Rinie van Est, *Winds of Change: A Comparative Study on the Politics of Wind Energy Innovation in California and Denmark* (New York: International Books, 1999); E. Beuse, et al. (eds), *Vedvarende Energi i Danmark 1975–2000* (Renewable Energy in Denmark 1975–2000), pp. 75–110 (Aarhus: Organization for Renewable Energy), available at www.ove.org.
28 P. Karnøe, and A. Buchhorn, 'Path Creation Dynamics and Winds of Change in Denmark', in *Promoting Sustainable Electricity in Europe*, edited by W. Lafferty et al. (Cheltenham: Edgar A. Elgar, 2008); M. Cohen, J. G. March, and J. P. Olsen, 'Garbage Can Model of Organizational Choice', *Administrative Science Quarterly* 17, no. 1 (1972): 1–25.
29 Peter Karnøe, *Danish Wind Turbine Industry – A Surprising International Success: On Innovations, Industrial Development and Technology Policy* (Copenhagen: Samfundslitteratur, 1991).
30 *Energikoncept 2030 – baggrundsrapport*, Energinet.dk, 2015, www.energinet.dk.
31 A. B. Antal, M. Hutter, and D. Stark, *Moments of Valuation: Exploring Sites of Dissonance*, 1st edn. (Oxford: Oxford University Press, 2015); D. Beunza, and R. Garud, 'Calculators, Lemmings or Frame-makers? The Intermediary Role of Security Analysts', *The Sociological Review* 55 (2007): 13–39; H. Mortensen, 'Calculating Wind Energy as Cost Burden or Cost Leader in Denmark' (working paper, Department of Development and Planning, Aalborg University, Aalborg, Denmark, 2015).
32 N. Nørgaard, and J. Tornbjerg, 'Finanslov: Sort dag for miljøet', *Politiken*, January 30, 2002, 7; J. Tornbjerg, 'Miljøet redder finansloven', *Politiken*, March 20, 2002, 4.
33 K. Vogt-Nielsen, and L. Husmer, *Dansk Miljøpolitik 2002: En analyse af VK-regeringens første måneder set med miljøbriller* (Copenhagen, Denmark: Center for Alternativ Samfundsanalyse, 2002).
34 K. Asdal, 'Enacting Things through Numbers. Taking Nature into Accounting', *Geoforum* 38, no. 1 (2008): 123–132.
35 Ritzau Note, *Information Newspaper*, 27 January 2011.
36 Ritzau Note, *Politiken Newspaper*, 2 November 2011.
37 D. Stark, *The Sense of Dissonance: Accounts of Worth in Economic Life* (Princeton, NJ: Princeton University Press, 2009).
38 Danish Economic Council, *Vismandsrapport, Dansk Økonomi*, Forår 2002, De Økonomiske Råd (Copenhagen: Danish Economic Council, 2002).
39 Only one of these two new offshore wind farms stipulated in the compromise of 2004 came online after extensive political delays, www.ens.dk/undergrund-forsyning/vedvarende-energi/vindkraft-vindmoller/havvindmoller/idriftsatte-parker-nye#Horns Rev II.
40 Authors' translation of the original quote from Danish: 'jeg har meget længe hørt til dem der var sådan lidt i tvivl om det der med klimaet … der er nok mange hos os der har været lidt forsigtige, for ikke at

sige fodslæbende i alt det her … Jamen, det var måske også forkert … man kan nok sige vi har ikke været den energipolitiske avantgarde.' Full text of the talk can be found in this source: http://ing.dk/artikel/fakta-laes-statsministerens-miljotale-93417-17.11.2008, retrieved June 17, 2016.

41 KEBMIN, Energiaftalen i korte træk, Klima-, Energi- og bygningsministeriet, 2002, 1.

42 H. P. Dejgaard, 'Miljø og klima rammes særligt hårdt i den store nedskæring af Danida Budget', *Ingeniøren*, 4 October 2015; D. Rehling, 'Grønt set er finansloven sort', *Information Newspaper*, 19 January 2015; 'Regeringen skærer 1,5 milliarder mere på ulandsbistand', *Information Newspaper*, 19 November 2015.

43 PSO stands for Public Service Obligation and PSO is in the context of EU law, that allows a governing authority to offer a special subsidy for a public service to a winning company, in this case renewable energy services. The PSO is offered because the so-called free-market cannot create incentives for the socially desired development. However, the so-called free market for fossil, nuclear and renewable energy is better seen as the political economy of energy markets as all fuels and technologies exist and are favoured by various state subsidies and regulations.

44 B. V. Mathiesen, H. Lund, D. Connolly, et al., 'Smart Energy Systems for Coherent 100% Renewable Energy and Transport Solutions', *Applied Energy* 145 (2015): 139–154.

45 Bijker, Hughes, and Pinch, *The Social Construction of Technological Systems.*

46 Margrethe Vestager, EU Commissioner for Competition, interview on Danish Radio, 22 January 2016, P1 Orientering, 9 minutes in.

47 DEFU, Vindkraft i elsystemet, Energiministriets og Elværkernes Vindkraftprogram, rapport EEV 83–02, September 1983, 2a.

48 Ibid., 2.3.

49 Danish Energy Agency, Omkostninger ved CO_2-reduktion for udvalgte tiltag – Midtvejsrapport, Energistyrelsen, May 2001.

50 Ibid., 8.

51 Ibid.

52 H. Lund, and E. Münster, 'Management of Surplus Electricity Production from a Fluctuating Renewable-energy Source', *Applied Energy* 76, no. 1–3 (2003): 65–74.

53 Mathiesen et al., 'Smart Energy Systems'; H. Lund, *Renewable Energy Systems: A Smart Energy Systems Approach to the Choice and Modeling of 100% Renewable Solutions*, 2nd ed. (Cambridge, MA: Academic Press, 2014).

54 H. Lund, et al., *Coherent Energy and Environmental System Analysis* (Aalborg, Denmark: Aalborg University, 2011).

55 www.energinet.dk/DA/El/Engrosmarked/Ny%20markedsmodel/Sider/default.aspx.

56 Pöyry, *Wind Energy and Electricity Prices: Exploring the 'Merit Order Effect'* (Brussels: European Wind Energy Association, 2010).

57 Henrik Mortensen, interview with author, 21 January 2010.

58 Klimarådet. 'Omstilling med omtanke – Status og udfordringer for dansk klimapolitik', Klimarådet.dk 2015.

59 'Dansk Fjernvarme: Værdien af vindmøllestrøm kan øges med milliarder med varmepumper', *Energy Supply, Online Newspaper*, June 9, 2015.

60 T. Færgeman, 'Der skal mere strøm på varmen', *Politiken Newspaper*, 28 September 2015, 9.

61 P. Pierson, 'Increasing Returns, Path Dependence, and the Study of Politics', *The American Political Science Review* 94, no. 2 (2000): 251–267.

62 P. Karnøe, and S. Møller, 'En analyse af udviklingsbetingelserne for udvalget energiteknologier – en magt og teknologianalyse' (master's thesis, Copenhagen University, 1985); H. Nielsen, K. Petersen, and Hans Siggaard Jensen, *Til Samfundets Tarv – Forskningscenter Risø's historie* (Risø: Danmarks Tekniske Universitet, Risø Nationallaboratoriet for Bæredygtig Energi, 1998), 560.

63 T. Færgeman, interview with CEO Mads Nipper, *Politiken Newspaper*, 22 October 2015.

64 T. Færgeman, interview with Siemens Global off-shore director, *Politiken Newspaper*, 25 October 2015.

65 D. Saietz, *Politiken*, 4 November 2015.

66 Dansk Energi, '3F: Op til 55.000 arbejdspladser i grøn omstilling', Dansk Energi, 2015, www.danskenergi.dk/Aktuelt/Arkiv/2015/September/15_09_24B.aspx.

67 M. Borup, et al., 'The Sociology of Expectations in Science and Technology', *Technology Analysis & Strategic Management* 18, nos. 3/4, July–September 2006, 285–298.

27

Twins of 1713

Energy security and sustainability in Germany

R. Andreas Kraemer

Introduction, background and context

Energy security and sustainability are twins, born in 1713. As key concepts they were explained and made practicable through management rules in the seminal 'Sylvicultura oeconomica' by Hans Carl von Carlowitz, and have guided policy-making ever since.[1] As conceptual foundations of Germany's present-day '*Energiewende*' – the country's green energy shift away from fossil and nuclear energy towards renewable energy supply – energy security and sustainability are topical and ever more relevant for many countries.

The thinking in Germany, or rather its precursor states, is much broader and predates the modern concept of energy security as it was developed notably in the OECD in response to the politically induced oil crises of the 1970s. Furthermore, energy security is part of a broader agenda of rational resource management, including the principles of sustainability.[2]

Germany's energy endowments

Germany is well endowed, to varying degrees, with all forms of energy, starting with relatively abundant forest cover over about one-third of its territory. The potential for hydropower is dispersed: there are few options for large impounding dams, but thousands of small dams exist, mostly in mountain regions of the South and the South-East, as do run-of-river plants integrated into weirs to increase river depth and assist navigation along the large rivers. Windmills were prominent as they were in other developed countries, but gave way to fuel-powered engines and later electric motors. Modern-day wind turbines for electricity generation harvest a wind resource of medium quality (if compared to other countries), notably in the North and the East of the country, but also increasingly on the high plains and ridges of the South. The strength of the sun, the insolation, in Germany is similar to that of Alaska, and yet solar energy, notably photovoltaic cells, have been installed on millions of rooftops as well as in large arrays on agricultural and other land. The farming sector produces enough waste for a dynamic bio-energy industry to develop and thrive. Although Germany has no notable present-day volcanic activity, there is reasonable potential for geo-thermal energy in some areas where hot rock is found at accessible depth, and heat-pumps can tap near-surface strata for both heating and cooling. Other than offshore wind power, which is being developed, the potential for ocean or marine energy is limited because of the physical characteristics of Germany's coasts. This energy endowment is enough, models show, to provide 100% clean, safe, and secure renewable energy even for a densely populated and heavily industrialized country like Germany.

During industrialization, Germany used up a large part of the stock of fossil energy on its territory. Remaining hard coal is in deep mines and uneconomic to extract. Surface mining of soft brown coal continues in various parts of Northern Germany, but is now threatened by competition from ever cheaper renewable power. Very small oil resources have been exploited in the past, and what is left would be uneconomic to extract. Natural gas, a fossil form of the greenhouse gas methane the Germans call *Erdgas* (earth gas), is found in some locations, mostly in Northern Germany, but the quantities extracted are declining fast. In fact, the remaining gas fields may be used better to store imported gas as a strategic reserve. There are also small, almost symbolic, uranium mines of local historical significance. Uranium mining was uneconomic and ended in the 1990s after German unification.

The most abundant energy resource in Germany is perhaps ingenuity, with high standards in the natural sciences, engineering and materials sciences, and effective training for technical professions and trades. The German national innovation system, if that concept is still valid for any country deeply embedded in the European Union, is generally good at continuous incremental improvement, rather than disruptive innovations of the kind associated with Silicon Valley. This innovation system, surrounded by a political system that forms ruling coalitions around the political centre and provides for comparatively high levels of policy continuity, permits industries to grow around key technologies and along the value chains. The strength of this system is perhaps demonstrated best, but also perversely in the larger context of energy policy, by the breadth and depth of nuclear technology in Germany, a country that had no significant uranium resources and no nuclear weapons of its own.[3] In spite of this, and in the absence of any sustainable economic rationale, Germany has practically all the technologies needed to manage the full nuclear fuel cycle for nuclear power, and the capacity to avail itself of all technologies needed to build, maintain and operate nuclear weapons. Yet, there are no plans to do so, and the decision to phase-out nuclear power is final; a reversal is politically unimaginable.

The current energy system transformation builds on energy security as a key objective and is motivated by sustainability concerns. It is a 'grass-roots', socially desired and politically directed shift towards energy efficiency, renewable energy supply and energy storage, with a smart electricity grid enabling demand flexibility. It increases energy security by most measures, if not all. It does so in part by creating novel links between the energy and transport systems, between electricity and gas, and between the electricity and the heating systems. These links all contribute to higher system reliability and resilience of the whole.

Contemporary context: global overheating

For years, the acute effects of climate change within Germany were not the main concern, but rather climate-change driven events around the world that might harm political stability in other countries, result in a loss of trade, induce migration, and ultimately cause conflict. Promoting good climate policies abroad was seen as being in Germany's best interest and as good global citizenship. The contribution of global warming to the various crises around the world is generally assumed if not always understood in Germany.

With its place, geographically and politically, at the heart of Europe, with all neighbours being Member States of the European Union (EU)[4], Germany is in a favourable position. Some

EU countries like Belgium, the Netherlands, Britain, or Denmark will likely suffer more from rising sea levels, while others around the Mediterranean will feel stronger effects from changing rainfall patterns.[5] Germany has comparatively strong, well-organized, and efficient government and can respond to emerging threats more effectively than countries with more limited statehood, especially countries outside the EU, from where refugees now come.

Germany is most vulnerable to the effects of climate change along the North Sea and Baltic coasts, but these are not densely populated. However, many houses, businesses, and much transport infrastructure is located along the rivers. Seasonally low flow already forces the occasional shutdown of nuclear plants and other installations. Recent record floods in all large rivers are seen as a consequence of changing climate, with a warmer atmosphere carrying more water and triggering stronger rainfall or snowfall. In time, a partial retreat from vulnerable areas will become necessary, yet there is no sense of urgency now.

Evolution of *Energiesicherheit* in Germany

While fossil energy sources were known from antiquity, their large-scale industrial use only developed in the past 200 years. Previously, wood had been the most important source of storable and transportable energy, not only in domestic heating and small trade activities such as blacksmithing but also in the industry of the time.

Since the Middle Ages, the region that is known as Germany today experienced episodes of energy scarcity with the consequence of industries collapsing and regimes weakening. The typical pattern was the harvesting of old-growth forest for mining purposes, without regard for the need to replant forests and consider the speed of tree growth to assess future energy availability. In some regions, hydropower was used, for instance for driving air bellows in metal processing, but that resource was always limited such that excessive energy needs were met by harvesting ever larger parts of the energy stock represented by standing trees.

One example is the history of the Harz mountains in the centre of Northern Germany. Mining and metal processing on an industrial scale developed since the Middle Ages, resulting in unsustainable pressure on the forest very early on. The Harz mountains became increasingly denuded, and many of the foresters and miners moved on, eastwards and southwards into other areas of Saxony.

The story is similar to that of the Eifel mountain region in the West of Germany, near Luxemburg, which has been mined since antiquity, through the Bronze and Iron Ages, and into Roman times. Its scale was relatively small, and mining did not change the landscape much. However, that changed in the 17th and 18th centuries, during which the hills were cleared of trees and not much more than a grass-covered highland was left. The foresters and miners mostly moved away.

By now the pattern was recognized, and stronger states with longer time-horizons, better government and administration had emerged, so that the dangers of unsustainable energy and resource management could be understood and guarded against through policy, law, and management rules.

Wood and its sustainable resource management (1713)

In 1713, Hans Carl von Carlowitz first spelled out the foundations for sustainability as a concept and sustainable development as an objective by establishing principles and practical rules for forest management and the rational use of forest resources. He hailed from a family of forest masters, was a lawyer and public administrator by training, and had travelled widely to study industry, notably mining, as well as governmental systems and forestry practices.

As a long-time mining administrator, he was given responsibility, in 1711, for the mining industry at the court in Freiberg, a mining town in Saxony known for mining silver. Mining was the most important industry at the time, employing a very large work force, because metals were needed in the manufacture of weapons and minting of money. In modern terms, the defence sector and national security, and the money supply and thus economic stability were all directly dependent on mining output, and so energy security was at the heart of policy considerations.

The most pressing threat to the mining industry of the time was the supply of timber, which was used not only for structural purposes inside the mines and at the mine heads, but also for setting up hot fires underground in the mines at the lodes with the metal ores. The heat would induce splitting in the rock to make the ore easier to mine. Wood was also used to fuel fires that needed to be hot enough for metal to melt from the ores. Mining and metal processing were similarly dependent on a continuous supply of good timber, wood that would burn hot enough to reach smelting point.

The seminal work 'Sylvicultura oeconomica' is the essence of von Carlowitz's understanding of forestry, which he understood to be about much more than the harvesting of timber. He laid out the principles of sustainable management of natural resources, addressed the optimal and sustainable management of forests and the rational use of forest products under the threat of overexploitation. On the basis of the understanding and management rules provided by von Carlowitz, reforestation of denuded areas and protection of forests were so successful that Germany could maintain mobility and fuel a large part of its vehicle fleet from wood gas derived from special burners mounted on cars and trucks under fuel oil embargo during World War II.

Von Carlowitz's legacy includes a present-day legal framework for forestry in Germany on the basis of the 1975 Federal Law for the Protection of the Forest and the Promotion of Forestry (*Gesetz zur Erhaltung des Waldes und zur Förderung der Forstwirtschaft (Bundeswaldgesetz)*). The stated purpose of the law is primarily:

> To protect and maintain the forest, to enlarge it as necessary, and to safeguard its rational management in a sustainable manner, and to do so in view of the forest's economic use (use function) and its importance for the natural environment, notably for sustaining its ability to provide ecosystem services, for the climate, water resources, air quality, soil fertility, the landscape, agriculture and infrastructure, and the recreation of the population (protection and recreation function). (own translation)

As a consequence of the law and the philosophy behind it, about one-third of land in Germany is forest, and the share of forest in total land use is rising, if slowly. The protection of the nation's forests is embedded deeply in German culture. When long-range air pollution threatened the forests' survival in the 1970s, '*Waldsterben*', forest die-back, became a rallying cry for the environmental movement and motivator for strong national and EU policies on air pollution control, which eventually achieved a turn-around in pollution and forest health. The impact of those policies was to force notably coal-fired power stations to retrofit with flue-gas desulphurization and controls to reduce other pollutants, including nitrogen oxides.

Harvest of eons: coal, the idea of autarchy, and its demise

Limitations to the availability and supply of wood in Germany would eventually have resulted in industrial decline, had it not been for the rise of coal as a consequence of James Watt's invention of the steam engine. The steam engine allowed water to be pumped from deeper mines, so that more hard coal of higher quality could be extracted. Hard coal is relatively easy and cheap to transport and store, and could thus be brought to locations that were otherwise attractive for industry.[6] Coal differs from wood in that it is not a growing resource but a fossil stock of energy built up over millions of years. As such, it held the promise of abundance, of economic growth without limits, at least as long as there was no end to coal supplies in sight, and the negative consequences of its use could be ignored.

Germany had a number of hard-coal-producing areas, where the last pits are now being closed, and still has several lignite-mining areas, where mines are co-located with lignite-burning power plants, as electricity is easier and cheaper to transport than heavy lignite coal. The timing of phasing out lignite mining is currently being debated, with end dates falling maybe in the 2035 to 2045 range. The political motivation for the coal phase-out is climate protection, but cost reductions in renewable energies and advances in information and communication technologies also make coal-fired power increasingly uncompetitive.

Past energy security concerns that would argue in favour of keeping lignite mines running are losing weight as a stable electricity supply from a portfolio of renewable energy systems and storage in a smart grid becomes ever more credible, especially as the European energy grids are increasingly integrated.

Its reserves of first cheap-to-mine hard coal close to the surface and later deeper hard coal gave Germany, for a while, the option of achieving autarchy in energy supply as well as a supply of carbon-molecules as feed-stock for the chemical industry. Until the end of World War II and the subsequent restructuring of the chemical industry, Germany built this industry on carbochemistry using coal (with ring-shaped molecules containing carbon) as feed stock.

Coal liquefaction, deriving liquids from coal using the Bergius (1913) and Fischer-Tropsch (1925) processes, provided an important option for Germany to achieve energy autarchy on the basis of domestic coal resources. Coal gasification was as important for maintaining gas supply for industries and cities in Germany as it was in other countries. The practice ended in Germany when the country switched to mostly imported natural gas and retired its gas works.

After 1945, the chemical industry in West Germany changed from carbochemistry to petrochemical processes using long-chained molecules containing carbon (and more hydrogen than in coal-derived molecules); the switch in East Germany with its lignite mines occurred after the fall of the Berlin Wall in 1989, in the context of a rapidly shrinking, even collapsing industry.

Collective energy security: oil and sea-lanes (post-1945)

The German Federal Republic (founded in 1949) changed its outlook from national energy security (with the option of autarchy) toward collective energy security, or energy policy in the context of the collective security framework provided by the West.

The development of (West) Germany's coal industry was placed under the High Authority of the European Coal and Steel Community (ECSC). The ECSC built on an earlier precedent in the Act of Mannheim (1868) on navigation on the Rhine as a shared river,[7] and laid the foundation for the adoption of the European Economic Community (EEC) and the European Atomic Energy Treaty (Euratom Treaty) in 1957. The 1951 Paris Treaty establishing the ECSC essentially ended national autonomy in energy policy as it was conceived before, not only for West Germany, but also for France, Italy and the three Benelux countries Belgium, Luxembourg, and the Netherlands. These countries, and later all members of the EEC and now the EU increasingly shared political and regulatory oversight over their energy industries in a process that has resulted in the European Energy Community Treaty and the European Energy Union now being developed.

The switch from carbochemistry to petrochemistry resulted in Germany becoming more reliant on the openness and security of sea-lanes (bringing crude oil) and thus served to tie Germany into the Western alliance. The switch was helped by the fact that the locations of major chemical works in Germany were along the Rhine river and its tributaries. Shipping costs on the Rhine were low enough to make the use of oil advantageous in economic terms.

The market share of oil rose until the oil-crises in the 1970s and early 1980s. The 'oil shocks' induced by the Organization of Petroleum Exporting Countries (OPEC) revealed an economic component of energy security that had been neglected in the decades before. The response of the oil-consuming, industrialized countries to OPEC's challenge was coordinated in the OECD and resulted in the establishment of the International Energy Agency (IEA). The IEA was given the mandate to orchestrate the building of strategic oil reserves and coordinate the release of those reserves at times of crisis.

In (West) Germany, the national response was manifold: oil-fired power generation was phased out, nuclear power generation was promoted (until it was stopped again by the nuclear catastrophe at Chernobyl in 1986), and energy efficiency including fuel efficiency in the transport sector was given more emphasis. The promotion of diesel that ultimately resulted in the Volkswagen scandal of 2015, can be seen as a consequence of the desire to make the German economy as fuel efficient as possible, and thus to increase economic energy security as an important factor in the competitive position of Germany as a trading nation.

Gas, the North Sea, and the embrace of Russia (1980–2014)

One of the key elements in reducing Germany's dependence on imported oil was to develop the small domestic gas fields and begin importing from neighbouring countries. The Netherlands, the UK and Norway, which developed oil and gas fields on land and in the North Sea, were close and reliable partners in Western Europe, and they provided additional energy security to Germany. The challenge at the time lay to the East.

From the 1970s, with the beginning of the '*Ostpolitik*' – ushered in by the then Chancellor Willy Brandt of the Social Democratic Party and a cornerstone of German foreign policy ever since – a new approach began to shape relations with the Soviet Union and the countries of the Warsaw Pact. In the 1980s, establishing good trading relations was seen to be key to reducing the risk of armed conflict in Europe, when any conflict at the time would have left Germany as a radioactive wasteland. The Soviet Union had gas but needed technological and financial support to develop the industry and build the pipelines necessary to deliver its gas to markets in Western Europe.[8]

The resulting 'gas for pipelines'-deal between (West) Germany and the Soviet Union resulted in increased diversification of fuels and their sources and thus improved energy security as it was understood at the time, and improved the geopolitical environment for Germany and thus, in a limited way, enhanced national security through energy cooperation. The deal was upheld by both sides through the collapse of the Soviet Union and the re-establishment of Russia, the new independence of some Soviet Republics that would become transit countries, and the political changes in the Warsaw Pact countries, including German unification. It even holds up after Russia's attack on and invasion of (part of) Ukraine in 2014.

Nuclear power and the dead-end military option (1957–1986)

Like the US with the 'Atoms for Peace' programme in the 1950s and together with the other founding members of the Euratom Treaty, Germany was swept along in the illusion that nuclear power would be a safe source of energy and a fountain of peace and prosperity.[9] The Euratom Treaty, Europe's quasi-constitutional commitment to subsidize nuclear power, was signed in March 1957, the year of the nuclear catastrophes in Windscale, UK, and the Mayak reactor near Kyshtym in the Soviet Union.

Plans to build a nuclear reactor in Wyhl in the South-West of Germany near France and Switzerland, aroused local resistance which stopped construction and triggered the development of an organized and increasingly influential anti-nuclear environmental movement. It included scientists and technical experts that could question and contradict the pro-nuclear narratives of the plant's promoters.

The movement found strength in the German tradition of providing energy supply through local utilities, usually controlled by municipalities, and a general suspicion of central government and big business involvement in the sector. Another contributing factor was the fact that Germany is not a nuclear weapons state with two important consequences: the links between the nuclear industry and the military and security policy community were considerably weaker than in other countries, notably neighbouring France, and it was easier to criticize nuclear technology or question the dominant nuclear scientists without being accused of sedition or treason. The political balance was more level than in other countries, and there was significantly more room for open political debate.

The movement established the Institute for Applied Ecology (Öko-Institut) in Freiburg,[10] the first of a cluster of privately initiated, independent policy institutes and think tanks.[11] In 1980, experts from the Öko-Institut published the book *Energiewende – Growth and Prosperity without Oil and Uranium* as a blueprint for the energy transformation that is now underway.[12] The energy system analysis and transformational strategy developed at the time was influenced by Amory Lovins's work at the Rocky Mountain Institute and notably his 1976 article in *Foreign Affairs*,[13] which opened eyes to the wider international implications of domestic energy policy choices.[14]

With the oil crisis in 1980/1981, German energy policy focused on reducing its dependence on imported oil. But the 1986 nuclear catastrophe in Chernobyl, Ukraine (then part of the Soviet Union) focused German minds again on the need to invest in 'anything but nuclear' and develop technologies, regulatory frameworks and political strategies to phase out this dangerous and ultimately uneconomic technology. The ideas in the 1980 book *Energiewende*, and the critical mass attained by the anti-nuclear movement and the community of anti-nuclear scientists and experts were available at this critical moment when German energy policy looked for new orientation.

Feed-in tariff: innovation of an effective policy instrument (1990)

In 1990, a seminal element of the political strategy was adopted: the German Power Feed-in Law (*Stromeinspeisegesetz*) guaranteeing grid access and establishing a feed-in tariff for renewable power. Influential, land-owning and politically conservative owners of hydropower dams in Germany's South obtained this federal law mandating power utilities to buy renewable electricity from them at stable rates. It is the achievement of an early alliance of conservative and progressive or 'green' political forces,[15] and it triggered innovation and business development in renewable energies.

The then newly elected red-green federal government consisting of the Social Democratic and the Green parties, in 1999 and 2000, negotiated a phase-out of nuclear power with the industry and upgraded the *Stromeinspeisegesetz* to become the Renewable Energy Act (*Erneuerbare-Energien-Gesetz* or EEG), the much-copied law that has undergone a number of upgrades and revisions since then.

The negotiated phase-out of 2000 guaranteed nuclear power plant operators residual operating time for each of their plants, roughly in line with their expected technical life and safety records. Newer plants would run longer than older ones. Power sales from the plants were expected to be sufficient to cover depreciation and allow operators to build sufficient reserves for future decommissioning and legacy costs.

It was also agreed that regulators could not demand safety retrofits at the expense of operators, but would have to reimburse the cost of any retrofits. This in effect avoided any 'expropriation in kind' (through regulatory action) by government fiat. Residual running time could be transferred from older plants, which presented the highest risk profile, to new plants, which were regarded as safer, in 'trades' that allowed operators to consolidate operating time for economic gain. The end of the nuclear age in Germany would have come sometime between 2021 and 2023, depending on the rate of power production over the two decades of the phase-out.

The futile attempt to reverse the nuclear phase-out (2010)

Legislative ambush by government on parliament is the best way to describe the ill-fated 2010 decision to delay the closure of Germany's nuclear power plants beyond the agreed phase-out date. It was adopted after the law was presented and pushed through in a very short time and without normal scrutiny by ministerial officials and legislators in parliament. As a consequence, the law was challenged as being unconstitutional for several reasons and may well have threatened the survival of the government of the day. It was ultimately defeated, for good reason, because it broke a long-standing cross-party consensus on nuclear phase-out, formed after the tragedy in Chernobyl in 1986.

At least since the 1970s, when the country was divided in a frozen conflict during the Cold War, nuclear technology and its ostensibly 'civilian' use in electricity generation was controversial in Germany. The controversy evolved into a Great Societal Conflict (*Gesellschaftlicher Grosskonflikt*) that involves large segments of society, and defines not just one generation but several. It called into question the four-sided relationship between citizens and ratepayers, municipalities as democratic local governments engaged in energy services, industry as operators of nuclear plants, and the central or federal government as regulator and promoter of energy policy choices. Pitched battles were fought in legislative chambers, before courts, and in the streets, until a phase-out of nuclear power was agreed between operators and the federal government in 2000. The Grand Societal Conflict subsided. Some mistakenly saw it as resolved, others knew it was dormant and ready to break out again if the phase-out were ever revoked.

What then drove the German parliament in the autumn of 2010 to extend the allowed operating time for the country's nuclear power plants? Lobbying by the plant operators, the four big incumbent utilities, seeking to overturn the nuclear phase-out they agreed to in 2000 is the obvious answer. Extending required pretending the untenable, when the truth was already widely known that nuclear power is too expensive, polluting, wasteful, and dangerous even for a rich, technologically advanced country like Germany.

Germans saw through the false pretences and took to the streets in ever-larger numbers to protest for a continuation and acceleration of the *Energiewende*, Germany's energy transition towards clean, safe, and increasingly cheap renewable energies. The 2010 extension of nuclear running time made no sense in the larger framework of energy policy, economic policy, or technology policy. The vote in the German parliament, the Bundestag, on 28 October 2010 is the most prominent and in many ways surprising outlier in an otherwise smooth process, of envisioning, preparing, planning, initiating and managing a grand transformation of an industrial country away from costly, dirty and dangerous fossil and nuclear power towards a future based on increasingly cheap, clean and safe renewable energies.

Fukushima and the restoration of the nuclear phase-out (2011)

Into 2011, the anti-nuclear demonstrations continued with increasing numbers of participants. The anti-nuclear Green Party was rising in opinion polls, approaching 30 per cent, and threatening to overtake the ruling conservative Christian Democratic Party (CDU) of the Chancellor, Angela Merkel. Among the demonstrators were grandparents carrying grandchildren on their shoulders; they were not young anti-industrialist or anti-capitalist rebels, but from the conservative core of society. Young people, in their teens, had their political awakening in that context.

The CDU faced not just the ignominy of having a high-profile law overturned by the courts or the embarrassment of losing upcoming elections at state (*Land*) or federal level, it faced the erosion of its electoral prospects for a generation. It was clear to most observers that the CDU-led federal government would have no choice but to annul the contested extension of the nuclear power plants' running time.

The tragic nuclear catastrophe at the Fukushima nuclear plant that started on 11 March 2011 and has been unfolding ever since, provided Angela Merkel with a face-saving opportunity for a deft U-turn to restore the nuclear phase-out. She used it without hesitation, and a new nuclear phase-out law was adopted by the German parliament on 31 July 2011.

The 2011 nuclear phase-out law differed from the negotiated nuclear phase-out of 2000 in important nuances. While the earlier agreement fixed the number of residual run-time hours for each plant, and made provisions for them to be transferred among plants to improve the economics and safety of operations, the new law set fixed end-dates for each nuclear plant to shut down. Overall, the plant owners actually gained operating hours, as long as they did not have to interrupt operations for lengthy retrofits. The new law sets incentives to skimp on maintenance and run the plants without interruption.

In practice, because of the low electricity prices on the German market, nuclear plants are unlikely to operate to the deadline in the law, but be taken out of service when the last refuelling would be due. Refuelling incurs large costs, both in the fuel and in the legacy cost of dealing with spent fuel later. In the prevailing market conditions, even existing nuclear power plants are uneconomic to run, even though they still benefit from subsidies and privileges.

Return of autarchy? Home-grown renewable energies (1990–)

Since the 1950s, the German energy economy opened up and was increasingly integrated into the world markets (for oil and coal), the emerging integrated European economy (for traded fossil energy products, gas and later electricity) and with the Soviet Union, later Russia (mainly for gas, but also for other fossil energy carriers). Economic integration and diversification of fossil fuels and their sources were considered viable paths to improve energy supply security, while also producing wider strategic and security-policy benefits.

In the aftermath of the OPEC-induced oil crises of the 1970s and early 1980s, and with a living memory of energy autarchy and local control over energy services, the vision of an *Energiewende* (as in the 1980 book title) rekindled the desire to develop sufficient energy supply capacity at home and in the European internal market, and to do so without relying on fossil energies that harm the climate, or dangerous and costly nuclear power.

Scenarios and modelling have shown that combinations of renewable energies plus storage in a smart grid can supply Germany based on current and projected patterns of demand. The perhaps best-known simulation has been developed by the *Kombikraftwerk* project which documented that a combination of existing renewable plants and storage facilities connected in a virtual power utility can be operated to match a fixed percentage of the grid demand curve. Scenarios then demonstrated how the renewable energy industries can be scaled up, and the percentage of the load curve covered by a larger virtual 'combination power plant' be increased until it reaches 100% (or more) by 2050, with some imports from and exports to neighbouring countries in the EU's internal market.[16]

This simulation does not take account of the potential to reduce energy demand through efficiency, and notably the dynamic efficiency that can be provided by demand flexibility and response, including the possibility that large industrial power users can become 'swing consumers' of electricity and ramp production up and down to match the availability of fluctuating but highly predictable renewable energies.[17] It also does not yet take into account the potential storage becoming available as a consequence of a shift towards electric mobility, which would reduce the current dependence of the transport sector on liquid fossil fuels and thus enhance energy security.

Grand co-transformation of power, gas, heat and transport

The German energy transformation has already created interesting new links between the electricity and the gas market. At times when renewable energy is abundant and the spot-market price of electricity is low (or even negative), operators of gas heating systems switch to electric resistor heating units. The conversion of noble electricity into low-temperature heat is considered a sin against entropy, but it makes economic sense for the operators. The gas thus saved simply remains in the distribution pipeline system, which acts as a gas storage of very large capacity, to be burnt another day. The use of gas in combined heat-and-power plants in effect converts the gas back into electricity, as a welcome by-product of the heating process.

The next step in the German *Energiewende* vision is the creation of new inter-linkages among various energy sectors, as is depicted in Figure 27.1.

Some of the most promising emerging technologies that would provide new options for energy storage, transport and conveyance are (various approaches in) the conversion of power-to-gas or power-to-liquid fuels or feed-stock for the chemical industry. The idea at the centre of the image is simple: when there is insufficient demand, renewable wind or solar power, which is fluctuating in availability irrespective of demand, is used to split water (2 H_2O) into hydrogen (2 H_2) and O_2. The H_2 is then combined with carbon from CO_2, which is taken from the atmosphere or the off-gas from combustion processes, to form first methane (CH_4) and then longer chains of alkanes and their derivatives.

The products are gaseous and liquid fuels, which can be integrated in existing distribution or 'down-stream' infrastructure that is a legacy of the fossil oil and gas industries. Power-to-gas and power-to-liquid technologies are expected to be cost-competitive as soon as the penetration of renewable energies is such that there is 'surplus' renewable power often and long enough to operate the conversion plants thousands rather than hundreds of hours in a year.

The implications for strategic energy security are clear: synthetic gas, transport and heating fuels, and feed-stock for industry derived from domestic renewable energies can substitute for fossil energy that otherwise have to be imported. Each step in the conversion also improves the storability of the product, so the technology allows build-up of reserve stocks for times of no wind during a dark and cold winter period, when the derived gaseous and liquid fuels can be used to power efficient heat-and-power plants.

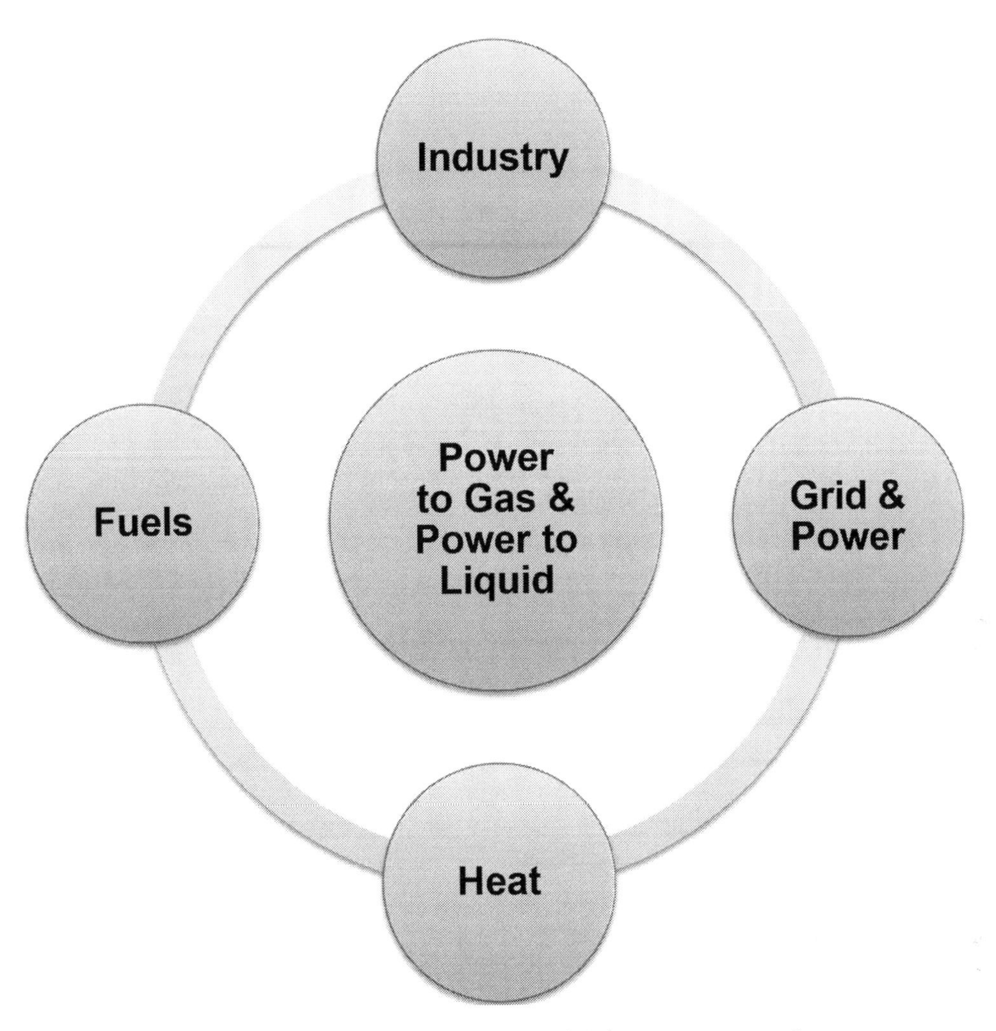

Figure 27.1 Conversion of power to gas and liquid fuels is key to sector coupling
Source: UBA, Umweltbundesamt (ed.) (2015, for English translation, p. 76), *Germany in 2050 – a greenhouse gas-neutral country*. Climate Change | 07/2014. Dessau-Rosslau, Umweltbundesamt. April 2014 (for original German publication). http://www.umweltbundesamt.de/en/publikationen/germany-in-2050-a-greenhouse-gas-neutral-country.

Security evaluation of the *Energiewende*

The 'green power shift' underway in Germany already produces manifold and large economic-policy benefits in terms of innovation, business creation and growth, qualification and employment especially in rural areas where there are few job alternatives, additional tax revenue and social security contributions, import substitution and added strength in the balances of trade and payments. The renewable energy sector even acts as an automatic economic stabilizer in a general downturn of the business cycle. Further, the policy adds to options in dealing with an aggressive and exploitative Russia, the importance of which became obvious when Russia invaded, occupied and annexed Ukraine's Crimea and involved itself in a 'civil war' in parts of the Donbas region since early 2014.

Economic energy security and other economic benefits

Since the 1980s, a new energy industry with an annual turnover of 40 billion euros and employing about 370,000 people has developed, of which perhaps 150,000 are net additional employment that would not exist in the absence of the *Energiewende*. This business and job creation in Germany is a consequence of domestic value generation in lieu of importing fossil energy commodities. The employment is across skill levels – from highly specialized technicians to farm hands – and geographically spread. This is particularly useful to stop the economic decline of rural areas and the migration to towns and cities, and thus to enhance the economic and social cohesion of Germany, where some regions are still struggling with the effects of industrial decline that came with the fall of the Berlin Wall and the collapse of the centrally planned economy of the former German Democratic Republic (or East Germany).

Substituting imported fuels with domestically produced energy strengthens the balances of trade and payments, and serves to isolate the German economy from price volatility on international energy markets. In fact, costs and prices for domestically produced renewable energy are not only predictable, but they are locked in by fixed investment in renewable generating plants. The *Energiewende* simply removes the uncertainties associated with factor prices from a growing segment of its industry and national economy. In addition to enhancing economic security of energy supply, the policy reduces vulnerability to physical supply disruptions as a consequence of natural disaster or enemy action.[18] This 'import substitution' implies the development of a broad and deep value chain on energy efficiency, renewable energies, smart grids and storage within Germany. The development of this industry requires a one-off investment over one or two generations, but then pays off for much longer. One of the benefits is that the domestic renewable energy industry, because it is largely disconnected from international markets, acts as an automatic stabilizer in economic crises that result from external shocks.

The growth of the domestic energy industry drives up tax revenue and stabilizes the social security systems. In the international discussion, the German *Energiewende* policy is often described (falsely) as being based on generous government subsidies. In fact, the financial mechanism does not involve any public funds, but a legally mandated pool financing system paid for mainly by households and small businesses. Because the policy does not cost public funds but increases tax and social security revenue, it is 'fiscally positive' and enlarges the government's ability to deal, for instance, with the legacy costs of nuclear and coal power generation. The lesson that well-designed policies for energy transformation can help governments with high deficits and debts is sadly often lost in the debate about public finance in the euro-zone.

Wholesale electricity prices, the prices paid by large industrial power users and utilities that buy electricity to distribute it to their customers, are very low in Germany – at around 3 cents per kWh – and projected to remain there for years to come. This is attracting inward investment, or the expansion of some electricity-intensive industries, such as aluminium recycling. Internationally, the German *Energiewende* is often criticized as being expensive, with high and rising electricity prices (in cents per kWh) for household users being cited in evidence. In such criticisms, no mention is made of the fact that prices for industrial users and wholesale prices for distribution utilities are declining, which is to be expected as the share of renewable electricity with zero or near-zero marginal cost is increasing its share of the market.

Grid stability and the frequency and duration of electricity supply interruptions, and the number of people and businesses affected, are further important considerations in energy security (of supply). The experience in Germany (as well as in Denmark) shows that grid stability improves and interruptions become shorter and less frequent as the share of distributed renewable electricity supply rises.

Nuclear power plants are large and tend to go off, more often than their operators like to admit, within seconds and without warning. The loss of any such plant must then be compensated on the grid, which requires a sufficient spinning reserve of generating capacity to step in, and a long-distance, high-capacity, high-voltage transmission grid capable of redirecting power flows in seconds. The physical back-up and technical competence and preparedness are costly to maintain, and they are themselves vulnerable to disturbances.

Distributed renewable generation does not impose the same risk and cost for preparedness and response capacity on the grid. Even if a renewable generator suddenly fails, which they or the substations through which they feed into the grid occasionally do, the effect on the grid is not large and can be compensated in most cases automatically and within one distribution grid cell (for solar) or one mid-voltage regional grid (for wind). There are not risks of knock-on effects on the grid as a whole as are presented by nuclear power plants.

Energy security and security implications of energy

As a consequence of the nuclear phase-out, Germany is eliminating the un-insurable high risks of nuclear power, first on its own territory and, by showing that a nuclear phase-out is technically possible and makes economic sense, perhaps also on the territory of its neighbours. Currently, the builders, owners and operators of nuclear power plants benefit from generous caps and waivers that limit their liability for damages to others resulting from nuclear accidents. International agreements furthermore limit liability for transboundary damages. The economic value of those caps and waivers is disputed. It would be revealed in insurance premiums, but there is no commercial market for unlimited liability insurance for nuclear plants. Whatever the value of the risk or the insurance of it, Germany will no longer be exposed when its last nuclear power plant shuts down at the end of 2022 (at the latest).

Once the last nuclear power plant has gone cold, Germany will no longer be adding to the already high (and largely unfunded) legacy costs of nuclear power, and can address the issue of long-term nuclear waste storage. This is not only an important economic security aspect of the nuclear energy system, it is also an issue of security in broader terms. As long as nuclear waste remains stored at nuclear power plant sites, as opposed to specialized, dedicated, high-security waste management and storage facilities, those waste sites are targets for attacks by potential enemies in international armed conflict or terrorists, whether or not they are sponsored by a state.

The overall, macro-economic assessment shows that the total cost of electricity supply to end users in Germany, expressed as a percentage of gross domestic product (GDP), representing the size of the German economy, has not changed much as a consequence of the *Energiewende*. In essence, the benefits listed above are being obtained at low net cost to the German economy.

International security implications of the Energiewende

Domestic renewable energies reduce the need for imported fossil fuels thus enhance energy supply security, in economic and, as noted by Frank-Walter Steinmeier, in strategic terms. Their 'non-deniability' makes domestic renewable energy supply an asset for national security.

Developing countries and emerging economies that have large fossil (or mineral) resources very often suffer from the resource curse, while its milder cousin, the Dutch disease, affects developed countries with relatively strong institutions of government. While the latter is not necessarily of security concern, the resource curse leads to more autocratic and repressive government, internal strife between regions or communities for control over the resources, slower development generally, and sometimes strong militaries that can threaten neighbouring countries. Some of the regimes built on oil and gas tend to export conflict.

The growth of ever cheaper renewable energies effectively reduces the market for fossil energies, in terms of market revenue even more than in physical quantities. Autocratic regimes that benefited from fossil energy revenue are weakened and this in turn leaves room for the renewal of societies as the resource curse is partially lifted. The renewal may not be peaceful, but it is unavoidable. The ability of the regimes to export conflict is greatly reduced; there is a peace dividend from lifting the resource curse.

A similar argument can be made concerning nuclear power and the security policy price of nuclear proliferation, the spreading of nuclear technology in the form of equipment and fissile materials. The cost of containing the ambitions of autocratic regimes, in some cases even 'rogue regimes', to avail themselves of nuclear explosive devices, is very high in terms of diplomatic effort, trade sanctions as well as overt and covert operations. While, in the past, nuclear technology spread mostly among state actors or state controlled businesses, there are an increasing number of private businesses in the nuclear industry. This tends to accelerate proliferation and increase the risk that non-state actors, such as terrorists, obtain material to make nuclear bombs, including 'dirty bombs'.

The German *Energiewende* shows that nuclear power can be phased out without harming the economy or energy security. This is not surprising, with hindsight, given the dismal economics of nuclear power generation, a fact that is also revealed by the current negotiations for the construction of the new nuclear plant at Hinkley Point in the UK.[19] Absent any rational business case for nuclear power, only one or more of the following factors can explain the building of new plants: collective economic delusion; corruption, as large-scale investments make large kick-backs possible; or (unstated) military purpose and intent.

Germany's experience and the lessons that can be drawn from it, show that nuclear power and proliferation were costly mistakes.[20] Instead of continuing to promote the 'civilian' use of nuclear power, as Germany is still committed to do under the Euratom Treaty, research and development, policy design and investment should focus on renewable energies and smart grids. International frameworks that support or protect nuclear power should be abolished, or redirected to facilitate a speedy phase-out of nuclear power. A first step would be to initiate negotiations to amend the Nuclear Non-Proliferation Treaty (NPT) and the Statute of the International Atomic Energy Agency (IAEA), and thus, in the interest of reducing the high security policy price of nuclear power, change the context for dealing with governments with aggressive nuclear programmes.[21]

In addition, the various international agreements that limit cross-border liability for damages resulting from nuclear accidents should be annulled. They provide an unjustifiable privilege to the builders, owners, operators and insurers of nuclear power plants at the expense of the victims of any accidents.[22]

Finally, the *Energiewende* is also a response to the growing security threats that are caused or aggravated by climate change.[23] Germany has helped bring down the cost of energy efficiency, renewable energy and smart-grid technologies to the point that they are cost-competitive with new and in many cases even existing fossil and nuclear energy supply. As a consequence, the German energy system transformation is self-sustaining and self-accelerating by now, and government policy has shifted to slow it in order to protect incumbent interest groups that are important for electoral reasons.

Energy transformation, German style, is likely to be replicated all over the world, for economic reasons alone, albeit from different starting points.[24] Generally, the transformations in other countries are likely to be much faster, however, as the relative costs and risks of energy technologies are now very clearly in favour of renewable energy supply. Economic rationality indicates that, therefore, no new investment will go into fossil or nuclear energy, a development that will improve energy security and lower the security policy price of energy at the same time.

Conclusions and outlook

Germany coined a new word, which the world is adopting into many languages. It is a fascinating idea: a developed country known for incremental and not radical or disruptive innovation commits to transform its energy infrastructure. To that end, it realigns its resources to research, develop, deploy and commercialize clean energy technologies and thus changes the way it powers its homes, businesses, and automobiles. If Germany succeeds, and all the indications so far point to likely success, no one can deny any more that the world is in the Age of Energy Transformation, a term found in the language agreed to at the US-EU Summit held in April 2007.

The original word *Energiewende* was made popular as the book title for a blueprint, published in 1980, of how to move beyond oil and uranium. Experts working with the Oeko-Institut, the Institute for Applied Ecology, building on ideas first spelled out by Amory Lovins of the Rocky Mountain Institute, defined a policy vector for the country, first for West Germany and later for the larger unified Germany, on which large majorities across the political spectrum agreed ever since.

Viewed in its entirety and with its evolution spanning large geopolitical changes as well as technological developments, the true nature of the *Energiewende* becomes plain to see: a slow transformation over decades. It is reasonably well monitored and managed as an adaptive policy process. Overall, it is an economically sound, low-risk strategy. With the benefit of hindsight we can now say that it could have been accelerated. Even today, its ambition is held back by what might be called an overabundance of caution.

Today, Germany has two issues that strengthen its soft power around the world: the 'Dual System' of vocational training in the arts and crafts, of technicians and managers, and the *Energiewende*. Both play well towards other perceived strengths of Germany and the Germans: technological prowess built on research as well as practical education, economic and fiscal solidity, and continuity in realizing long-term projects. It is not surprising, therefore, that a new 'Energy Foreign Policy' (*Energieaussenpolitik*) is emerging, coupled with the external climate policy developed over the past 20 to 25 years.[25] Both are strong on public diplomacy as a complement to more traditional foreign policy instruments.

The most important contribution Germany may be able to make for the success of international climate negotiations, and the regular revisions agreed to at the Climate Summit in Paris at the end of 2015, is to show that transforming the energy system, once begun in earnest, is much easier than it first appears. Many decision-makers around the world still cling to the belief that protecting the climate involves sacrifice today for the benefit of future generations. The example of Germany's *Energiewende* shows that the energy transformation produces short-term benefits that outweigh its costs, that those costs are not higher than maintaining the old, non-sustainable energy system, and that the costs are coming down as experience accumulates. Saving the planet is getting cheaper by the day.

In countries that are rich in fossil resources, the 'resource curse' tends to result in centralization, increasing corruption of elites and, ultimately, the emergence of autocratic, repressive government.[26] The experience with nuclear technology is no better, and the security policy price paid for the proliferation of nuclear technology through the energy sector, is high and rising. This is in stark contrast to the German experience, where the *Energiewende* is a contagious democratic happening driven by citizens across the country. People all around the world are fascinated by the readiness of Germans to engage and invest in the *Energiewende*, and their ability to do so as a consequence of changes in policy and regulation.

Notes

1 Carlowitz, Hans-Carl von; Joachim Hamberger (ed.), 2013 [1713], 'Sylvicultura oeconomica oder Haußwirthliche Nachricht und Naturmäßige Anweisung zur Wilden Baum-Zucht' (Sylvicultura oeconomica or economic information and natural-science-based instruction on the care of trees in the wild). München: Oekom, ISBN [illegible] (annotated re-edition of the historic book establishing sustainability as a principle).

2 This chapter is based on reflections begun in 2015 as Visiting Scholar at the Center for Energy and Environmental Policy (CEEPR) of the Massachusetts Institute of Technology (MIT). I am grateful to Michael Mehling, MIT CEEPR's Executive Director, and colleagues for probing and stimulating discussions. The main writing was part of a research fellowship at the Institute for Advanced Sustainability Studies (IASS) in Potsdam, building on earlier work at Ecologic Institute, Berlin, Germany. Jane A. Johnston helped with many comments and edits on drafts.

3 Germany still has a small number (of around one dozen) nuclear warheads provided by the USA with dual key controls, legally making Germany a nuclear weapons state. This is symbolic and of no practical or political relevance.

4 See Chapter 6 in this volume.

5 Kraemer, R. Andreas, 'Security Through Energy Policy: Germany at the Crossroads'. *eJournal USA* 14, no. 9 (2009): 19–21, http://ecologie.ev/2984.

6 Soft, brown coal, or lignite, is relatively heavy for the energy it contains and is best used close to the mine. Industrial development based on lignite would have been very different to what has been.

7 See Kraemer, R. Andreas, 'Dissolving the "Westphalian System": Transnationalism in transboundary water management', in *Strategic Review* 2, no. 4 (2012): 43–47. http://ecologic.eu/7434.

8 For an interesting contemporary assessment see Director of Central Intelligence (1982), 'The Soviet Gas Pipeline in Perspective'. CIA document SNIE 3–11/2–82 of 21 September 1982. URL: www.foia.cia.gov/sites/default/files/document_conversions/89801/DOC_0000273322.pdf. For a thorough treatise of the shift from oil to gas in Germany, see Duffield, John S. (2014), 'Germany: From Dependence on Persian Gulf Oil to Russian Gas', in John S. Duffield, *Fuels Paradise – Seeking Energy Security in Europe, Japan, and the United States* (Baltimore, MD: Johns Hopkins University Press), 151–194.

9 For the history of the nuclear industry in Germany see Radkau, Joachim, and Lothar Hahn, *Aufstieg und Fall der deutschen Atomwirtschaft* (The Rise and Fall of the German Atomic Industry) (München: Oekom Verlag, 2013).

10 See www.oeko.de/en.

11 The eight leading institutes form the Ecornet Ecological Research Network, see www.ecornet.eu.

12 Krause, Florentin, Hartmut Bossel, and Karl-Friedrich Müller-Reißmann, *Energiewende – Wachstum und Wohlstand ohne Erdöl und Uran* (Frankfurt am Main: S. Fischer, 1980).

13 Lovins, Amory (1976), 'Energy Strategy: The Road Not Taken', *Foreign Affairs*, October. www.foreignaffairs.com/articles/united-states/1976-10-01/energy-strategy-road-not-taken. Also available as a reprint (with an introduction) at www.rmi.org/Knowledge-Center/Library/E77–01_EnergyStrategyRoadNotTaken.

14 For his contribution to the *Energiewende*, Amory Lovins was awarded the Officer's Cross of the Order of Merit of the Federal Republic of Germany on 17 March 2016.

15 Stefes, Christoph H., 'Bypassing Germany's Reformstau: The Remarkable Rise of Renewable Energy', *German Politics* 19, no. 2 (2010): 148–163. http://dx.doi.org/10.1080/09644001003793222.

16 See www.kombikraftwerk.de/100-prozent-szenario/power-flow-animation.html.

17 Morris, Craig (2015), 'Germany's cheap, green energy keeps aluminium sector healthy', *Renewables International*, 30 March. http://reneweconomy.com.au/2015/germanys-cheap-green-energy-keeps-aluminium-sector-healthy-58772.

18 Frank-Walter Steinmeier, the German Foreign Minister, identified domestic renewable energy as being 'non-deniable' by enemy action (short of invading the territory of Germany), in his speech on

26 March 2015 at the Berlin Energy Transition Dialogue (BETD). www.energiewende2015.com/wp-content/uploads/2015/03/150326-Berlin-Energy-Transition-Dialogue-Rede-BM-Steinmeier.pdf (in German).

19 See Chapter 15 in this volume.

20 'We will look back and think that nuclear was an expensive mistake', Paul Massara, chief executive of RWE NPower, one of the UK's big power generators, as quoted in Pratley, Nils, and Sean Farrell (2015), 'Planned Hinkley Point nuclear power station under fire from energy industry', in *The Guardian*, 9 August 2015. www.theguardian.com/environment/2015/aug/09/planned-hinkley-point-nuclear-power-station-energy-industry. (The original article was published in the *Sunday Times*.)

21 See Kraemer, R. Andreas (2010), 'A Transatlantic Agenda for Global Nuclear Governance', summary of the 7th Transatlantic Energy Governance Dialogue convened by the Global Public Policy Institute (GPPI) and the Brookings Institution, Potsdam, Germany, 4–5 March. http://ecologic.eu/3303.

22 See Kraemer, R. Andreas, 'Germany, Fukushima and global nuclear governance', *Strategic Review* 2, no. 4, (2012): 143–152. http://ecologic.eu/7436.

23 See Chapters 3 and 4 in this volume.

24 Kraemer, R. Andreas, and Christoph H. Stefes, 'The changing energy landscape in the Atlantic Space', in Jordi Bacaria and Laia Tarragona (eds), *Atlantic Future. Shaping a New Hemisphere for the 21st century: Africa, Europe and the Americas* (Barcelona: CIDOB, 2016), 87–102.

25 See, for instance, Müller-Kraenner, Sascha, *Energy Security* (London; Sterling, VA: Earthscan, 2008), 141–156.

26 See Chapters 8 and 10 in this volume.

28

Energy transitions and climate security in Italy

Morena Skalamera and Fabio Farinosi

Introduction

The socio-economic context

Italy is characterized by peculiar natural resources, geography, socio-cultural and economic factors. The Italian energy mix has for many years consisted of a dominant role for oil (until 2012), a much higher share of gas and hydro as compared to other European countries, and a limited use of coal. Furthermore, Italy heavily relies on foreign energy supplies. Italian reliance on imported fuels (particularly oil and gas, but also coal and electricity) has remained very high: above 80% until recently. By comparison, the EU-28 has a rate of import dependency of about 53%.[1] Italy has one of the highest dependence rates in Europe, which causes concern when energy prices are high or in case of supply disruptions. On the other hand, stagnating and aging populations and low economic growth have in the last few years signaled a somewhat reduced demand for hydrocarbons.

The last years' fall in economic growth raises important questions about Italy's declared energy priorities: how quickly will the country develop the necessary technology to improve energy efficiency and lower carbon intensity? How will it seek to reconfigure its energy mix subsequent to the eurozone debt crisis?

The 2008 financial crisis and the subsequent shocks, such as the eurozone debt crisis, have disrupted the global economic order of the prior decade. In this context, Italy is still struggling with the double crisis of high sovereign debt and stagnant growth. Stabilizing the debt burden, while stimulating growth and employment, is a tough balancing act. It calls for adjusting the old social welfare state model and deregulating labor markets, steps that will be socially disruptive before paying off. This requires extensive market and political reforms, which, however, face strong headwinds from vested interest groups and infrastructure constraints. Especially in large and lucrative sectors such as energy, there is a general will for more business-friendly policies but also firm opposition from vested interests. All these factors combined with rising demand for political participation inspired by the harder economic realities have eventually forced the government to start wider political and economic reforms. Reform would need to set the agenda toward completely breaking up the privileges of ex-state-owned monopolies (i.e. ENI, the Italian multinational oil and gas corporation) and interest groups, spurring investment growth and building functional social safety nets, which are the prerequisites for sustainable domestic demand driven growth. Italy is definitely on the right track, but judging by the slow reform speed of the past decade, entrenched corruption and powerful industrial elites, the transition may take well into the 2020s. On the other hand, assuming that reforms and fiscal consolidation are implemented and paid off by the 2020s, Italy may enjoy much-improved prospects. Yet there is considerable uncertainty about how these drivers will impact energy demand over the coming years and decades.

The energy mix

The remainder of this decade will most likely be shaped by slow but profound transition processes toward an energy mix largely consisting of a combination of renewables and natural gas. Regarding renewables, policies have been influenced by the EU's 20–20–20 targets[2] which Italy is not only expected to meet but highly exceed, especially after the demand decrease due to the economic crisis. Italy's energy mix has historically experienced a higher share of gas and oil products and a lower share of coal. In the period 1995–2013, in fact, the fuel mix showed a continuous decrease in consumption of oil and oil products, a steady increase in gas use (peaking in 2005), and the sustained growth of renewables. Currently, 10–11% of gas is produced domestically. The remainder is imported, mainly through pipelines.[3]

Today Italian economics and, consequently, energy consumption is depressed by the eurozone crisis (Figure 28.1). Barring major and, for the moment unexpected, changes in attitudes to nuclear, the contributions of wind and solar to Italian power generation will continue growing. There are however signs that the current economic malaise is eroding the government's resolve to continue subsidizing these technologies as generously as they have in the recent past.

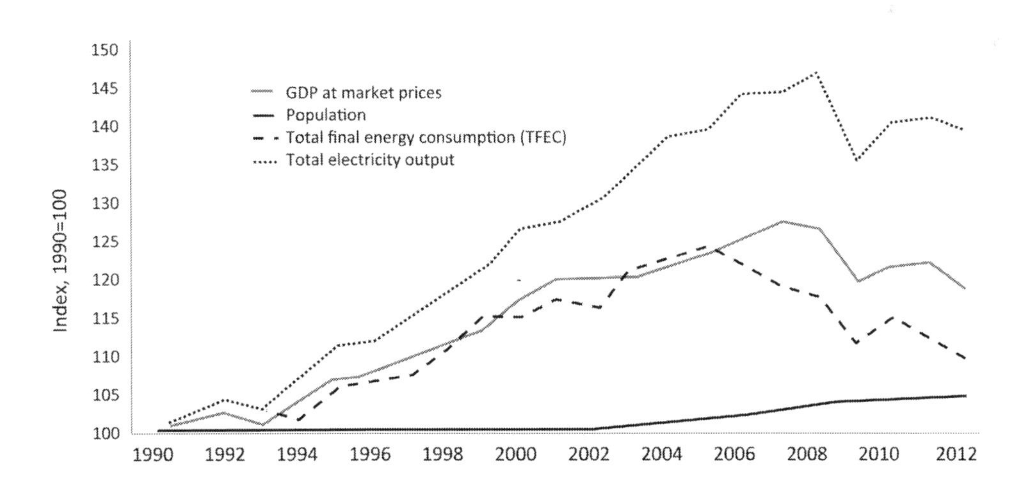

Figure 28.1 Population, income and energy consumption in Italy 1990–2012, 1990=100.
Source: World Bank, World Development Indicators (2015). Retrieved from http://data.worldbank.org/data-catalog/world-development-indicators.

Identifying Italian tradeoffs between energy security, competitiveness and climate security

In such a complex contest, this article seeks to disentangle the following questions: how has Italy historically dealt with satisfying the three key policy objectives of energy security, sustainable climate (climate security), and economic competitiveness? Were all the three key policy goals approached simultaneously, or, rather, was precedence given to one or two elements of the trilemma at the expense of the others? What changes do we observe over time? Can we consider Italy successful in managing the nexus between energy security, competitiveness and climate-related goals?

A comprehensive empirical analysis of the "paradigm shifts" in the Italian energy policy panorama warrants an investigation of the battle unfolding at the high-policy level between "conservative" constraints favoring hydrocarbons, vis-à-vis inputs toward a more sustainable, modern energy environment. Such an overview requires a time frame of approximately 40 years. However, the period between 2000 and 2015 will be essential to understand, first, why Italy appeared to have favored energy security and climate security at the expense of economic competitiveness, and second, why Italy has encountered so many obstacles in devising more sustainable national energy plans. Conversely, what has been the role of the EU in soliciting Italy to adapt to the new environment, and with what results?

The particular contexts in which changes have been taking place will be of paramount importance in explaining the predominance of certain agendas over others at the national level. The chapter focuses on the country's approach toward four crucial energy sources: coal, oil, natural gas, and renewables. Each section employs in-depth historical analysis to assess the critical importance of the particular energy source in the country's overall energy strategy. In-depth analysis requires great attention to detail in order to unravel complex pathways of cause and effect. In that sense, each section's analytical focus lies on constructing a strong explanatory model of Italy's overall energy priorities, while identifying the key players and tracing what drives them – a pursuit of energy security, climate security, or rather, economic competitiveness.

Energy security

The country's energy mix

In 2010, a large share (84%) of the Italian primary energy supply came from imports, resulting in a strong dependence on foreign fossil fuels,[4] while the European average was much lower at 53%.[5] Indigenous production from renewables, gas and crude oil covers only 10%, 4%, and 3% respectively of the national primary needs.[6]

The current energy-mix makes the Italian economy more exposed to the global geopolitical instabilities of the oil and gas-producing countries, as compared to the northern European countries. This is due to the limited availability of domestic mineral resources, combined with electricity production's strong dependence on fossil fuels. Such a situation should be also viewed in the light of the decision to put an end to the nuclear program, following the referendum of 1987, in turn subsequent to the Chernobyl accident. In June 2011, in the aftermath of the Fukushima Daiichi tragedy and as part of another referendum, Italy reconfirmed its refusal of nuclear power. The reasons also included the perceived risk of nuclear technology in a landslide, flood and earthquake prone country, and the risk of pollution by nuclear waste.[7] In any event, the outcome is an absence of nuclear power generation in the energy mix and a stronger attention to renewables.

Regarding oil, Italy is highly dependent on external sources of supply, importing over 90% of its oil needs.[8] This fossil fuel supplies only 5.5% of electric power, while 54.7% is employed by the Italian transportation sector. However, unlike most of its European counterparts, Italy has made large efforts in the form of compressed natural gas (CNG) vehicles in the transport sector, in order to reduce the use of oil.[9]

In 2011, the Italian generation mix was very different than the European one,[10] featuring gas as 48% of total power production, followed by 27% from renewables (including hydro at 18%), only 15% from coal, 2.6% from old oil plants and 7.4% from other sources. In the early 1990s, the increase in the use of natural gas largely happened at the expense of oil (Figure 28.2, panel a). Italy has a twofold interest in natural gas: gas is used in industries and for domestic heating, but is also widely employed to produce electricity. In 2010, power generation accounted for almost 40% of total gas demand in Italy. The residential and commercial sector is the second biggest source of demand growth for natural gas. As a result, even though Italy has indigenous production of natural gas, over the past decades imports from abroad have grown rapidly (Figure 28.2, panel b). Renewables, on the other side, were still low on the political agenda (Figure 28.3). Nevertheless, especially since 2009, the development of renewables seems to be somewhat eroding the share of gas.[11]

In 2010 still about three-quarters of Italy's supply mix came from oil and natural gas (only slightly down from 88% in 1973). The remaining shares were split between coal (9.2%), hydro (2.4%) and other energies, such as renewable sources (12.3%), rapidly rising.[12] The potential for further reduction in coal in the foreseeable future is large, owing to the extended growth of renewable energy and the local resistance to coal. Thanks to severe regulatory restrictions coal's

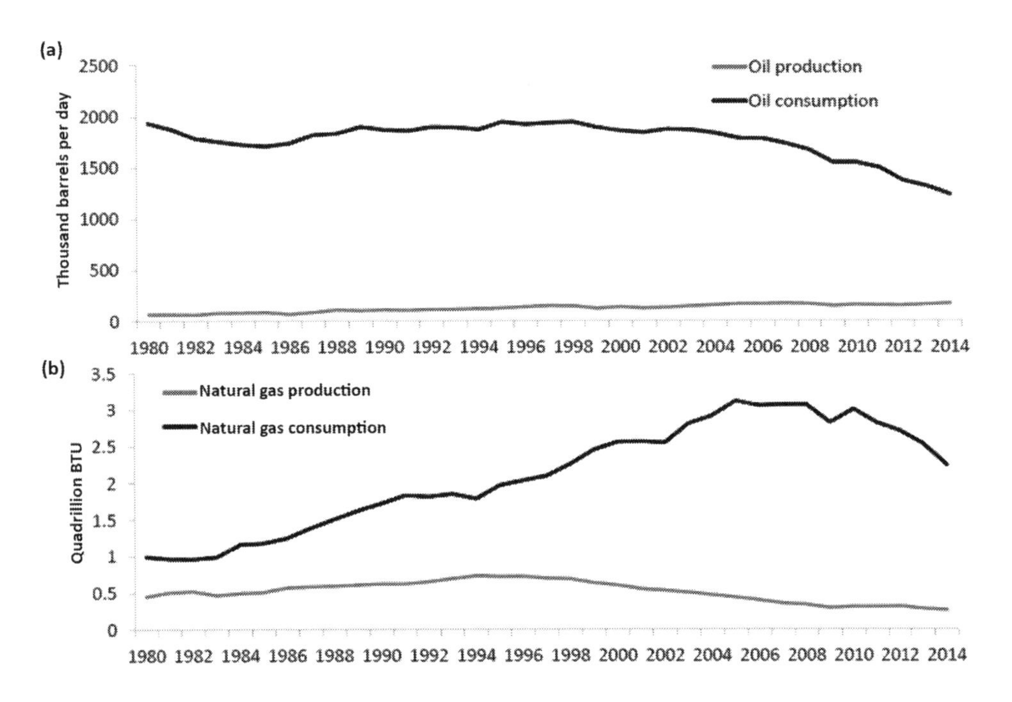

Figure 28.2 Oil (a) and natural gas (b) demand and internal production 1980–2014
Source: Authors' elaboration based on US Energy Information Administration (US EIA), International Energy Statistics, US EIA Database, 2015, http://www.eia.gov/cfapps/ipdbproject/IEDIndex3.cfm#.

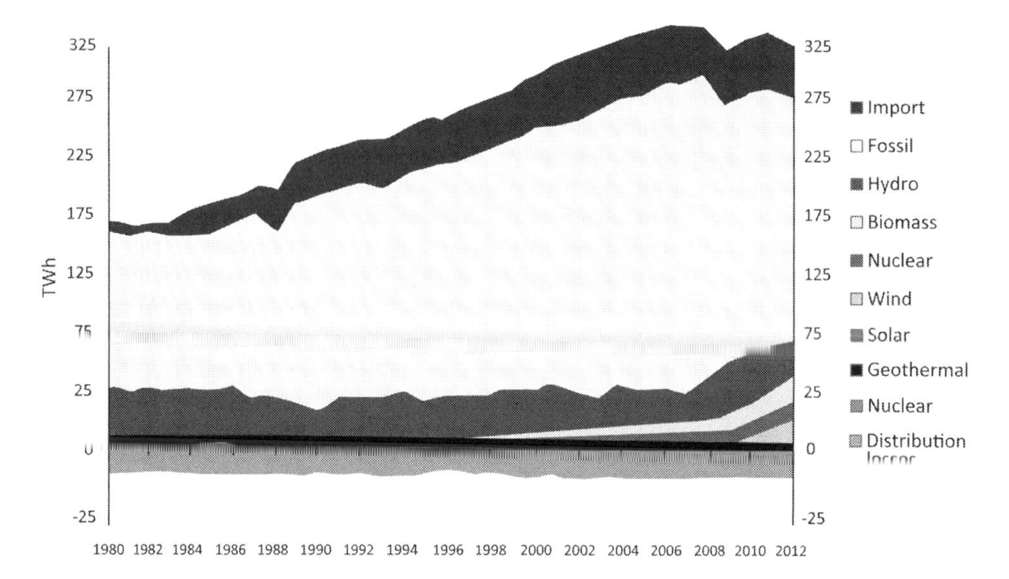

Figure 28.3 Italy's electricity mix 1980–2012
Source: Authors' elaboration based on US EIA, International Energy Statistics database (2015). Retrieved from http://www.eia.gov/cfapps/ipdbproject/IEDIndex3.cfm#.

share of Italy's generation mix is steadily declining, being replaced by natural gas and renewables. Recently coal fired power has captured market share in the EU because coal imports have become cheaper and given that the EU carbon price is not sufficiently high to impact relative economics. This has raised a flurry of speculation over whether the pace of Italy's gas demand growth may be checked by coal price developments eroding gas's recent competitiveness in the power sector. A renaissance for coal based power generation is, however, unlikely. Coal use is under sustained regulatory attack. It is feasible that the combination of moderate natural gas prices and gradually tougher climate and other environmental policies, combined with higher carbon prices, will contribute to lower – in parts of the world negative – growth in coal consumption from 2020 onwards. This development will also be fostered by development of electricity production based on new renewables.

Historical background

While today's share of oil in the Italian energy mix is shrinking, this was not always so. During the second half of the 1930s the Italian government exerted considerable pressure on AGIP (*Azienda Generale Italiana Petroli*), for a rapid exploitation of national mineral resources to achieve self-sufficiency.[13] In the interwar period, due to strong political pressure oil's contribution to the national energy mix gradually grew. Quite obviously, oil's share in the national energy mix had grown even more dramatically after World War II, passing from 22.1% in 1950, to 44% in 1960, 72.6% in 1970 and about 75.3% in 1973.[14]

Italian economic growth, and thus competitiveness in the 1950s and 1960s was mainly powered by oil; in 1973 the share of oil in primary energy consumption reached 79%.[15] In absolute terms, oil consumption has remained relatively static since 1970, but its primary energy share has decreased significantly, steadily replaced by natural gas. In fact, the oil shocks in the 1970s reinforced the country's emphasis on gas rather than oil. Moreover, environmental policies made natural gas the primary fuel for power generation in the 1990s and, to a lesser extent, in the 2000s. In the 2010s, dependency on fuel oil of the 1980s gave way to a new dependency on both gas and renewables.[16] Although gas became the uncontested champion of Italy's energy supply mix, the start of Italy's preference for gas is somewhat accidental. While searching for oil during World War II, the state company AGIP found large quantities of natural gas in the Po Valley, in Northern Italy, where the majority of promising oil and gas fields were located. After the war, AGIP developed the resources and by 1960 Italy was the largest consumer and producer of gas in Europe.[17]

In 1953 ENI *(Ente Nazionale Idrocarburi)* was created with the mission to provide energy to the rapidly growing economy.[18] ENI was given the exclusive right to look for and exploit hydrocarbon deposits and the exclusive right to build and run gas and oil pipelines in the Po Valley. As a result, ENI (and the companies controlled by ENI) had a monopoly (de jure or de facto) in all segments of the gas chain.

ENI's vertical integration and monopolistic position contributed to the expansion of the gas network to other parts of the country, including the Southern regions. Yet gas production was driven by rapid industrial development concentrated in the North. To speed up the use of gas, in 1949 the first Italian gas-fired power station was set up. In 1974, the first Russian gas flowed in the direction of Italy. By 1980, the gas national network had reached 15,000 km and covered almost the entire country. Since the early 1990s the length of the Italian gas network has tripled, driven by the remarkable growth in power generation from gas.[19] In the early 2000s, Italy had cemented the predominance of natural gas over all other primary sources. At that time the country became the fourth major world importer after the USA, Germany and Japan, while Algeria and Russia were supplying most of the imported methane.

As for oil, indeed, the first wave in the decrease of consumption (1980–2008) is due to the increasing incidence achieved by natural gas. The additional energy demand, compared to 1980, was almost completely satisfied by natural gas, whose emissions per energy unit are about 30% lower.[20] Concerning the recent part of the trend, the sharp decrease in oil consumption is due to the expansion of renewables, whose contribution almost doubled in the period 2008–2010 thanks to the incentives created by the Italian government.[21] The phasing out of fuel-oil in the power sector and the rapid success of gas-fired power plants (albeit, still, under long-term oil-indexed contracts) was part of the national program to lessen the dependence on oil imports due to growing environmental concerns.

The key role of refining

Still today Italy plays an important role as Europe's largest refining center, and is a net exporter of refined products, providing finished products to other countries (Figure 28.4). While ENI had the largest share of the market in 2008 (around 30%), it intends to reduce its presence in the retail market and will focus on upstream and refining activities. There are three non-OECD companies operating in Italy: Tamoil Italia (Libya), Petroleum Italiana (Kuwait) and Lukoil (Russia). The three companies have refining and marketing operations.[22] The country's refining capacity grew rapidly: in 1951 it was four times the value of 1940 and more than twice the 1948 figure.[23] Although the internal supply of crude oil soon turned out to be practically non-existent and that of natural gas insufficient, for many years Italy played an important role as supplier of refined products to foreign countries. The discovery of new rich oil fields in the Middle East and the shift of refining activities toward safer locations, closer to the consumption areas, gave new importance to Italy's position in the middle of the Mediterranean Sea and gave rise to the creation of a number of independent refining companies.

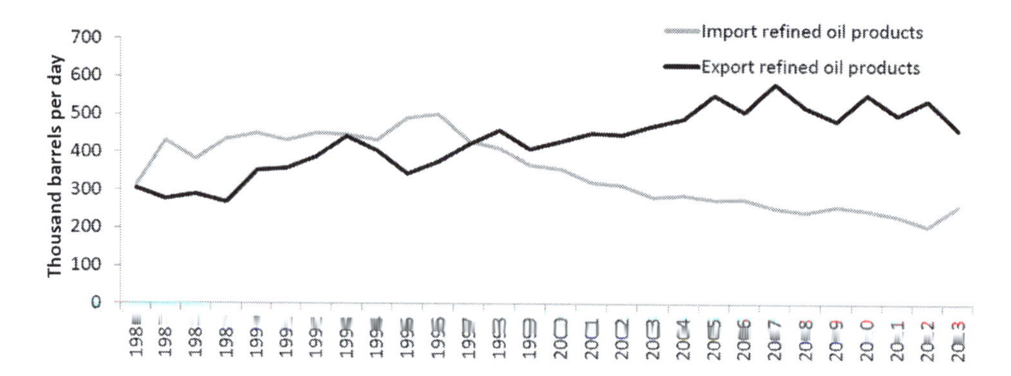

Figure 28.4 Refined oil products import and export 1986–2013
Source: Authors' elaboration based on US EIA, International Energy Statistics database (2015). Retrieved from http://www.eia.gov/cfapps/ipdbproject/IEDIndex3.cfm#.

Substantial investments have been carried out in order to adapt the refineries to the decline in heavy fuel oil demand in the power sector and the growth of cleaner fuel consumption in the transport sector. Further investment in conversion capacity and fuel quality is expected in light of ever-growing demand for diesel fuels and greater availability on the market of sour (rather than sweet) crude oils. Nevertheless, the refining sector is currently undergoing a painful crisis along with the entire downstream oil sector, mainly due to lower demand and competition from cheaper Asian refining facilities. Therefore, a large share of older refineries will likely be scrapped.

Entrenched interests

The entrenched interests of big conglomerates, such as ENI, Enel and Edison lie behind policies of protection and adaptation of Italy's refining capacity. Especially former state oil company ENI retains a dominant position in the Italian oil and gas sector, although a number of Italian and foreign private companies have established a significant presence. This is due to path-dependent policies carried out in the 1980s. At that time, under a general global climate of low oil prices following the shocks of the 1970s, Italy continued to rely on hydrocarbons and, increasingly, on imported electricity, thereby giving scant attention to alternative renewable sources. The two oil shocks in 1973 and 1979 only partially affected this state of affairs, leading to a decrease in energy intensity (i.e. the amount of energy consumed per unit of GDP produced) but not to a permanent decline in energy imports (from 82% to 83% in the same period).[24] Due to the scarcity of domestic energy resources and high energy costs, energy intensity in Italy has historically been lower than the European average. In such a climate, the two public energy oligopolies, ENI and ENEL (oil and gas corporation the first, electricity the second), gained further power. ENI, which had entered the 1980s as the eighth largest oil company, in 1990 was the fifth largest for sales and the third for assets and employment.[25] By the early 1980s, the state company had definitively entered the elite of the world oil market. Political and economic considerations, both at domestic and international levels, thus, influenced the defining of Italian strategy on oil consumption. As for gas, import dependency for natural gas is very high, standing at around 90%.[26] In 2013, 45% of gas imports came from Russia, 20% from Algeria, 9% from Libya, and 8.6% from Qatar, with the rest coming mostly from EU countries and Norway.[27] Yet, still, two countries alone – Algeria (22.7 bcm) and

Russia (20.0 bcm) – account for two-thirds of Italy's imports.[28] Italy, along with Germany and the UK, is one of the largest European markets for natural gas. However, unlike the other two, Italy sources most of its gas from Russia and Algeria and, despite the country's overwhelming dependence on imported gas it greatly relies on gas for electricity generation. Such a situation causes economic hardship when imported energy prices are high or, even worse, in cases of supply disruptions.

As noted by Luciani and Mazzanti, historically the two incumbent companies (ENI and ENEL), have taken major decisions affecting national energy security, with little input from the government.[29] The decision to build a strong interdependence with Russia was definitely informed by both commercial and geopolitical considerations, reaching a peak during the Berlusconi governments,[30] epitomized by the ENI-Gazprom strategic partnership agreement of 2006 that launched common projects in all segments of the gas chain. With the government's support, ENI has completed some of its most spectacular gas projects: Blue Stream, a deep under-water pipeline connecting Italy with Turkey and then all the way to Russia, Green Stream connecting Libya to Sicily, and Transmed, the pipeline between Tunisia and Italy, which was completed in 1983.[31]

Concentrated foreign supplies

Both in oil and in natural gas Italy relies on a concentrated set of supplies. While oil supply sources rely on over 30 countries, Libya and Russia are the dominant sources, each accounting for almost a quarter of all Italian crude oil imports. Saudi Arabia, Iraq and Iran together represent an additional quarter of oil imports.[32] The Persian Gulf as a whole is the main region that supplies Italy's hydrocarbons, while Russia is the largest supplier of both oil and gas.[33] Expensive infrastructure, where Italy's ENI pioneered deep underwater gas pipelines transport, stays at the center of the Italian approach to energy security, in which Russia, the Middle East and North Africa remain the main points of reference.

Today Italy still relies on fossil fuels for a quite large portion of its electricity generation: IEA, US-EIA and TERNA data for 2012 show that Italy relied on traditional sources for about 67% of its electricity generation. This situation is largely due to Italy's long-lasting decision to avoid nuclear power at a national level and exposes Italy to the turbulent geopolitics of its supplying regions, the Middle East and Eurasia.

For instance, the spread of the Arab Spring to new countries in the Middle East and North Africa led to large supply losses in 2011 and 2012. The main losses in 2012 were from South Sudan, Syria, Libya and Iran.

Furthermore, the ongoing war in Ukraine raises doubts over gas supplies, too. Given that Gazprom is not on the list of sanctioned companies and there have been no problems on gas pipelines, supplies have thus far continued to flow. However, the elevated supply risk associated with the Ukraine war will probably last for several years.

As already noted, Italy's strategy on hydrocarbons in the past has made the country reliant on a limited number of unstable countries. Moreover, it largely reflects the preferences and priorities of the main energy company: ENI. Path dependent practices that ENI established since its inception explain why Italy may have had a more difficult time adapting to the new energy-climate nexus within the EU.

Despite several measures to restrain ENI's dominant position, the latter remains crucial throughout the gas chain. Such a situation clearly impedes the attainment of Italy's national goals for 2020, such as greater energy efficiency, competition in the gas market and the creation of the Southern European gas hub.[34]

Hydrocarbons lose but show resilience

In sum, the brief picture that we have sketched illustrates that oil largely powered Italy's path to industrialization following World War II. Technical improvements in production and refining went side by side with a renewed oil nationalism, which included in its ranks the new and dynamic Italian national agency for energy, ENI, created by Enrico Mattei. Subsequently, as already noted, gas expansion was promoted by a national program to alleviate the country's dependence on oil imports.

However, the Italian government did not pursue an interventionist policy to reduce oil's contribution to primary energy (compared with other European countries). Indeed, Italy plays an important role as Europe's largest refining center, and is a net exporter of refined products. While recognizing that oil contracts may represent a barrier to environmental security, considering the market power of incumbents such as ENI and ENEL which due to their refining and marketing operations guard the status quo, Italy also considered the importance of oil for Italian competitiveness and relied on market mechanisms for the evolution of oil contracts vis a vis other sources. Moreover, due to its availability and environmentally friendly characteristics compared to oil, natural gas became an increasingly important fuel in the Italian energy mix. At least until the 2000s, renewables retained a negligible role, except for hydropower that supplies about 15% of the national electricity.

In 2010, the power produced by hydroelectric power plants, including pumped stations, covered about 18% (54.4 TWh) of the total. More generally, in 2010 12.2% of the Italian energy mix consisted of renewables, in turn constituted mainly by hydroelectric power (67.6%).[35]

In general, fluctuations in the price of oil (acting slightly on energy efficiency in times of high oil prices) did not, however, significantly affect Italy's energy dependency that stabilized around 80% starting from the 1980s.[36] This strategy is consistent with profit maximization by holding companies, as profits obtained, say, by an ad-hoc program encouraging larger use of renewables would have not been of the same magnitude as gains obtained by incumbents in the refining business.

Italy's trajectory in hydrocarbon consumption also shows that oil and gas became appealing primary energy sources due to a strong incumbent (i.e. ENI), which has de facto led the Italian national energy policy and has favored energy security and economic competitiveness at the expense of climate security.

A key question then becomes, how has the former state energy company ENI's dominant position in the Italian gas sector influenced the tackling of Italy's conundrum between competitiveness, security and environmentally-driven concerns? High dependence on imported gas, for the residential and power generation sectors, and complex access to storage, triggered risks of periodic gas shortages. In sum, all this has meant that Italy hardly features high on the gas security parameter.

Energy security: Italy's position within the EU

Due to such a situation, Italy was severely affected by a disruption of gas supplies over the winter of 2005–2006 (partly due to a Russia-Ukraine gas dispute), and has since taken significant measures to better prepare for another similar situation. Starting from 2007, the Ministry of Economic Development has adopted and updated its legislation regarding specific emergency procedures. The update establishes the roles of the actors involved, the system monitoring procedures, and the measures to be taken by the Ministry in the case of a crisis.[37]

However, prolonged gas shortages in 2006 and 2009 have not dramatically changed the strategy of Italy's most important authority for gas security, ENI. Bellicose statements from Gazprom and the Russian authorities against the pillars of the EU's gas liberalization process have been quietly supported by ENI. Moreover, despite the EU's persistent goals at gas market liberalization at the EU level, the path towards a common internal gas market has been fraught with obstacles (see below).

In Europe, Gazprom's big clients are Germany, France and Italy, which together account for more than half of the EU's consumption of Russian gas.[38] These countries have a long history of friendly "special relations" with the Kremlin with policies focused on promoting the interests of their respective "national champions" through bilateral relations with Russia, rather than an EU-wide strategy on security of supply. These three large member states have traditionally been characterized by national or regional monopolies, supported by the respective governments, or have directly delegated energy governance to their "national champions," as in the Italian case. Historically, the three biggest Russian clients who enjoy close relations with Moscow – Germany, France, and Italy in particular – have blocked steps toward a real EU energy integration. The issue of sovereignty over energy policy has cropped up repeatedly in the history of European integration to justify individual and varied approaches to energy security. Energy specialists are divided between those who advocate a truly common European internal and external energy policy[39] and those who believe that one step at a time should be taken instead; i.e. the EU should make the internal gas market integration a priority, and only when that goal is achieved an eventual external energy policy should be pursued.[40] In sum, the current weakness of energy policy at the EU level is due to a complex mix between: an internal market and competition policy, a nascent sustainable energy policy and a still empty EU-level security of supply policy. At the core of this problem is the "mandate" issue, which is the EU's lack of legitimacy over its member states' energy policies. With the Lisbon Treaty now the European Union has a mandate to establish a common energy policy based on "solidarity, sustainability, security of supply and economic efficiency."[41]

Most recently, subsequent to the Ukraine crisis, former Polish PM Donald Tusk has used the very same concept, sovereignty, to argue for a common EU external energy policy against Russia. On 2014, writing an editorial for the *Financial Times*, Tusk warned, "Gas security is a fundamental prerequisite of sovereignty."[42] Yet, Italy, France and Germany are simply not convinced that a truly common external European energy policy is in their interest. The UK, somewhat ironically considering its traditional Euroscepticism, having already liberalized its own markets, is pushing for fully integrated EU energy markets.[43] Comprehensive and definitive views on the best way forward are beyond the scope of this chapter, yet only in the context of this dilemma can we understand the lingering challenges moving forward. With this in mind, we now turn to the second pillar affecting Italy's energy strategy, economic competitiveness.

Economic competitiveness

Falling indigenous oil and gas production, continuous economic crisis, low or negative energy demand growth, planned and unplanned fuel mix changes driven by policies and changing economic and energy realities, have made an overarching analysis of Italian competitiveness a daunting task. Moreover, emerging doubts about the consistency of Brussels's 2020 targets, and divergent policy priorities within the EU put a question mark over Italy's strength as a global climate policy frontrunner.

According to the National Energy Strategy, Italy intends to double its domestic production of oil and gas by 2020 and boost renewable power generation as it moves to cut consumers' energy costs and boost faltering economic growth.[44] This optimistic scenario entails the reduction of hydrocarbon imports to 67%, although it remains to be seen whether indigenous production will be cheaper than imported fuels.

In any event, the Italian economic situation is currently complex and uncertain. After a decade of very slow growth, the economic crisis reduced GDP by more than 5%. Sustainable growth is the government's declared goal,[45] however, its attainment is hindered by a series of structural factors, most notably energy prices much higher than the European average (especially on electricity). According to the National Energy Strategy there are four main structural factors impeding Italian competitiveness:[46]

1 The Italian energy mix and in particular the electricity mix mainly consists of gas and renewables and largely differs from the European one for the absence of nuclear and much lower volumes of coal.
2 The wholesale price of gas in Italy is higher than the European average despite the web of "special relationships" that ENI has established with a series of foreign gas suppliers, such as Russia (see section 1). For example, in 2011 Italian gas was on average 25% more expensive than the gas sold at North European hubs.[47] Even the price of the long-term take-or-pay Italian gas contracts is on average higher than the same contracts elsewhere in Europe. Such a situation negatively affects the final electricity price.
3 Italy has Europe's highest incentives for the production of renewable energy (for example, the incentives for the production of solar energy – photovoltaic – have been twice as generous as the German ones) despite the dire state of Italian finances and with a sharp negative effect on the general cost of energy: more than 20% of the average Italian energy bill consists of incentives to the production of renewables.
4 There is also a series of other costs related to public policies on tariffs and widespread inefficiencies.

These factors, compounded with Italy's still insufficient resiliency due to overwhelming dependence on foreign hydrocarbon supplies and the inability to respond effectively during crisis periods (as the February 2012[48] crisis has revealed), diminish Italian flexibility and therefore have a negative impact on the competitiveness of the system. The much-needed structural reforms are still incomplete, largely because of vested interests and to a lesser degree because they would be socially painful and unpopular with voters.

In order to enhance competitiveness by 2020 the Italian Ministry of Economic Development has put forth the following goals: aligning the Italian electricity costs with European ones; reducing the large gap between energy costs for businesses and residential consumers while making sure that the long-term energy transition does not hinder industrial competitiveness. Other first-tier priorities are growth in the "green economy," investments in gas storage and regasification facilities, and domestic production of hydrocarbons. The Government plans to allocate a sum of 170–180 billion euros to implement such measures up to 2020.[49]

On the other hand, the high energy prices led the Italian productive sector to extremely high energy efficiency. The energy intensity of GDP in the period 2000–2014 (unit of energy per unit of GDP) was one of the 10 lowest in the world.[50]

In sum, the government promotes positive economic development by combining more effectively environmental sustainability and economic competitiveness. In that respect, the development of a liquid natural gas market is considered key to position Italy as a gas hub in the Mediterranean, thereby gaining both in economic competitiveness and in secure energy supplies.

Diversification

In the last decade Italy has tried to diversify its energy sources in an attempt to redress the excessive reliance on certain supplying countries and to lower the import price of its hydrocarbons.

Therefore, Italy's interest in broadening its hydrocarbon foreign sources is driven not only by security of supply-related worries but also by competitiveness-driven considerations. Italy is largely reliant on pipeline imported oil and gas supplies. High oil prices and the fact that major shares of these come from unstable regions raise both economic and fuel supply security issues. Italy needs to import more and more of its gas supplies too, but faces a different set of suppliers and trade routes for gas with respect to oil, and therefore hopes that European shale gas in the long term will add materially to indigenous supply.

In order to boost both its competitiveness and energy supply, Italy aims to broaden its base of gas supply sources and import routes. It is also taking steps to strengthen gas and power grids, improve interconnectedness with the rest of Europe and facilitate gas and power exchanges across borders. Efforts to make more supply available to Italy and the rest of Europe have focused on, among other things, opening a "Southern corridor" for Caspian and potentially Central Asian and/or Middle Eastern gas imports. A decade-long rivalry between various consortia aiming to build pipeline systems from Azerbaijan via Turkey to Southeast Europe came to an end in 2013, with the Shah Deniz field owners declaring the Trans-Adriatic Pipeline (TAP) proposal the winner.

With TAP coming to fruition, Italy has taken a first step to diversify its still very concentrated supplies of gas. To bring Azeri gas to Italy and the rest of Europe, the TAP will connect Greece, Albania and Italy, and will join the domestic gas network in Brindisi (Puglia). There are also many projects for LNG terminals, at different levels of planning, for a total capacity of about 24–32 billion cubic meters (bcm).[51] Additional infrastructure is seen as a way to develop further competition, add flexibility to the system, and transform the country into a Southern European hub.

In this regard, there now is action in addition to the decades-long diversity talk. Rome seems to be largely interested in an economic cooperation with Azerbaijan: the Southern Gas Corridor has been agreed upon and it has now started its implementation phase.[52] And so, in a couple of years, a southern corridor should be taking fuel from the Caspian Sea through Azerbaijan, Georgia and Turkey, and into Europe, bypassing Russia. Those advances combined with other moving parts – such as liquefied gas plants off the Adriatic and Tyrrhenian coasts and greater gas integration with the rest of Europe, mean the Italian government will be getting closer to its aspirations of setting itself as a gas hub in the Mediterranean, thereby also weakening Russia's grip. Nonetheless, Russia is not standing idly by. The EU's attempts to get Caspian gas and mobilize support for the Southern corridor are complicated by Russia's plans to build a giant pipeline of its own across the Black Sea to Southeast Europe, called Turkish Stream. Given Southeast Europe's prolonged economic downturn, the market hardly supports the construction of several major new import pipelines into this region in the same timeframe.

Most recently, in Africa, the recent discoveries of huge gas fields in Mozambique and Tanzania have opened prospects of future oil and LNG shipments from these countries. In July 2014 ENI CEO, Claudio Descalzi, asserted that the proven reserves in the area could cover the gas consumption of a nation like Italy for the next thirty years.[53]

Yet, due to still incomplete liberalization, ENI's supremacy over national storage, and an insufficient number of LNG terminals to balance ENI's supplies, in the foreseeable future Italy will remain exposed to the vagaries of Russian, Libyan and Algerian foreign policy, and in turn to ENI's relationship with these countries' corporate counterparts. Furthermore, ENI's sustained opposition to domestic liberalization in the gas market has contributed to Italy paying among the highest gas prices in Europe, and thereby hindered competitiveness.

Domestic gas liberalization – fraught with obstacles

In Italy, ENI – the former monopoly and the main player in the market – is also the largest operator in terms of sales to final users. Moreover, ENI controls SNAM that owns 94% of nearly 34,000 km of grid. ENI's dominant position in the Italian market, in particular its control of the grid, has generally meant less transparency and convenience for end-users. Nowadays this is a highly debated issue, subsequent to the reforms of the Monti government, which has committed to unbundling SNAM from ENI. Yet, only in mid-2012 Italian gas prices started aligning with those of the rest of Western Europe.

The majority of storage is managed by STOGIT, completely owned by SNAM (i.e. ENI). Almost every winter Italy experiences a shortage in the supply of Russian gas (because of severe climatic conditions) that tests the Italian storage capacity, which explains why there are multiple projects to increase the storage capacity.

According to the National Energy Strategy published in 2013, Italy requires an increase of about 75 million cubic meters of gas supply a day and about 5 billion cubic meters of storage capacity – which represents an increase of almost 50% compared to the current commercial capacity.[54] This increased storage will secure the system in case of emergency situations similar to those of February 2012, gradually reducing the need for measures to limit fuel consumption. This storage capacity will also contribute to enhance the liquidity and competitiveness of the market, representing a potential for modulation of streams for export.

In fact, market liberalization in the 2000s failed to achieve levels of competition in the mid and downstream sectors to the extent seen in North West European markets. This resulted not only in one of the highest European end-user gas prices, but also delayed the development of a liquid natural gas hub. Today, the gas industry is fully liberalized but competition has yet to reach its full potential with a few players still dominating the upstream and wholesale sectors. And although the retail sector is more diversified, market concentration is still significant.[55] Gas imports are delivered mostly via long-term oil-indexed contracts, which have come under pressure since 2008, due to a wave of renegotiations for price reductions in the contracts and a revision of take-or-pay (TOP) clauses. Even gas renegotiations, which were initiated by ENI, did not, however, fundamentally change the preceding situation. A real reduction of the cost of natural gas for final users would require an opening of the gas market that can only be fully achieved after the unbundling of the gas network, run by ENI. Such strategic management of the import infrastructure and the full liberalization of the market could, in principle, make Italy a European hub for natural gas, as hoped by the government. However, due to infrastructure constraints but also firm opposition from vested interests, this goal has not been achieved yet.

Climate security

The EU 2020 targets and their implications for Italy

After the adoption of the Kyoto Protocol, the European Commission set an ambitious plan to achieve higher environmental sustainability for the EU's energy sector. A set of targets was calculated for Europe as a whole and national targets were assigned to each member state.[56] In Italy these targets entail increasing the share of renewable energy sources in gross energy consumption to 17.0%, with a share of 26.4% in electricity generation by 2020. National greenhouse gas emissions were set to be reduced by 13% in 2020, as compared to their 2005 levels. The target for gross energy consumption was set at a level of 158 Mtoe. A recent study, published by the European Energy Agency, stated that Italy is one of the 13 member states currently considered on course to achieve all the three targets by 2020.[57] In the period 2005–2013, efforts to increase energy efficiency and stagnating economic conditions decreased the gross final energy consumption by 10%. Currently, Italian energy intensity is lower than the EU-28 average. The relative contribution of renewable sources to the final energy mix increased by 158% in the same period, reaching a share of 11% in 2013.[58] Between 1990 and 2004, Italy recorded an increase in emissions due to economic growth. In more recent years, the combined effect of the economic crisis and the higher share of renewables in the energy mix led to a notable reduction of carbon emissions. Over the same period, oil use in power generation was replaced by natural gas. Energy efficiency and a rapid growth of production from renewables had a positive effect on dealing with GHG emission reductions (Figure 28.5). Italy over-achieved the intermediate target for the year 2013 and is considered on track for the achievement of the 2020 goal.

To achieve the objectives defined in the Kyoto Protocol (in terms of CO_2 emissions) and to meet the ambitious targets of EU directive 2009/28/EC, Italy adopted several policy instruments. These included tradable certificates for economically subsidizing energy efficiency and renewable sources, feed-in tariffs, investment subsidies, and tax deductions.

In Italy the green energy certificates system was introduced in 1999. It targeted large plants, while small installations (mainly micro-hydro and small photovoltaic for household use) were subsidized through feed-in tariffs. From that moment onwards both domestic energy production and energy imports were bound by obligatory quotas of renewable energy. The quotas were first set to 2% and later increased by an annual rate. The producers or importers of traditional energy had two options: either directly produce a growing amount of energy from renewable sources or cover part or all of their requirements by buying green certificates on the compliance market. Producers of renewable energy benefit from financial and pricing support: the electricity price and the revenue from GECs sales.[59] The system worked rather well at the beginning, but after a few years, under the combined effect of mismanagement in the compliance market and economic crisis, the price of the certificates fell considerably, aggravating the

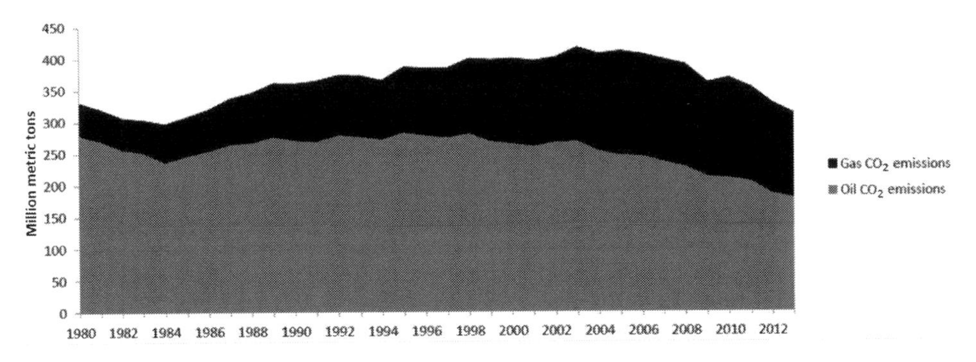

Figure 28.5 Carbon dioxide emission from the two main fossil fuels in the Italian energy mix 1980–2013.

Source: Authors' elaboration based on US EIA, International Energy Statistics database (2015). Retrieved from http://www.eia.gov/cfapps/ipdbproject/IEDIndex3.cfm#.

private investments on RES. The tradable certificate system was progressively phased out between 2013 and 2015 and replaced by feed-in tariffs.

In sum, the incentive scheme's impact on renewable sources installation was substantial, especially for wind and solar, which already account for significant shares of power generation, and for the refurbishment of medium and large hydropower plants. On the other hand, as in the case of new hydropower installations, it also created controversial dynamics. From 2000 to 2010, the number of plants increased by 39.3%, from 1,958 to 2,729, but installed capacity grew by only 0.7%, from 15,641MW to 17,826 MW.[60]

Moreover, many downside risks remain. Increasing shares of intermittent power put power grids to severe tests. Since the wind does not always blow and the sun does not always shine, and since electricity cannot easily be stored, there will be a need for massive investments to ensure demand-side flexibility.

In any event, barring major and, for the moment unexpected, changes in attitudes to nuclear, the contributions of renewables to Italy's power generation will continue growing. Wind and solar power have already captured substantial market shares and upon economic recovery will most likely pave the way for an even stronger support to renewables in the power sector.[61]

Conclusions

In recent years, Italian perceptions over the availability, reliability and affordability of energy sources have all shifted considerably. In terms of affordability, a disruptive economic recession in parallel with policies geared toward increased use of renewable energy left policy-makers with uncomfortable choices. The dilemma revolved around positive, albeit inevitably costly, climate and environmental policies and the competitiveness-driven imperative to reduce public expenditures, increase jobs, and foster growth. As for reliability, both the geopolitical tumult between Russia and Europe over Ukraine and the turbulent geopolitics of the Middle East significantly modified Italian views on the perceived supply risk associated with dependence on these regions.

Looking forward, Italian energy policies will be shaped by slow but profound transition processes, naturally accompanied by high levels of uncertainty. In the late 2010s Italy is expected to gradually climb out of recession and return to positive, albeit unimpressive, growth. However, should long-term commitment to reforms continue slacking, Italy runs the risk of sliding into a prolonged stagnation, with serious social unrest. In sum, the 2010s do not seem to be optimistic for Italy's economic growth. Meanwhile Italy's economic woes not only dampen overall energy demand, but also dilute efforts on the renewables build-up, potentially in favor of coal and gas. In fact, the national energy portfolio management will be heavily influenced by these developments. Gas will be increasingly used to back intermittent and unpredictable power generation by renewable energy and will remain in steady demand. The Italian gas market is the third largest in Europe with strong demand growth, especially from the power generation sector. Yet market competition has failed to achieve levels of competition already present in Northern European markets. This resulted not only in some of the highest European end-user gas prices, but also delayed development of a liquid trading gas hub. In short, liberalization failed to dismantle a rather rigid market structure organized around the incumbent, ENI.

The picture, however, is not uniformly bleak. As noted above, to achieve the objectives defined in the Kyoto Protocol (in terms of CO_2 emissions) and to meet the EU's ambitious 20–20–20 targets, Italy adopted several policy instruments that led to a notable reduction of carbon emissions in the energy mix. Given the combined effect of economic crisis and higher shares of renewables, Italian energy intensity is currently lower than the EU-28 average.

Oil demand already peaked and growth will be limited by its relatively high price and efficiency improvements in the transport sector, in addition to environmental policies. Although there are pressures to revise nuclear policy, Italy's long-term decision to do without nuclear power is not likely to be altered in the foreseeable future.

Natural gas will continue to account for the bulk of power generation although the new renewables – mainly wind and solar – are on the rise and already accounting for significant shares of power generation. This trend is driven mainly by climate and environmental policies, continued technological improvements, but also Italy's desire to mitigate local pollution and diversify energy supply. Hydrocarbons' dominance, however, remains high. As already noted, in the 1970s hydrocarbons peaked to over 80% and have never since fallen below that level. In addition, Italy is becoming increasingly reliant on imported oil and gas supplies. Nevertheless, there are encouraging developments on the hydrocarbons front, too.

To achieve the "diversification of gas sources" as defined by the National Energy Strategy, Italy played an instrumental role in bringing to fruition a southern corridor that will be taking fuel from the Caspian Sea through Azerbaijan, Georgia and Turkey, and into Europe, bypassing Russia. This policy represents a substantial shift from Italy's traditional position within the EU. The current fiscal crisis has dramatically exposed the dilemmas of an EU torn between federalist and nationalistic pressures. In that respect, European euro-sceptics and European federalists alike feel vindicated in arguing that no monetary union could have worked without a fiscal union. The same dilemmas and differences in opinion have surrounded the EU-led energy policy. As for Italy, despite a preference for sovereignty in security of supply seen as critical for national security, the government has recently moderated its behavior. Subsequent to a geopolitical standoff with Russia over Ukraine, it has cautiously supported a more muscular Brussels-led energy policy.

In conclusion, Italy has started a painful but necessary reform process, which has triggered significant social turmoil. The "bite" of these much-needed political and economic reforms will largely determine the pace and the success of Italy's approach vis-à-vis its three key policy goals – energy security, climate security, and economic competitiveness.

By recently approving of a sweeping constitutional reform and a new electoral law, Italy is proving itself more flexible than many observers could have hoped. How vigorously the government is poised to improve the relationship between overall energy use and carbon intensity, however, remains to be seen.

Notes

1 Eurostat, "Eurostat Information Society Statistics," Eurostat, 2015, http://ec.europa.eu/eurostat.
2 Reduction of 20% carbon dioxide emissions and 20% increase in energy efficiency with respect to 1990, 20% of energy coming from renewable sources by 2020.
3 M. R. Virdis, M. Gaeta, E. De Cian, et al., *Pathways to Deep Decarbonization in Italy*, Deep Decarbonization Pathways Project, 2015, http://deepdecarbonization.org/countries/.
4 C. Cammi, and M. Assanelli, *An Overview of Italy's Energy Mix* (Paris: Institut français des relations internationales (IFRI), 2012).
5 Italian Ministry of Economic Development, *Stratergia Energetica Nazional* (Rome: Italian Ministry of Economic Development, 2013), www.sviluppoeconomico.gov.it/images/stories/normativa/20130314_Strategia_Energetica_Nazionale.pdf.
6 Italian Ministry of Economic Development, *Stratergia Energetica Nazionale*.
7 Virdis et al., *Pathways to Deep Decarbonization in Italy*.
8 International Energy Agency (IEA), *Oil and Gas Security: Emergency Response of IEA Countries – Italy* (Paris: International Energy Agency, 2010), www.iea.org/publications/freepublications/publication/oil-and-gas-emergency-policy–italy-2010-update.html.

9 A. Honoré, *The Italian Gas Market: Challenges and Opportunities* (Oxford: Oxford Institute for Energy Studies (OIES), 2013).
10 Italian Ministry of Economic Development, *Stratergia Energetica Nazionale.*
11 Ibid.
12 Honoré, *The Italian Gas Market.*
13 International Energy Agency (IEA), and Organization for Economic Co-operation and Development (OECD), *Development of Competitive Gas Trading in Continental Europe: How to Achieve Workable Competition in European Gas Markets?* (Paris: OECD, 2008), www.iea.org/publications/freepublications/publication/gas_trading.pdf.
14 G. Pastori, "Between Continuity and Change: The Italian Approach to Energy Security," in *Energy Security: Visions from Asia and Europe*, ed. A. Marquina (New York: Palgrave Macmillan, 2008).
15 IEA, and OECD, *Development of Competitive Gas Trading*, 2008.
16 Honoré, *The Italian Gas Market.*
17 International Energy Agency (IEA), *World Energy Outlook 2012* (Paris: IEA, 2012).
18 Honoré, *The Italian Gas Market.*
19 Ibid.
20 Cammi, and Assanelli, *An Overview of Italy's Energy Mix.*
21 Ibid.
22 IEA, *Oil and Gas Security.*
23 P. A. Toninelli, "Energy and the Puzzle of Italy's Economic Growth," *Journal of Modern Italian Studies* 15, no. 1 (2010): 107–127.
24 Pastori, "Between Continuity and Change."
25 Toninelli, "Energy and the Puzzle of Italy's Economic Growth."
26 IEA, *Oil and Gas Security.*
27 Virdis et al., *Pathways to Deep Decarbonization in Italy.*
28 IEA, *Oil and Gas Security.*
29 G. Luciani, and M. R. Mazzanti, "Italian Energy Policy: The Quest for More Competition and Supply Security," *The International Spectator* 41, no. 3 (2006): 87.
30 M. Skalamera, "Italy's Path to Gas Liberalisation: Corporate Power, Monopoly Distortions and the Russia Factor," *Contemporary Italian Politics* 7, no. 2 (2015): 161–184.
31 IEA, and OECD, *Development of Competitive Gas Trading.*
32 IEA, *Oil and Gas Security.*
33 Pastori, "Between Continuity and Change," 91.
34 Italian Ministry of Economic Development, *Strategia Energetica Nazional.*
35 Cammi, and Assanelli, *An Overview of Italy's Energy Mix.*
36 Pastori, "Between Continuity and Change."
37 Honoré, *The Italian Gas Market.*
38 P. Noel, *Beyond Dependence: How to Deal with Russian Gas*, European Council on Foreign Relations, 2008, www.ecfr.eu/publications/summary/beyond_dependence_how_to_deal_with_russian_gas.
39 K. Rosner, "The European Union: On Energy, Disunity," in *Energy Security Challenges for the 21st Century: A Reference Handbook*, ed. G. Luft and A. Korin (Portsmouth, NH: Greenwood Publishing Group, 2009).
40 C. van der Linde, *Turning a Weakness into Strength. A Smart External Energy Policy for Europe* (Paris: Institut Français des Relations Internationales (IFRI), 2008).
41 The new powers are embodied in the Article 194 of the Lisbon Treaty, see www.lisbon-treaty.org/wcm/the-lisbon-treaty/treaty-on-the-functioning-of-the-european-union-and-comments/part-3-union-policies-and-internal-actions/title-xxi-energy/485-article-194.html.
42 D. Tusk, "A United Europe Can End Russia's Energy Stranglehold," *Financial Times*, April 21, 2014.
43 R. Youngs, *Energy Security: Europe's New Foreign Policy Challenge* (New York: Routledge, 2009).
44 Honoré, *The Italian Gas Market.*
45 Italian Ministry of Economic Development, *Stratergia Energetica Nazionale.*
46 Ibid.
47 Ibid.
48 Oxford Institute for Energy Studies, Oxford Institute for Energy Studies, 2012, www.oxfordenergy.org/wpcms/wp-content/uploads/2012/04/Lessons-from-the-February-2012-gas-crisis.pdf.
49 Italian Ministry of Economic Development, *Stratergia Energetica Nazionale.*

50 Enerdata, 2015, "Energy intensity of GDP at constant purchasing power parities," https://yearbook.enerdata.net/energy-intensity-GDP-by-region.html.
51 While current capacity is 12 bcm. Italian National Strategy, 2013, 66.
52 The Southern Gas Corridor will supply the EU with Caspian gas from the Shah Deniz 2 field. Two pipeline networks, TANAP and TAP, will transport Azeri natural gas through Turkey, Greece, and Albania to the final destination, Italy.
53 www.ilmessaggero.it/PRIMOPIANO/POLITICA/renzi_mozambico_eni_investimento_50_miliardi_descalzi_assicura_non_andremo_via_gela/notizie/806148.shtml.
54 Italian Ministry of Economic Development, *Stratergia Energetica Nazionale*.
55 Honoré, *The Italian Gas Market*.
56 European Commission, *Europe 2020 A Strategy for Smart, Sustainable and Inclusive Growth* (COM(2010)), European Commission, 2010, http://eur-lex.europa.eu/legal-content/EN/TXT/?uri=URISERV:em0028.
57 European Environmental Agency, *Trends and Projections in Europe 2015: Tracking Progress towards Europe's Climate and Energy Targets* (Copenhagen: European Environmental Agency, 2015), www.eea.europa.eu/publications/trends-and-projections-in-europe-2015.
58 Ibid.
59 F. Farinosi, L. Carrera, J. Mysiak, et al., "Tradable Certificates for Renewable Energy: The Italian Experience with Hydropower," in *2012 9th International Conference on the European Energy Market*, 1–7, IEEE, 2012, http://ieeexplore.ieee.org/lpdocs/epic03/wrapper.htm?arnumber=6254695.
60 Ibid.
61 Italian Ministry of Economic Development, *Stratergia Energetica Nazionale*.

29

Energy transitions in the Caribbean and Central America

Robert E. Looney

Introduction

If one had to pick one area in the Western Hemisphere where an energy transition away from imported oil and coal would be the most beneficial, it would no doubt be the Caribbean/ Central American region. Currently both are beginning to experience the effects of global warming in the form of more powerful storms, shore erosion and severe droughts.

Both regions also have extremely low levels of energy security with many countries totally dependent on imported oil to meet their energy needs, and thus most exposed to external shocks stemming from rapid changes in the international price of oil. Only one country in the region, Trinidad and Tobago, has sufficient reserves of energy to be a net exporter. In part, the region's energy situation has resulted in many of the hemisphere's lowest per capita income countries being located there.

A successful energy transition at this time is even more necessary because the current system is unsustainable. The two regions have historically bridged the gap between low income and high energy costs with subsidized access to oil from regional producers. Subsidized oil was first provided by Mexico and Venezuela through the Acuerdo de San José beginning in 1980, and more recently by Venezuela alone through Petrocaribe beginning in 2001.

Petrocaribe has enabled member states to take advantage of a deferred payment system for up to 50% of market prices. The deferred payment was due in 25 years at 1–2% interest. Cuba, Nicaragua and Haiti were three of the larger beneficiaries. Steady access to cheap Venezuelan oil gave the member countries little incentive to diversify away from oil, only compounding their dependence on foreign energy imports. Intentionally or not, the loans through the Petrocaribe program have resulted in a number of the Central American and Caribbean becoming heavily indebted[1] to Venezuela in addition to becoming dependent on Venezuela for their energy security. Countries like Haiti, Nicaragua and Cuba have thus become quite vulnerable to Venezuelan political pressure.[2] However, with the decline in Venezuelan oil production together with the sharp drop in oil prices beginning in the fall of 2014, Venezuela has begun to cut its oil subsidies[3] and it may only be a matter of time before it is forced to withdraw the program completely.

As noted, another compelling reason for initiating an energy transition is the increasing cost of inaction.[4] The two regions, particularly the Caribbean are likely to be the most adversely affected parts of the world by climate change. Increased hurricane intensity, hotter temperatures and increased ocean levels threaten livelihoods[5] and lives throughout the region.[6]

Fortunately, several fortuitous developments have the potential of lowering the costs of transitioning away from imported oil. First while it is always extremely difficult to forecast future oil prices there is the distinct likelihood that prices will remain relatively low for some time.[7] Lower oil prices should result in subdued energy prices even as the transition to alternative fuels takes place. Second, both regions are potential beneficiaries of energy initiatives such as authorizing the export[8] of crude oil on the part of the United States, together with the energy revolution in that country and the potential for low cost liquefied natural gas (LNG) exports[9] to the region. Finally, rapid progress in green energy technologies[10] along with the low price of oil are significantly reducing the costs of transitioning away from fossil fuels

Even though both the motives to and means for initiating major efforts at energy transitions exist it is apparent that success is far from assured. While there are a number of common themes that run between the Caribbean and Central America, broad generalizations are difficult. Significant differences exist between the regions and for individual countries within each region. The next section examines some of the more relevant differences for the purpose of identifying where gains towards transitioning away from imported energy have occurred, and where progress has been more elusive.

Energy trends

While both the Central American and Caribbean countries, with the notable exception of Trinidad and Tobago, are net importers of energy, significant differences exist. Due largely to reduced opportunities for hydro-power, dependence on imports for energy use is on average about 10% higher for the Caribbean group (Table 29.1). The Central American countries have developed an integrated electricity system[11] linking the six countries, a development impossible to implement in the island states of the Caribbean.

The Caribbean region has a population of about 40 million (about 12% of the US population) and consumes roughly 50,000 gigawatt-hours of electricity per year (about 1% of US electricity demand). The World Bank projects the Caribbean region's electricity demand nearly to double by 2030.[12]

The region is dependent on fossil fuels, with oil and natural gas supplying roughly 90% of electricity. The transportation sector is also almost exclusively dependent on oil, leaving the region suffering from energy prices that are not only high but also extremely volatile.

Within the Caribbean sample of countries both the Dominican Republic and Jamaica have the highest rates of energy imports, both countries averaging use percentages in the 80s. However, with the exception of Cuba, all of the sample countries achieved reductions in their energy import percentages between 2000 and 2012 (the latest date of available data). Energy import percentages were the lowest for Haiti, and that country also had the greatest percentage reduction in imports (22.1%) between 2000 and 2012. In sharp contrast, during this period, Cuba increased its percentage of imported energy use by 16.4%.

Costa Rica experienced the greatest reduction between 2000 and 2012 in the percentage (16.9%) of its energy coming from imports, while Panama experienced the greatest increase (10.8%) in its imported energy. While El Salvador, Honduras and Guatemala all had slight increases in their imports as a share of total energy use, Nicaragua's fell by nearly 6%.

In the area of fossil fuel shares in total energy consumption the picture is one of gradually increasing shares in the Caribbean, and falling shares in Central America (Table 29.2). On average fossil fuels account for about 15–20% more in total energy consumption in the

Table 29.1 Energy imports, net (% of energy use)

Country	*2000*	*2005*	*2010*	*2011*	*2012*	*2000–2012 % change*
Central America						
Costa Rica	57.6	46.9	47.6	48.2	47.9	−16.9
El Salvador	46.6	44.8	46.2	47.6	47.5	1.9
Guatemala	25.1	30.8	26.4	25.6	25.6	1.8
Honduras	49.1	55.0	51.2	51.4	51.7	5.4
Nicaragua	46.4	48.0	46.8	48.5	43.7	−5.8
Panama	70.6	73.2	77.3	79.8	78.2	10.8
Average	*49.2*	*49.8*	*49.2*	*50.2*	*49.1*	*-0.3*
Caribbean						
Cuba	44.3	46.2	54.0	51.3	51.5	16.4
Dominican Republic	90.7	88.3	89.5	89.7	89.4	−1.4
Jamaica	84.6	88.9	82.8	82.6	82.2	−2.9
Haiti	23.3	19.5	17.8	18.0	18.2	−22.1
Average	60.7	60.7	61.0	60.4	60.3	−0.6
Energy exporter						
Trinidad and Tobago	−93.2	−116.6	−112.0	−104.9	−107.0	14.8

Source: Data from database: World Development Indicators, http://data.worldbank.org/data-catalog/world-development-indicators. Last updated: 10/14/2015.

Table 29.2 Fossil fuel energy consumption (% of total)

Country	*2000*	*2005*	*2010*	*2011*	*2012*	*2000–2012 % change*
Central America						
Costa Rica	59.0	46.9	47.5	48.3	47.8	−19.0
El Salvador	44.9	44.1	46.0	47.4	47.3	5.3
Guatemala	42.3	44.2	32.7	30.9	30.9	−27.0
Honduras	48.3	54.9	51.2	51.6	51.6	6.9
Nicaragua	46.0	48.0	46.9	48.6	43.7	−5.1
Panama	70.2	73.4	77.2	79.7	78.3	11.5
Average	*51.8*	*51.9*	*50.3*	*51.0*	*49.9*	*-4.6*
Caribbean						
Cuba	69.9	80.1	88.8	86.7	86.5	23.8
Dominican Republic	90.7	88.3	89.5	89.7	89.4	−1.4
Jamaica	84.6	88.9	82.8	82.6	82.2	−2.9
Haiti	23.3	19.5	17.8	48.0	18.2	−22.1
Average	*67.1*	*69.2*	*69.7*	*69.2*	*69.1*	*2.9*
Energy exporter						
Trinidad and Tobago	99.8	99.8	99.9	99.9	99.9	−5.9

Source: Data from database: World Development Indicators, http://data.worldbank.org/data-catalog/world-development-indicators. Last updated: 10/14/2015.

Caribbean, with fossil fuels making up over 80% of energy consumption in Cuba, Dominican Republic and Jamaica. In contrast the highest percentage of fossil fuel consumption occurs in Panama where it averages in the 70s.

Between 2000 and 2012, the largest reductions in fossil fuel shares in energy consumption were in Guatemala (27%), Haiti (22.1%) and Costa Rica (19%). Cuba accounted for the largest increase in fossil fuel energy shares (23.8%) followed by Panama at 11.5%.

The development of alternative energies such as solar, wind and other natural sources has also varied across the two regions (Table 29.3). In general, alternative energies account for a much greater share of total energy use in Central America, with these sources averaging in the high teens. In contrast alternative sources have never accounted for 1% of total energy use in the Caribbean. Percentage-wise, both regions had a considerable expansion in these types of energy during 2000 to 2012, with Central American alternative sources increasing by 37.5% and those in the Caribbean countries at 33.1%, albeit from a very low base. In absolute shares, the biggest gains have occurred in Costa Rica, El Salvador and Nicaragua. At the same time significant reductions occurred in Honduras and Haiti.

In contrast to the sharp differences in energy patterns between the Caribbean and Central America noted previously, combustible renewables such as biogas, industrial waste, biomass, and municipal waste average around 30–35% of total energy use in both the Caribbean and Central America (Table 29.4). Also both regions had fairly significant reductions in this source of energy with its share of total energy falling by 8.8% between 2000 and 2012 in Central America and 6.7% in the Caribbean. Also between 2000 and 2012 significant reductions in this fuel source occurred in Cuba (55.4%), El Salvador (44.9%) and Panama (40.9%). The only major increase occurred in Costa Rica (55.7%).

Table 29.3 Alternative and nuclear energy (% of total energy use)

Country	*2000*	*2005*	*2010*	*2011*	*2012*	*2000–2012 % change*
Central America						
Costa Rica	33.8	36.6	35.2	36.0	38.7	14.6
El Salvador	19.6	23.2	35.3	34.8	33.8	72.6
Guatemala	3.3	5.1	5.5	5.2	5.4	61.4
Honduras	6.5	3.6	5.8	5.3	5.3	−18.5
Nicaragua	5.3	9.5	10.7	9.5	15.5	193.2
Panama	11.4	11.1	9.7	8.7	11.2	−2.4
Average	*13.3*	*14.8*	*17.1*	*16.6*	*18.3*	*37.5*
Caribbean						
Cuba	0.1	0.1	0.1	0.1	0.1	69.8
Dominican Republic	0.9	2.5	1.7	1.8	2.0	120.5
Jamaica	0.3	0.5	0.7	0.7	0.8	206.5
Haiti	1.2	0.7	0.4	0.3	0.3	−72.1
Average	*0.6*	*0.9*	*0.7*	*0.7*	*0.8*	*33.1*
Energy exporter						
Trinidad and Tobago	0.0	0.0	0.0	0.0	0.0	0.0

Source: Data from database: World Development Indicators, http://data.worldbank.org/data-catalog/world-development-indicators. Last updated: 10/14/2015.

Table 29.4 Combustible renewables and waste (% of total energy)

Country	*2000*	*2005*	*2010*	*2011*	*2012*	*2000–2012 % change*
Central America						
Costa Rica	8.6	16.5	17.3	15.8	13.4	*55.7*
El Salvador	34.0	32.1	18.5	17.6	18.7	*−44.9*
Guatemala	55.3	51.1	61.6	63.6	63.7	*15.3*
Honduras	44.4	41.4	43.0	43.3	43.0	*−3.3*
Nicaragua	48.3	42.5	42.5	42.0	40.8	*−15.6*
Panama	18.0	15.7	13.0	11.5	10.6	*−40.9*
Average	*34.8*	*33.2*	*32.6*	*32.3*	*31.7*	*−8.8*
Caribbean						
Cuba	30.0	19.8	11.1	13.2	13.4	*−55.4*
Dominican Republic	8.4	9.2	8.7	8.5	8.5	*1.7*
Jamaica	15.1	10.6	16.6	16.7	17.0	*12.5*
Haiti	75.5	79.8	81.8	81.8	81.5	*8.0*
Average	*32.3*	*29.9*	*29.6*	*30.0*	*30.1*	*−6.7*
Energy exporter						
Trinidad and Tobago	0.22	0.25	0.07	0.07	0.07	*−67.0*

Source: Data from database: World Development Indicators, http://data.worldbank.org/data-catalog/world-development-indicators. Last updated: 10/14/2015.

While there is some controversy as to how best to measure energy efficiency, a simple ratio of GDP per unit of energy use (Table 29.5) provides a rough picture of improvements in the use of energy. On this basis both the Caribbean and Central American countries have been improving their use of energy, with the Caribbean countries as a whole seeing a 61.7% improvement between 2000 and 2010, compared with a 15.6% improvement in Central America. Gains have been across the board with significant increases in Cuba (99.8%), Dominican Republic (55.6%) and Jamaica (47.5%). In Central America, Panama had a gain in energy efficiency of 75.1% during this time. On the other hand, two other Central American countries, Guatemala (−5.3%) and Honduras (−4.5%) saw the amount of GDP supported by a unit of energy fall during this period.

The patterns noted above illustrate the difficulty of drawing broad conclusions over the energy situation in the region as a whole, or even for the Caribbean or Central America separately. Clearly several of the patterns can be explained by differential resource endowments, especially the limited alternative energy sources in the Caribbean. Electricity generation is a prime example. In a sample of Caribbean countries 94.2% of electricity generation comes from conventional thermal plants (Table 29.6). In contrast conventional thermal accounts for only 38.1% of generation in Central America.

With the potential for hydro-power much greater, Central America generates 43.4% of its electricity from this source compared with only 4.1% in the Caribbean. Clearly however other factors are at work. While one might anticipate a greater role for renewable energy in the Caribbean, this source generates only 1.7% of our regional sample's electricity compared with 18.5% in Central America.

Table 29.5 GDP per unit of energy use (constant 2011 PPP $ per kg of oil equivalent)

Country	*2000*	*2005*	*2010*	*2011*	*2012*	*2000–2012 % change*
Central America						
Costa Rica	13.1	11.9	12.4	12.9	13.4	2.3
El Salvador	9.4	9.3	10.7	10.7	10.7	13.8
Guatemala	10.0	10.5	9.6	9.4	9.5	−5.3
Honduras	7.3	6.6	7.1	7.1	6.9	−4.5
Nicaragua	6.9	7.1	7.8	8.0	7.8	12.2
Panama	9.4	11.6	14.5	15.0	16.4	75.1
Average	*9.4*	*9.5*	*10.4*	*10.5*	*10.8*	*15.6*
Caribbean						
Cuba	9.7	14.8	18.1	19.1	19.4	99.8
Dominican Republic	10.0	12.4	15.6	15.8	15.5	55.6
Jamaica	5.5	6.2	8.4	8.0	8.1	47.5
Haiti	7.3	8.1	8.6	9.0	9.4	30.1
Average	*8.1*	*10.4*	*12.7*	*13.0*	*13.1*	*61.7*
Energy exporter						
Trinidad and Tobago	2.3	2.1	2.0	2.0	2.1	**−10.3**

Source: Data from database: World Development Indicators, http://data.worldbank.org/data-catalog/world-development-indicators. Last updated: 10/14/2015.

Table 29.6 Diversity of electricity generation (% by source)

	Conventional thermal	*Hydroelectric*	*Other renewables*
Central America			
Costa Rica	7.8	71.2	21.0
El Salvador	39.3	29.8	30.9
Guatemala	31.9	48.1	20.0
Honduras	54.9	37.7	7.3
Nicaragua	58.9	9.9	31.3
Panama	35.9	63.8	0.3
Average	*38.1*	*43.4*	*18.5*
Caribbean			
Barbados	100.0	0.0	0.0
Dominican Republic	86.2	12.8	0.9
Jamaica	90.7	3.7	5.6
Trinidad and Tobago	99.8	0.0	0.2
Central America	*94.2*	*4.1*	*1.7*

Source: World Energy Council, *Energy Trilemma Index*, 2015, https://www.worldenergy.org/data/trilemma-index/country/

Other differences no doubt stem from policy choices made by individual governments. If one accepts the basic premise of the energy trilemma – that a country cannot simultaneously have marked improvements in energy security, energy affordability and environmental quality, then movements in these three outcomes should in part reveal changes in national priorities in the energy area.

For the Caribbean sample (Table 29.7) Barbados appears relatively strong in energy equity, and in recent years has shifted towards energy security at the expense of environmental sustainability. The Dominican Republic on the other hand scores well in environmental sustainability and has given energy equity priority over energy security. Jamaica's highest priority has been energy equity. The country has traditionally favored environmental stability over energy

Table 29.7 The Energy Trilemma: Caribbean

	2011	*2012*	*2013*	*2014*	*2015*	*2011/2015*		*Score*
						Difference	*Average*	
Barbados								
Energy performance	*32*	*36*	*59*	*63*	*91*	−59	56.2	
Energy security	68	70	118	117	114	−46	97.4	D
Energy equity	51	45	41	34	47	4	43.6	B
Environmental sustainability	19	24	25	40	84	−65	38.4	C
Dominican Republic								
Energy performance	*112*	*106*	*104*	*92*	*100*	12	102.8	
Energy security	120	119	114	111	119	1	116.6	D
Energy equity	111	107	106	87	88	23	99.8	C
Environmental sustainability	56	54	55	54	56	0	55.0	B
Jamaica								
Energy performance	*120*	*119*	*123*	*117*	*124*	−4	120.6	
Energy security	128	127	116	121	126	2	123.6	D
Energy equity	78	76	81	79	82	−4	79.2	C
Environmental sustainability	96	98	110	90	105	−9	99.8	D
Trinidad								
Energy performance	*82*	*86*	*113*	*64*	*75*	7	84.0	
Energy security	66	74	79	50	61	5	66.0	B
Energy equity	50	49	95	30	33	17	51.4	B
Environmental sustainability	116	116	115	112	111	5	114.0	D

Source: World Energy Council, *Energy Trilemma Index*, 2015, https://www.worldenergy.org/data/trilemma-index/country/

security, although that may be changing with the country's recent drop in country ranking on environmental stability and improvement in the energy security area. As an oil producer Trinidad automatically scores well in the energy security dimension, with energy equity a close second. For the energy importers, the general pattern therefore is one of energy equity and environmental sustainability coming at the expense of energy security.

Energy priorities for the Central American countries are a bit more difficult to pin-point. Instead of the expected trilemma pattern of two energy dimensions dominating a third, the situation is generally one of a single dimension dominating the other two. As a virtue of their use of hydro-power all score the highest on the environmental sustainability dimension (Table 29.8). This doesn't necessarily signify that environmental sustainability was a goal in and of itself, but more likely that hydropower was an attractive source of electricity.

For these countries therefore, marginal changes in their rankings are likely to provide a more accurate picture of national energy priorities than that revealed through an examination of absolute levels. From this perspective, in recent years El Salvador has given preference to energy equity over energy security. The same applies to Guatemala, although that country continues to make progress in improving its environmental sustainability.

In recent years Honduras has seen a deterioration along all three energy dimensions. The decline was the greatest in the area of energy equity, suggesting a policy shift towards energy security. Nicaragua continues to make good progress in environmental sustainability while energy equity appears to take precedence over energy security. Costa Rica has one of the top scores in environmental sustainability, with the last few years signaling a shift towards energy security over energy equity. Finally, Panama continues to score extremely well in environmental sustainability with a shift towards energy equity over energy security in recent years.

Summing up, the general pattern revealed by country performance on the three dimensions of the energy trilemma is one where the Caribbean countries have given preference to energy equity and energy security over environmental sustainability. A more varied pattern exists in Central America with countries displaying continued interest in environmental sustainability, but with some attaching more preference to energy security and others showing concern for energy equity.

US initiatives towards the region

The Caribbean had an average cost of 0.33 dollars per kilowatt-hour (kWh)[13] in 2012, nearly three times the US cost of electricity – a considerable economic burden for most of the region's economies. High electricity rates together with the extreme vulnerability to external developments faced by most of the energy poor Caribbean countries, and as noted, with the prospect that Venezuela may have to begin withdrawing its energy subsidy, Petrocaribe program, the United States has developed several energy related initiatives for the region.[14] Given its concerns over global warming and the underdevelopment of renewable energy sources throughout the region, the Obama Administration has focused its efforts on expanding this source of energy. Several highlights include a $20 million financing facility for green energy projects in the region. The US Overseas Private Investment Corporation (OPIC) will also be involved providing financing and insurance for new energy projects.

In addition to the programs initiated by the Obama Administration there are a number of public and private programs:

- The Ten Island Renewable Challenge/Operation Smart Island Economies. The Carbon War Room and Rocky Mountain Institute provide various services, including technical assistance to create roadmaps for Caribbean countries and financing.

Table 29.8 The Energy Trilemma: Central America

	2011	*2012*	*2013*	*2014*	*2015*	*2011/2015*		*Score*
						Difference	*Average*	
El Salvador								
Energy performance	*33*	*41*	*37*	*36*	*56*	−23	40.6	
Energy security	61	71	68	61	96	−35	71.4	C
Energy equity	63	67	64	71	66	−3	66.2	C
Environmental sustainability	15	11	11	11	18	−3	13.2	A
Guatemala								
Energy performance	*41*	*47*	*38*	*26*	*48*	−7	40.0	
Energy security	47	31	40	31	73	26	[illegible]	C
Energy equity	69	72	75	73	76	−7	73.0	C
Environmental sustainability	30	35	36	29	16	14	29.2	A
Honduras								
Energy performance	*97*	*94*	*95*	*105*	*117*	−20	101.6	
Energy security	116	116	111	114	125	−9	116.4	D
Energy equity	86	83	90	102	102	−16	92.6	D
Environmental sustainability	57	53	52	55	68	−11	57.0	B
Nicaragua								
Energy performance	*114*	*115*	*105*	*99*	*109*	5	108.4	
Energy security	102	105	100	100	111	−9	103.6	D
Energy equity	101	101	91	101	101	0	99.0	D
Environmental sustainability	85	89	87	59	67	18	77.4	B
Costa Rica								
Energy performance	*22*	*25*	*18*	*17*	*22*	0	20.8	
Energy security	71	77	57	51	64	7	64.0	B
Energy equity	46	47	45	56	128	−82	64.4	B
Environmental sustainability	2	2	2	2	2	0	2.0	A
Panama								
Energy performance	*25*	*27*	*28*	*43*	*51*	−26	34.8	
Energy security	55	54	53	86	104	−49	70.4	D
Energy equity	59	60	58	50	54	5	56.2	B
Environmental sustainability	11	14	18	17	15	−4	15	A

Source: World Energy Council, *Energy Trilemma Index*, 2015, https://www.worldenergy.org/data/trilemma-index/country/

- Caribbean Climate Innovation Center. The World Bank's Climate Technology Program has created eight Climate Innovation Centers (CIC) throughout the world. The Caribbean CIC supports ventures through services such as proof of concept funding and access to technical facilities and information.
- Canada's Caribbean Program. Through their Foreign Affairs, Trade, and Development division, Canada's Caribbean Program serves eleven island and three continental nations, although not specifically addressing the energy sector.

There is a logic to focusing on renewables, and also for government financial assistance in developing projects in the area. Despite numerous physical opportunities for the development of renewables, particularly solar and wind, as noted above, there has been insignificant development throughout the region. The main constraint has been the limited opportunities for commercial profitability. In many Caribbean countries, state-owned electricity producers enjoy entrenched monopolies and have therefore not felt enough pressure to accommodate alternative sources of energy.

Currently, low oil prices and the prospects for them remaining low for some time are working to discourage investments in alternative energies. High upfront costs for renewable systems will likely continue to be the primary limiting factor to the industry. Still a number of other factors have tended to inhibit investment in renewables throughout the region.

Most importantly foreign investors have been overly cautious over financing projects owing to concerns that existing rules favor local firms or are changed capriciously for political reasons. On a broader level, there is also the perception held by many investors that many countries throughout the region have weak governance structures and retain outdated anti-business sentiments. In short, potential investors in renewable energy projects perceive a high level of resistance to private capital and regulatory reform by many countries. They are not convinced national institutional structures exist to the extent that even expanded initiatives by the United States and other countries will gain much traction in domestic markets. These are important considerations given the fact that nearly all the investments in renewables will have to be undertaken by the private sector.

There is good reason for concern. A quick glance at the World Bank's Ease of Doing Business data base shows that the Caribbean in particular ranks very poorly in terms of providing an environment conducive to large scale private sector participation. The Caribbean sample of Barbados, Dominican Republic, Haiti, Puerto Rico, Jamaica and Trinidad had an average rank of 100.5 out of 189 at the start of 2016 (Table 29.9). This was 12.3% higher than that found in Central America. While it was easier to start a business in the Caribbean, the average rank of Central American countries was 64.7% for credit and 20.3% for electricity.

With regard to concerns over governance throughout the region, at a January 2015 summit in Washington, DC as part of the "Caribbean Energy Security Initiative," the US vice-president, Joe Biden, told the region's leaders that they ought to harmonize regulatory frameworks and ensure that dispute-resolution systems were "predictable." Progress in this area while necessary is only a start at what has to be done to bring the region up to competitive standards with other potential investment destinations.

The World Bank's measure of regulatory quality (Table 29.10) illustrates the magnitude of the problem, particularly in the Caribbean region. Regulatory quality is generally low by international standards in both the Caribbean and Central America with the Caribbean sample of countries averaging in the high 30s, and Central America in the low 50s. Furthermore, the gap between the two regions has increased considerably over time from 25% in 1996 to 43% in 2014 as a result of fairly steady progress in Central America and retrogression in the Caribbean.

Table 29.9 Caribbean/Central America ease of doing business

Countries	*Ease of doing business*		*Starting a business*		*Getting electricity*		*Getting credit*		*Enforcing contracts*	
	2016	*2015*	*2016*	*2015*	*2016*	*2015*	*2016*	*2015*	*2016*	*2015*
Caribbean										
Barbados	119	116	100	95	87	82	126	118	164	164
Dominican Republic	93	90	110	116	140	148	97	90	115	114
Haiti	182	179	188	187	136	133	174	171	123	120
Puerto Rico	57	56	51	45	57	57	7	6	100	100
Jamaica	64	71	9	17	80	67	7	12	107	107
Trinidad	88	85	72	68	27	24	42	36	167	167
Average	*100.5*	*99.5*	*88.3*	*88.0*	*87.8*	*85.2*	*75.5*	*72.2*	*129.3*	*128.7*
Central America										
El Salvador	86	97	125	120	107	106	107	106	109	109
Honduras	110	115	150	139	143	143	143	143	150	150
Guatemala	81	81	101	99	21	18	21	18	173	173
Nicaragua	125	123	123	119	94	90	94	90	94	94
Costa Rica	58	79	121	116	23	30	23	30	124	121
Panama	69	66	44	37	32	31	32	31	148	147
Average	*88.2*	*93.5*	*110.7*	*105.0*	*70.0*	*69.7*	*70.0*	*69.7*	*133.0*	*132.3*
Caribbean/CA % difference	*12.3*	*6.0*	*-25.3*	*-19.3*	*20.3*	*18.2*	*20.3*	*18.2*	*-2.8*	*-2.8*

Source: World Bank, *Doing Business*, http://www.doingbusiness.org/rankings.

Table 29.10 Regulatory quality (percentile)

Country	*1996*	*2000*	*2005*	*2010*	*2013*	*2014*	*1996–2014 % change*
Central America							
Costa Rica	72.1	70.6	66.7	68.9	68.9	70.2	−2.6
El Salvador	52.0	57.4	53.4	62.7	60.3	64.4	24.0
Guatemala	43.0	48.5	41.7	48.8	45.5	47.6	10.7
Honduras	26.5	37.3	30.4	45.5	45.9	38.9	47.1
Nicaragua	37.3	48.0	42.2	43.1	43.1	38.5	3.2
Panama	68.6	69.1	58.3	63.2	62.7	65.4	−4.7
Average	*49.9*	*55.1*	*48.8*	*55.3*	*54.4*	*54.2*	*8.6*
Caribbean							
Cuba	11.3	10.8	6.9	3.4	3.8	7.2	−36.0
Dominican Republic	41.7	44.1	42.7	47.9	48.8	51.4	23.4
Jamaica	60.3	57.8	59.3	58.9	58.4	56.7	−5.9
Haiti	17.2	15.2	11.3	17.7	19.1	14.4	−16.0
Trinidad and Tobago	69.1	72.1	67.7	68.4	59.3	59.6	−13.7
Average	*39.9*	*40.0*	*37.5*	*39.2*	*37.9*	*37.9*	
Caribbean/CA % difference	*25.0*	*37.9*	*29.9*	*41.1*	*43.5*	*43.0*	

Source: World Energy Council, *Energy Trilemma Index*, 2015, http://data.worldbank.org/data-catalog/worldwide-governance-indicators

In 2014 average scores in Central America were 8.6% higher than in 1996, whereas those in the Caribbean were 5.1% lower.

The main declines (in percentile terms) in regulatory quality between 1996 and 2014 occurred in Cuba (36%), Haiti (16%), Trinidad and Tobago (13.7%). On the other hand, the Dominican Republic experienced an improvement of 23.4%. In Central America despite slight declines in already high levels in Costa Rica and Panama, major gains were registered by Honduras (47.2%) and El Salvador (24%).

A related World Bank measure, government effectiveness, also has great relevance for implementing successful strategies in the renewable energy area. In contrast to regulatory quality government effectiveness does not show great differences between the Caribbean countries and those in Central America. However, rankings are generally quite low (35–40 percentiles), although they have been gradually improving in both regions (Table 29.11). Cuba is a bright spot in the Caribbean with a 193% improvement over its 1996 percentile, moving from the 18.1st percentile in 1996 to the 52.9th in 2014. On the other hand, Haiti's percentile ranking fell from 9.3 in 1996 to 1st by 2014 or a decline of 89.6%. During this period, government effectiveness also declined in neighboring Dominican Republic, but at the much lower rate of 12.8%.

El Salvador has led the way in Central America with its percentile increasing by 114.5% between 1996 and 2014, at the same time however both Guatemala (−28.2%) and Nicaragua (−8.3%) suffered declines.

A final index of interest for investors is that of corruption (Table 29.12). The World Bank's control of corruption measure shows another interesting contrast between the Caribbean and Central American countries. As with government effectiveness, both regions have poor records

Table 29.11 Government effectiveness (percentile)

Country	*1996*	*2000*	*2005*	*2010*	*2013*	*2014*	*1996–2014 % change*
Central America							
Costa Rica	62.0	62.9	59.0	65.1	67.9	69.2	11.8
El Salvador	24.9	35.1	44.4	56.0	49.8	53.4	114.5
Guatemala	34.1	35.6	27.8	28.7	27.3	24.5	−28.2
Honduras	*19.5*	*33.7*	*30.7*	*31.1*	*25.8*	*20.7*	5.9
Nicaragua	21.0	30.2	24.4	16.7	23.0	19.2	−8.3
Panama	57.6	62.4	**37.3**	59.8	62.7	63.0	9.4
Average	*36.5*	*43.3*	*37.3*	*42.9*	*42.7*	*41.7*	*14.2*
Caribbean							
Cuba	18.1	45.9	36.6	42.6	39.7	52.9	193.0
Dominican Republic	42.4	42.9	35.1	29.7	36.8	37.0	−12.8
Jamaica	60.5	56.6	54.6	62.7	55.0	60.1	−0.6
Haiti	9.3	3.9	5.9	2.9	3.8	1.0	−89.6
Trinidad and Tobago	56.1	70.2	60.5	63.6	63.6	64.4	14.8
Average	*37.3*	*43.9*	*38.5*	*40.3*	*39.8*	*43.1*	15.6

Source: World Energy Council, *Energy Trilemma Index*, 2015, http://data.worldbank.org/data-catalog/worldwide-governance-indicators.

Table 29.12 Control of corruption

Country	*1996*	*2000*	*2005*	*2010*	*2013*	*2014*	*1996–2014 % change*
Central America							
Costa Rica	73.7	79.0	66.8	72.4	71.8	75.0	1.8
El Salvador	21.5	36.1	42.4	51.4	47.8	43.3	101.6
Guatemala	22.9	28.3	32.7	37.6	33.0	28.4	23.7
Honduras	**12.7**	**19.5**	**26.8**	**19.5**	**17.2**	**23.6**	85.7
Nicaragua	32.2	10.7	34.1	23.3	25.4	19.2	−40.3
Panama	42.4	43.4	44.8	45.7	45.9	46.2	8.8
Average	*34.2*	*36.2*	*41.3*	*41.7*	*40.2*	*39.3*	*14.7*
Caribbean							
Cuba	70.7	71.2	63.9	69.5	62.2	58.7	−17.0
Dominican Republic	54.6	31.2	35.1	21.9	21.1	23.1	−57.8
Jamaica	49.3	53.7	43.9	44.3	46.5	43.8	−11.2
Haiti	10.2	4.9	2.9	7.1	11.0	7.7	−24.9
Trinidad and Tobago	83.4	62.0	55.1	45.2	47.4	33.7	−59.7
Average	*53.7*	*44.6*	*40.2*	*37.6*	*37.6*	*33.4*	*−37.8*
Caribbean/CA % difference	*−36.2*	*−18.9*	*2.7*	*10.8*	*6.9*	*17.7*	

Source: World Energy Council, *Energy Trilemma Index*, 2015, http://data.worldbank.org/data-catalog/worldwide-governance-indicators.

in this area with average scores generally in the 40s. However as in the case of regulatory quality, scores in Central America have increased over time (14.7% between 1996 and 2014), while those in the Caribbean sample have fallen (37.8% between 1996 and 2014). As a result, the Caribbean's higher average percentile differential of 36.2% in 1994 had fallen to a lower differential of 17.7% by 2014.

Between 1996 and 2014, increased corruption was a particular problem in the Dominican Republic (57.8%) and Trinidad and Tobago (59.7%). Nicaragua in Central America also saw its corruption ranking fall considerably (40.3%). In Central America, however the rest of the countries showed remarkable gains in this area. These were led by El Salvador (101.6%) and Honduras (85.7%). While these gains are impressive, the overall average scores in Central America still remain quite low by international standards.

No doubt the institutional deficiencies in both regions will severely limit interest in renewable development, especially that undertaken by foreign firms. The main concern however is with the Caribbean where not only are key governance measures low, by international standards and in relation to Central America, but many are currently in sharp decline.

Country experiences

While the prospects for a successful energy transition in the Caribbean appear bleak, there are bright spots. A closer examination of several countries' efforts and progress shows that it is possible to make progress despite a number of institutional and financial impediments.

Haiti

Even prior to the devastating January 2010 earthquake, Haiti's energy sector suffered from a number of deficiencies[15] that prevented the sector from assisting the country's growth and development. The main difficulties included (1) very limited access to electricity, especially in rural areas, (2) over-dependence on fossil fuels for generating electricity, (3) significant energy losses from the national grid, (4) high cost and unreliable supplies, and (5) household reliance on wood and charcoal for home cooking. The latter practice had led to extensive deforestation.[16]

The earthquake was a dramatic reminder of the extent to which the country's energy sector was unsustainable. In particular, it demonstrated the urgent need for effecting a transition to a more accessible, efficient and ecologically sustainable system. However, effecting broad changes in Haiti's energy and sector is a major challenge. Not only is the country's power system one of the weakest[17] in the Western Hemisphere, but the government's capacity to effect constructive change in the country's power sector is extremely limited.[18]

Of the four country case studies examined here, Haiti stands out in a number of important regards. Of the energy importers, Haiti derives the lowest share of its energy usage from imported fuels. Between 2000 and 2012 Haiti also had the largest percentage decrease in the share of fossil fuels in its energy mix (Table 29.2). A similar pattern exists with regard to the fuels in the country's total fuel consumption, and the decline (2000–2012) in the share of energy consumption from this source was second only to that of Guatemala. Haiti also had one of the lowest shares of energy derived from alternative sources and this share declined by far the most percent-wise (72.1%) between 2000 and 2012 (Table 29.3). In sharp contrast Haiti leads other countries by a large margin in the share of its energy derived from combustible renewables and waste (Table 29.4).

These patterns are largely reflective of the fact that most Haitian households rely on biomass, particularly in the form of charcoal, for their cooking.[19] With the wide spread earthquake

-related destruction to the country's power grid, many households were forced to revert almost exclusively to biomass for their energy.

As noted the Haitian government currently has a very limited capability of undertaking a major expansion of the energy sector or even an upgrading of the current system. Of the Caribbean/ Central American countries included in the World Bank's Governance Indicators,[20] Haiti receives by far the lowest scores in government effectiveness.[21] Even worse, of the Caribbean/ Central American countries, the country's ranking in government effectiveness has been declining at the most rapid rate (Table 29.11) since 1996, the starting date of the governance measures.

Next to Cuba, Haiti has by far the lowest levels of regulatory quality.[22] Again, this key governance indicator has been declining over time, although not as rapidly as those in Cuba (Table 29.10). Finally, Haiti has by far the highest levels of corruption. This governance dimension has also been deteriorating over time. Finally, the country also scores the lowest in many aspects of the World Bank's ease of doing business index (Table 29.9). These patterns imply that, at least for some time, it is unlikely the Haitian government will be able to take the lead in modernizing and expanding the country's energy sector and power generating capacity.

The World Energy Council[23] did not include Haiti in its energy trilemma index. However, a similar exercise was undertaken by the World Economic Forum (WEF). The WEF's Energy Architecture Performance Index (EAPI)[24] is a trilemma-type construct measuring national energy performance. The index's three sub-components[25] bear a close resemblance to the classic trilemma. Equivalent to the trilemma's energy affordability is the WEF's Economic Growth and Development, a measure assessing the extent to which a country's energy architecture adds to or detracts from economic growth. In a manner similar to the trilemma's environmental sustainability, the WEF's Environmental Sustainability measures the environmental impact of energy supply and consumption. Finally, as with the trilemma's energy security the WEF's Energy Access and Security sub-index measures the extent to which an energy supply is accessible, secure, and diversified.

Of the four countries chosen for closer examination, Dominican Republic, Haiti, Costa Rica and Haiti, Haiti has by far the lowest overall EPI score. The country also trailed the others in energy access and security, but despite past deforestation for charcoal production, faired fairly well in environmental sustainability. Haiti and Nicaragua shared the lowest rankings in the economic growth and development component. Finally, Haiti had the lowest improvement in its EPI between 2013 and 2015. These scores further confirm the poor state of the Haitian power sector as well as that sector's ability to play a vital role in the country's economy and economic growth.[26]

Yet despite the many obstacles, some progress has been made. In 2001 32% of the country's inhabitants had access to energy – electricity, solar, and generators. By 2012 this had increased to just 36%. For the urban regions the corresponding change was 62% to 63% while rural populations stayed at 11%. By 2012 and in terms of income group, for the country as a whole, 58.3% of the non-poor households had access to a sustainable source of energy. This declined to 28.2% for the poor and 7.9% for the extremely poor. Corresponding percentages for the urban population were 73.0% (non-poor), 51.3% (poor) and 32.4% (extremely poor). Access was much lower in the rural areas with 26.1% of the non-poor having access to a sustainable source of energy. For the poor this fell to 9.8% and further to 2.8% for those in a state of extreme poverty.[27]

A World Bank assessment[28] found that much of the country's power difficulties stem from poor management and regulatory oversight of the country's electricity utility, Electricité d'Haïti

(EDH). Worsening commercial performance has led to a lack of infrastructure maintenance and rapidly deteriorated quality of electricity service, including frequent service interruptions and large voltage fluctuations.

EDH's weak grid infrastructure, poor commercial performance, and inadequate controls over subcontracted electricity generation by independent power producers (IPPs) have led to a financial drain on government resources. Due to its inability to meet electricity demand and in an attempt to expand electricity availability, EDH has subcontracted part of the production of electricity to IPPs. Unable to cover its operating expenses, including fuel costs and power purchases in part because of low bill collection rates, EDH has relied on fiscal transfers from the Treasury averaging US$200 million annually in recent years (equivalent to 10% of the national budget and 1–2% of gross domestic products (GDP)).[29]

For the near term there are numerous opportunities for donors,[30] the private sector, and even individuals to collaborate on some of the renewable technologies – from solar lights and charging stations for cell phones, to development of seed oil crops for biodiesel. With the spread of renewable electricity to rural areas many new jobs will be created directly and indirectly by new light industries taking advantage of local talents and resources.[31] But these are still not at a stage that provides the entire country with a reliable low-cost energy supply. The potential to use natural gas, given the revolution underway in the US and the development of LNG export terminals along the Gulf coast, is now an increasingly realistic option. The fact that the Dominican Republic has a natural gas docking station has the potential of opening up a number of opportunities to begin the shift towards a more sustainable energy.

Realistically, for the near term there is little reason to expect Haiti to make significant progress towards transitioning to a sustainable energy mix. The country's capacity and business climate will have to be effectively dealt with before the transition towards improved energy security, sustainable environment and access to affordable power can begin in full force. In the longer term, assuming progress in institutional reforms occurs, the country has great promise for renewables, and the agreements reached in the Paris conference should help move both foreign and domestic investments and finance in this direction.

Dominican Republic

The Dominican Republic provides a contrasting case to the energy situation found in Haiti. Until 1950 the history of both countries followed a similar course of protracted power struggles, frequent periods of domestic instability and global commodity price instability. As a result, by 1950 their GDP per capita were nearly identical. Today however, Haiti has taken on many of the characteristics of a failed state, while the Dominican Republic enjoyed an income per head almost five times larger than its island neighbor.[32]

Contrasts in the energy sector and related critical areas of government competency are just as pronounced. Whereas in our sample of ten Caribbean and Central American countries, Haiti had one of the lowest ratios of energy imports to total energy use, the Dominican Republic has one of the highest, averaging around 90% (Table 29.1). The difference occurs largely as a result of percentage use of fossil fuel in the national energy mix. Haiti has by far the lowest percentage, while the Dominican Republic has the highest, again around 90% (Table 29.2). The same pattern exists with combustible renewables and waste as a percentage of total energy with Haiti having the highest ratio and the Dominican Republic the lowest.

Because the Dominican Republic has little in the way of fossil energy reserves the country is extremely dependent on imports for the bulk of its energy. Currently the country's power is generated from natural gas (31%), coal (15%), diesel (6%), fuel oil (29%), isolated systems (8%), wind (2%), and hydropower (9%). GDP per unit of energy use is considerably higher in the Dominican Republic and has increased at a much more rapid rate since 2000 (Table 29.5). As in the case of Haiti, the Dominican Republic has been receiving preferential oil-financing terms under Venezuela's Petrocaribe program.

Contrasts also abound in the area of government effectiveness where the Dominican Republic has averaged in the low 40th/high 30th percentile since 1996. On the other hand, Haiti has never reached the 10th percentile and is currently at the first percentile (Table 29.11). While not quite as striking, differences in regulatory quality (Table 29.10) are notable with the Dominican Republic averaging in the high 40th percentile and increasing by 23.4% since 1996. Haiti's, on the other hand, averaged around the 15th percentile while declining by 16% since 1996.

Corruption differences are also significant (Table 29.12) although both countries have seen a marked increase in corruption in recent years. The Dominican Republic also has developed a much more business friendly environment except for the ability of firms to obtain electricity – here both countries score, along with Honduras (Table 29.9) the lowest among the sample countries.

As noted earlier, Haiti was not included in the World Energy Council's trilemma database. The Dominican Republic on the other hand was included and posted relatively low scores on the energy security and energy equity dimensions, with environmental sustainability consistently receiving the highest score (Table 29.7). However, trilemma differences between the two countries are very apparent in the World Economic Forum's Global Energy Architecture Performance Index. Here both countries score fairly evenly in environmental stability. However, the Dominican Republic leads Haiti by a wide margin in the critical areas of economic growth and development and energy access and security.

Despite the Dominican Republic's lead over Haiti in many critical energy indices and government effectiveness the country's energy sector is still beset by a number of problems stemming from many years of low output capacity, poor sector management, rising demand and widespread theft, non-payment and technical (transmission and distribution) losses. In addition, the public energy transportation network is poorly developed, which leads to frequent energy blackouts.

According to a recent executive survey conducted by the World Economic Forum, within Latin America only Haiti and Venezuela have a worse-functioning electricity service. The sector's problems have resulted in regular blackouts, lost productivity, and an erosion in the competitiveness of manufacturing and other sectors.[33]

Although lower international fuel costs should ease pressures somewhat, the power sector will remain one of the major constraints on the country's business climate. The country was in 123rd position in terms of quality of the electricity supply in the World Economic Forum's Global Competitiveness Report for 2015–2016,[34] above only Haiti (138th) and Venezuela (131st) in the Latin American region. At the start of 2016 the country dropped three places in the World Bank's Ease of Doing Business[35] ranking, to 93rd out of 189 economies, due in part to lengthy and expensive electricity procedures – it takes an average of 82 days to get the service running, compared with 65 days on average in the region. At the start of 2016 getting electricity in the Dominican Republic was even harder than in Haiti.

While many of the country's power sector problems are well known and have been present for years, little progress has been made in addressing them. Inefficiencies and budget deficits have been covered over through public subsidies. This practice has placed considerable pressure on the country's public finances. For example, in 2014 the government's budget allocated US $1,200m for subsidies to the electricity sector while in 2015 lower global oil prices reduced this to US $869m, or 1.4% of GDP.

The Medina administration's energy strategy has largely centered on expanding the number of conventional coal burning thermal plants. In trilemma terms there is a clear priority towards energy affordability and to a lesser extent energy security at the expense of environmental sustainability

The administration has also advanced the idea of an "energy pact."[36] The idea is to approach a consensus on energy reforms through a dialogue between representatives of various sectors of society. The process was begun in 2005, but as of early 2016 had not arrived at a final reform agenda, and a set of milestones to monitor progress. It is anticipated that the pact will be finalized later in 2016, although many issues remain unresolved.[37]

There is widespread hope that the country's energy transition will be structured to conform to international standards set at the UN's Climate Change Conference in early December 2015. However, with the planned expansion of coal-fired plants, it is difficult to see how the country will be able to conform to the new set of international standards. This is especially the case given limited government support in recent years for renewable energy.

Further hindering the country's energy transition, in 2012 a series of fiscal reforms rescinded many of the tax advantages for investing in renewables. Given the country's past record on renewables and the fact that the energy pact appears likely to focus on more conventional power sources, a notable expansion of renewable energy in the country's energy mix appears problematic in the medium term.[38]

Nicaragua

Just as Haiti and the Dominican Republic present an interesting contrast in energy transitions in the Caribbean. Nicaragua and Costa Rica have followed somewhat different paths in moving away from fossil fuels to a more sustainable and secure energy mix. As with Haiti's relatively lower capacity (low government effectiveness, poor business climate) Nicaragua has made less progress in this transition than Costa Rica.

As with the two Caribbean countries Nicaragua and Costa Rica have pressing reasons for reaching a more sustainable energy path. In Nicaragua's case, the country is finding climate change is having a devastating effect on the country's coffee crop, forcing many farmers to shift to cocoa,[39] while in Costa Rica's case not only the coffee crop, but the country's booming eco-tourism industry are also adversely affected by temperature changes.[40] Both countries are heavily dependent on Venezuela's Petrocaribe program which, as noted, is in danger of being scaled back or eliminated at any time.

Some major differences in capacity in implementing an energy transition stand out. Costa Rica is slightly below the 70th percentile (2014) in the World Bank's measure of government effectiveness whereas Nicaragua is slightly below the 20th percentile. Not only that, but Costa Rica's government effectiveness percentile has increased by 11.8% since 1996, whereas Nicaragua's has declined by 8.3% during this time. Similarly, Costa Rica is in the 75th percentile (2014) in the World Bank's control of corruption index, whereas Nicaragua's percentile is 19.2. Nicaragua's ranking on this index has fallen by 40.3% since 1996, whereas Costa Rica's has increased by 1.8%. Finally Costa Rica ranks 58th (2016) in the World Bank's Ease of Doing Business, while Nicaragua comes in at 125.

Capacity differences are no doubt at least partly responsible for the big gap in energy efficiency between the two countries, and may be an element affecting energy trilemma rankings. Costa Rica gets nearly twice the GDP generated per unit of energy use (Table 29.5), while the World Energy Council gives Costa Rica grades of A for environmental stability and B for both energy security and energy equity. Nicaragua in contrast received a B for environmental sustainability but D for both energy security and energy equity (Table 29.8). Similarly, Costa Rica ranks 11th in the World Economic Forum's Global Energy Architecture Performance Index, while Nicaragua ranks 82nd (Table 29.13).

Other differences are subtler. For both countries, the percentage of energy imports in total energy use is in the high 40s although between 2000 and 2012 Costa Rica reduced (Table 29.1) these by nearly 17%, while Nicaragua has had a reduction of nearly 6%. Similarly, Costa Rica has had a fairly dramatic reduction in fossil fuel as a share of the country's energy mix, reducing the percentage of fossil fuels from 59% in 2000 to 47.8% in 2012 or an overall reduction of 19% (Table 29.2). Nicaragua reduced its share of fossil fuels from nearly 52% in 2000 to 43.7% in 2012, for a 5.1% reduction overall.

Larger differences are seen in the area of alternative energy where these sources account for 38.7% (2012) of total energy use in Costa Rica, but only 15.5% (2012) in Nicaragua (Table 29.3). As in Haiti combustibles and renewables account for a large share of Nicaraguan energy use (40.8% in 2012), while in Costa Rica (Table 29.4), as in the case of the Dominican Republic, a relatively low share (13.4% in 2012).

Similarly, the two countries vary considerably in their energy mix for electricity generation. Hydroelectric accounts for 71.2% (Table 29.6) of Costa Rica's electricity followed by other renewable (21.0%) and conventional thermal 7.8%. Nicaragua on the other hand relies largely on conventional thermal (58.9%) followed by other renewable (31.3%) and hydroelectric (9.9%).

Both countries share a commitment to improved energy security and a sustainable environment. Costa Rica has been more successful in this regard and, as noted earlier, is one of the world leaders in moving towards a completely carbon neutral environment, now targeted for 2021. Both countries have been important developers of hydropower and geothermal energy, and have considerable scope for expansion in these areas. Both are also focused on expanding the amount of energy they will receive from solar, wind, agricultural waste and biofuels. For both countries however billions of dollars in new renewable energy investments will be necessary to take advantage of their renewable resource potentials.

The Nicaraguan case is particularly interesting in that not only is Nicaragua the poorest country in Central America, it also has until recent years had one of the highest energy costs in Central America.[41] The country also relies the most on conventional thermal to generate its electricity (Table 29.6).

Over the years and despite many impediments, the country has made significant progress in addressing its energy and security needs. In the 1990s, the dilapidated state of the country's power distribution system was responsible for the loss of up to 25% of all power generated and supply could cover just one-half of consumer demand. Daily blackouts, often lasting up to 6–10 hours, were the norm as late as 2006.[42]

In 2007 the new Government of President Daniel Ortega signed several cooperation agreements with the Venezuelan Government of Hugo Chávez. Venezuela agreed to supply discounted oil to Nicaragua under its Petrocaribe initiative, as well as to build a refinery in the country.

In late 2008 the state power company, Empresa Nicaragüense de Electricidad (ENEL), announced that, thanks to assistance primarily from Venezuela and Cuba, and the installation of an additional 120 MW generating capacity, prolonged power cuts in Nicaragua would be prevented. By 2011 the Nicaraguan energy system had a surplus with a total installed capacity of approximately 1,000 MW and an average demand of 550 MW.[43]

Renewable energy has become an integral part of the Sandinistas' economic development strategy.[44] In recent years this has involved attempts to expand the country's hydroelectric

Table 29.13 Global Energy Architecture Performance Index 2013–2015 (scores)

Country	*Country rank*		*EPI index*		*Economic growth and development*		*Environmental sustainability*		*Energy access and security*	
	2015	*2013*	*2015*	*2013*	*2015*	*2013*	*2015*	*2013*	*2015*	*2013*
Caribbean										
Dominican Republic	62	52	0.59	0.56	0.61	0.53	0.51	0.61	0.65	0.56
Haiti	119	94	0.44	0.43	0.44	0.44	0.67	0.64	0.22	0.20
Average	*90.5*	*73*	*0.52*	*0.50*	*0.53*	*0.49*	*0.59*	*0.63*	*0.44*	*0.38*
% change: 2013–2015										
Dominican Republic			5.36		15.09		-16.39		16.07	
Haiti			2.33		0.00		4.69		10.00	
Central America										
Costa Rica	11	19	0.72	0.65	0.69	0.62	0.69	0.61	0.77	0.72
Nicaragua	82	77	0.54	0.48	0.46	0.37	0.61	0.60	0.54	0.45
Average	*46.5*	*48*	*0.63*	*0.57*	*0.58*	*0.50*	*0.65*	*0.61*	*0.66*	*0.59*
% change: 2013–2015										
Costa Rica			10.8		11.3		13.1		6.9	
Nicaragua			12.5		24.3		1.7		20.0	

Source: Global Energy Architecture Performance Index, Report 2015, December 2014.

Note: Scores are on a scale of 0 to 1.

generating capacity. In 2015 hydroelectricity provided 9.9% of Nicaragua's total production of electrical energy.[45] Expansion plans include the 253-MW Tumarín hydroelectric project[46] in the South Atlantic Autonomous Region to be undertaken by Centrales Hidroeléctricas de Nicaragua at an estimated cost of some US $1,000m and, on completion, will be one of Central America's largest hydroelectric dams. The dam, the construction of which was scheduled to be completed in 2019, was expected to generate an annual average of 1,184 GWh of energy. The Nicaraguan Government planned drastically to cut the cost of energy imports by generating 80% of the country's energy requirements from renewable sources by 2017.

Nicaragua's efforts in transitioning to a more energy and environmental sustainable mix are impressive given concerns, noted earlier, over the country's capacity to undertake such major changes. For most countries the national investment climate is critical in determining the flow of capital into renewable energy. Nicaragua has defied this general rule by ranking fourth in Latin America after Uruguay, Brazil, and Chile, but before Costa Rica in Climatescope's Enabling Framework Parameter[47] which assesses the mechanisms in place that can facilitate investment in renewables.

Despite its limitations in areas such as government effectiveness, corruption and a poor business climate, Nicaragua has developed a policy framework to support renewables and to improve legal security for investors in the sector. In addition, a highly professional microfinance industry has helped finance local businesses that specialize in assisting households and small businesses in introducing and drawing on renewable sources of energy.

Costa Rica

Costa Rica represents one of the more successful efforts in transitioning to a sustainable environment with improved energy security. The country has long been a leading advocate of hydropower and geothermal energy, and is now actively branching out into other renewables such as solar, wind, agricultural waste and biofuels.[48]

These actions are all key components of the country's strategy to become carbon neutral (C-neutral) by 2021. Costa Rican President Oscar Arias observed that, "we do this with the hope that, eventually, we will be able to show the world that what ultimately needs to be done, can be done. As a small country, this is Costa Rica's important contribution to the climate change issue."[49]

In part, Costa Rica's energy strategy is designed to take advantage of the country's geographical setting, and natural resource base. The presence of active volcanoes has opened up the opportunity for large developments in geothermal power generation. In addition, Costa Rica has abundant sunlight and wind, offering numerous opportunities for renewable development.

Electricity is provided by the Instituto Costarricense de Electricidad (ICE), a state controlled, vertically-integrated utility. The Costa Rican market includes private players, in the form of cooperatives and independent power producers selling electricity to ICE.

To meet the country's energy objectives, Costa Rican government offers a spectrum of incentives including import, value added and income tax breaks on select renewable energy materials and equipment. In addition, ICE periodically holds tenders to contract new clean energy capacity. A five-year net metering pilot program was in place in Costa Rica until February 2015.[50] The project was cancelled when the capacity limit of 10MW was achieved, but consumers that had already joined the program will remain connected for 15 years.

As of 2014 Costa Rica had 2.8GW of installed power capacity. Large hydro accounted for 44% of this while oil and diesel contributed 21%. The remaining 35% was clean energy of which 1.4% was from biomass and waste, 8% geothermal, 7% wind and 19% small hydro.[51] As a result of its successful energy transition, by 2015, Costa Rica was able to operate the country's grid for the first 75 days only using renewable energy.[52]

While remarkable, Costa Rica's energy transition has not been without setbacks. By placing emphasis on energy security (less dependence on Venezuela) and environmental sustainability, the third dimension of the energy trilemma, energy affordability has suffered[53] with electricity tariffs increasing by 142% between 2005 and 2015.[54]

The expected decline in the costs of renewable energy should help bring prices down in the future. Another factor behind the increases in tariffs stems from electricity demand simply outrunning supply. This problem can be addressed through regulatory reform[55] removing restrictions on the amount of electricity smaller producers can put into the grid. Currently the private sector's generation is limited to 30% of the total power generated in the National Electric System. This highly regulated market unfortunately stifles competition. Since, ICE is not required to purchase from these generators, that means that these private generators must wait until the capacity demand increases.

With a high dependence on hydropower, the country is quite vulnerable to droughts and thus sudden surges in electricity prices. However, expanding other renewable sources over time will alleviate this problem. In short, the country has a number of options at its disposal for improving the energy affordability dimension.

In the near term, Costa Rica now has the option of importing some of its energy needs thus dampening price increases. The country is part of the Central American Regional Market (Mercado Eléctrico Regional, MER), which interconnects seven countries in the region. Due to severe droughts in the last several years Costa Rica has relied on the regional market to help meet its domestic electricity needs. In 2014, for example, Costa Rica imported 251GWh from the Central American Regional Market, 18% of all the imports that year.[56]

The Costa Rica story holds out hope to many smaller countries wishing to transition to a more environmentally friendly and secure energy mix. Just knowing that this is possible will no doubt help build public support in these countries for pressing ahead. With the falling costs of renewable energies, the means are there for most countries to make significant progress on these two dimensions of the trilemma. For many however, increases in short run higher energy bills are a price they will face.

However, the Costa Rican model is not likely to take hold in larger more industrialized countries, particularly ones with heavier infrastructures and developed coal and/or oil resources. Here, a consensus on the path of the transition will be much more difficult to form.

Conclusions

The Caribbean and Central American regions illustrate the difficulty of making sweeping generalizations concerning energy transitions. Even in a regional setting where countries share a number of similarities such as small domestic markets with limited fossil energy reserves, a great variety of energy mixes are possible.

Still several patterns prevail. Progress or lack of movement toward a secure sustainable energy mix is largely related to government capacity, especially at the extremes as illustrated by Haiti with the least capacity and Costa Rica with the most. However, explaining progress in intermediate cases like Dominican Republic and Nicaragua requires additional insights. Are leftist regimes more concerned with the environment and energy security? Is there a political economy effect where democracies tend to place more emphasis on energy affordability? These may be avenues worth exploring in future research.

Notes

1 Andres Schipani, "Petrocaribe: a legacy that is both blessing and curse," *Financial Times*, April 17, 2015, www.ft.com/cms/s/0/04c55724-bea7-11e4-8036-00144feab7de.html#axzz3y0SFtZGz.

2 Harold Trinkunas, "Changing Energy Dynamics in the Western Hemisphere: Impacts on Central America and the Caribbean," Brookings Institution, Washington, DC, April 2014; www.brookings.edu/research/papers/2014/04/04-energy-central-america-caribbean-trinkunas.

3 Jackie Northam, "Venezuela Cuts Oil Subsidies to Caribbean Nations," *NPR*, March 30, 2015; www.npr.org/sections/thetwo-way/2015/03/30/396399497/venezuela-cuts-oil-subsidies-to-caribbean-nations.

4 Ramon Bueno, Cornelia Herzfeld, Elizabeth Stanton, and Frank Ackerman, *The Caribbean and Climate Change: The Costs of Inaction*, Tufts University, Medford, MA, May 2008, http://ase.tufts.edu/gdae/pubs/rp/caribbean-full-eng.pdf

5 Christopher Pala, "Study finds fish stocks in Caribbean declining fast," *Caribbean 360*, January 22, 2016, www.caribbean360.com/news/study-finds-fish-stocks-in-caribbean-declining-even-faster-urgent-reversal-needed?utm_source=Caribbean360%20Newsletters&utm_campaign=d986cdb171-Vol_11_Issue_015_News1_22_2016&utm_medium=email&utm_term=0_350247989a-d986cdb171-39418289.

6 Troy Lorde, Charmaine Gomes, Dillon Alleyne, et al., *An Assessment of the Economic and Social Impacts of Climate Change on the Coastal and Marine Sector on the Caribbean*, United Nations, New York, 2013.

7 Georgi Kantchev, "Banks Slash Oil-Price Forecasts Again: Predictions of a Recovery have been Dialed Back," *Wall Street Journal*, January 12, 2016.

8 Alison Sider, "U.S. Exports First Freely Traded Oil in 40 Years: Energy Companies Race to Ship Abroad After Ban is Lifted," *Wall Street Journal*, January 13, 2016.

9 Jason Bordoff, "How Exporting U.S. Liquefied Natural Gas Will Transform the Politics of Global Energy," *Wall Street Journal*, November 17, 2015.

10 Harold Trinkunas, "Changing Energy Dynamics," op. cit.

11 *Energy Integration in Central America: Full Steam Ahead*, Inter-American Development Bank, Washington, DC, June 23, 2013.

12 Rigoberto Ariel Yepez-Garcia, Todd M. Johnson and Luis Alberto Andres, *Meeting the Balance of Electricity Supply and Demand in Latin America and the Caribbean* (Washington, DC, World Bank, June 2011).

13 "U.S.-Caribbean Clean Energy Will See Venezuela Pushback," *Oxford Analytica*, May 11, 2015.

14 "Fact Sheet: U.S.-CARICOM Summit: Deepening Energy Cooperation," The White House, Washington, April 9, 2015.

15 "HA-T1130: Towards a Sustainable Energy Sector Haiti – White Paper," Inter-American Development Bank, Washington, 2012.

16 Nathan C. McClintock, *Agroforestry and Sustainable Resource Conservation in Haiti: A Case Study*, www.ncsu.edu/project/cnrint/Agro/PDFfiles/HaitiCaseStudy041903.pdf.

17 *Haiti Energy*, USAID, Washington, DC, January 26, 2016, www.usaid.gov/haiti/energy.

18 Fabrice Guerrier, *Haiti National Energy Policy Reform: Implementation of a Bioregional Framework to Achieve Sustainable National Electrification*, April 1, 2014, www.academia.edu/6940125/Haiti_National_Energy_Policy_Reform_Implementation_of_a_Bioregional_Framework_to_Achieve_Sustainable_Nationwide_Electrification.

19 Chemonics, *Sustainable Energy Solutions for Haiti*, www.chemonics.com/OurWork/OurProjects/Pages/Haiti-Improved-Cooking-Technology-Program.aspx.

20 *Worldwide Governance Indicators*, World Bank, Washington, DC, 2016, http://data.worldbank.org/data-catalog/worldwide-governance-indicators.

21 *Government Effectiveness Indicator*, Millennium Challenge Corporation, Washington, DC, www.mcc.gov/who-we-fund/indicator/government-effectiveness-indicator.

22 *Regulatory Quality*, World Bank, Washington, DC, http://info.worldbank.org/governance/wgi/pdf/rq.pdf.

23 *World Energy Council, Energy Trilemma Index, 2015*, www.worldenergy.org/data/trilemma-index/.

24 *Global Energy Architecture Performance Index Report 2015* (Geneva: World Economic Forum, Geneva, October 2014).

25 *Global Energy Architecture Performance Report 2015*, op. cit., p. 7.

26 *Haiti Energy, USAID*, op. cit., www.usaid.gov/haiti/energy.

27 *Rural Development in Haiti: Challenges and Opportunities,* World Bank, Washington, DC, September 2014.

28 Raju Jan Singh and Mary Barton-Dock, *Haiti: Toward a New Narrative*, Washington, DC: World Bank, 2015.

29 Raju Jan Singh and Mary Barton-Dock, *Haiti: Towards a New Narrative*, op. cit., p. 68.

30 *Powering Haiti with Clean Energy*, Clinton Foundation in Haiti, www.clintonfoundation.org/our-work/clinton-foundation-haiti/programs/powering-haiti-clean-energy.
31 Matthew Lucky, Katie Auth et al., *Haiti Sustainable Energy Roadmap: Harnessing Domestic Energy Resources to Build a Reliable, Affordable and Climate-Compatible Electricity System*, Worldwatch Institute, Washington, November 2014, www.worldwatch.org/bookstore/publication/haiti-roadmap.
32 Aleksander Laszek, "Why is Haiti Poorer than the Dominican Republic?" In Leszek Balcerowica and Andrzej Rzonca, *Puzzles of Economic Growth* (Washington, DC: World Bank, 2015).
33 "Dominican Republic economy: Power-sector reform discussions get under way," *EIU ViewsWire*, February 20, 2015.
34 *The Global Competitiveness Report 2015–2016* (Geneva: World Economic Forum, 2015).
35 *Doing Business: Dominican Republic 2016* (Washington, DC: World Bank, 2015).www.doingbusiness.org/data/exploreeconomies/dominican-republic/.
36 "Energy in the Dominican Republic: All Eyes on the Pacto Electrico," Institute of the Americas, Washington, www.iamericas.org/en/recent-articles-sp-619120327/2142-energy-in-the-dominican-republic-all-eyes-on-the-pacto-electrico.
37 "Bishop denies gridlock on Dominican Republic Electricity Pact," *Dominican Today*, January 15, 2016, www.dominicantoday.com/dr/economy/2016/1/15/57858/Bishop-denies-gridlock-on-Dominican-Republic-Electricity-Pact.
38 "Dominican Republic economy: Government sets ambitious targets for renewable," *EIU ViewsWire*, December 18, 2015.
39 Luc Cohen and Ivan Castro, "As climate change threatens CentAm coffee, a cocoa boom is born," *Reuters*, January 18, 2016, www.reuters.com/article/us-climatechange-cocoa-coffee-idUSKCN0UW1AV.
40 "Costa Rica's Environment: The Social Perspectives," http://blogs.nelson.wisc.edu/es112-308-4/climate-change/.
41 Tim Rogers, "Electricity costs to increase 7.78 % next week," *The Nicaragua Dispatch*, April 12, 2013, http://nicaraguadispatch.com/2013/04/electricity-costs-to-increase-7-78-percent-next-week/.
42 Tim Rogers, "The next 'revolution' for Nicaragua: energy independence," *Christian Science Monitor*, February 7, 2012, www.csmonitor.com/World/Americas/2012/0207/The-next-revolution-for-Nicaragua-energy-independence.
43 "Nicaragua Announces 4 Billion Dollar Renewable Energy Investment Plan," *Eurasia Review*, September 24, 2014, www.pronicaragua.org/en/newsroom/news-on-nicaragua/1301-nicaragua-announces-4-billion-dollar-renewable-energy-investment-plan.
44 "Nicaragua: Poverty Reduction Strategy Paper," International Monetary Fund, Washington, May 2010, www.imf.org/external/pubs/ft/scr/2010/cr10108.pdf.
45 World Energy Council, *Dynamic Data*, worldenergy.org/data.
46 "The Tumarin hydro project in Nicaragua with its 253 MW will generate half the energy consumed in the country," *EN EnergyNews: Todo Energia*, November 4, 2014, www.energynews.es/english/the-tumarin-hydro-project-in-nicaragua-with-its-253-mw-will-generate-half-of-the-energy-consumed-in-the-country/.
47 *Climatescope 2015*, http://global-climatescope.org/en/.
48 *Renewable Energy in Central America*, op. cit., p. 9.
49 Roberto Dobles Mora, "Costa Rica's Commitment: On the Path to Becoming Carbon-Neutral," *UN Chronicle*, June 2007, http://unchronicle.un.org/article/costa-rica-s-commitment-path-becoming-carbon-neutral/.
50 Cham Brownell, "The battle over distributed generation of electricity in Costa Rica," *The Tico Times*, April 11, 2015, www.ticotimes.net/2015/04/11/the-battle-over-distributed-generation-of-electricity-in-costa-rica.
51 "Costa Rica," *Climatescope 2015*, http://global-climatescope.org/en/country/costa-rica/#/details.
52 Maria Gallucco, "Costa Rica Used 100% Renewable Energy in the First 75 Days of 2015," *International Business Times*, March 23, 2015, www.ibtimes.com/costa-rica-used-100-renewable-energy-first-75-days-2015-1855654.
53 "Costa Rica economy: Quick View – Energy costs set to remain high," *EIU ViewsWire*, September 5, 2014.
54 "Costa Rica: Electricity Rate Rose 142% in 10 Years," CentralAmericaData.com, August 31, 2015, www.centralamericadata.com/en/article/home/Costa_Rica_Electricity_Rate_Rose_142_in_10_years.
55 "Costa Rica economy: Electricity market backlog pushes up rates," *EIU ViewsWire*, October 25, 2012.
56 "Costa Rica," Global Climate Scope, http://global-climatescope.org/en/download/reports/countries/climatescope-2015-cr-en.pdf.

Appendix

Empirical patterns[1]

Do those countries that give priority to environmental stability/climate change have a common set of characteristics that set them apart from those countries that give priority to energy security/energy affordability? To answer this question a discriminant analysis was performed[2] on the two groups of countries in Table A.2.

The independent variables used in the analysis were from: (a) the various elements of the EIU's Democracy Index,[3] (b) World Bank Governance Indicators,[4] voice and accountability, political stability, government effectiveness, regulatory quality, rule of law, and control of corruption, and (c) World Energy Council,[5] the other two dimensions of the trilemma, energy security and energy affordability.

Of these variables the discriminant analysis found three to be statistically significant in defining a unique environment for each group. In relative order of importance these were: (a) the World Energy Council's energy security ranking, (b) the World Bank's measure of voice and accountability, and (c) the EIU's measure of political participation (Table A.1).

Most countries were correctly classified with a high degree of probability, for example, Russia, a country that does not prioritize energy sustainability, had a 99% chance of being placed in that group based on its energy security levels, voice and accountability scores and the extent of political participation in that country.

Only three countries were misclassified: India, Germany and Denmark. For example, India, based on its energy trilemma dimension rankings was originally grouped with countries that did not prioritize energy sustainability. The discriminant analysis suggest the country has more in common with those countries that do prioritize energy sustainability

What the analysis suggests is that countries which prioritize energy sustainability/climate security are more likely to have relatively low levels of energy security, and relatively high levels of democracy (voice and accountability). Similarly, countries that do not prioritize climate security have relatively high levels of energy security, and low levels of democracy. The average ranking on the EIU's political participation was about the same for both groups.

Since the discriminant analysis provides a composite profile of each group one can infer that even lower levels of political participation are more effective in raising the priority of climate security in relative democracies, but that even higher levels of participation may not be effective in this regard in relative non-democracies.

A final statistical test, regression analysis,[6] asked whether there were variables within each group that might improve a particular country's commitment to energy sustainability. Drawing on the same set of variables used in the previous exercise the EIU's Democracy Index was the only statistically significant variable to the total sample of countries. However, this variable accounted for only around 38% of the differences in country rankings in environmental sustainability.

However, looking at each sub-group of countries separately (those which do prioritize climate security and those which do not), the analysis found that the EIU's democracy ranking could account for around 88% of the differences in energy sustainability rankings for countries that prioritized energy sustainability. On the other hand, no variables were statistically significant in accounting for the observed levels of energy sustainability in those countries that did not prioritize energy sustainability.

This analysis has several policy implications. First, while many observers are focusing on the amount of assistance certain countries might require to meet their internationally agreed upon voluntary carbon reduction commitments, efforts to improve democracy might be even more successful in assuring significant levels of decarbonization.

Second, for those countries that have not prioritized energy sustainability, financial assistance is unlikely to produce meaningful results. Nor are efforts simply focused on nurturing democratic institutions. Instead assistance in improving energy security, along with assistance in improving democratic institutions and political participation are likely to pay much higher dividends if these efforts are sufficient to elevate the country to that group of countries that do prioritize energy sustainability.

Table A.1 Profiles of country groups based on environmental sustainability

Country	*Initial grouping*	*Probability of correct placement*	
		Probability Group 0	*Probability Group 1*
Saudi Arabia	0	0.93	0.07
Canada	0	0.57	0.42
Russia	0	0.99	0.01
Mexico	0	0.97	0.03
South Africa	0	0.88	0.12
China	0	0.99	0.01
United States	0	0.66	0.34
United Kingdom	1	0.28	0.72
Brazil	1	0.42	0.58
Indonesia	0	0.94	0.05
Egypt	0	0.85	0.15
Japan	1	0.02	0.97
Thailand	0	0.77	0.23
Pakistan	1	0.01	0.99
India	0	0.47	0.53*
Jordan	1	0.10	0.90
Turkey	1	0.49	0.51
France	1	0.26	0.74
Denmark	1	0.60	0.40*
Germany	0	0.23	0.77*
Italy	1	0.10	0.90
Dominican Republic	1	0.01	0.99
Costa Rica	1	0.23	0.76
Nicaragua	1	0.08	0.92

Note: * Signifies not classified in original group.

Table A.2 Group discriminating variables

Country	*Energy security*	*Voice/accountability*	*Political participation*
Group I			
UK	8.00	1.30	6.67
Brazil	33.00	0.41	5.56
Japan	64.30	1.04	6.11
Pakistan	61.70	0.74	2.78
Jordan	114.70	−0.77	3.89
Turkey	66.00	0.32	3.00
France	42.00	1.22	7.78
Denmark	3.70	1.55	8.33
Italy	[illegible]	0.90	7.22
Dominican Republic	114.70	0.14	5.00
Costa Rica	57.30	1.13	8.04
Nicaragua	103.70	−0.37	4.44
Average	61.37	0.59	5.90
Group II			
Saudi Arabia	54.00	−1.78	2.22
Canada	1.00	1.43	7.78
Russia	6.30	−1.04	5.00
Mexico	32.00	−0.05	7.22
South Africa	38.30	0.65	8.33
China	19.30	−1.54	3.33
US	7.70	1.05	7.22
Indonesia	17.00	0.13	6.67
Egypt	53.70	−1.19	3.33
Thailand	94.30	−0.85	5.56
India	68.30	0.42	7.22
Germany	27.70	1.46	7.78
Average	34.97	−0.11	5.97

Note: Political participation = EIU political participation index values, 10, highest, 0, lowest.

Notes

1 Copies of the computer print-outs are available from the author upon request.
2 Using IBM SPSS Statistics 22. A description of the output can be found at www.ats.ucla.edu/stat/spss/output/SPSS_discrim.htm.
3 The Economist Intelligence Unit, *Democracy Index 2015: Democracy in an age of anxiety* (London: EIU, 2016), www.eiu.com/public/topical_report.aspx?campaignid=DemocracyIndex2015.
4 World Bank, Governance Indicators 2015, http://data.worldbank.org/data-catalog/worldwide-governance-indicators.
5 World Energy Council, *World Energy Trilemma 2015*, www.worldenergy.org/work-programme/strategic-insight/assessment-of-energy-climate-change-policy/.
6 Again using IMB SPSS Statistics 22, www.ats.ucla.edu/stat/spss/webbooks/reg/chapter1/spssreg1.htm.

Index

Made in the USA
Columbia, SC
26 October 2023

24893309R00289